Nutrient Requirements and Feeding of Finfish for Aquaculture

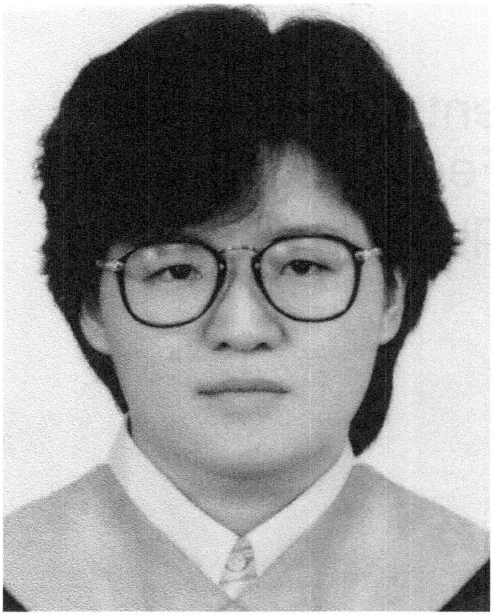

This book is dedicated to the memory of

Dr Tippawan Paripatananont,

a colleague and good friend, whose efforts and dedication to complete the chapter on snakehead and *Pangasius* catfish inspired us all to accomplish the task.

Nutrient Requirements and Feeding of Finfish for Aquaculture

Edited by

CARL D. WEBSTER

Aquaculture Research Center
Kentucky State University
Frankfort, KY 40601, USA

and

CHHORN LIM

USDA-ARS
Aquatic Animal Health Research Laboratory
PO Box 952
Auburn, AL 36831, USA

CABI *Publishing*

CABI is a trading name of CAB International

CABI Head Office
Nosworthy Way
Wallingford
Oxfordshire OX10 8DE
UK

Tel: +44 (0)1491 832111
Fax: +44 (0)1491 833508
Email: cabi@cabi.org
Web site: www.cabi.org

CABI North American Office
875 Massachusetts Avenue
7th Floor
Cambridge, MA 02139
USA

Tel: +1 617 395 4056
Fax: +1 617 354 6875
Email: cabi-nao@cabi.org

A catalogue record for this book is available from the British Library, London, UK.

Library of Congress Cataloguing-in-Publication Data
Nutrient requirements and feeding of finfish for aquaculture / edited by Carl D. Webster and Chhorn Lim
p. cm.
Includes bibliographical references
ISBN 0-85199-519-5 (alk. paper)
1. Fishes—Feeding and feeds. 2. Fishes—Nutrition—Requirements.
I. Webster, Carl D. II. Lim, Chhorn.

SH156.N82 2002
639.3—dc21 2001037511

ISBN-13: 978-0-85199-519-9
ISBN-10: 0-85199-519-5

First published 2002
Transferred to print on demand 2009

Printed and bound in Great Britain by
CPI Antony Rowe, Chippenham and Eastbourne.

Contents

Contributors

Geoff L. Allan, *New South Wales Fisheries, Port Stephens Fisheries Centre, Taylors Beach, NSW 2316, Australia.*

Frederic T. Barrows, *US Fish and Wildlife Service, Bozeman Technology Center, Bozeman, MT 59715, USA.*

Mali Boonyaratpalin, *Department of Fisheries, Jatujak, Bangkok 10900, Thailand.*

Ilda G. Borlongan, *Aquaculture Department, Southeast Asian Fisheries Development Center, Tigbauan, Iloilo, the Philippines.*

Paul B. Brown, *Department of Forestry and Natural Resources, Purdue University, 1159 Forestry Building, West Lafayette, IN 47907-1159, USA.*

Chris Carter, *FRDC-CRC, Southern Bluefin Tunic Subprogram, PO Box 120, Henley Beach, SA 5020, Australia.*

Steven Clarke, *FRDC-CRC, Southern Bluefin Tunic Subprogram, PO Box 120, Henley Beach, SA 5020, Australia.*

Shawn D. Coyle, *Aquaculture Research Center, Kentucky State University, Frankfort, KY 40601, USA.*

Sergiusz Czesny, *Illinois Natural History Survey, Lake Michigan Biological Station, Zion, IL 60099, USA.*

Konrad Dabrowski, *School of Natural Resources, Ohio State University, 2021 Coffey Road, Columbus, OH 43210, USA.*

Harry V. Daniels, *Department of Zoology, North Carolina State University, 207 Research Station Road, Plymouth, NC 27962, USA.*

Dong-Fang Deng, *Department of Biochemistry and Molecular Biology, Mississippi State University, PO Box 9650, Mississippi State, MS 39762, USA.*

Terry A. Dick, *Department of Zoology, University of Manitoba, Winnipeg, Manitoba R3T 2N2, Canada.*

Débora M. Fracalossi, *Departamento de Aquicultura, Centro de Ciências Agrárias, Universidade Federal de Santa Catarina, Caixa Postal 476, 88040-900 Florianópolis, SC, Brazil.*

Margie L. Gallagher, *Institute for Coastal and Marine Resources, East Carolina University, 125 Ragsdale, Greenville, NC 27858-4353, USA.*

Delbert M. Gatlin III, *Department of Wildlife and Fisheries Sciences and Faculty of Nutrition, Texas A&M University System, College Station, TX 77843-2258, USA.*

Brett Glencross, *FRDC-CRC, Southern Bluefin Tunic Subprogram, PO Box 120, Henley Beach, SA 5020, Australia.*

Barbara Grisdale-Helland, *AKVAFORSK (Institute of Aquaculture Research AS), N-6600 Sunndalsøra, Norway.*

John Gunn, *FRDC-CRC, Southern Bluefin Tunic Subprogram, PO Box 120, Henley Beach, SA 5020, Australia.*

Ronald W. Hardy, *Hagerman Fish Culture Experiment Station, University of Idaho, 3059F National Fish Hatchery Road, Hagerman, ID 83332, USA.*

Ståle J. Helland, *Department of Animal Science, Agricultural University of Norway, N-1432 Ås-NLH, Norway.*

Silas S.O. Hung, *Department of Animal Science, University of California, One Shield Ave., Davis, CA 95616, USA.*

S.J. Kaushik, *Fish Nutrition Research Laboratory, Unité Mixte INRA-IFREMER, 64310 Saint Pée-sur-Nivelle, France.*

Kotaro Kikuchi, *Abiko Research Laboratory, Central Research Institute of Electric Power Industry, 1646 Abiko, Abiko, Chiba 270-1194, Japan.*

V. Kiron, *Department of Aquatic Biosciences, Tokyo University of Fisheries, Konan, Minato, Tokyo 108-8477, Japan.*

Shunsuke Koshio, *Faculty of Fisheries, Kagoshima University, 4-50-20, Shimoarata, Kagoshima 890-0056, Japan.*

William Koven, *National Center for Mariculture, Israel Oceanographic and Limnological Research, POB 1212, Eilat 88112, Israel.*

Menghe H. Li, *Thad Cochran National Warmwater Aquaculture Center, Mississippi State University, PO Box 197, Stoneville, MS 38776, USA.*

Chhorn Lim, *USDA-ARS, Aquatic Animal Health Research Laboratory, PO Box 952, Auburn, AL 36830, USA.*

Rebecca Lochmann, *Department of Aquaculture and Fisheries, University of Arkansas at Pine Bluff, PO Box 4912, Pine Bluff, AR 71611, USA.*

Toshiro Masumoto, *Laboratory of Fish Nutrition, Kochi University, Nankoku, Kochi, 783-8502, Japan.*

Malgorzata Matusiewicz, *Department of Biochemistry, Medical Academy, Wroclaw, Poland.*

H. Shivananda Murthy, *Department of Aquaculture, College of Fisheries, University of Agriculture Sciences, Mangalore 575002, India.*

Tippawan Paripatananont, *Institute of Agricultural Technology, Walailak University, Tasala, Nakhon Si Tamarat 80160, Thailand.*

Felicitas P. Pascual, *Aquaculture Specialists, Inc., Nanga, Guimbal, Iloilo, the Philippines.*

Harold Phillips, *Department of Aquaculture and Fisheries, University of Arkansas at Pine Bluff, PO Box 4912, Pine Bluff, AR 71611, USA.*

Edwin H. Robinson, *Thad Cochran National Warmwater Aquaculture Center, Mississippi State University, PO Box 197, Stoneville, MS 38776, USA.*

Kirsten Rough, *FRDC-CRC, Southern Bluefin Tunic Subprogram, PO Box 120, Henley Beach, SA 5020, Australia.*

Stuart J. Rowland, *New South Wales Fisheries, Grafton Fisheries Centre, Grafton, NSW 2460, Australia.*

Shuichi Satoh, *Department of Aquatic Biosciences, Tokyo University of Fisheries, Konan, Minato, Tokyo 108-8477, Japan.*

Shi-Yen Shiau, *Department of Food Science, National Taiwan Ocean University, Keelung, Taiwan 202, Republic of China.*

Trond Storebakken, *AKVAFORSK, Institute of Aquaculture Research AS, N-6600 Sunndalsøra, Norway.*

Toshio Takeuchi, *Department of Aquatic Biosciences, Tokyo University of Fisheries, Konan, Minato, Tokyo 108-8477, Japan.*

James H. Tidwell, *Aquaculture Research Center, Kentucky State University, Frankfort, KY 40601, USA.*

Robert van Barneveld, *FRDC-CRC, Southern Bluefin Tunic Subprogram, PO Box 120, Henley Beach, SA 5020, Australia.*

Carl D. Webster, *Aquaculture Research Center, Kentucky State University, Frankfort, KY 40601, USA.*

Kevin Williams, *CSIRO Division of Marine Research, PO Box 120, Cleveland, Qld 4163, Australia.*

X. Yang, *Department of Cancer Immunology and AIDS, Dana–Faber Cancer Institute and Harvard University Medical School, Boston, MA 02115, USA.*

Preface

Aquaculture constitutes a vital and growing segment of agriculture worldwide. The increased demand for fish as a result of rapid population growth, increased disposable income and preferences for fish over other animal proteins for personal, cultural or health reasons further accelerate the industry's growth. For the foreseeable future, aquaculture operations will need to be expanded and production increased. The industry will need to be efficient and profitable and have a minimal environmental impact. Feeds and feeding practices are crucial elements in the sustainability, profitability and well-being of the modern aquaculture industry because feed costs represent between 30 and 70% of the total operating expenses. Further, it is becoming evident that nutrition plays an important role in immune system function and disease resistance in fish. As a result, the quality of feed and feeding management have become very critical. Feeds must be highly digestible, contain an optimal nutrient and energy balance and be economical for a given species cultured under a particular production system.

The benefits of having high-quality feed, however, can only be realized if fish are properly fed. For maximum production and profits, farmers are interested in a high rate of feed consumption. However, loss of nutrients if feed is not consumed immediately and uneaten or excess feed represent an economic loss as well as a possible source of environmental pollution. Thus, feed should be fed in such a manner as to minimize waste, optimize growth, allow for efficient conversion, minimize stress and maintain fish health. Economics, environmental issues and fish health are important factors vital to the sustainability of the aquaculture industry. These factors increase the pressure on fish nutritionists to develop more cost-effective, nutritionally balanced and low-pollution diets as well as to improve feeding strategies.

This book is a compilation of the currently available information on nutrition, feeds and feeding practices of various fish species of economic importance to regional and/or global aquaculture industries. Each chapter describes the nutrient requirements (protein and amino acids, energy, lipids and fatty acids, carbohydrates, vitamins and minerals), practical diet formulations and feeding practices by species.

This book should be useful to students, researchers, fish nutritionists, aquaculturists, feed formulators and manufacturers, extension specialists, producers and persons interested in fish nutrition and fish production. We hope this book will fulfil its intended purpose of serving as an important and valuable reference resource in nutrition of the species listed. The book may also serve as a guide for species for which little nutritional information is available. It is hoped that this book will also stimulate research on nutrition for species where the information is lacking.

Carl D. Webster
Chhorn Lim

Acknowledgements

The editors gratefully acknowledge the contributions made by the chapter authors, and thank Laura Muzinic and Kenneth Thompson for assistance with graphics, and Michelle Coyle for her typing and proofreading prowess. The preparation of this book has involved the cooperative efforts of many people, of whom we are extremely appreciative. Our gratitude is also extended to our families for their enormous patience and support during the preparation of this book.

Carl D. Webster: this book is dedicated to my wife Caroline, our daughter NancyAnn and our other 'children', and our beloved Darwin, Misty, KC, Poppins, Michael, Ignatz and Barley (all dogs go to heaven).
Chhorn Lim: this book is dedicated to my wife Brenda and our children Chheang Chhun, Chhorn, Jr and Brendan.

Introduction to Fish Nutrition

1

Carl D. Webster[1] and Chhorn Lim[2]
[1]Aquaculture Research Center, Kentucky State University, Frankfort, KY 40601, USA; [2]USDA-ARS, Aquatic Animal Health Research Laboratory, Auburn, AL 36830, USA

Proteins and Amino Acids

Proteins are organic compounds comprised of conjugated amino acids in proportions that are characteristic of each protein. This nutrient always contains carbon, hydrogen, oxygen and nitrogen; however, some proteins also contain sulphur, phosphorus and iron. Since most proteins contain about 16% nitrogen, crude protein of a product can be obtained by determining the nitrogen content and then multiplying that value by 6.25 (100% ÷ 16% = 6.25). Plants generally contain their protein in actively growing portions, such as leaves and seeds, while animals tend to distribute protein in a wider variety of tissues, such as bones, skin, organs, feathers, scales and muscle. Body protein is constantly undergoing two processes: protein synthesis (anabolism) and protein breakdown (catabolism). Protein deposition (or loss) equals the amount of protein synthesis minus the amount of protein breakdown.

The basic structural component of proteins consists of amino acids. Ten amino acids cannot be synthesized by vertebrates, including fish, and must be supplied in the diet. These essential (indispensable) amino acids are: arginine, histidine, isoleucine, leucine, lysine, methionine, phenylalanine, threonine, tryptophan and valine (Fig. 1.1). There are other amino acids that fish can synthesize. These amino acids are termed non-essential (dispensable) amino acids and consist of: alanine, asparagine, aspartic acid, cysteine, cystine, glutamic acid, glutamine, glycine, hydroxyproline, proline, serine and tyrosine. Non-essential amino acids can be synthesized by transfer of an amino group to α-keto acids which can be derived from non-protein sources, such as glucose.

When a particular essential amino acid is deficient in a diet, it is referred to as a limiting amino acid because it limits the synthesis of protein. All of the amino acids needed to synthesize a protein must be available or no synthesis can occur. This is why protein quality is important in fish nutrition. Protein synthesis involves a series of reactions that are specific for each protein. Each protein is

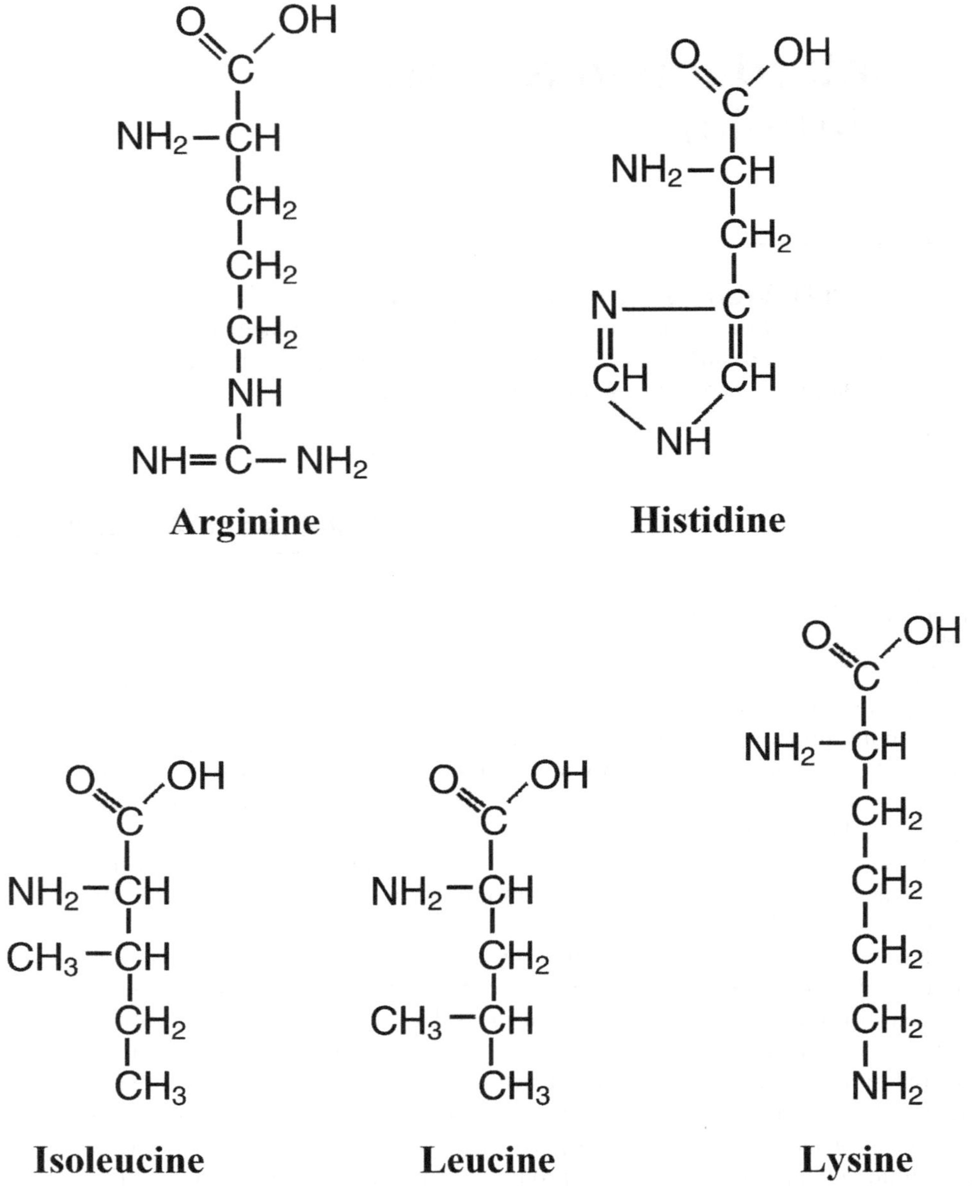

Fig. 1.1. Chemical structures of the ten essential amino acids (continued opposite).

manufactured according to the code derived from deoxyribonucleic acid (DNA), a chromosomal component of the cell. Messenger ribonucleic acid (mRNA) transcribes the information from the DNA and brings it to the cytoplasm of the cell as a single strand; transfer RNA (tRNA) carries the specific amino acids to

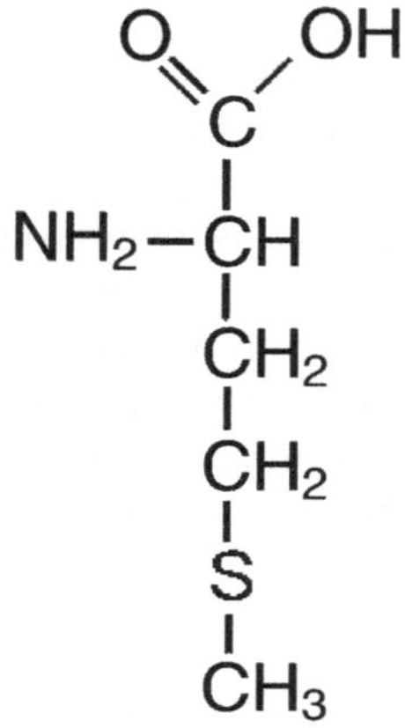

Methionine

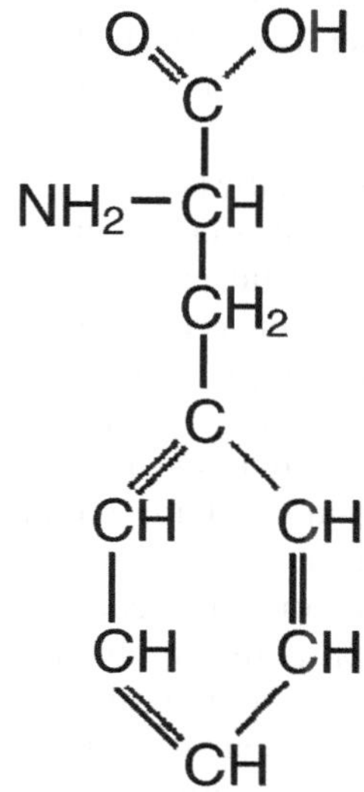

Phenylalanine

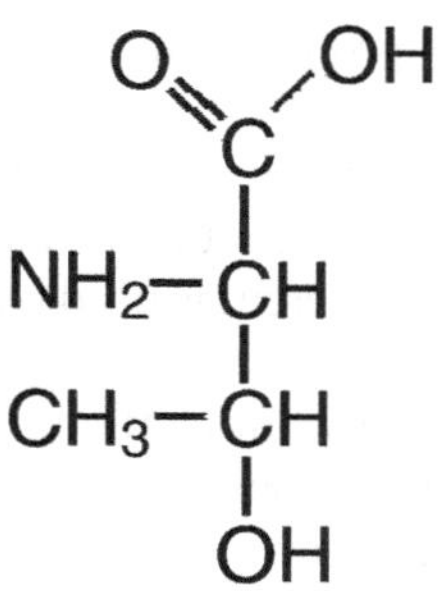

Threonine

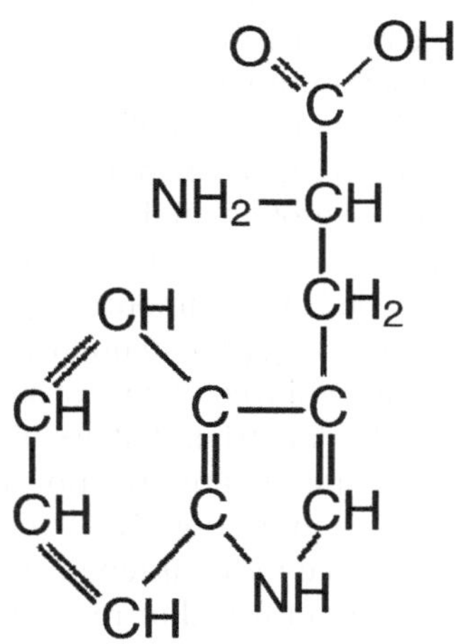

Tryptophan

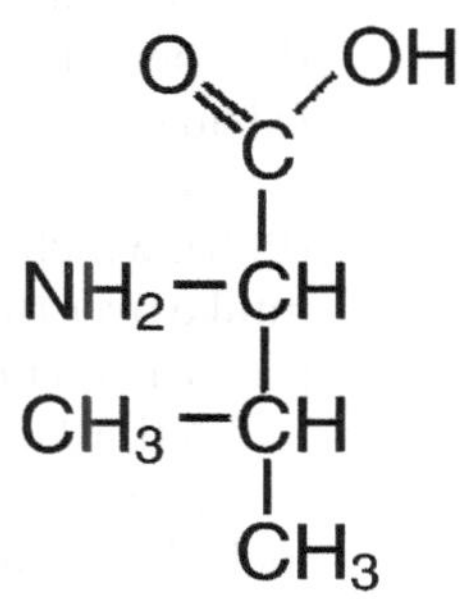

Valine

the ribosome, where they interact with mRNA; ribosomal RNA (rRNA) is a component of the ribosomes in the cell and is the site of protein synthesis. As the DNA message is decoded, tRNA brings the proper amino acids to be conjugated. The amino (NH_2) portion of one amino acid will combine with the carboxyl (COOH) portion of another amino acid, forming a peptide linkage and releasing water (H_2O). This process of joining amino acids is continued until the specific molecule (protein) encoded by the DNA is produced. Each sequence of amino acids is a different protein and these different proteins accomplish different functions in the body.

Proteins occur in nature in a number of forms that possess unique chemical properties and can be divided into simple proteins and conjugated proteins. Simple proteins consist of only amino acids or their derivatives, while conjugated proteins are joined to other non-protein substances. Simple proteins consist of the albuminoids, albumin, globulins, hormones and blood proteins, contractile proteins, such as actin–myosin, and keratins. Conjugated proteins include the chromoproteins (combination of a protein and a pigmented substance), lecithoproteins (combination of a protein and lecithin), lipoproteins (combination of a lipid and a protein), metalloproteins (combination of a metal and a protein), nucleoproteins (combination of a nucleic acid and a protein) and phosphoproteins (combination of phosphorus and a protein; the phosphorus is in a form other than phospholipid or nucleic acid).

Protein requirement is somewhat of a misnomer in that it is actually the requirement for essential amino acids that is important for the proper growth, development and health of fish. While crude protein values are important to a nutritionist formulating a fish diet, they may not allow the nutritionist to determine the quality of the protein. Gelatin is a high-protein source; however, it is virtually devoid of tryptophan. Feather meal, while high in protein, is very poorly digested and utilized by fish, resulting in reduced growth if incorporated at high levels in the diet. Thus, it is important to know the amino acid composition and digestibility of a protein source, as well as its total protein content, when formulating a diet. There are some measurements and mathematical equations that one can utilize to quantify protein quality. One such method is to determine the apparent protein utilization (APU). APU is the amount of protein gain of fish fed an experimental diet, divided by the amount of protein fed.

A second means of determining protein quality is the protein efficiency ratio (PER), which is defined as the total weight gain of fish divided by the total protein intake over the period of the feeding trial. PER assumes that all protein is used for growth and no allowance for maintenance (turnover) is made.

A third method, net protein utilization (NPU) (or net protein value), attempts to take into account protein used for maintenance. This is accomplished by the inclusion of a treatment group of fish that receive diets containing no dietary protein. The NPU is the protein gain of a group of fish fed the experimental diet minus the protein loss of a similar group fed a protein-free diet, divided by the weight of the protein consumed.

Most proteins are highly digestible. In fish possessing a stomach, pepsins and hydrochloric acid (HCl) initiate proteolysis. Once in the intestine, luminal enzymes, such as trypsins, chymotrypsins and carboxypeptidase A and B, further break down the protein into amino acids. Amino acids are absorbed across the brush border of the intestines in fish. In agastric fish (fish without stomachs) proteolysis begins directly in the anterior intestine.

There are a number of factors that affect the protein requirement of fish. One is the size of the fish; small fish require more protein than larger fish. As mentioned previously, protein quality is important. A third factor is water temperature; when water temperature is at or near a fish's optimum temperature for growth, a higher dietary protein level is needed for optimum growth. A fourth factor is feeding rate; if fish are fed all they consume, the lower the protein level required compared with fish fed a restricted (not all the diet a fish will consume) amount of diet. A fifth factor is the presence of natural food items in the culture system. Fish stocked in ponds at low density may have access to more natural foods than those stocked at high density or in a raceway. Lower dietary protein level or poorer-quality dietary protein may be used for fish with access to natural foods as compared with fish cultured where natural foods are absent or limiting. Lastly, the dietary energy level affects the dietary protein requirements of fish. If the non-protein energy level of a diet is low, the fish will utilize protein to meet metabolic energy needs. This is inefficient and financially wasteful. If the energy level is too high, it may suppress food intake and the fish will not consume enough diet to meet protein requirements. This may reduce fish growth.

Energy

Energy is not a nutrient but is a property of nutrients that is released during metabolic oxidation of protein, lipids and carbohydrates. Thus, energy is an abstraction; it can only be measured as it is converted from one form to another. Energy is defined as the capacity to do work, but in a biological sense it is muscle activity and energy needed to run chemical reactions in the body, to move molecules against a concentration gradient and for other bodily processes. Fish, like other animals, must obtain their energy from oxidizing chemical bonds. Energy is measured in heat units called calories and is defined as the amount of heat needed to raise the temperature of 1 g of water by 1°C (from 14.5°C to 15.5°C).

The energy of ingested diets is partitioned into various fractions in terms of animal utilization (Fig. 1.2). Gross energy (GE) is the energy released as heat when an organic substance is completely oxidized to carbon dioxide (CO_2) and water using a bomb calorimeter. Intake energy (IE) is the gross energy in the diet (food) consumed and can be calculated as: IE = GE (amount of food consumed). Faecal energy (FE) is the gross energy of the faeces. Digestible energy (DE) is the intake energy of the diet (food) consumed that is not excreted in faeces and is calculated as: DE = IE – FE. Urinary energy (UE) and gill excretion energy (GEE) can be measured but are generally negligible in fish. Metabolizable energy (ME) is the intake energy that is not lost in the faeces or urine or through the gills, and can be calculated from the equation: ME = IE – (FE + UE + GEE). Net energy (NE) is the intake energy that is left after FE, UE, GEE and heat energy (HE) are deducted. The NE is the portion of IE that is available for maintenance (NEM) and various productive purposes (NEP), such as growth and reproduction.

Fish, like all animals, need energy to live. The citric acid (Krebs) cycle is the primary producer of adenosine triphosphate (ATP) and hydrolysis of ATP is the

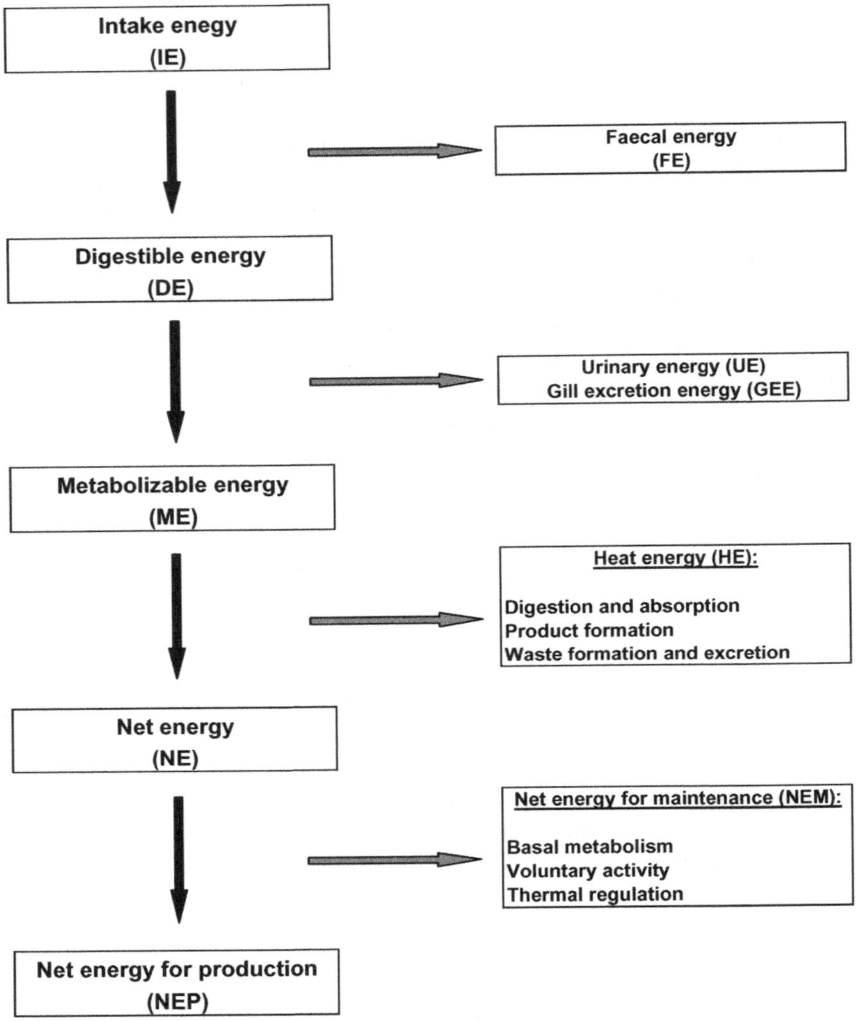

Fig. 1.2. The energy of ingested food is comprised of various fractions. The breakdown of the energy intake occurs in the course of the metabolic process.

source of energy at the cellular level. Complete oxidation of 1 mol of glucose yields 686 kcal and hydrolysis of ATP yields 8 kcal mol^{-1}. Thus, theoretically, the hydrolysis of 1 mol of glucose could produce 85 mol of ATP ($686 \div 8 = 85$); however, in reality, only about 39 mol of ATP are produced, the rest being lost as heat to the surrounding water.

While it appears that fish may have a high protein requirement compared with other animals, in reality, fish have a low requirement for energy. One reason for the lower energy requirement of fish is that they do not need energy for maintenance of body temperature as do mammals. A second reason is that fish live in water and are neutrally buoyant. This minimizes the energy required by the fish's muscles to maintain their position in space. Land animals must maintain their body position against the force of gravity. Since many fish have swim-bladders to assist in maintaining their position in the water column, they require less muscle activity to maintain their position in space. A third reason for the lower energy needs of fish is that fish excrete approximately 85% of metabolic wastes as ammonia (NH_3) directly through the gills into the surrounding water with little energy cost. In contrast, mammals must expend energy to produce urea, while birds expend energy to produce uric acid.

There are a number of factors that affect the energy requirements of fish, including: (i) physical activity; (ii) temperature; (iii) fish size; (iv) growth rate; (v) species; and (vi) food consumption. Physical activity includes swimming, fleeing from a predator and stress. Water temperature has the greatest effect of any of the factors that influence the energy requirement of fish. Since the body temperature of the fish is similar to that of the surrounding water, water temperature near the optimal temperature of a particular fish species leads to an increase in its metabolic rate and increases in appetite, growth rate and activity. Conversely, as the water temperature declines below the optimal range, metabolic rate is reduced. The size of the fish influences energy requirement because small fish tend to have higher energy requirements per unit weight than do larger fish. Similarly, fast-growing fish will require more energy than slow-growing fish.

For most land animals, carbohydrates are the major sources of energy; however, fish generally do not utilize carbohydrates as well as land animals. Pigs can utilize (digest) carbohydrates at about 90%, while the digestibility of carbohydrates for carps can range from 55 to 60%. Carbohydrates contain 4.1 kcal of GE g^{-1}, and are the least expensive dietary energy source. Thus, they are important constituents in the diets of herbivorous and omnivorous fish species. Further, carbohydrates allow for pellet expansion during extrusion of pellets, which controls pellet density and buoyancy. Proteins contain 5.6 kcal of GE g^{-1} of energy and are readily utilized by fish. However, as protein is the most expensive component in a diet, it is uneconomical to use protein as an energy source. Thus, lipid and carbohydrate sources are added to diets to increase the necessary energy. Lipids are the most concentrated sources of energy added to aquaculture diets; lipids contain approximately 9.4 kcal of GE g^{-1}. Lipids contain more energy because they contain more carbon and hydrogen. Oxidation of 1 g of hydrogen produces 34.5 kcal, while oxidation of 1 g of carbon produces 8 kcal. Carbohydrates have their hydrogens oxidized and only the carbons are not yet oxidized. However, there are many unoxidized hydrogens and carbons in a lipid molecule.

Fish must be fed diets containing appropriate amounts of energy. Since fish eat to satisfy their energy requirement, too much energy in the diet may result in fatty fish, which may decrease dressing percentage or the fish will not consume enough diet to satisfy their protein and other nutrient requirements, and growth rate will be reduced. If there is too little energy in the diet, fish will utilize part of the protein as an energy source, which is economically wasteful.

Lipids and Fatty Acids

Lipids include fats, oils and waxes. Fats are esters of glycerol and fatty acids that are solid at room temperature. Oils are glycerol esters that are liquid at room temperature. Waxes are esters of fatty acids with alcohols other than glycerol. Lipids are essential nutrients in fish diets and contain the three elements carbon, hydrogen and oxygen; however, unlike carbohydrates, a larger proportion of carbon and hydrogen comprise lipid molecules. Because of the larger proportion of carbon and hydrogen atoms, lipids liberate approximately 9.4 kcal of GE g^{-1} and are the best sources of energy in terms of kcal g^{-1} compared with carbohydrates (4.1 kcal of GE g^{-1}) and proteins (5.6 kcal of GE g^{-1}).

Biosynthesis of lipids involves the combining of two carbon units, called acetyl coenzyme A (CoA), into long chains and then adding hydrogen. Most lipid formation occurs in the microsomes of adipose tissue, the liver and the mitochondria of the cell (to a lesser degree). Approximately 98% of the lipids in diet ingredients (foods) and 90% of the lipid in the body occur as triacylglycerols (triglycerides). Triglycerides are a combination of three fatty acid molecules (chains) attached to a glycerol backbone.

Fatty acids are the key components of all lipids. The amount of hydrogen on each carbon (degree of saturation) and the length of the carbon chain determine the physical and nutritional characteristics of lipids. Saturation refers to the number of hydrogen atoms on carbon atoms. When a single bond joins two carbon atoms together, the carbon atoms within the chain have two hydrogen atoms associated with each carbon while the carbon at one of the ends of the fatty acid chain has three hydrogens. The other terminal carbon has an acid group (COOH) with a double bond between the carbon (C) and the oxygen (O), and a single bond between the C and the hydroxyl group (OH). If all the bonds connecting carbon atoms together in the fatty acid chain are single bonds, the fatty acid is said to be saturated with hydrogen (Table 1.1). When carbon atoms within the chain are joined by one or more double bonds, those carbon atoms are only able to have one hydrogen atom bonded to them. Fatty acids that have one or more double bonds are said to be unsaturated. A fatty acid with one double bond is called a monoenoic fatty acid (monounsaturated); a fatty acid which has two double bonds is a dienoic fatty acid; a fatty acid with three or more double bonds is called a polyenoic (polyunsaturated) fatty acid.

The second factor that influences the characteristics of a fatty acid is the length of the chain or number of carbon atoms in the fatty acid molecule. Fatty acid chains are formed by the addition of acetyl CoA (two-carbon) units. Thus, the majority of fatty acids in nature have an even number of carbons and, of these, the most biologically important to fish are those having between 16 and 22 carbon atoms.

There are other important lipids, including phospholipids, lipoproteins and cholesterol. Phospholipids are the structural components of cell membranes. Lipoproteins are the principal vehicle for lipid transport in the blood and there are four main types: chylomicrons, very low-density lipoproteins (VLDL), low-density

Table 1.1. Structure of some common fatty acids.

Common name	No. of carbons (class)	Structure
Saturated		
Caproic	6	$CH_3(CH_2)_4$ COOH
Caprylic	8	$CH_3(CH_2)_6$ COOH
Capric	10	$CH_3(CH_2)_8$ COOH
Lauric	12	$CH_3(CH_2)_{10}$ COOH
Myristic	14	$CH_3(CH_2)_{12}$ COOH
Palmitic	16	$CH_3(CH_2)_{14}$ COOH
Stearic	18	$CH_3(CH_2)_{16}$ COOH
Arachidic	20	$CH_3(CH_2)_{18}$ COOH
Lignoceric	24	$CH_3(CH_2)_{22}$ COOH
Monoenoic		
Palmitoleic	16 (n-7)	$CH_3(CH_2)_5CH{=}CH(CH_2)_7$ COOH
Oleic	18 (n-9)	$CH_3(CH_2)_7CH{=}CH(CH_2)_7$ COOH
Dienoic		
Linoleic	18 (n-6)	$CH_3(CH_2)_4CH{=}CHCH_2CH{=}CH(CH_2)_7$ COOH
Polyenoic		
Linolenic	18 (n-3)	$CH_3CH_2CH{=}CHCH_2CH{=}CHCH_2CH{=}CH(CH_2)_7$ COOH
Arachidonic	20 (n-6)	$CH_3(CH_2)_4CH{=}CHCH_2CH{=}CHCH_2CH{=}CHCH_2$ $CH{=}CH(CH_2)_3$ COOH
Eicosapentaenoic	20 (n-3)	$CH_3CH_2CH{=}CHCH_2CH{=}CHCH_2CH{=}CHCH_2CH{=}$ $CHCH_2CH{=}CH(CH_2)_3$ COOH
Docosahexaenoic	22 (n-3)	$CHCH_2CH{=}CHCH_2CH{=}CHCH_2CH{=}CHCH_2CH{=}$ $CHCH_2CH{=}CHCH_2CH{=}CH(CH_2)_2$ COOH

lipoproteins (LDL) and high-density lipoproteins (HDL). Cholesterol is synthesized in the body, but some species, notably crustaceans, require cholesterol in the diet.

Digestion of lipids differs from digestion of carbohydrates and proteins in that lipids are not water-soluble. Thus, before enzymes can begin to break down lipids, they are first solubilized by bile secreted from the gall-bladder. Bile emulsifies the lipid and allows for digestion to occur. When the emulsified lipids come into contact with lipases (enzymes that hydrolyse lipids), the lipids are broken down into monoacylglycerols (monoglycerides), diglycerides, free fatty acids and glycerol. Micelles, emulsified mixtures of monoglycerides, fatty acids and bile salts, attach to the surface of the intestinal epithelial cells and are absorbed. Once inside the cells, long-chain fatty acids are re-esterified (joined to glycerol) to form triglycerides. The triglycerides combine with cholesterol, phospholipids and lipoproteins to form chylomicrons, which are small lipid droplets. Chylomicrons pass into the lymphatic system and eventually enter the blood, where they are transported to body tissues. Adipose tissue is the major site for the removal and storage of chylomicron triglycerides. However, fatty acids in phospholipids are vital for maintaining cell and subcellular membrane integrity, and fatty acids play a role in cholesterol metabolism and are the precursors of prostaglandins.

Lipids perform four main functions in the body: (i) provide energy; (ii) provide essential fatty acids; (iii) serve as structural components; and (iv) serve in regulatory functions. As sources of energy, lipids are the most concentrated source, with approximately 9.4 kcal of GE g^{-1}. An excess of energy, whether derived from carbohydrates, protein or lipid, is stored as triglycerides within the adipose tissue (cells) of the body. Adipose cells are found beneath the skin, between the muscle fibres and around the abdominal (visceral) organs and the membranes that support the organs. Free fatty acids (FFA) are the sources of immediate energy for fish. As they are oxidized from the fatty acid chains into two carbon segments, they form acetyl CoA, which releases energy during the Krebs cycle (tricarboxylic acid (TCA) or citric acid cycle).

Fish require dietary lipids to meet their essential fatty acid requirements. Each species of fish may have a specific fatty acid requirement, but, in general, it appears that cold-water fish require highly unsaturated fatty acids (HUFA) of the n-3 class of lipids, while warm-water fish require HUFA from either the n-3 or n-6 classes, or a mixture of both. Some fish, such as rainbow-trout (*Oncorhynchus mykiss*) can chain-elongate and desaturate fatty acids with 18 carbons, specifically linolenic acid (18:3n-3), to HUFA with 20 and 22 carbons of the n-3 class, specifically eicosapentaenoic acid (EPA; 20:5n-3) and docosahexaenoic acid (DHA; 22:6n-3). This ability to synthesize EPA and DHA from linolenic acid allows nutritionists to formulate diets containing less expensive plant oils that contain linolenic acid (such as linseed oil) instead of using more expensive marine fish oils, which are rich sources of EPA and DHA (Table 1.2). However,

Table 1.2. Table of fatty acids of various fats and oils.

Common name	Numerical name	A	B	C	D	E	F	G	H	I	J	K
Myristic	14:0				1					8.0	7.6	3.0
Palmitic acid	16:0	13	26	13	48	12	3	12	8	29	18	27
Palmitoleic acid	16:1 (n-7)		1	1						8	8	
Stearic acid	18:0	4	3	2	4	6	2	2	5	4	2	21
Oleic acid	18:1 (n-9)	29	18	75	38	55	58	24	21	13	17	40
Linoleic acid	18:2 (n-6)	54	48	9	9	26	19	54	66	1	2	2
Linolenic acid	18:3 (n-3)						10	8		1	1	0.5
Eicosapentaenoic acid	20:5 (n-3)									10	9	
Docosahexaenoic acid	22:6 (n-3)									13	8	

A, maize oil; B, cottonseed oil; C, olive oil; D, palm oil; E, groundnut oil; F, canola oil; G, soybean oil; H, sunflower oil; I, menhaden oil; J, herring oil; K, beef tallow.

most marine fish have a limited ability to convert linolenic acid into EPA and DHA, and thus the dietary sources of these essential fatty acids must be supplied, usually through the addition of marine fish oil (such as menhaden fish oil, cod-liver oil or herring oil) to the diet. The striped bass (*Morone saxatilis*) and its hybrids are examples of fish that require a dietary source of essential EPA and DHA.

Carbohydrates

Carbohydrates are organic compounds composed of carbon, hydrogen and oxygen. They are one of the major classes of nutrients besides proteins and lipids. Carbohydrates are abundant in plants because they are the storage form of energy in plants, in contrast to animals, which store energy as lipid (fat). Carbohydrates typically fall into three main groups: monosaccharides, oligosaccharides and polysaccharides.

Monosaccharides

Monosaccharides are simple sugars and are rarely found in the natural world. Rather, they occur as components of complex carbohydrate molecules. Simple sugars are categorized by the number of carbon atoms in the molecule. A sugar containing three carbons is a triose, four a tetrose, five a pentose and six a hexose. Pentoses and hexoses are the most abundant groups of monosaccharides.

Pentoses

All pentoses have the same chemical formula, $C_5H_{10}O_5$, but each has a slightly different structure. There are three pentose sugars: arabinose, ribose and xylose. Arabinose is found in gums and, when several arabinose molecules are joined together, a pentosan (pentose sugars polymerized into large molecules) is formed. Ribose is a very important pentose sugar because, when it is joined with pyrimidines and purines, nucleosides are formed. Ribose is also a component of riboflavin (a vitamin). Xylose is produced from the breakdown of woody substances and, when several xylose molecules are joined together, the pentosan xylan is formed.

Hexoses

As with the pentoses, all hexoses have the same chemical formula, $C_6H_{12}O_6$, but slightly different structures. There are four hexoses: fructose, galactose, glucose and mannose. In nature, only fructose and glucose occur in free form. Galactose joined with glucose forms the disaccharide lactose (also called milk sugar). Mannose is found in the polysaccharide mannan.

Derivatives of monosaccharides

Monosaccharides that have been reduced by having a hydroxyl group (OH) replace the ketone or aldehyde group (C-O) are known as sugar alcohols. One such sugar alcohol, glycerol, is important in the metabolic pathway, while other sugar alcohols, such as galactitol, inositol, mannitol, sorbitol and xylitol, are present in some foods. Amino sugars have an amino (NH_2) group substituted for a hydroxyl group (OH). Galactosamine and glucosamine are two important members of this group of amino sugars. Sugar acids are a third group of monosaccharide derivatives and these compounds contain an acid group (COOH) in their chemical structure. Some familiar sugar acids are gluconolactone, glucuronic acid and ascorbic acid (vitamin C).

Oligosaccharides

Oligosaccharides contain between two and ten monosaccharides that are chemically bonded together. Most oligosaccharides are the result of catabolism (breakdown) of polysaccharides. There are three categories of oligosaccharides: disaccharides ($C_{12}H_{22}O_{11}$), trisaccharides ($C_{18}H_{32}O_{16}$) and tetrasaccharides ($C_{24}H_{42}O_{21}$).

Disaccharides

Disaccharides are compound sugars comprised of two monosaccharides and include lactose, maltose, sucrose and trehalose. Lactose (milk sugar) and sucrose (table sugar) are two of the most important carbohydrates in human foods; however, they are not important in fish diets as most fish tend to utilize monosaccharides and oligosaccharides poorly.

Trisaccharides

There are three trisaccharides: maltotriose, melezitose and raffinose; however, these are found only in limited quantities in some plants. Maltotriose is a trisaccharide formed during starch digestion and is comprised of three glucose molecules. Melezitose is found in sap from some coniferous plants and contains two molecules of glucose and one fructose molecule. Raffinose is found in sugar beets, molasses, beans and cottonseed meal, and consists of one molecule each of glucose, fructose and galactose.

Tetrasaccharides

There are two sugars in this group: stachyose and maltotetrose. Stachyose is composed of two molecules of galactose, one molecule of glucose and one molecule of fructose. Stachyose and raffinose, two sugars found in beans, cannot be split into monosaccharides by enzymes in the digestive tract of some monogastric animals and must be broken down by microbial action. This results in fermentation, which produces gas in the digestive tract and results in flatulence in many mammals, including humans. Maltotetrose is formed during the digestion of starch and consists of four glucose molecules.

Polysaccharides

Polysaccharides are large sugar complexes that contain repeating chains of monosaccharides. There are three categories of polysaccharides: pentosans $(C_5H_8O_4)_n$, hexosans $(C_6H_{10}O_5)_n$ and mixed polysaccharides.

Pentosans

There are two pentosans, araban and xylan, which yield pentose sugars when hydrolysed. Araban is a chain of arabinose molecules while xylan is a chain of xylose molecules. Both are widely found in plant polysaccharides.

Hexosans

Hexosans are polysaccharide sugars that contain hexose as their repeating sugar units. There are six hexosans: cellulose, starch (amylose and anylopectin), dextrins, glycogen, inulin and mannan. For fish, these are the most important carbohydrate sources, except for cellulose, which is indigestible.

Mixed polysaccharides

There are several mixed polysaccharides found in nature, many of which serve structural or protective roles. These include agar, carrageenan, chitin, hemicelluloses, pectins and gums.

Carbohydrates have several important roles in fish nutrition and are useful in fish diets for numerous contributions. Carbohydrates, like proteins, provide approximately 4.1 kcal of GE g^{-1}. Following digestion and absorption, carbohydrates can be used to meet the immediate energy needs of cells, converted to glycogen and stored in the liver and muscle for later energy needs or converted to fat and stored in adipose tissue for energy reserves. Carbohydrates can spare protein in some fish species by supplying metabolic energy, which spares protein for anabolic purposes, such as tissue synthesis. However, not all carbohydrates yield nutrients that can be utilized by fish. Cellulose, gums, pectins and hemicelluluses are commonly referred to as fibre. Generally, fibre is undesirable in fish diets since it may hinder nutrient absorption and results in the production of diet pellets that have low binding properties and water stability.

Energy is derived from the catabolism of glycogen (or glucose) to pyruvate with the generation of ATP. In the glycolytic pathway, glycogen is converted to glucose-1-phosphate; glucose-6-phosphate; fructose-6-phosphate; fructose-1, 6-diphosphate; glyceraldehyde-3-phosphate; 1, 3-diphosphoglycerate; 3-phosphoglycerate; 2-phosphoglycerate; and ultimately pyruvate. The pyruvate is then converted to ATP in the citric acid (TCA) cycle. Steps in the TCA cycle are pyruvate to citrate, isocitrate, oxalosuccinate, α-ketoglutarate, succinyl-CoA, succinate, fumarate, malate and oxaloacetate. When 2 mol of pyruvate are oxidized, 30 mol of ATP are produced. These 30 mol of ATP are added to the 6–8 mol of ATP produced in the glycolytic pathway. Since each mole of ATP yields 8 kcal, approximately 300 kcal are produced from oxidation of each mole of glucose.

Some carbohydrates and their derivatives have special functions. Glucuronic acid detoxifies chemicals and bacterial by-products; heparin prevents blood clotting; and carbohydrates that bind to lipids occur in nerve tissues.

Vitamins and Minerals

Vitamins

Vitamins are organic substances that are essential for growth, health, reproduction and maintenance, but required in small amounts. Since fish cannot synthesize vitamins at all or can only synthesize in insufficient quantity for normal development, growth and maintenance, they must be supplied in the diet. Each vitamin performs a specific function in the body and one vitamin cannot substitute for or replace another vitamin. Vitamins can be classified into two groups: fat-soluble and water-soluble. The fat-soluble vitamins include vitamin A, vitamin D, vitamin E and vitamin K; water-soluble vitamins consist of biotin, choline, folic acid, niacin, pantothenic acid (B_3), riboflavin (B_2), thiamine (B_1), pyridoxine (B_6), cyanocobalamin (B_{12}), and vitamin C (ascorbic acid). The fat-soluble vitamins contain carbon, hydrogen and oxygen, while water-soluble vitamins contain these three elements plus nitrogen, and cyanocobalamin contains cobalt. Fat-soluble vitamins are absorbed from the intestinal tract along with lipids, and any condition or factor that increases lipid absorption will increase the absorption of fat-soluble vitamins. Absorption of water-soluble vitamins is simpler because water is constantly absorbed from the intestine into the bloodstream.

Water-soluble vitamins are generally not stored in the body and excess vitamins are excreted; however, excess fat-soluble vitamins can be stored in the body. Excesses of fat-soluble vitamins in fish diets may cause physiological or health problems. Thus, adding excess amounts of vitamins not only may be financially wasteful, but may compromise the well-being of the fish.

Not all the vitamins present in diet ingredients are in available forms. For instance, niacin in many cereal grains is bound to protein and cannot be absorbed unless the ingredient is treated with an alkali. Further, vitamins are destroyed during diet processing and storage, due to heating, oxidation, sunlight or mould growth. Thus, some vitamins may need to be added to the diet in excess of requirements due to the anticipated losses during diet processing, production and storage. The requirements and deficiency signs for various vitamins will not be presented in this chapter as this information is provided in detail in the following chapters.

Vitamin A

Vitamin A can exist in several forms as vitamin A alcohol, vitamin A aldehyde and vitamin A acid. The alcohol is the most common form and is generally referred to as retinol; the aldehyde is called retinal or retinene; the acid is called retinoic acid. Retinol (traditionally called vitamin A) is an ester (retinyl palmitate) and is biologially active as an alcohol, or aldehyde, and an acid (Fig. 1.3). Dehydroretinol, also called vitamin A_2, differs from retinol in that it has an extra double bond and has only 40% of the biological activity.

Vitamin A is a fairly colourless fat-soluble vitamin. It is insoluble in water and thus almost no absorption of vitamin A occurs in the stomach. In the intestine, vitamin A is emulsified with bile salts and absorbed into the intestinal mucosa. Vitamin A is essential for a number of physiological processes: vision, growth, reproduction, and coenzyme and hormone roles. The best understood role of vitamin A is related to vision. When light hits the retina, the pigment rhodopsin (which contains vitamin A) is converted to another pigment, called retinaldehyde. As a result of this change, images are sent to the brain via the optic nerve. Rhodopsin is reconverted in the dark, but some vitamin A is lost in the reaction and vision would be impaired if sufficient vitamin A were not supplied in the diet or by body stores to replace that which is lost.

Vitamin D

Vitamin D consists of approximately ten different sterol compounds with vitamin activity; however, only two are of practical importance: ergocalciferol (vitamin D_2 or calciferol) (Fig. 1.4) and cholecalciferol (vitamin D_3) (Fig. 1.5). Cholecalciferol is the form added to fish diets. This is transported by a specific vitamin D carrier protein to the liver, where it is converted to 25-hydroxycholecalciferol (25-OH-D_3) and is eventually converted to 1, 25-(OH_2)-D_3. This compound is then transported and stored in fatty tissues, skeletal muscles and bones. Vitamin D is associated with calcium and phosphorus metabolism, influencing the absorption of these minerals and their deposition in bones; vitamin D increases calcium and phosphorus from the intestine. However, excessive dietary amounts of vitamin D can lead to hypercalcaemia and can cause reduced growth and abnormal deposition of calcium in soft tissues (such as the heart and blood-vessels) which

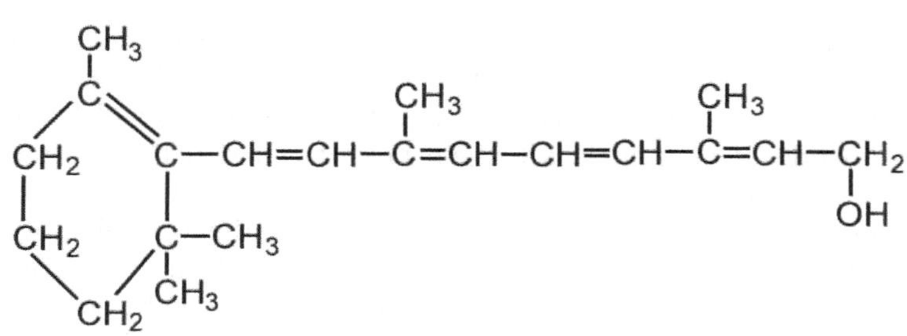

Fig. 1.3. Retinol (vitamin A_1).

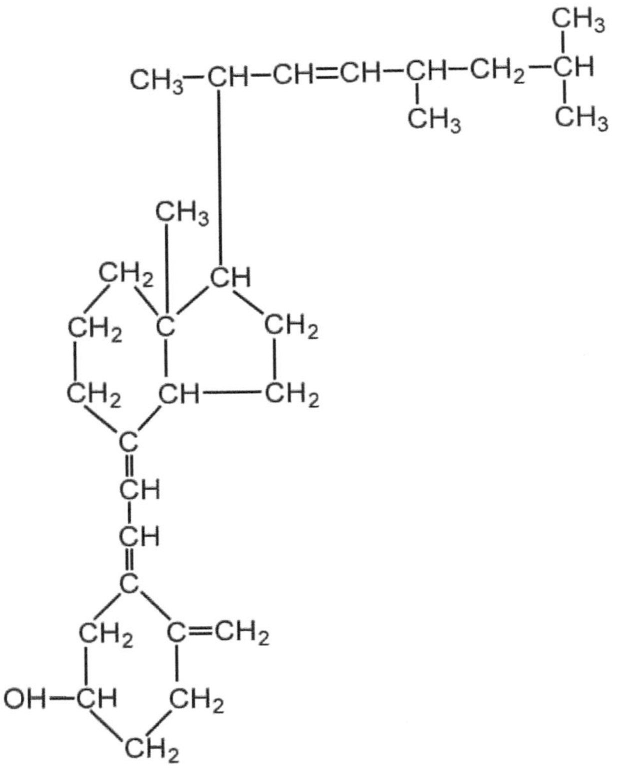

Fig. 1.4. Vitamin D_2.

may result in these tissues calcifying so that their normal functioning is impaired or adversely affected.

Vitamin E

There are eight tocopherols and tocotrienols that have been identified as having vitamin E activity and all are called vitamin E. The molecule with the most vitamin E activity is α-tocopherol (Fig. 1.6). Absorption takes place in the intestine and the presence of bile and lipid is required for proper absorption. Once absorbed, vitamin E is transported to and stored in adipose tissue, liver and muscle. The tocopherols and tocotrienols are soluble in alcohol and lipid solvents, but, as with the other fat-soluble vitamins, are insoluble in water. Vitamin E is stable to heat, but destroyed by exposure to oxygen, iron, copper and ultraviolet light. The primary function of vitamin E is to help protect cell structure, intracellular components and enzymes from degradation. It is a powerful antioxidant that not only prevents the rancidity of lipids in the diet and the fish's digestive tract, but also interrupts the oxidation of highly unsaturated fatty acids in tissue membranes, especially cellular membranes. In an animal's body, vitamin E also protects vitamin A, carotene, vitamin C and ATP from being oxidized. Vitamin E also works with selenium to protect cell membranes.

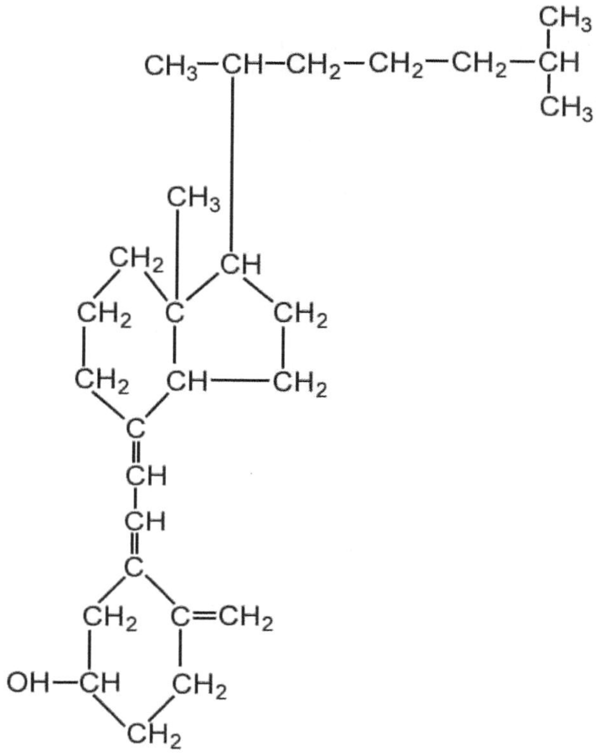

Fig. 1.5. Vitamin D_3.

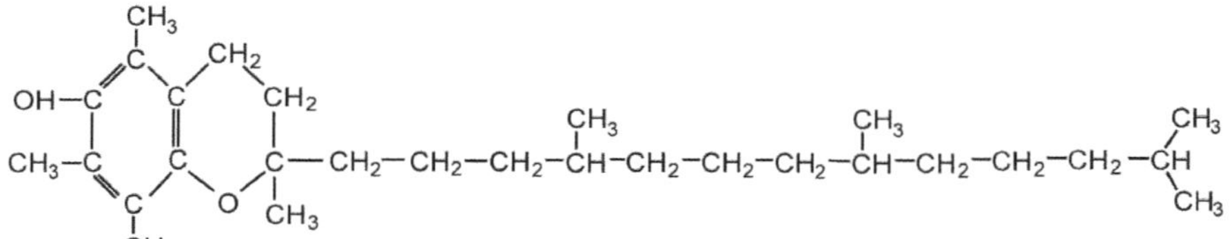

Fig. 1.6. Vitamin E (α-tocopherol).

Vitamin K

There are two naturally occurring forms of vitamin K: vitamin K_1 (phylloquinone), which occurs in green plants, and K_2 (menaquinone), which is synthesized by microorganisms. There are also several synthetic compounds that possess vitamin K activity; the best known and commonly utilized vitamin K is menadione (zimethyl-1, 1, 4-naphthoquinone), also known as K_3 (Fig. 1.7). Menadione is converted to vitamin K_2 in the body and is two to three times more potent than either K_1 or K_2. Vitamin K is absorbed from the intestine and, since it is fat-soluble, it requires bile for maximum absorption. Vitamins K_1 and K_2 are stored only in small amounts in the liver and, unlike the other fat-soluble vitamins, any excess can be excreted.

Biotin

Biotin is a sulphur-containing vitamin and is a cyclic derivative of urea with an attached thiophene ring (Fig. 1.8). Biotin is absorbed from the intestine, but, if avidin, a protein in raw egg-white, is present, absorption is prevented because avidin binds the biotin. Heating egg-whites inactivates the avidin, thus eliminating its ability to bind to biotin. Some fish, such as tilapia, have bacteria in the

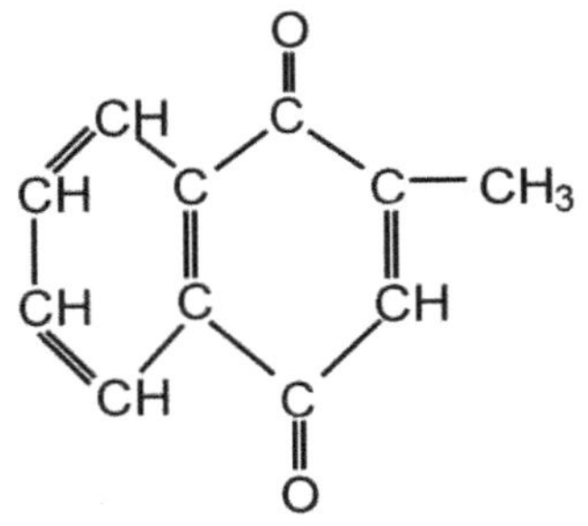

Fig. 1.7. Menadione (vitamin K_3).

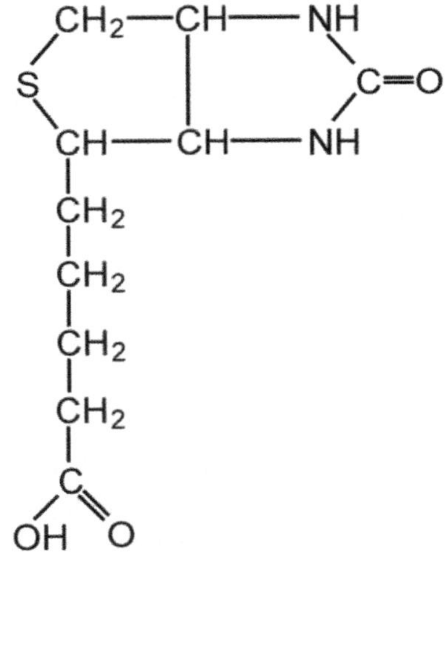

Fig. 1.8. Biotin.

intestine that can synthesize biotin, but even this synthesis may occur too late in the intestine for it to be maximally absorbed. Thus, a dietary source of biotin should be added to fish diets to meet their dietary requirements. Biotin is a coenzyme for transferring CO_2 from one compound to another (decarboxylation and carboxylation). These reactions are involved in carbohydrate, lipid and protein metabolism and include the interconversion of pyruvate and oxaloacetate, the interconversion of succinate and propionate and the conversion of malate to pyruvate.

Vitamin C

Vitamin C, also known as ascorbic acid, is closely related in structure to the monosaccharide sugars. It is synthesized from glucose and other simple sugars by plants and many animal species. However, humans, monkeys, guinea-pigs, fruit-eating bats and teleost fish cannot convert glucose to ascorbic acid because they lack the enzyme L-gulonolactone oxidase. Thus, a dietary source of vitamin C must be provided to meet their requirement. Two forms of vitamin C occur in the body: ascorbic acid and dehydroascorbic acid (Fig. 1.9). While most of the vitamin C exists as ascorbic acid, it can be oxidized to dehydroascorbic acid. Vitamin C is readily absorbed in the intestine and is taken up by several tissues, including the retina of the eye, liver, spleen, brain and kidney. Oxidation of vitamin C is accelerated by air, light, heat, copper and iron. Thus, when adding ascorbic acid to fish diets, processing losses must be factored in so that extra vitamin C is added to compensate for processing losses.

Vitamin C is important in the formation and maintenance of collagen; metabolism of the amino acids proline, lysine, tyrosine and tryptophan; absorption and transport of iron; metabolism of lipids and cholesterol; as an antioxidant in the protection of vitamins A and E and highly unsaturated fatty acids; in the development of strong bones; and in the metabolism of folic acid.

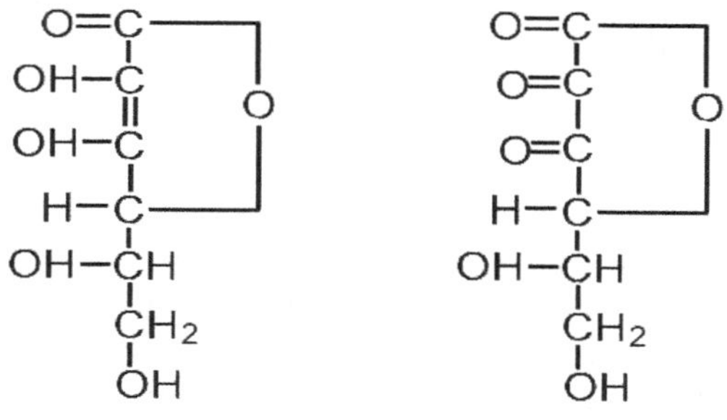

Ascorbic acid **Dehydroascorbic acid**

Fig. 1.9. Ascorbic acid (vitamin C).

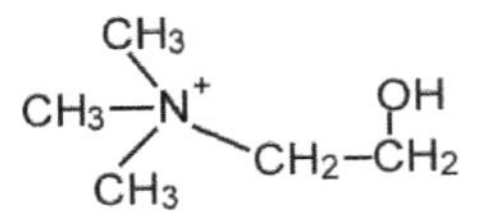

Fig. 1.10. Choline.

Choline

Choline ($C_5H_{15}NO_2$) is a simple molecule with three methyl (CH_3) groups joined to a nitrogen (Fig. 1.10). Choline is absorbed in the small intestine and is water-soluble. It is extremely hygroscopic (absorbs water when exposed to air) and forms stable crystalline salts, such as choline chloride. It is fairly stable to heat, but unstable in strong alkali. Choline is a constituent of phospholipids (primarily lecithin) and prevents fatty livers through the transport of lipids. It is also a component of acetylcholine and is important in nerve transmission, and it acts as a methyl donor.

Cyanocobalamin

Cyanocobalamin, also called vitamin B_{12}, is the largest and most complex of all the vitamins in terms of molecular structure (Fig. 1.11). The main component of the molecule is a porphyrin ring containing cobalt. A cyanide (CN) group may be attached to the cobalt, producing the compound cyanocobalamin, which is the commercially available form of the vitamin. The cyanide group can be replaced by a hydroxyl group (OH), giving hydroxycobalamin (the naturally occurring form of the vitamin) and can be replaced by a nitrite group (NO_2) producing nitritocobalamin. Vitamin B_{12} is absorbed in the intestine and is stable to heat but destroyed by light and strong acid or alkaline solutions.

In fish, vitamin B_{12} is converted to one of two active coenzyme forms: coenzyme B_{12} (adenosylcobalamin) or methyl B_{12} (methylcobalamin). Coenzyme B_{12} has an adenosine ribonucleoside attached to the cobalt atom in place of the cyanide group. Vitamin B_{12} is essential for blood formation and the synthesis of myelin (a lipoprotein of the nervous system), is required for normal carbohydrate and lipid metabolism and serves as a coenzyme in the biosynthesis of methyl groups (CH_3).

Folic acid (folacin)

There is no single vitamin with the name 'folacin', but the term is used to describe a group of closely related substances having the biological activity of folic acid (Fig. 1.12). Folic acid is absorbed by active transport and by diffusion in the intestine and is slightly soluble in water, unstable to heat and destroyed by exposure to light. After it has been absorbed, folic acid is changed into several active coenzyme forms, including tetrahydrofolic acid. These coenzymes transfer single carbon units from one compound to another and are involved in the formation of haem (the protein containing iron in haemoglobin), formation of tyrosine from

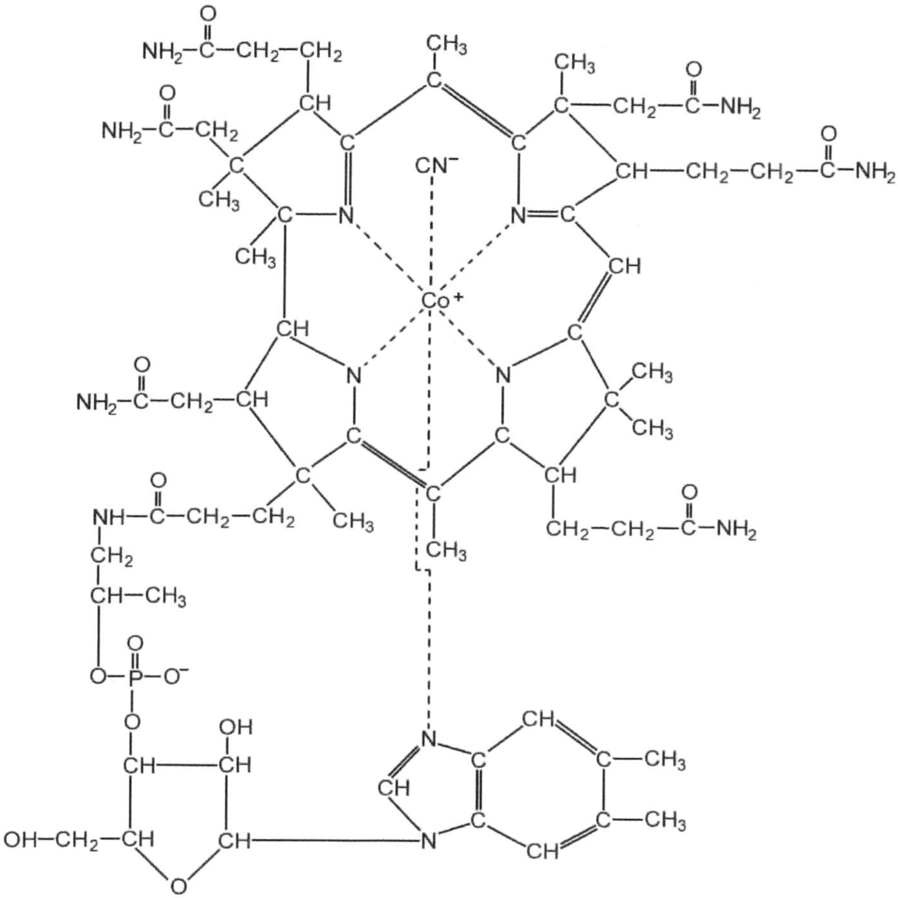

Fig. 1.11. Vitamin B_{12}.

phenylalanine and glutamic acid from histidine, synthesis of choline from ethanolamine and formation of methionine from homocysteine.

Niacin

Niacin is a term used to include several B-complex vitamins, including nicotinic acid and nicotinamide (Fig. 1.13). Both of these natural forms have equal niacin activity. Both nicotinic acid and nicotinamide are derivatives of pyrimidine. Niacin is readily absorbed from the intestine and transported to the liver, where it is converted to nicotinamide adenine dinucleotide (NAD). Niacin is found in the body tissues as either NAD or NAD phosphate (NADP). Nicotinic acid and nicotinamide are soluble in water and are not destroyed by light, oxidation or heat. The principal role of niacin is as a component of NAD and NADP. These coenzymes function in systems necessary for cell respiration. Along with the riboflavin and thiamine-containing coenzymes, NAD and NADP serve as hydrogen acceptors and donors in oxidation–reduction reactions that release energy. Both NAD and NADP are also included in fatty acid synthesis, protein synthesis and DNA synthesis.

Pantothenic acid

Pantothenic acid, also known as vitamin B_3, is composed of pantoic acid and β-alanine (Fig. 1.14). Like other B vitamins, pantothenic acid is readily absorbed from the intestine and is transported to body tissues, with high concentrations in the liver. Calcium pantothenate, the commercially available form of the vitamin, is water-soluble and very stable. Pantothenic acid is a component of two enzymes: CoA and acyl carrier protein (ACP). CoA and ACP are required by cells to synthesize fatty acids. Pantothenic acid (as a part of CoA) is also required in the citric acid cycle (TCA cycle, also called the Krebs cycle), for the formation of acetylcholine and in the metabolism of proteins, lipids and carbohydrates.

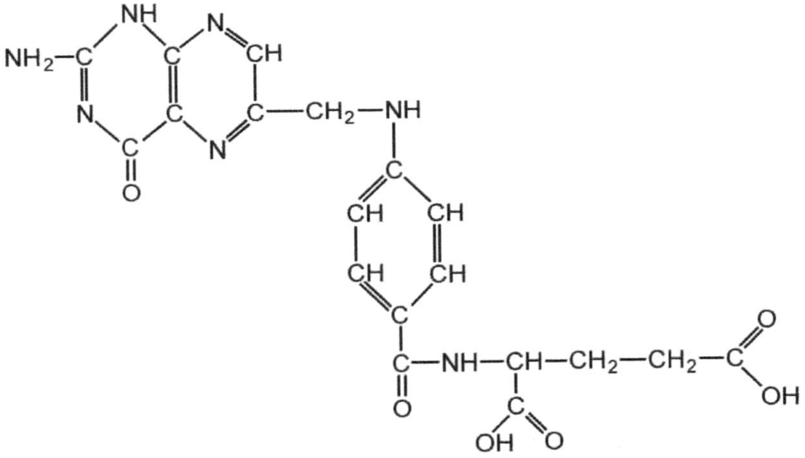

Fig. 1.12. Folic acid.

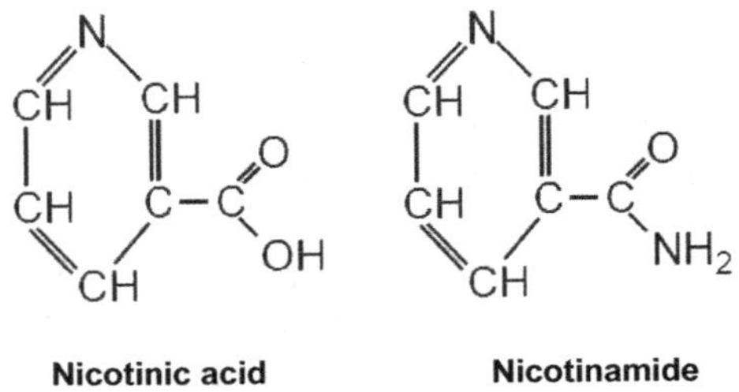

Fig. 1.13. Niacin.

Pyridoxine

Pyridoxine, also known as vitamin B_6, is found in three readily interconvertible forms: pyridoxine (Fig. 1.15), pyridoxal phosphate and pyridoxamine phosphate. In the free form, pyridoxine is absorbed from the intestine and is readily soluble in water and stable to heat, but is destroyed by oxidation and ultraviolet light. In its coenzyme forms (usually as pyridoxal phosphate or pyridoxamine phosphate), it is involved in a large number of physiological processes, especially protein metabolism, but also lipid and carbohydrate metabolism. In protein metabolism, pyridoxine is involved in: (i) shifting an amino group (NH_2) from a donor amino acid to an acceptor acid so that another amino acid can be formed; this process is called transamination; (ii) removal of the carboxyl groups (COOH) (decarboxylation) from certain amino acids to form other compounds, such as serotonin, noradrenalin and histamine; and (iii) assisting in the removal of amino groups from amino acids not needed for growth (deamination) so that carbon residues can be utilized for energy.

Riboflavin

Riboflavin, also called vitamin B_2, is composed of an alloxine ring that is linked to an alcohol derivative of ribose (Fig. 1.16). Riboflavin is absorbed in the intestine by tissue diffusion. It is not very soluble in water and is heat-stable in neutral and acid solutions, but is destroyed by light (especially ultraviolet light). Riboflavin functions as a part of the enzyme group called flavoproteins. Flavin mononucleotide (FMN) and flavin adenine dinucleotide (FAD) are two coenzymes in several important oxidation–reduction reactions by which energy is released to the cell. Riboflavin is involved in fatty acid, amino acid and carbohydrate metabolism. Riboflavin and pyridoxine are also required to produce niacin from tryptophan.

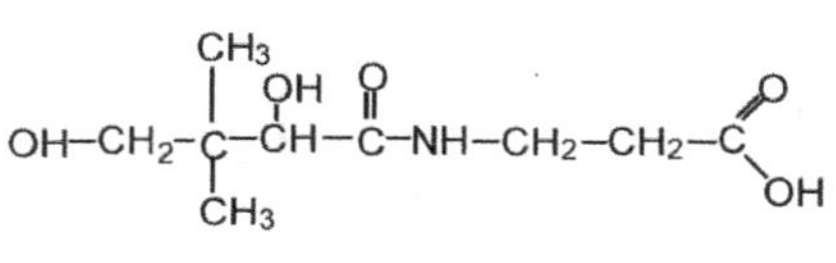

Fig. 1.15. Pantothenic acid.

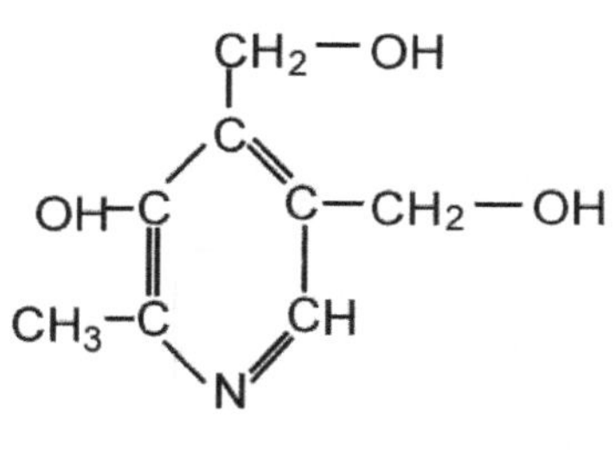

Fig. 1.14. Pyridoxine.

Thiamine

Thiamine, also called vitamin B_1, was the first of the B-complex vitamins to be isolated in pure form (hence the designation B_1). Thiamine (Fig. 1.17) is composed of carbon, hydrogen, oxygen, nitrogen and sulphur and consists of a molecule of pyrimidine (pyrimidine ring) and a molecule of thiazole (thiazole ring) connected by a methylene (CH_2) bridge. Thiamine is water-soluble and heat-stable in acid solutions, but auto-oxidation and ultraviolet light destroy it. Thiaminase, an enzyme present in the raw flesh of some species of fish, also destroys thiamine, so fish diets should not be formulated to contain large amounts of raw fish unless extra thiamine is added. Thiamine is essential as a coenzyme (thiamine diphosphate) in the conversion of pyruvic acid and the formulation of acetyl CoA. Thiamine diphosphate is also a coenzyme in the conversion of glucose to lipid (transketolation).

Minerals

A mineral is an element found in ash when a food or body tissue is burnt. Minerals are classified into two groups based on the relative amounts needed in the diet: macrominerals, elements that are required in large amounts, and microminerals or trace minerals, elements that are required in very small amounts. The general functions of minerals include: structural components of the skeletal system (such as calcium (Ca), phosphorus (P), magnesium (Mg), sodium (Na), and potassium

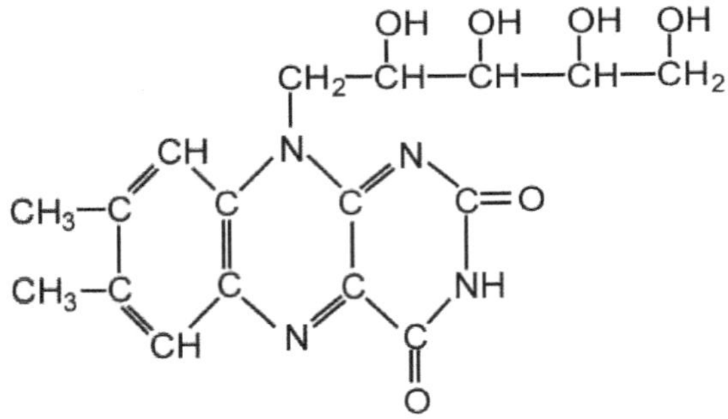

Fig. 1.16. Riboflavin.

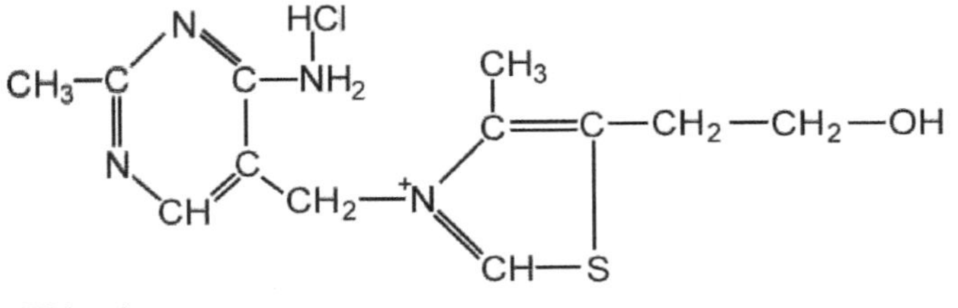

Fig. 1.17. Thiamine.

(K)), components of organic compounds (such as proteins and lipids), enzyme-system activators (such as zinc (Zn) and copper (Cu)) and maintaining acid–base and osmotic balances (such as Na, K and chloride (Cl)).

Macrominerals

CALCIUM The major function of calcium is to give strength to bones, but it is also important in controlling the heartbeat, used in the transmission of nerve impulses and necessary for blood clotting, and calcium controls various enzymes. Calcium is absorbed in the intestine and several factors enhance calcium uptake by the body, including vitamin D, dietary protein and an acid medium. Conversely, a vitamin D deficiency reduces the absorption of calcium, as does an excess of dietary phosphorus, excessive lipid or fibre in the diet and the presence of phytic acid. Phytic acid, which is found in cereal grains and oil-seed meals, prevents the absorption of calcium by forming an insoluble salt (calcium phytate).

CHLORIDE Chloride is an essential mineral because it is an electrolyte and is found in extracellular fluids in the body. The chloride content of the blood is higher than that of any other mineral and the negatively charged chloride ion plays a major role in the regulation of osmotic pressure and acid–base balance.

MAGNESIUM Nearly 70% of magnesium in the body is located in bones, while approximately 30% is found in various fluids and soft tissues (such as the liver and muscle). Magnesium is a constituent of bones and teeth; it is essential for cellular metabolism, often as an enzyme activator involved in high-energy phosphate transfer of adenosine diphosphate (ADP) and ATP; and it is involved in certain peptidase activations for protein digestion. If not enough magnesium is added to a diet, calcium could be deposited in soft tissues, forming calcified lesions, while excess of magnesium may disrupt calcium and phosphorus metabolism.

PHOSPHORUS Phosphorus is readily absorbed by fish from the water via the gills so a dietary supply of phosphorus is theoretically not needed. However, since phosphorus is limiting in aquatic systems, being taken up by plants and algae, a dietary source is needed for proper fish growth and well-being. Phosphorus is essential for bone formation and maintenance, for muscle-tissue function, as a component of nucleic acids, maintaining osmotic and acid–base balances, phospholipid synthesis, protein formation and enzyme systems. Phosphorus is involved in several chemical relationships. If the diet contains an excess of calcium relative to phosphorus, free calcium will be present and form insoluble tricalcium phosphate. An excess of dietary phosphorus over calcium will decrease the absorption of both calcium and phosphorus. Further, excesses of iron, aluminium and magnesium may bind phosphorus to form insoluble salts, which inhibit phosphorus absorption.

Much of the phosphorus in many plant-protein sources (such as soybean meal) and plant ingredients is in the form of phytate, which is poorly utilized and may depress the absorption of iron and calcium. Highly digestible sources of phosphorus, such as monocalcium and dicalcium phosphate, are often added to fish diets to meet the phosphorus requirement of fish.

POTASSIUM Potassium is the third most abundant element in the body, after calcium and phosphorus. Potassium and sodium are closely interrelated in the maintenance of proper osmotic pressure within cells, and these minerals are involved in the maintenance of proper acid–base balance. Potassium ions relax muscles and are used in enzyme reactions. Excessive levels of potassium could interfere with magnesium absorption.

SODIUM Sodium is the major positively charged ion (cation) in the fluid outside the cell (extracellular fluid), where it assists in the maintenance of osmotic and acid–base balance. Sodium is also associated with muscle contraction, nerve function and carbohydrate absorption. Sodium, potassium and chlorine (chloride) all control osmotic pressure and the acid–base equilibrium.

Microminerals

CHROMIUM Chromium is an essential element, but it does not work alone; instead it acts with other substances to control metabolism, such as insulin, several enzymes and DNA and RNA. Chromium, in the form of glucose tolerance factor (GTF), is released into the blood whenever there is a dramatic increase in the glucose and/or decrease in insulin levels in the blood. GTF and insulin both act to allow amino acids, fatty acids and sugars to pass more easily from the blood into the cells of tissues. Chromium, as well as other minerals, can also activate the digestive protease trypsin.

COBALT Cobalt is a component of vitamin B_{12} (cyanocobalamin), which has an essential function in red blood cell formation.

COPPER Copper is a cofactor for several enzyme systems used in energy metabolism, and is required for normal functioning of the brain, skeleton and spinal cord. Copper is also involved in iron metabolism, as it facilitates the absorption of iron from the intestine and releases iron from storage in the liver. However, excesses of cadmium, iron, lead and zinc reduce the utilization of copper. Copper in the diet is relatively non-toxic to most monogastric species; however, fish are very susceptible to copper toxicity if exposed to copper in the water.

IODINE Iodine is an essential component of the thyroid hormones, thyroxine and tri-iodothyronine, which regulate the rate of oxidation with cells. In so doing, the thyroid influences physical growth, nerve- and muscle-tissue functions, circulatory activity and metabolism. Some aquaculture diet ingredients, such as canola meal and mustard seed, can contain goitrogens, which interfere with the production of thyroxine and may produce goitre.

IRON Iron is an essential mineral in the formation of red blood cells. Iron combines with protein to make haemoglobin, the iron-containing compound of red blood cells. Iron is involved in transporting oxygen via red blood cells and is also a component of enzymes that are utilized in energy metabolism. A deficiency of iron may cause microcytic anaemia, characterized by small, pale red blood cells.

An excess of iron in the diet can bind with phosphorus to form an insoluble iron–phosphate complex, which could lead to a phosphorus deficiency. Since free iron is toxic, the iron molecule is transported along with a protein. Two atoms of ferric iron are bound to one molecule of transferrin, a β-globulin protein. If the level of iron ions exceeds the binding capacity of transferrin, iron toxaemia may occur.

MANGANESE Manganese is involved in the formation of bone, blood clotting, insulin function and cholesterol synthesis, and is an activator in several enzymes in the metabolism of proteins, lipids, carbohydrates and nucleic acids.

SELENIUM Selenium is a rare, non-metallic element that appears to be very important in maintaining the health of animals under stressful conditions. Vitamin E and the amino acids cysteine and methionine may act as partial substitutes for selenium in some of its functions. The biological availability of selenium from different sources varies. Selenium in wheat appears to be 100% available; however, only about 33–50% is available in fish-meal, possibly due to selenium being bound to mercury and other heavy metals. Fish diets with high percentages of polyunsaturated fatty acids (PUFA) but reduced levels of vitamin E might increase the requirement for selenium, because PUFA may be converted to toxic peroxides unless there is sufficient vitamin E to prevent this process, and selenium is required to activate the enzyme glutathione peroxidase, which destroys peroxides. Thus, like vitamin E, selenium acts as an antioxidant to protect the cell against oxidative damage. However, these antioxidants act differently; selenium functions throughout the cytoplasm to destroy peroxides, while vitamin E is present in the membrane constituents of the cell and prevents peroxide formation. Thus, the requirements for each nutrient can only be partially offset by addition of the other. Further, when diets are deficient in the amino acids methionine and cysteine, the selenium requirement may increase. This is due to two sulphur amino acids being converted to glutathione, which has a limited ability to carry out the functions of glutathione peroxidase.

ZINC Zinc is poorly absorbed, with approximately 10% of zinc being taken up by body tissues. Further, excesses of calcium, phytic acid or copper inhibit zinc absorption. Zinc is required for normal bone calcification, in the transfer of carbon dioxide in red blood cells, and for the synthesis and metabolism of proteins and nucleic acids.

European Sea Bass, *Dicentrachus labrax*

2

S.J. Kaushik
Fish Nutrition Research Laboratory, Unité Mixte INRA-IFREMER, 64310 Saint-Pée-sur-Nivelle, France

Introduction

European sea bass (*Dicentrarchus labrax*) is a major marine fish species reared in Europe, particularly in the Mediterranean region, and production is expected to continue to increase in the near future. Current production is estimated to be about 45,000 tons, the major producing country being Greece, followed by Turkey, Italy and Spain. About 180 million juveniles are produced annually in hatcheries across Europe and transferred to grow out in marine cage farms in the Mediterranean. Very much appreciated by consumers, the price is about 5.5 Euros kg^{-1}. It is also a reasonable forecast that cultured sea bass will shortly predominate over wild-captured sea bass. The species has been fully domesticated for two or three decades, with all the different phases of production (controlled reproduction, larval culture, grow-out of juveniles to market size) well under control. Despite considerable research effort in recent years on the nutrition and feeding of these species, knowledge of the nutritional requirements is still limited.

Nutrient Requirements

Protein and amino acids

Initial studies (Alliot *et al.*, 1978; Métailler *et al.*, 1981) reported that European sea bass would have very high dietary protein requirements, ranging from 52 to 60%. Subsequently, it was shown that the dietary crude protein levels can be decreased to between 48 and 54% (Hidalgo and Alliot, 1988; Ballestrazzi *et al.*, 1994). More recent works (Pérez *et al.*, 1997; Dias *et al.*, 1998a) show that the optimum growth of sea bass can be achieved with diets having 43–45% protein. Dias *et al.* (1998a) found no growth differences in sea bass fed diets with either

43% or 52% protein, provided the dietary digestible energy levels are maintained above 21 MJ kg^{-1}.

Métailler *et al.* (1973) showed that European sea bass require the same ten essential amino acids (EAA) as the salmonids. Quantitative data on requirements for some of these EAA have been obtained in recent years (Table 2.1), based on dose–response curves using semipurified diets, for arginine (Tibaldi *et al.*, 1994), lysine (Tibaldi and Lanari, 1991), methionine and cysteine (Thébault *et al.*, 1985), tryptophan (Tibaldi *et al.*, 1993), and threonine (Tibaldi and Tulli, 1999). For other EAA, indirect estimations have been made using methods such as those based on overall body protein accretion and an ideal protein composition has been provided (Kaushik, 1998b).

The effect of water temperature on the protein requirements has been studied (15°C and 20°C by Hidalgo and Alliot, 1988; 18°C and 25°C by Peres and Oliva-Teles, 1999). In both studies, protein requirements were estimated to be 48–50% of the diet and found not to be affected by water temperature.

Data on the apparent digestibility coefficients (ADC) of some common feedstuffs for sea bass have been made available in recent years (Santinha *et al.*, 1996; da Silva and Oliva-Teles, 1998). Much as in other teleosts, the ADC of protein is generally high, while that of energy is slightly lower in some of the plant-protein sources, reflecting the variability in the availability of carbohydrates (Table 2.2).

Spyridakis *et al.* (1988) also determined the true and apparent digestibility coefficients for some ingredients in sea bass and found few differences between sources. However, the ADC of some amino acids were found to be different from that of the crude protein, suggesting the need for more systematic work on amino acid availability.

Currently, there are very few data on the energy requirements of European sea bass, either in terms of maintenance or for growth. Data based on oxygen consumption rates of sea bass under unfed conditions suggest that the maintenance requirements would be approximately 14–16 kJ kg^{-1} body weight (BW) per day. In terms of dietary protein-to-energy ratios, although early data indicated values above 22 mg digestible protein (DP) kJ^{-1} digestible energy (DE), the data from Dias *et al.* (1998a) suggest that the optimum protein-to-energy ratio of the diets for sea bass should be 19 mg kJ^{-1}, in diets with at least 21 MJ kg^{-1} DE, with the major portion of non-protein energy being supplied by dietary fats.

Table 2.1. Available data on the quantitative essential amino acid requirements (expressed as g per 16 g nitrogen) of European sea bass.

Amino acid	Requirement
Arginine	4.1
Lysine	4.8
Threonine	2.6
Tryptophan	0.5
Methionine + cystine	4.4

Table 2.2. Apparent digestibility coefficients (ADC) of selected protein sources in European sea bass.

		ADC (%)		
Ingredient	CP (%)	Protein	Energy	Reference
Fish-meal, Denmark	72.3	94.6	95.0	1
Fish-meal, Norway	74.9	96.0	94.5	2
Fish-meal, Portugal	70.1	89.5	86.2	2
Fish-meal, Peru	62.1	94.4	94.0	1
Fish soluble protein concentrate (CPSP G), France	71.7	96.5	94.6	2
Poultry-meat meal	62.2	97.3	96.5	1
Meat meal, defatted, steam-cooked	75.1	92.0	86.4	2
Blood meal	97.1	90.6	92.1	2
Dry brewer's yeast	48.0	88.6	85.0	1
Soybean meal	46.2	88.9	82.2	1
Soybean meal (solvent-extracted)	51.6	89.8	69.3	2
Soy protein concentrate	64.8	97.3	87.9	1
Rapeseed meal	35.6	93.4	78.1	1
Maize gluten meal	54.5	91.3	86.9	1
Wheat gluten	77.0	98.4	93.3	1
Extruded pea meal	24.5	92.5	84.2	1
Extruded lupin-seed meal	42.0	93.7	87.6	1

1, F. Tulli and E. Tibaldi (personal communication); 2, Da Silva *et al.* (1998); CP, crude protein; CPSP G, Concentré Protéique de Solubles de Poissons-Gras (Fish soluble protein concentrate – fat rich)

Lipids and fatty acids

Alliot *et al.* (1979b) observed higher growth rates in sea bass fed diets containing a dietary lipid level of 12% compared with fish fed a diet containing 8% lipid. Two later studies (Métailler *et al.*, 1981; Pérez *et al.*, 1997) with sea bass, however, did not show any beneficial effect from increasing the lipid level. Data of Dias *et al.* (1998a) and Lanari *et al.* (1999) suggest that an increase in dietary lipid up to 18–19% improves protein utilization in sea bass. However, Peres and Oliva-Teles (1999) suggest that very high levels (30% lipid) would lead to a growth depression in sea bass.

Although it is generally recognized that long-chain polyunsaturated fatty acids (PUFA) of the n-3 series, eicosapentaneoic acid (EPA) (20:5n-3), and docosahexaenoic acid (DHA) (22:6n-3), are considered essential for a marine fish such as the European sea bass, precise quantitative data have not been clearly established. Given the importance of PUFA during early embryonic and larval development (Bell *et al.*, 1996), much work has been undertaken with larval stages of sea bass fed PUFA-enriched live prey (Fukusho *et al.*, 1985). An attempt was made to determine the essential fatty acid (EFA) needs of sea bass using formulated dry diets (Coutteau *et al.*, 1996) and their data suggest a requirement of about 1% of n-3 PUFA for sea bass.

Supplementation with between 1 and 2% of phospholipids was also found by Geurden *et al.* (1997a,b) to bring about a significant improvement in the growth of juvenile sea bass, as well as in the retention of fatty acids. Ballestrazzi and Lanari (1996) observed that sea bass is capable of utilizing fatty acid calcium salts of herring oil as efficiently as fish oil.

The fatty acid composition of the neutral lipids in sea bass, as in many other teleosts, reflects that of the dietary fatty acid profile. However, sea bass are prone to have a high fat deposition in the liver. Given that marine oils rich in n-3 PUFA are susceptible to oxidation, the consequences in terms of hepatic enzyme activities as well as muscle peroxidation characteristics (Alvarez *et al.*, 1998) are worth taking into account.

Carbohydrates

European sea bass do not appear to have high digestive amylolytic activity. Digestibility of carbohydrates in sea bass is affected by the nature and complexity of the carbohydrate supplied. Extruded starch, maltose or glucose is better digested than crude starch. Use of precooked or pretreated plant products then becomes necessary for improving starch digestibility and hence DE supply. At the postabsorptive level, a prolonged hyperglycaemia is also reported in sea bass (Peres *et al.*, 1999), with postprandial patterns comparable to those observed in other species. Very early on, Alliot *et al.* (1979b) suggested a possible protein-sparing potential of dietary glucose in European sea bass. While some authors (Pérez *et al.*, 1997) found no growth differences in sea bass fed diets containing either crude or extruded starch, Dias *et al.* (1998a) observed better growth and feed utilization in sea bass fed extruded starch than in those fed crude starch. However, very high levels (above 30%) appear to induce some growth depression.

Vitamins and minerals

Data on quantitative vitamin requirements of European sea bass are extremely scarce. A dietary requirement for vitamin C below 50 mg kg^{-1} diet has been reported for sea bass. A recent work with juvenile sea bass (Fournier *et al.*, 2000) has indicated a much lower requirement for growth but higher needs for maximizing ascorbate concentration in the liver. There is also indirect but clear evidence that the requirements of European sea bass for most water-soluble vitamins would not differ from those established for the salmonids. Data from a recent study (Kaushik *et al.*, 1998) suggest that the recommendations on requirements for water-soluble vitamins established for salmonids (NRC, 1993) can be applied to European sea bass fed practical diets. Although no quantitative requirement data are available as regards fat-soluble vitamins, a dietary supply of vitamin E at levels of 500 mg kg^{-1} diet has been recommended as a measure of defence against peroxidative damage (Messager *et al.*, 1992).

There is a definite lack of data on the mineral requirements of European sea bass. Dietary phosphorus availability has received some attention (da Silva and Oliva-Teles, 1988; Santinha *et al.*, 1996), the bioavailability being higher in feedstuffs of animal origin than in those of vegetable origin. Unpublished data suggest that the requirements would be in the range of 0.6%. The possible improvement of the availability of phosphorus in plant feedstuffs by dietary exogenous phytase has also been demonstrated (Oliva-Teles *et al.*, 1998).

Practical Diets

Larval diets

As with most marine fish, first-feeding larvae of European sea bass are fed live prey items, such as the rotifers *Brachionus* sp. and *Artemia*. There is abundant literature on the beneficial effects of enrichment of these live prey with different nutrients (n-3/n-6 PUFA, vitamins, amino acids). Most of the recommendations as regards nutrient supply to larval sea bass have been obtained using variably enriched live prey (Bell *et al.*, 1996; Merchie *et al.*, 1996; Navarro *et al.*, 1997, 1998). After a period ranging from 25 to 40 days of growing the larvae with such live prey, they are gradually transferred to formulated dry diets. Although all hatcheries still follow the sequential supply of the live prey, there have been several research efforts to develop formulated diets for first-feeding larvae.

These works have shown that digestive enzyme activities are present even during early development of sea-bass larvae and that they are affected by dietary factors (Zambonino Infante and Cahu, 1994a,b; Péres *et al.*, 1996). Other studies have shown the beneficial effects of the use of protein hydrolysates instead of whole proteins in larval diets (Carvalho *et al.*, 1995; Zambonino Infante *et al.*, 1997; Cahu *et al.*, 1999). Similarly, studies by Geurden *et al.* (1997a,b) have identified the importance of dietary phospholipids for growth and tissue lipid composition of sea-bass larvae after transfer from live prey to compound dry diets. Cahu *et al.* (1998) were thus able to rear sea-bass larvae with a compound diet from first feeding onwards, although growth rates of these larvae were significantly lower than those fed *Artemia*. Our own, more recent, unpublished data indicate that it is possible to rear the larvae of European sea bass with formulated dry diets right from the first feeding onwards with equal growth rate and survival to those fed live prey. Given the continued interest and progress in this field of developing artificial diets as a total replacement for live prey, one can expect precise data on the requirements of sea-bass larvae for different nutrients in a short time, which will also require development of semipurified diets for first-feeding larvae.

Juvenile and grow-out stages

Diets for juvenile stages are generally rich in protein and those up to marketable size generally receive extruded dry diets with crude protein levels ranging from 43 to 50% and fat levels between 12 and 25%. Fish-meal remains the major protein source (Table 2.3), although even very early on studies were undertaken by Alliot *et al.* (1979a) to find alternatives to fish-meal. Current diets contain terrestrial plant products, such as soybean meal, maize gluten meal or wheat gluten, supplying anywhere between 30 and 40% of dietary protein. Partial or total replacement of fish-meal by plant ingredients in the diets have been attempted (Dias, 1999; Gouveia and Davis, 2000); however, poor growth performance of sea bass fed such diets has been linked to a reduction in voluntary diet intake. An improvement in voluntary diet intake by the addition of an amino acid mix as an attractant appears to hold promise (Gomes *et al.*, 1997). Addition of plant oils to a commercial diet for sea bass has reportedly decreased growth performance compared with addition of fish-oil (Yildiz and Sener, 1997). As with salmonids, culture of sea bass in the marine environment and how to minimize effects on environmental discharges need to be investigated more thoroughly (Ballestrazzi *et al.*, 1994). Sea bass exhibit postprandial ammonia excretion patterns comparable to those of other teleosts (Dosdat *et al.*, 1996) and the nutritional strategies for the reduction of nutrient loads into the environment are also similar (Kaushik, 1998a).

Brood-stock diets

A reduction of the dietary protein level from 51% to 34% was shown to reduce brood-stock performance in sea bass (Cerdâ *et al.*, 1994). Much emphasis has also been placed on the importance of both n-3 and n-6 highly unsaturated fatty acids (HUFA) for the reproductive performance of sea bass (Bell *et al.*, 1997; Navas *et al.*, 1998; Bruce *et al.*, 1999). Data indicate that marine fish-oils rich in long-chain HUFA are essential for improving the quality of eggs, much as in most other marine fish.

Table 2.3. Example of a practical diet based on fish-meal for European sea bass.

Ingredients	%
Fish-meal, CP > 70%	52.0
Extruded wheat	19.6
Soybean meal, CP 48%	9.5
Fish-oil	15.9
Binder	1.0
Mineral premix	1.0
Vitamin premix (NRC, 1993)	1.0

CP, crude protein.

Blazquez *et al.* (1995) found that a dietary supply of 17-α-methyltestosterone at 10 mg kg^{-1} diet from 126 to 226 days postfertilization was effective for masculinization of sea bass. More recent work by Chatain *et al.* (1999) demonstrated that even lower doses of a synthetic analogue, 17-α-methyldehydrotestosterone (0.5, 3.0 or 5.0 mg kg^{-1} diet), are sufficient to induce complete and permanent masculinization of sea bass, although a small proportion of males (10%) showed testicular deformities. Very few studies have reported that feeding purified diets to sea bass has led to growth rates comparable to those with fish-meal-based practical diets (Table 2.4).

Feeding Practices

Hatching and larval rearing of European sea bass are undertaken on land-based farms with a very good control of the physical and microbial environment. As mentioned above, enriched live prey (rotifers, *Artemia*) are used for rearing larvae during the first 4–6 weeks in circular, self-cleaning tanks. In order to avoid problems related to swim-bladder inflation, surface skimmers are necessary. They are slowly 'weaned' on to formulated dry diets. Juveniles (1–5 g) are transferred from hatcheries to rearing tanks and, after periods ranging from 3 to 6 months, transferred to grow-out cages in the sea. Cage culture is the major production system used for sea-bass culture in the Mediterranean area. The size of the cages is variable, with stocking densities ranging between 6 and 20 kg m^{-3}. There is also an increasing interest in land-based, thermoregulated, totally recirculated systems.

Some of the basic bioenergetic principles as applicable to marine fish feeding were provided recently (Kaushik, 1998a). Dias *et al.* (1998b) showed that, when there is a dilution of dietary nutrients by bulk agents, sea bass are capable of compensating by increasing their diet intake to a certain extent. However, as with salmonids, use of nutrient-dense diets for this species may be advisable. We have also recently shown that voluntary diet intake and feed : gain ratio are inversely related to dietary DE content (Boujard *et al.*, 2000).

Table 2.4. Example of a semipurified diet providing about 8% nitrogen (N) for studies on nutrient requirements of European sea bass.

	Diet	
Ingredients	1	2
Casein, vitamin-free	15.0	25.0
Fish-soluble protein concentrate	1.0	1.0
Dextrin	38.0	20.4
Agar	1.0	1.0
Vitamin mix	5.0	5.0
Mineral mix	5.0	5.0
Fish-oil + lecithin	11.0	11.0
Cellulose	5.0	5.0
Amino acid mix	19.0	26.6

Sea bass generally feed under daylight conditions. A feeding frequency of two to three meals during the light hours of the day is considered most efficient (Tsevis *et al.*, 1991). Recent studies (Boujard *et al.*, 1996; Azzaydi *et al.*, 1998) have also demonstrated that sea bass are capable of using demand feeders and that they exhibit a diel pattern of feeding activity rhythms. Sea bass also seem to exhibit a certain plasticity in their feeding rhythms in that they are capable of adjusting their feeding activity in response to diet availability. Compared with manual or automatic feeding, the use of demand feeders has the definite advantage of avoiding diet wastage and better feed efficiency (Coves *et al.*, 1998; Paspatis *et al.*, 1999), as well as reduced environmental loads (Paspatis *et al.*, 2000). Sea bass are able to adjust their diet intake in accordance with their nutritional needs.

References

Alliot, E., Pastoureaud, A. and Patrois, J. (1978) Etude de l'efficacité protéique de quelques farines de poissons pour l'alimentation du bar, *Dicentrarchus labrax*. *Tethys* 8, 335–338.

Alliot, E., Pastoureaud, A. and Nedelec, J. (1979a) Utilization des farines végétales et des levures cultivées sur alcanes pour l'alimentation du bar *Dicentrarchus labrax*. Influence sur la croissance et la composition corporelle. In: Halver, J.E. and Tiews, K. (eds) *Proceedings of the World Symposium on Finfish Nutrition and Fishfeed Technology*, Vol I. Heenemann, Berlin, pp. 229–238.

Alliot, E., Pastoureaud, A. and Nedelec, J. (1979b) Etude de l'apport calorique et du rapport calorico-azoté dans l'alimentation du bar *Dicentrarchus labrax*. Influence sur la croissance et la composition corporelle. In: Halver, J.E. and Tiews, K. (eds) *Proceedings of the World Symposium on Finfish Nutrition and Fishfeed Technology*, Vol. I. Heenemann, Berlin, pp. 241–255.

Alvarez, M.J., Lopez-Bote, C.J., Diez, A., Corraze, G., Arzel, J., Dias, J., Kaushik, S.J. and Bautista, J.M. (1998) Dietary fish oil and digestible protein modify susceptibility to lipid peroxidation in the muscle of rainbow trout (*Oncorhynchus mykiss*) and sea bass (*Dicentrarchus labrax*). *British Journal of Nutrition* 80, 281–289.

Azzaydi, M., Madrid, J.A., Zamora, S., Sanchez-Vazquez, F.J. and Martinez, F.J. (1998) Effect of three feeding strategies (automatic, ad libitum demand-feeding and time-restricted demand-feeding) on feeding rhythms and growth in European sea bass (*Dicentrarchus labrax* L.). *Aquaculture* 163, 283–294.

Ballestrazzi, R. and Lanari, D. (1996) Growth, body composition and nutrient retention efficiency of growing seabass (*Dicentrarchus labrax* L.) fed fish oil or fatty acid Ca salts. *Aquaculture* 139, 101–108.

Ballestrazzi, R., Lanari, D., D'Agaro, E. and Mion, A. (1994) The effect of dietary protein level and source on growth, body composition, total ammonia and reactive phosphate excretion of growing seabass (*Dicentrarchus labrax*). *Aquaculture* 127, 197–206.

Bell, J.G., Farndale, B.M., Bruce, M.P., Navas, M.N. and Carillo, M. (1997) Effect of broodstock dietary lipid on fatty acid composition of eggs from sea bass (*Dicentrarchus labrax*). *Aquaculture* 149, 107–119.

Bell, M.V., McEvoy, L.A. and Navarro, J.C. (1996) Deficit of didocosahexaenoyl phospholipid in the eyes of larval seabass fed an essential fatty acid deficient diet. *Journal of Fish Biology* 49, 941–952.

Blazquez, M., Piferrer, F., Zanuy, S., Carrillo, M. and Donaldson, E.M. (1995) Development of sex control techniques for European seabass (*Dicentrarchus labrax* L.) aquaculture: effects of dietary 17 alpha-methyltestosterone prior to sex differentiation. *Aquaculture* 135, 329–342.

Boujard, T., Jourdan, M., Kentouri, M. and Divanach, P. (1996) Diel feeding activity and the effect of time-restricted self-feeding on growth and feed conversion in European seabass. *Aquaculture* 139, 117–127.

Boujard, T., Gélineau, A., Corraze, G., Kaushik, S., Coves, D., Dutto, G. and Gasset, E. (2000) Effect of dietary lipid content on circadian rhythm of feeding activity in seabass. *Physiology and Behaviour* 68, 683–689.

Bruce, M., Oyen, F., Bell, G., Asturiano, J.F., Farndale, B., Carrillo, M., Zanuy, S., Ramos, J. and Bromage, N. (1999) Development of broodstock diets for the European seabass (*Dicentrarchus labrax*) with special emphasis on the importance of n-3 and n-6 highly unsaturated fatty acid to reproductive performance. *Aquaculture* 177, 85–97.

Cahu, C., Zambonino Infante, J., Escaffre, A.M., Bergot, P. and Kaushik, S. (1998) Preliminary results on seabass (*Dicentrarchus labrax*) larvae rearing with compound diet from first feeding comparison with carp (*Cyprinus carpio*) larvae. *Aquaculture* 169, 1–7.

Cahu, C., Zambonino Infante, J.L., Quazuguel, P. and Le Gall, M.M. (1999) Protein hydrolysate vs fish meal in compound diets for 10-day old seabass *Dicentrarchus labrax* larvae. *Aquaculture* 171, 109–119.

Carvalho, A.P.C., Ferreira, M.L., Oliva Teles, A. and Bergot, P. (1995) Utilization of fish protein hydrolysate in diets for seabass (*Dicentrarchus labrax* L.) post larvae. In: Castello I Orvay, F. and Calderer I Reig, A. (eds) *Actas V Congresso Nacional Acuicultura*. University of Barcelona, Spain, pp. 415–419.

Cerdâ, J., Carillo, M., Zunuy, J.R. and Higuera, M. (1994) Influence of nutritional composition of diet on seabass, *Dicentrarchus labrax* L., reproductive performance and egg and larval quality. *Aquaculture* 128, 345–361.

Chatain, B., Saillant, E. and Peruzzi, S. (1999) Production of monosex male populations of European seabass, *Dicentrarchus labrax* L. by use of the synthetic androgen 17 alpha-methyldehydrotestosterone. *Aquaculture* 178, 225–234.

Coutteau, P., Stappen, G.V. and Sorgeloos, P. (1996) A standard experimental diet for the study of fatty acid requirements of weaning and first ongrowing stages of the European seabass *Dicentrarchus labrax* L.: comparison of extruded and extruded/coated diets. *Archives of Animal Nutrition* 49, 49–59.

Coves, D.G., Lemarie, G. and Dutto, G. (1998) A simple way of avoiding feed wastage in European seabass, *Dicentrachus labrax*, under self-feeding conditions. *Aquatic Living Resources* 11, 395–401.

da Silva, J.G. and Oliva-Teles, A. (1998) Apparent digestibility coefficients of feedstuffs in seabass (*Dicentrarchus labrax*) juveniles. *Aquatic Living Resources* 11, 187–191.

Dias, J. (1999) *Lipid Deposition in Rainbow Trout (*Oncorhynchus mykiss*) and European Seabass (*Dicentrarchus labrax*): Nutritional Control of Hepatic Lipogenesis.* University of Porto, Porto, Portugal, and University of Bordeaux I, Bordeaux, France, 190 pp.

Dias, J., Alvarez, M.J., Diez, A., Arzel, J., Corraze, G., Bautista, J.M. and Kaushik, S.J. (1998a) Regulation of hepatic lipogenesis by dietary protein/energy in juvenile European seabass (*Dicentrarchus labrax*). *Aquaculture* 161, 169–186.

Dias, J.H., Huelvan, C., Dinis, M.T. and Metailler, R. (1998b) Influence of dietary bulk agents (silica, cellulose and a natural zeolite) on protein digestibility, growth, feed intake and feed transit time in European seabass (*Dicentrarchus labrax*) juveniles. *Aquatic Living Resources* 11, 219–226.

Dosdat, A., Servais, F., Métailler, R., Huelvan, C. and Desbruyères, E. (1996) Comparison of nitrogenous losses in five teleost fish species. *Aquaculture* 141, 107–127.

Fournier, V., Gouillou-Coustans, M.F. and Kaushik, S. (2000) Hepatic ascorbic acid saturation is the most stringent response criterion for determining the vitamin C requirement of juvenile European seabass (*Dicentrarchus labrax*). *Journal of Nutrition* 130, 617–620.

Fukusho, K., Okauchi, M., Nuraini, S., Tsujigado, A. and Watanabe, T. (1985) Preliminary results on the nutritional evaluation of omega 3-HUFA-enriched *Artemia* nauplii for larvae of the seabass, *Dicentrarchus labrax*. *Aquaculture* 49, 223–229.

Geurden, I., Coutteau, P. and Sorgeloos, P. (1997a) Effect of dietary phospholipid supplementation on growth and fatty acid composition of European seabass (*Dicentrarchus labrax* L.) and turbot (*Scophthalmus maximus* L.) juveniles from weaning onward. *Fish Physiology and Biochemistry* 16, 259–272.

Geurden, I., Coutteau, P. and Sorgeloos, P. (1997b) Increased docosahexaenoic acid levels in total and polar lipid of European seabass (*Dicentrarchus labrax*) post larvae fed vegetable or animal phospholipids. *Marine Biology* 129, 689–698.

Gomes, E., Dias, J. and Kaushik, S.J. (1997) Improvement of feed intake through supplementation with an attractant mix in European seabass fed plant-protein rich diets. *Aquatic Living Resources* 10, 385–389.

Gouveia, A. and Davies, S.J. (2000) Inclusion of an extruded dehulled pea seed meal in diets for juvenile European seabass (*Dicentrarchus labrax*). *Aquaculture* 182, 183–193.

Hidalgo, F. and Alliot, E. (1988) Influence of water temperature on protein requirement and protein utilization in juvenile seabass (*Dicentrarchus labrax*). *Aquaculture* 72, 115–129.

Kaushik, S.J. (1998a) Nutritional bioenergetics for the estimation of wastes in non-salmonids. *Aquatic Living Resources* 11, 211–217.

Kaushik, S.J. (1998b) Whole body amino acid composition of European seabass (*Dicentrarchus labrax*), gilthead seabream (*Sparus aurata*) and turbot (*Psetta maxima*) with an estimation of their IAA requirement profiles. *Aquatic Living Resources* 11, 355–358.

Kaushik, S.J., Gouillou-Coustans, M.F. and Cho, C.Y. (1998) Application of the recommendations on vitamin requirements of finfish by NRC (1993) to salmonids and seabass using practical and purified diets. *Aquaculture* 161, 463–474.

Lanari, D., Poli, B.M., Ballestrazzi, R., Lupi, P., D'Agaro, E. and Mecati, M. (1999) The effects of dietary fat and NFE levels on growing European seabass (*Dicentrarchus labrax*): growth rate, body and fillet composition, carcass traits and nutrient retention efficiency. *Aquaculture* 179, 351–364.

Merchie, G., Lavens, P., Storch, V., Ubel, U., Nelis, H., de Leenheer, A. and Sorgeloos, P. (1996) Influence of dietary vitamin C dosage on turbot (*Scophthalmus maximus*) and European seabass (*Dicentrarchus labrax*) nursery stages. *Comparative Biochemistry and Physiology* 114A, 123–133.

Messager, J.L., Stephan, G., Quentel, C. and Laurencin, F.B. (1992) Effects of dietary oxidized fish oil and antioxidant deficiency on histopathology, haematology, tissue and plasma biochemistry of seabass *Dicentrarchus labrax*. *Aquatic Living Resources* 5, 205–214.

Métailler, R., Febvre, A. and Alliot, E. (1973) Note préliminaire sur les acides aminés essentiels du loup ou bar. *Conseil Général des Pêches pour la Mediterranée* 52, 91–96.

Métailler, R., Aldrin, J.F., Messager, J.L., Mavel, G. and Stephan, G. (1981) Feeding of European seabass (*Dicentrarchus labrax*): role of protein level and energy source. *Journal of World Mariculture Society* 12, 117–118.

Navarro, J.C., McEvoy, L.A., Bell, M.V., Amat, F., Hontoria, F. and Sargent, J.R. (1997) Effect of different dietary levels of docosahexaenoic acid (DHA, 22:6 omega-3) on the DHA composition of lipid classes in seabass larvae eyes. *Aquaculture International* 5, 509–516.

Navarro, J.C., Hontoria, F. and Amat, F. (1988) Effect of alternate feeding with a poor long-chain polyunsaturated fatty acid *Artemia* strain and a rich one for seabass (*Dicentrarchus labrax*) and prawn (*Penaeus kerathurus*) larvae. *Aquaculture* 74, 307–317.

Navas, J.M., Mananos, E., Thrush, M., Ramos, J., Zanuy, S., Carrillo, M., Zohar, Y. and Bromage, N. (1998) Effect of dietary lipid composition on vitellogenin, 17 beta-estradiol and gonadotropin plasma levels and spawning performance in captive seabass (*Dicentrarchus labrax* L.). *Aquaculture* 165, 65–79.

NRC (National Research Council) (1993) *Nutrient Requirements of Fish*. National Academy of Sciences, Washington, DC, 114 pp.

Oliva-Teles, A., Pereira, J.P., Gouveia, A. and Gomes, E. (1998) Utilisation of diets supplemented with microbial phytase by seabass (*Dicentrachus labrax*) juveniles. *Aquatic Living Resources* 11, 255–259.

Paspatis, M., Batarias, C., Tiangos, P. and Kentouri, M. (1999) Feeding and growth responses of seabass (*Dicentrarchus labrax*) reared by four feeding methods. *Aquaculture* 175, 293–305.

Paspatis, M., Boujard, T., Maragoudaki, D. and Kentouri, M. (2000) European seabass growth and N and P loss under different feeding practices. *Aquaculture* 184, 77–88.

Péres, A., Cahu, C., Zambonino Infante, J.L., Le Gall, M.M. and Quazuguel, P. (1996) Amylase and trypsin response to dietary carbohydrate and protein level depends on the developmental stage in seabass (*Dicentrarchus labrax*) larvae. *Fish Physiology and Biochemistry* 15, 237–242.

Peres, H. and Oliva-Teles, A. (1999) Effect of dietary lipid level on growth performance and feed utilization by European seabass juveniles (*Dicentrarchus labrax* L.). *Aquaculture* 179, 325–334.

Peres, H., Gonçalves, P. and Oliva-Teles, A. (1999) Glucose tolerance in gilthead seabream (*Sparus aurata*) and European seabass (*Dicentrarchus labrax*). *Aquaculture* 179, 415–423.

Pérez, L., Gonzalez, M., Jover, M. and Fernandez, J.C. (1997) Growth of European seabass fingerlings (*Dicentrarchus labrax*) fed extruded diets containing varying levels of protein, lipid and carbohydrate. *Aquaculture* 156, 183–193.

Santinha, P.J.M., Gomes, E.F.S. and Coimbra, J.O. (1996) Effects of protein level of the diet on digestibility and growth of gilthead sea bream, *Sparus auratus* L. *Aquaculture Nutrition* 2, 81–87.

Spyridakis, P., Gabaudan, J., Metailler, R. and Guillaume, J. (1988) Digestibility of proteins and availability of amino acids in some feed ingredients for European seabass (*Dicentrarchus labrax*). *Reproduction, Nutrition and Development* 28, 1509–1517.

Thébault, H., Alliot, E. and Pastoureaud, A. (1985) Quantitative methionine requirements of juvenile seabass, *Dicentrarchus labrax*. *Aquaculture* 50, 75–87.

Tibaldi, E. and Lanari, D. (1991) Optimal dietary lysine levels for growth and protein utilisation of fingerling seabass (*Dicentrarchus labrax* L.) fed semipurified diets. *Aquaculture* 95, 297–304.

Tibaldi, E. and Tulli, F. (1999) Dietary threonine requirement of juvenile European seabass (*Dicentrarchus labrax*). *Aquaculture* 175, 155–166.

Tibaldi, E., Tulli, F. and Pinosa, M. (1993) *Quantitative Tryptophan Requirement of Seabass Fingerlings*. European Aquaculture Society Special Publication 19, Ostend, Belgium, 482 pp.

Tibaldi, E., Tulli, F. and Lanari, D. (1994) Arginine requirement and effect of different dietary arginine and lysine levels for fingerling seabass (*Dicentrarchus labrax*). *Aquaculture* 127, 207–216.

Tsevis, N., Klaoudatos, S. and Conides, A. (1991) Food conversion budget in seabass, *Dicentrarchus labrax*, fingerlings under two different feeding frequency patterns. *Aquaculture* 101, 293–304.

Yildiz, M. and Sener, E. (1997) Effect of dietary supplementation with soybean oil sunflower oil or fish oil on the growth of seabass (*Dicentrarchus labrax* L. 1758). International Center for Advanced Mediterranean Agronomic Studies, CIHEAM, Zaragoza (Spain). *Cahier Options Méditerranéens* 22, 225–233.

Zambonino Infante, J.L. and Cahu, C. (1994a) Influence of diet on pepsin and some pancreatic enzymes in seabass (*Dicentrachus labrax*) larvae. *Comparative Biochemistry and Physiology* 109A, 209–212.

Zambonino Infante, J.L. and Cahu, C. (1994b) Development and response to a diet change of some digestive enzymes in seabass (*Dicentrarchus labrax*) larvae. *Fish Physiology and Biochemistry* 12(5), 399–408.

Zambonino Infante, J.L., Cahu, C. and Péres, A. (1997) Partial substitution of di- and tripeptides for native proteins in seabass diet improves *Dicentrarchus labrax* larval development. *Journal of Nutrition* 127, 608–614.

Asian Sea Bass, *Lates calcarifer*

3

Mali Boonyaratpalin[1] and Kevin Williams[2]
[1]Department of Fisheries, Jatujak, Bangkok 10900, Thailand;
[2]CSIRO Division of Marine Research, PO Box 120, Cleveland, Qld 4163, Australia

Introduction

Asian sea bass (*Lates calcarifer*), also known as 'baramundi', is a primarily carnivorous finfish widely distributed in the tropical and subtropical areas of the western Pacific and Indian Ocean countries (Greenwood, 1976). Its natural habitat ranges between longitude 50° E and 160° E and latitude 24° N and 25° S. Under natural conditions, its life cycle is biphasic, involving an initial growth period of 2–3 years in the inland fresh waters and a subsequent migration into the sea for sexual maturation and spawning. The attributes of being adaptable to a wide range of water salinity and turbidity probably make sea bass a suitable species for culture under captive conditions.

Sea bass has been cultivated in both brackish-water and freshwater ponds as well as in marine cages. However, a gradual acclimatization of fry to new salinity is necessary. The cage culture remains the preferred system over that of the pond system. Floating or stationary types of cages may be used. The most common and manageable size of floating cage is 50 m^3 (5 m × 5 m with a depth of 2 m). Initial stocking density ranges from 40 to 50 fish m^{-3}. Cannibalism is one of the serious problems encountered during early stages of sea-bass culture. This may be related to the uniform size of the fry and heavy stocking density.

In the Asian Pacific region, Malaysia, Indonesia, Taiwan and Thailand remain the leading countries involved in commercial sea-bass production, with Hong Kong, Singapore and Australia also playing an important role. The difficulty in obtaining approval from the government to establish cages in the Great Barrier Reef Marine Park is a constraint in Australia. Large-scale sea-bass culture has become popular only recently. In Thailand, for example, the first successful spawning and seed production were achieved in 1975, whereas sea-bass farming in Australia began around 1986. Since then, Australia has recorded a rather rapid and continuing increase in sea-bass production. The production in Australia increased from around 22 Mt in 1988 (Lobegeiger and Barlow, 1991) to 310 Mt in 1996 (FAO, 1998) and is projected to reach 2000 Mt by the year 2005. In contrast, the rate of increase in sea-bass production in Taiwan, Thailand, Malaysia and Indonesia seems to have plateaued. The annual sea-bass production in these countries in 1994 was around 8800, 4000, 2700 and 2400 Mt, whereas in 1996 it was 7000, 3800, 2400 and 1700 Mt, respectively (FAO, 1998). The decrease in production in these countries during the 2-year period seems to be related to a rather limited market demand and relatively low prices. Therefore, future expansion of sea-bass culture will depend on market forces and whether culture technologies can be improved to sustain farm profitability in the face of reducing operating margins.

Nutrient Requirements

Protein and amino acids

A relatively high dietary protein level may be suggested for sea bass since the fish is primarily a carnivore under natural conditions. Sakaras *et al.* (1988) estimated the dietary crude protein (CP) requirement of juvenile sea bass to be 50%. The research from the same laboratory subsequently showed that the highest growth rate was achieved with a dietary level of 45% CP (Sakaras *et al.*, 1989). The optimal dietary protein level for grow-out sea bass has been reported to range between 40 and 45% (Wong and Chou, 1989). Few studies have been conducted to determine the essential amino acid requirements of sea bass. Coloso *et al.* (1993) reported the tryptophan requirement of juvenile sea bass to be 0.5% of dietary protein. The requirements for methionine, lysine and arginine have been determined to be 2.24, 4.5–5.2 and 3.8% of dietary protein, respectively (Millamena, 1994; Rimmer *et al.*, 1998). Additionally, Boonyaratpalin *et al.* (1990) found that an excessive concentration of tyrosine in the sea-bass diet may result in kidney malfunction. Clearly, much work needs to be done to understand the dietary amino acid concentrations and their interrelationships necessary for optimum sea-bass performance.

Energy

Catacutan and Coloso (1995) reported that sea bass can grow adequately on a diet containing 42.5% CP with an estimated protein-to-digestible energy (DE) ratio of about 30.7 mg CP kJ^{-1} DE, which corresponds to 25.2 mg CP kJ^{-1} gross energy (GE) or 27.1 mg digestible protein (DP) kJ^{-1} DE. Williams and Barlow (1999) demonstrated that a diet containing 42% DP and 15.5 kJ DE g^{-1} provided optimal feed conversion ratio (FCR) and nitrogen retention. They suggested an optimal DP : DE ratio of 26.7 mg kJ^{-1} or a CP : GE ratio of 24.5 mg kJ^{-1}. These data agree with the CP : GE value of 25.8 mg kJ^{-1} and 25.2 mg kJ^{-1} reported by Tubongbanua (1987) and Catacutan and Coloso (1995), respectively. In a more recent work, Williams *et al.* (1999) showed that 55% CP and 20 kJ DE g^{-1} were necessary to realize the growth potential of juvenile sea bass. These results indicate that diets with relatively high protein and energy density can support a higher rate of fish growth compared with that achieved by using standard diets containing 45% CP and 15.5 kJ DE g^{-1}.

Lipid and fatty acids

Sakaras *et al.* (1988, 1989) showed that the optimum dietary lipid level for sea-bass fingerlings is between 15 and 18%. Tucker *et al.* (1988) found that when 9–62 g sea bass were fed diets containing 9.3 and 12.9% fat, growth was similar but the FCR was significantly lower with the 12.9% fat diet. Borlongan and Parazo (1991) found that growth and survival of sea-bass fry fed a diet containing 9% supplemented lipid comprising 4.5% cod-liver oil and 4.5% soybean oil were higher compared with those fed cod-liver oil alone or soybean oil alone. In addition, the performance of the fry decreased when these dietary lipid sources were replaced with coconut oil and the poorest performance was observed when the diet contained no lipid supplement. These researchers also reported that the fatty acid profiles of the fry were influenced by the fatty acid composition of the dietary lipid.

Wanakowat *et al.* (1993) fed sea-bass fingerlings with diets containing 13% lipid and graded levels of n-3 highly unsaturated fatty acids (HUFA). They observed signs of essential fatty acid (EFA) deficiency after 2 weeks when the n-3 HUFA dietary level was 0.46%. Signs of mild EFA deficiency were also observed in fish fed a diet containing 0.88% n-3 HUFA and no such signs were detected in fish fed a diet containing 1.72% n-3 HUFA. The diet with the highest level of n-3 HUFA also gave the best performance in terms of growth. Early EFA deficiency signs were characterized as reddening of the fins and skin, which was followed by abnormal appearance of eyes, loss of appetite, poor growth and swollen and pale livers. Wanakowat *et al.* (1993) also fed sea bass six diets comprising fish-meal and casein with levels of n-3 HUFA ranging from 1 to 2% of the diet for a 10-week period. Levels of n-3 HUFA in the diets were adjusted using graded levels of squid liver oil. They observed no significant difference in growth, feed efficiency or mortality. Neither did they detect any signs of EFA deficiency. The results obtained by Wanakowat *et al.* (1993) indicate that dietary n-3 HUFA levels of 1% are adequate. Dhert *et al.* (1990) showed that the n-3 HUFA deficiency in sea-bass larvae led to a delayed metamorphosis or a complete lack of metamorphosis as well as a decreased tolerance to stress caused by a high salinity of 65 p.p.t. Rimmer *et al.* (1994) reported that 20 : 5n-3 is required by 20-day-old sea-bass larvae for survival. Ambient water temperature appears to influence EFA requirement and ratios. Williams *et al.* (1999) demonstrated that sea-bass growth was poor and unaffected by EFA dietary level at 20°C. At a water temperature of 29°C, however, the growth responded linearly with an increasing dietary ratio of n-3 : n-6 (0.58 : 1 to 2.24 : 1). The researchers recommended an n-3 : n-6 ratio of about 1.5–1.7 : 1 and a dietary level of eicosapentaenoic acid (EPA) and docosahexaenoic acid (DHA) of about 1.8% for juvenile sea bass.

Carbohydrates

Most fish have a limited capacity to assimilate and metabolize dietary carbohydrate (Cowey, 1988). Since the natural food of sea bass is high in protein, it is assumed that they utilize carbohydrate poorly. Catacutan and Coloso (1997) found that carbohydrate has a sparing effect for lipid as an energy source in practical diets for juvenile sea bass. They recommended a dietary carbohydrate level of 20% for diets containing lipid levels ranging from 6 to 18% and 42.5% CP. Carbohydrates are useful not only as a source of energy but also as a binder for pelleted feeds.

Vitamins and minerals

The vitamin requirements for sea bass have been established for young fish by supplementing both practical and semipurified diets with various levels of a specific vitamin. The addition of a vitamin mix to trash fish fed to sea-bass fingerlings increased body weight from 9.36 to 23.48 g and reduced the FCR from 7.44 to 3.83 during a 9-week rearing period (Phromkhuntong *et al.*, 1987). These requirements are based on dietary levels that support maximum growth or tissue storage and prevent deficiency signs. No differences in weight gain, feed efficiency and mortality were observed when sea bass were fed practical diets without the addition of choline, niacin, inositol or vitamin E (Boonyaratpalin *et al.*, 1988) and pyridoxine or pantothenic acid (Pimoljinda and Boonyaratpalin, 1989). Poor weight gain and feed efficiency were observed in fish fed diets without supplemental thiamine and riboflavin after 60 days (Pimoljinda and Boonyaratpalin, 1989).

In a study using semipurified diets, Wanakowat *et al.* (1989) demonstrated that 5 mg kg^{-1} of pyridoxine was required for normal growth, whereas 10 mg kg^{-1} was needed for normal lymphocyte levels. Boonyaratpalin and Wanakowat (1993) fed sea bass with semipurified diets deficient in several vitamins for 10 weeks and found that thiamine, pantothenic acid, inositol and vitamin E were essential for fish growth. The requirement of pantothenic acid for juvenile sea bass (3–30 g) was determined by Boonyaratpalin *et al.* (1994b) using semipurified diets containing 0, 15, 30, 60 and 90 mg kg^{-1} of calcium D-pantothenate. Growth ceased after 2 weeks in fish fed diets without supplemental calcium D-pantothenate and total mortality occurred within 6 weeks. The authors estimated the pantothenic acid requirement to be 15 mg kg^{-1} diet for normal growth, feed efficiency and survival. However, highest tissue storage was found to occur at a level of 90 mg kg^{-1} diet. Whether the tissue storage level continues to increase at dietary levels of calcium D-pantothenate higher than 90 mg kg^{-1} diet is not known since the highest level of this vitamin tested was 90 mg kg^{-1} diet.

The vitamin C requirement of sea-bass fingerlings was determined in a series of studies by Boonyaratpalin *et al.* (1989a,b, 1994a). The results indicated that the dietary level of L-ascorbic acid required for normal growth ranged from 500 to 700 mg kg^{-1} and a supplemental level of 1100 mg kg^{-1} was required for normal tissue storage. Fish fed diets without added ascorbic acid had normal growth rates for about 15 days. Thereafter, growth decreased and deficiency signs developed. Severe mortality occurred after 45 days and all vitamin C-deficient fish died within 60 days. However, with a stable form of vitamin C, such as ascorbyl-2-monophosphate-magnesium (Mg), a supplementation level of 30 mg kg^{-1} diet was sufficient for normal growth and prevention of deficiency signs. The bioavailability of other stable forms of vitamin C derivatives continues to be a subject of investigation. Also, the role of ascorbic acid in the disease resistance and immunocompetence of sea bass remains unclear to date. The dietary requirements for various vitamins with their respective signs of deficiency in sea bass are summarized in Table 3.1.

As sea bass do not readily accept purified diets, mineral requirements have not been adequately evaluated. Neither have the mechanisms that are involved in mineral absorption, utilization and homoeostasis been adequately defined. Boonyaratpalin and Phongmaneerat (1990) supplemented fish-meal-based diets with 0, 0.5, 1.0 and 2.0% monosodium phosphate and showed that the growth of sea bass fed a diet with 0.5% monosodium phosphate supplementation was superior to the growth of fish fed diets with higher phosphate levels. However, they found that 1.0% monosodium phosphate supplementation gave the best feed efficiency and protein efficiency ratio. They estimated the dietary available phosphorus requirement of sea bass to be about 0.55–0.65%. Growth rate and bone mineralization were not affected by feeding sea bass on a low-ash fish-meal diet that contained no additional phosphorus (Chaimongkol, 1998). Furthermore, utilization of other minerals was not compromised on this diet. Supplementation of the diet with 0.25% monosodium phosphate did result in higher bone concentrations of phosphorus and zinc. Zinc absorption in fish fed low-ash diet was significantly higher than those fed a high-ash diet.

Practical Diets

A potential difficulty in determining the digestibility of a feed ingredient for an aquatic animal is obtaining faeces samples that are unaffected by nutrient leaching or contamination. Table 3.2 summarizes the apparent digestibility of ingredients commonly used in sea-bass diets. The protein and energy digestibility of meals of marine origin is high and generally higher than meals of terrestrial ingredients. An exception to this is wheat gluten, which is found to be almost completely digestible. Digestibility of meat meal is lower than that of fish-meal, probably owing to its high ash content. Chaimongkol (1998) reported that digestibility of protein as well as absorption of calcium, magnesium, zinc and phosphorus decreased with increasing dietary ash. Variable ash concentration in meat meals also make their digestibility quite variable. The relatively low digestibility

Table 3.1. Vitamin requirements and deficiency signs for sea bass (from Boonyaratpalin *et al.*, 1989a,b, 1993, 1994a,b; Wanakowat *et al.*, 1989; Boonyaratpalin and Wanakowat, 1993).

Vitamins	Requirement (mg kg^{-1} diet)	Deficiency signs
Thiamine	R	Poor growth, substantial posthandling shock, high mortality
Riboflavin	R	Erratic swimming, cataracts
Pyridoxine	5–10*	Avoidance of schooling, erratic spiral swimming, surfacing, high mortality, convulsions
Pantothenic acid	15–90*	Ventral fin haemorrhage and erosion, haemorrhage around operculum and isthmus, clubbed gill, total mortality in 5–6 weeks
Nicotinic acid	n/a	
Biotin	n/a	
Inositol	R	Poor growth, abnormal bone formation
Choline	n/a	
Folic acid	n/a	
Ascorbic acid	700;† 25–30‡	Gill haemorrhages, distortion of gill filaments and hyperplasia, short operculum, short snout, exophthalmia, loss of equilibrium, scoliosis, lordosis, broken back, fatty liver, muscle degeneration and haemorrhage, low blood parameters and low tissue hydroxyproline
Vitamin A	n/a	
Vitamin D	n/a	
Vitamin E	R	Muscular atrophy, susceptible to disease
Vitamin K	n/a	

* For maximum tissue storage.
† For crystalline ascorbic acid.
‡ For ascorbyl-2-monophosphate-Mg or ascorbic acid glucose.
R, required but quantitative requirement not established; n/a, no information available.

of ingredients of plant origin is probably a result of the presence of indigestible fibre. Variability in digestibility of oil-seed meals is sometimes related to their inadequate processing and the presence of antinutritional factors. It appears that, similar to other carnivorous fish, sea bass is able to digest well a variety of feed ingredients of terrestrial origin (Hajen *et al.*, 1993; Gomes *et al.*, 1995; Gaylord and Gatin, 1996).

As world supplies of fish-meal are becoming limited, research continues to evaluate the usefulness of terrestrial feed ingredients as substitutes for fish-meal. Pongmaneerat and Boonyaratpalin (1995) found that a mixture of soybean meal and maize-gluten meal at a ratio of 5 : 3 could replace 25% of fish-meal protein in

Table 3.2. Apparent protein and energy digestibility coefficients of major protein feed ingredients in sea-bass diets (from Williams *et al.*, 1998).

	Digestibility coefficients (%)*	
Feed ingredient	Protein	Energy
Danish fish-meal	87.9 ± 1.0	83.3 ± 1.3
Tuna fish-meal	92.3 ± 1.0	69.3 ± 1.3
Poultry offal meal	78.8 ± 3.5	76.7 ± 5.6
Meat meal (high ash)	53.9 ± 3.9	58.2 ± 6.5
Meat meal (low ash)	63.5 ± 3.4	66.5 ± 3.4
Soybean meal (full-fat)	84.8 ± 3.8	75.9 ± 7.8
Soybean meal (solvent-extracted)	86.0 ± 0.8	69.4 ± 1.7
Canola meal	81.0 ± 2.3	56.1 ± 3.0
Lupin meal (dehulled)	98.1 ± 1.3	61.5 ± 1.8
Groundnut meal	91.9 ± 8.0	68.7 ± 5.0
Wheat-gluten meal	101.9 ± 1.6	98.8 ± 3.1

* Mean ± standard error. Data derived from faeces collected either by hand-stripping or by intestinal dissection.

the diet of juvenile sea bass without any adverse effects on performance. Similarly, Boonyaratpalin *et al.* (1998) have shown that nearly 37% of protein from fish-meal in sea-bass diet can be replaced with solvent-extracted soybean meal without compromising their performance. It was also demonstrated that soaked raw soybean was a relatively poor protein source for sea bass and extruded full-fat soybean meal was found to be less palatable for juvenile sea bass. Cuzon and Thouarde (1993) showed that diets containing 22% greaves meal supported good growth of sea bass. They found lipid infiltration of the liver cell but no other pathology was evident.

Williams *et al.* (1999) determined growth rate and nitrogen retention in sea bass subsequent to incremental replacement of up to 70% of nutritionally adequate summit diet with soybean meal, casein, dehulled lupins and meat meal. Nitrogen retention was more adversely affected by lupins and soybean meal than by casein and meat-meal inclusions. These results show the superior nutritional value of animal sources over the vegetable sources and their potential as replacements of fish-meal in practical diets of juvenile sea bass. Williams *et al.* (1999) also showed that meat meal could replace all of the fish-meal in sea-bass diet without causing any deleterious effects on growth, efficiency of feed utilization or eating quality of the fish. Williams and Barlow (1999) demonstrated that a mixture of meat meal and spray-dried blood meal at a ratio of 5.5 : 1 can be used to replace all fish-meal in grow-out sea-bass diets without altering growth, FCR or survival.

Despite investigations to explore alternative diet ingredients, practical sea-bass diets in South-East Asia remain primarily composed of trash fish made into moist pellets. In contrast, sea bass cultured in Australia are reared entirely on dry pelleted diets. Some examples of commonly used sea-bass diets are provided below. Each of these diets has its own advantages and farmers may select one, taking into account price, availability, acceptability and performance.

1. A moist weaning diet is made by adding vitamins to minced fish flesh. It is fed to fry (25 days old) as a weaning diet, the larvae being typically fed twice daily to satiation.

2. Crumbled pelleted diets that are dust-free and uniform in size may be used in conjunction with a moist weaning diet to facilitate the change-over to a dry diet. The size of the crumbs is increased as the fry grow. Crumbled shrimp diets top-dressed with 8–10% fish-oil are also used for fry.

3. Moist pellets are usually prepared at the farm site using a mincer by mixing a meal-form formulated diet and minced trash fish at a ratio of 1 : 1. The moist diet is made daily and fed on the same day. The formulated powder comprises a mixture of fish-meal, grains or grain by-products, vitamin and mineral premixes and fish-oil.

4. Another form of moist-pellet diet is prepared at the farm site by mixing a meal-formulated diet with fresh water at a ratio of 10 : 4.

5. Floating or slowly sinking extruded dry pellets are also used. In Australia extruded pellets are fed twice a day for fish weighing 20–100 g and once daily for fish larger than 100 g. These diets typically contain about 50% protein and 20% lipid.

6. Brood-stock diet is usually comprised of fresh or frozen trash fish fed at about 2% of body weight, three times a week. Vitamin and fish-oil supplements are usually added to the baitfish prior to feeding the brood-stock. In Thailand, sardine and yellowstripe are commonly used for the same purpose.

References

Boonyaratpalin, M. and Phongmaneerat, J. (1990) *Requirement of Seabass for Dietary Phosphorus.* Technical Paper No. 4, National Institute of Coastal Aquaculture, Department of Fisheries, Thailand, 20 pp. (in Thai).

Boonyaratpalin, M. and Wanakowat, J. (1993) Effect of thiamine, riboflavin, pantothenic acid and inositol on growth, feed efficiency and mortality of juvenile seabass. In: Kaushik, S.J. and Luguet, P. (eds) *Fish Nutrition in Practice.* Biarritz, France, pp. 819–828.

Boonyaratpalin, M., Unprasert, N., Kosutharak, P., Chumsungnern, S. and Sothana, W. (1988) *Effect of Choline, Niacin, Inositol and Vitamin E on Growth, Feed Efficiency and Survival of Seabass Fingerling in Freshwater.* Technical Paper No. 7, National Institute of Coastal Aquaculture, Department of Fisheries, Thailand, 22 pp. (in Thai).

Boonyaratpalin, M., Unprasert, N. and Buranapanidgit, J. (1989a) Optimal supplementary vitamin C level in seabass fingerling diet. In: Takeda, M. and Watanabe, T. (eds) *The Current Status of Fish Nutrition in Aquaculture.* Tokyo University of Fisheries, Tokyo, Japan, pp. 149–157.

Boonyaratpalin, M., Unprasert, N., Kosutharak, P., Sotana, W. and Chumsungner, S. (1989b) *Effect of Different Levels of Vitamin C Added in Diet on Growth, Feed Efficiency and Survival Rate of Seabass.* Technical Paper No. 6, National Institute of Coastal Aquaculture, Department of Fisheries, Thailand, 26 pp. (in Thai).

Boonyaratpalin, M., Boonyaratpalin, S. and Supamataya, K. (1994a) Ascorbyl-phosphate-Mg as a dietary vitamin C sources for seabass (*Lates calcarifer*). In: Chou, L.M., Munro, A.D., Lam, T.J., Chen, T.W., Cheong, L.K.K., Ding, J.K., Hooi, K.K.,

Phang, V.P.E., Shim, K.F. and Tan, C.H. (eds) *The Third Asian Fisheries Forum*. Asian Fisheries Society, Manila, the Philippines, pp. 725–728.

Boonyaratpalin, M., Wanagowat, J. and Hangsapreurke, K. (1994b) Pantothenic acid requirement of seabass. In: De Silva, S.S. (ed.) *Fish Nutrition Research*. Asian Fisheries Society Special Publication 9, Asian Fisheries Society, Manila, the Philippines, pp. 23–30.

Boonyaratpalin, M., Suraneiranat, P. and Tunpibal, T. (1998) Replacement of fish meal with various types of soybean products in diets for the Asian Seabass, *Lates calcarifer*. *Aquaculture* 161, 67–78.

Boonyaratpalin, S., Fryer, J.L., Hedrick, R.P., Supamattaya, K., Direkbusarakom, S. and Jadesadakraisorn, U. (1990) *Development of Kidney Disease and Occurrence of Pathogens in Seabass (*Lates calcarifer*) Fed Two Different Diets*. Technical Paper No. 8, National Institute of Coastal Aquaculture, Department of Fisheries, Thailand, 13 pp. (in Thai).

Borlongan, I.G. and Parazo, M.M. (1991) Effect of dietary lipid sources on growth, survival and fatty acid composition of seabass (*Lates calcarifer*, Bloch) fry. *Bamidgeh* 43, 95–102.

Catacutan, M.R. and Coloso, R.M. (1995) Effect of dietary protein to energy ratios on growth, survival, and body composition of juvenile Asian seabass, *Lates calcarifer*. *Aquaculture* 131, 125–133.

Catacutan, M.R. and Coloso, R.M. (1997) Growth of juvenile Asian seabass, *Lates calcarifer*, fed varying carbohydrate and lipid levels. *Aquaculture* 149, 137–144.

Chaimongkol, A. (1998) The utilization of phosphorus in fish meal by seabass *Lates calcarifer*. MS thesis, Kasetsart University, Bangkok, Thailand, 102 pp.

Coloso, R.M., Murillo, D.P., Borlongan, I.G. and Catacutan, M.K. (1993) Requirement of juvenile seabass *Lates calcarifer* Bloch, for tryptophan. In: *Program and Abstract of the VI International Symposium on Fish Nutrition and Feeding, 4–7 October 1993, Hobart, Australia.*

Cowey, C.B. (1988) The nutrition of fish: the developing scene. *Nutrition Research Reviews* 1, 255–280.

Cuzon, G. and Thouarde, E. (1993) Feeding of culture *Lates calcarifer*: replacement of fish meal by meat meal in the grow-out diet. In: Kaushik, S.J. and Lughet, P. (eds) *Fish Nutrition in Practice*. Paris, France, pp. 691–704.

Dhert, P., Lavens, P., Duray, M. and Sorgeloos, P. (1990) Improved larval survival at metamorphosis of Asian seabass (*Lates calcarifer*) using omega 3-HUFA enriched live food. *Aquaculture* 90, 63–74.

FAO (1998) *Aquaculture Production Statistics 1987–1996*. FAO Fisheries Department, Rome, Italy, 197 pp.

Gaylord, T.G. and Gatlin, D.M. (1996) Determination of digestibility coefficients of various feedstuffs for red drum (*Sciaenops ocellatus*). *Aquaculture* 139, 303–314.

Gomes, E.F., Rema, P. and Kaushik, S.J. (1995) Replacement of fish meal by plant proteins in the diet of rainbow trout (*Onchorynchus mykiss*): digestibility and growth performance. *Aquaculture* 130, 177–186.

Greenwood, P.H. (1976) A review of the family Centropomidae (Pisces Perciformes). *Bulletin of the British Museum of Natural History (Zoology)* 29, 1–81.

Hajen, W.E., Higgs, D.A., Beams, R.M. and Dosanj, B.S. (1993) Digestibility of various feed stuffs by post-juvenile chinook salmon (*Onchorynchus tshawytscha*) in sea water. 2. Measurement of digestibility. *Aquaculture* 112, 333–348.

Lobegeiger, R. and Barlow, C. (1991) *Aquaculture Production Survey Queensland 1990–91: Report to Farmers*. Miscellaneous Publication, Queensland Department of Primary Industries, Brisbane, Australia.

Millamena, O.M. (1994) Review of SEAFDEC/AQD fish nutrition and feed development researchs. In: Santiago, C.B., Coloso, R.M., Millamena, O.M. and Borlongan, I.G. (eds) *Feeds for Small-Scale Aquaculture. Proceeding of the National Seminar-workshop on Fish Nutrition and Feeds.* SEAFDEC Aquaculture Department, Iloilo, the Philippines, pp. 52–63.

Phromkhuntong, W., Supamattaya, K. and Jittione, W. (1987) *Effect of Water-soluble Vitamins on Growth, Body Composition and Histology of Seabass.* Report of the Aquatic Science Division, Faculty of Natural Resources, Prince of Songkhla University, Thailand, 37 pp.

Pimoljinda, T. and Boonyaratpalin, M. (1989) *Study on Vitamin Requirements of Seabass* Lates calcarifer *Bloch in Sea Water.* Technical Paper No. 3, Phuket Brackishwater Fisheries Station, Department of Fisheries, Thailand, 24 pp. (in Thai).

Pongmaneerat, J. and Boonyaratpalin, M. (1995) *Use of Plant Proteins in Diet for Seabass (*Lates calcarifer*).* Technical Paper No. 14, National Institute of Coastal Aquaculture, Department of Fisheries, Thailand, 12 pp.

Rimmer, M.A., Reed, A.W., Levitt, M.S. and Lisle, A.T. (1994) Effects of nutritional enhancement of live food organisms on growth and survival of barramundi, *Lates calcarifer* (Bloch), larvae. *Aquaculture Fisheries Management* 25(2), 143–156.

Sakaras, W., Boonyaratpalin, M., Unprasert, N. and Kumpang, P. (1988) *Optimum Dietary Protein Energy Ratio in Seabass Feed I.* Technical Paper No. 7, Rayong Brackishwater Fisheries Station, Thailand, 20 pp. (in Thai).

Sakaras, W., Boonyaratpalin, M. and Unprasert, N. (1989) *Optimum Dietary Protein Energy Ratio in Seabass Feed II.* Technical Paper No. 8, Rayong Brackishwater Fisheries Station, Thailand, 22 pp. (in Thai).

Tubongbanua, E.S. (1987) Development of artificial feeds for sea bass (*Lates calcarifer*). In: Copland, J.W. and Grey, D.L. (eds) *Management of Wild and Cultured Sea Bass/ Barramundi (*Lates calcarifer*).* ACIAR Proceedings No. 20, Canberra, Australia, pp. 186–188.

Tucker, J.W. Jr, Mackinnon, M.R., Russell, D.J., O'Brien, J.J. and Cazzola, E. (1988) Growth of juvenile barramundi (*Lates calarifer*) on dry feeds. *Progressive Fish Culturist* 50, 81–85.

Wanakowat, J., Boonyaratpalin, M., Pimoljinda, T. and Assavaaree, M. (1989) Vitamin B6 requirement of juvenile seabass *Lates calcarifer.* In: Takeda, M. and Watanabe, T. (eds) *The Current Status of Fish Nutrition in Aquaculture.* Tokyo University of Fisheries, Tokyo, Japan, pp. 141–147.

Wanakowat, J., Boonyaratpalin, M. and Watanabe, T. (1993) Essential fatty acid requirement of juvenile seabass. In: Kaushik, S.J. and Luquet, P. (eds) *Fish Nutrition in Practice.* Paris, France, pp. 807–817.

Williams, K.C. and Barlow, C. (1999) Nutritional research in Australia to improve pelleted diets for grow-out baramundi *Lates calcarifer* (Bloch). In: Cabanban, A.S. and Phillips, M. (eds) *Aquaculture of Coral Fishes and Sustainable Reef Fisheries.* Institute for Development Studies, Sabah, Malaysia, pp. 163–172.

Williams, K.C., Barlow, C. and D'Souza, F. (1998) Larval penaeid and grow-out finfish nutritional research in Australia. In: Rimmer, M. and Williams, K.C. (eds) *Proceedings of ACIAR–NACA Grouper Aquaculture Workshop.* Bangkok, Thailand, pp. 26–35.

Williams, K.C., Barlow, C., Rodgers, L., McMeniman, N. and Johnston, W. (1999) High performance grow-out pelleted diets for cage culture of barramundi (Asian seabass) *Lates calcarifer.* In: *Proceeding of the First International Symposium on Cage Culture in Asia.* Tungkang, Taiwan, 29 pp.

Wong, F.J. and Chou, F. (1989) Dietary protein requirement of early grow-out seabass (*Lates calcarifer* Bloch) and some observations on the performance of two practical formulated feeds. In: *Report of the Workshop on Shrimp and Finfish Feed Development.* Johor Bahru, Malaysia, pp. 91–102.

Red Sea Bream, *Pagrus major*

4

Shunsuke Koshio
Faculty of Fisheries, Kagoshima University, 4-50-20, Shimoarata, Kagoshima 890-0056, Japan

Introduction

The red sea bream, *Madai* in Japanesex, is one of the most popular fish species in Japanese aquaculture due to its economic feasibility and traditional food habits. The aquaculture production of red sea bream (*Pagrus major*) is the second largest in Japan, following the yellowtail (*Seriola quinqueradiata*). The aquaculture production of red sea bream in 1998 was 82,500 metric tonnes (Statistical and Information Department, 2000). This quantity is 1.8 times higher than that in 1988. Furthermore, nearly 81 million juveniles of this species were produced in 1998, of which 30 million were used for the stock enhancement and 51 million for aquaculture (Statistical and Information Department, 2000). Together with many nutritional studies, a variety of diets for the species have been introduced into the market in recent years. In 1998, the feeds for red sea bream reached 180,000 metric tonnes, which constitute 60% of total diets for marine fish (Aquaculture Feed Association, Tokyo, 1999, personal communication). Although most nutritional studies have focused on the growing stage of the species, studies on larval and brood-stock nutrition have also progressed in the past few years. Studies before the 1990s had been conducted using moist pellets, which contained more than 30% moisture. However, due to the recent development of dry diets, new findings are based on the data of studies with the dry diets.

Nutrient Requirements

Proteins and amino acids

Yone (1976) reported that the optimal dietary protein level was 55% with 10% dietary lipid in moist pellets for the species. However, a study using dry-type diets (Takeuchi *et al.*, 1991) indicated that, in 2 g fish, the growth performance of the 35-day trial seemed to improve with increased dietary protein from 37% to 52%, when fed diets containing white fish-meal and casein as major protein sources. In

8 g fish, the growth rate and feed conversion efficiency appeared to be better in fish fed the test diet containing 52% dietary protein than in fish fed the 42% dietary protein, irrespective of the dietary lipid content.

Since the quality of protein sources is very important in diet formulation as well as the determination of nutrient requirements, we conducted a study to evaluate the nutritional value of several protein sources (Koshio *et al.*, 1996). Five protein sources – casein, sardine meal, squid meal, soybean-protein isolate (SPI) and krill meal – were tested (Table 4.1). Protein (55%) and gross energy (4.5 kcal g^{-1}) contents of test diets were maintained constant in all diets. The fastest growth was found in fish fed the sardine meal-based diet, followed by squid meal, krill meal, casein and SPI diets (Fig. 4.1). Survival rates in all treatments were almost the same except the SPI-fed group, which showed significantly lower survival than other groups. The poor performance of fish fed SPI-based diet could be related in part to the lack of dietary methionine and lysine, poor palatability, high standard metabolism and lower specific dynamic action (Koshio *et al.*, 1996).

Recently, due to the high cost and limited supply of high-quality fish-meal, many studies on protein replacement have been conducted for cultured species in Japan (Koshio, 1999). Although soybean meals (SBM) have some negative effects on fish performance, the products have been widely used, due to their constant supply and economically reasonable cost. Ukawa *et al.* (1994) suggested that 14 g red sea bream could utilize commercially available SBM and the inclusion of 25% SBM in the diet did not show any negative growth performance compared with fish-meal-based diet. Moreover, we tested twin-screw extruded SBM, which has a very low level of trypsin inhibitors and antigenicities, for 3 g red sea bream

Table 4.1. Compositions of test diets (g 100 g^{-1}) for red sea bream.

	Diets				
Ingredients	1	2	3	4	5
Casein (vitamin-free)	45	–	–	–	–
Brown fish-meal	–	59	–	–	–
Squid meal	–	–	64	–	–
Soybean protein	–	–	–	54	–
Krill meal	–	–	–	–	66
Dextrin	4	4	4	4	4
Alpha-starch	4	4	4	4	4
Pollack liver oil	6	6	6	6	6
Soybean lecithin	3	3	3	3	3
Mineral mixture	4	4	4	4	4
Vitamin mixture	4	4	4	4	4
n-3 HUFA	1	1	1	1	1
Attractants	2	2	2	2	2
Activated gluten	5	5	5	5	5
Alpha-cellulose	22	8	3	13	1
Total	100	100	100	100	100

(Sumule *et al.*, 2000). The growth was not significantly different between groups fed the fish-meal-based diet and the diet containing 32% twin-screw extruded SBM.

The effects of possible combinations of protein sources on fish performances have also been investigated in recent years (Aoki *et al.*, 1997, 1998, 2000). Fish averaging 27 g and 550 g were fed diets containing soybean protein concentrate, SBM, maize-gluten meal, meat meal and poultry-feather meal with or without fish-meal (Aoki *et al.*, 2000). The growth performance of juveniles was not as good as that of fish fed a commercial diet, but feed efficiency and the haematochemical condition of fish fed the above diets together with low levels of fish-meal were similar to those of fish fed the commercial diet.

There are limited studies on amino acid, including essential amino acid (EAA), requirements of red sea bream. The lysine (Lys) requirement of red sea bream (1.7 g initial weight) was investigated by feeding semimoist diets (Forster and Ogata, 1998). When zein, fish-meal, gelatin and L-lysine hydrochloric acid (HCl) were used as ingredients, they found that the requirements were close to 3.6, 4.3 and 4.4% of dietary protein for the specific growth rate, feed efficiency and nitrogen retention, respectively. Forster and Ogata's (1998) estimated values for other EAA (% of dietary protein) based on the A/E ratio (amount of each EAA/total amount of EAA) were as follows: arginine (Arg), 3.5; histidine (His), 1.4; isoleucine (Ile), 2.2; leucine (Leu), 4.2; methionine + cystine (Met + Cys), 2.2; phenylalanine + tyrosine (Phe + Tyr), 4.1; threonine (Thr), 1.8; tryptophan (Trp), 0.6; valine (Val), 2.5%. Yone (1976) reported the requirement of the ten EAA (g 100 g^{-1} diet) according to the growth response of red sea bream as follows: Arg, 3.7; His, 1.7; Ile, 2.3; Leu, 3.4; Met, 1.1; Phe, 2.5; Thr, 1.7; Trp, 0.6; Val, 3.1; Lys, 4.3. A study on the arginine requirement of red sea-bream larvae (Lopez-Alvarado and Kanazawa, 1994) suggested that the requirement was

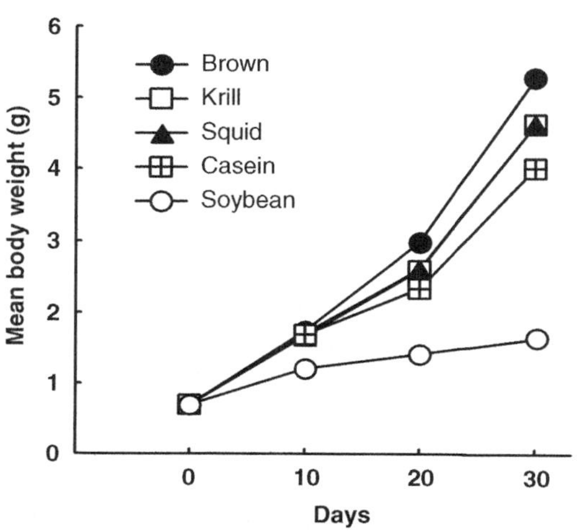

Fig. 4.1. Growth curve of red sea-bream juveniles fed diets with different protein sources. Brown, brown fish-meal.

4.2% of dietary protein (2.5% of a 60% protein diet). It is likely that larval red sea bream have a higher requirement for the amino acid than juveniles or adults for the optimal growth.

Energy

There are limited studies on the energy requirement of red sea bream. In a study on the search for suitable dietary levels of protein and lipid, four metabolizable energy (ME) levels with different dietary protein and lipid levels were tested (Takeuchi *et al.*, 1991). The study suggested that growth increased with increased dietary ME up to 3.7 kcal g^{-1} diet under the 42% protein level. With a 52% protein level, growth increased with increased dietary ME up to 4.3 kcal g^{-1} diet. The optimum calorie-to-protein ratio (kcal kg^{-1}/% protein) was 87 for the 42% protein group and 73–80 for the 52% protein group.

We conducted a study using the response surface analysis method (RSM) to determine the appropriate combination of energy-yielding nutrients for the optimal growth of red sea-bream juveniles (Koshio *et al.*, 1998). Based on the concept of the triangle graph (Toyomizu *et al.*, 1982), 24 fish-meal-based dry diets were formulated. The protein energy (PGE) intake ranged from 30 to 70%, fat energy (FGE) from 10 to 60% and carbohydrate energy (CGE) from 0 to 60%. Second-order polynomial equations were applied to construct the RSM and contour map of the growth performance. Figure 4.2 presents the results for the specific growth rate of juvenile red sea bream. The growth of the fish depended largely on the PGE. It can be concluded that red sea-bream juveniles required almost 60% of energy from protein to obtain the optimal growth. At the same time, we found that the ratio of CGE and FGE is one of the important factors that affect fish performances. As shown in Fig. 4.2, red sea bream required more energy from lipids than from carbohydrates for maximum growth and the ratio of CGE and FGE was less than 1.0 to obtain the optimal range of growth. It is also confirmed that red sea bream had a very poor ability to utilize carbohydrates since slow growth was obtained when fish took more CGE than FGE. Even when dietary CGE was not supplied, the growth did not decrease as long as enough PGE and FGE were available (Fig. 4.2).

Lipids and fatty acids

The optimal dietary lipid levels for maximum growth were examined using 30–40 g fish (Yone *et al.*, 1974; Yone, 1975). These studies suggested that the best performance was obtained in fish fed the diet containing about 10% fish-oil. They also indicated that fish fed a diet containing soybean oil showed a very poor growth rate due to the lack of eicosapentaenoic acid (EPA), 20:5n-3, and docosahexaenoic acid (DHA), 22:6n-3. Takeuchi *et al.* (1991) reported that increasing dietary energy by increasing dietary lipid resulted in improvement of the growth rate. The optimal combination seemed to be 15% dietary lipid and 52% dietary protein, together with 4 kcal g^{-1} diet of ME when red sea-bream

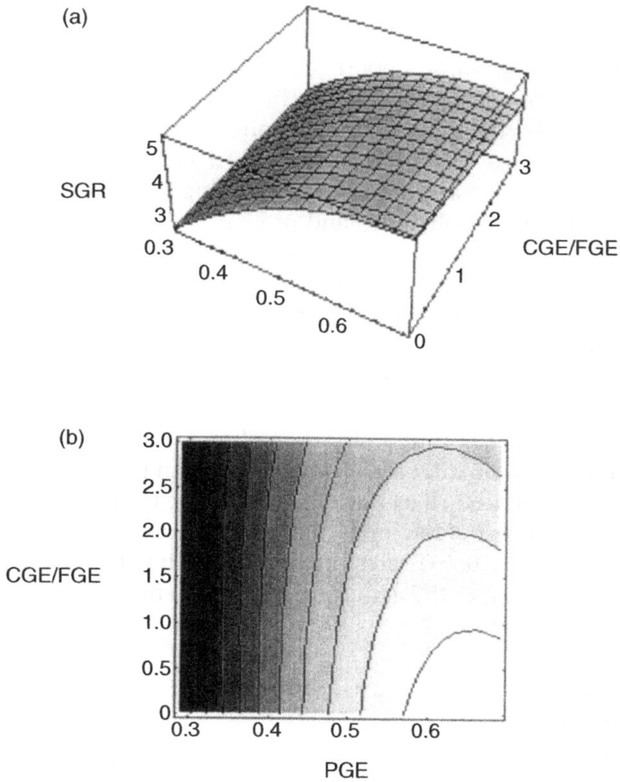

Fig. 4.2. Response surface (a) and contour map (b) for the specific growth rate (SGR) of red sea-bream juveniles fed for 30 days with 24 test diets containing a wide range of energy from protein, lipid and carbohydrate. PGE, protein energy; CGE, carbohydrate energy; FGE, fat energy.

juveniles were fed the diet containing either 1.2% or 2.2% n-3 highly unsaturated fatty acids (HUFA). A further increase of dietary lipids had no effect on the growth enhancement. They also suggested that the highest growth rate was obtained at 15% dietary lipid together with 3.7 kcal g^{-1} diet of ME. However, the study showed that the growth decreased when the ME increased up to 4.1 kcal g^{-1} diet with 20% dietary lipid. Furthermore, it was suggested that the growth performance depended on the dietary level of n-3 HUFA and that the n-3 HUFA requirement increased with increasing dietary lipid content (Takeuchi *et al.*, 1991). It was also reported that a suitable level of n-3 HUFA was approximately 20% of dietary lipid regardless of lipid levels (Takeuchi *et al.*, 1992). Among n-3 HUFA, EPA and DHA are essential fatty acids (EFA) for marine species. Dietary EPA and DHA requirements for red sea-bream juveniles were 1% and 0.5%, respectively, and the efficiency of DHA as an EFA was about twice that of EPA for 3–12 g red sea bream (Takeuchi *et al.*, 1990).

Carbohydrates

Marine fish utilize carbohydrates poorly compared with freshwater fish, and red sea bream is no exception. It was reported that the growth of 70 g red sea bream was highest in fish fed the glucose-free diet and growth indices decreased with increased dietary glucose levels (Furuichi *et al.*, 1971). The authors suggested that poor performance of fish fed the diet containing high glucose resulted from the low digestibility of carbohydrate and protein and poor utilization of carbohydrate by red sea bream. Furuichi and Yone (1980) also indicated that, when 30 g red sea bream were fed diets containing different dextrin levels (0 to 40%), the following were found in fish fed more than 30% dietary dextrin: (i) poor growth and feed efficiency; (ii) high hepatic glycogen storage; (iii) low muscle and liver lipid contents; and (iv) decreased protein and dextrin absorption with increased dietary dextrin.

In the study on the utilization of carbohydrates, Furuichi and Yone (1982) reported that the utilization of α-starch was better than that of dextrin or glucose in terms of growth and feed efficiency when fish were fed twice daily. They suggested that the utilization of low-molecule carbohydrates could be improved by multiple feeding since carbohydrates that were consumed during the successive feedings were absorbed after the activity of metabolic enzymes increased.

Vitamins and minerals

The following ten water-soluble vitamins are probably essential for red sea bream, as for other fish species: vitamins B_1, B_2, B_6, B_{12} and C, nicotinic acid, pantothenic acid, folic acid, inositol and choline. Furthermore, although there are very limited studies on the essentiality and requirement for fat-soluble vitamins namely A, D, E and K, vitamin A is probably essential.

The requirement for vitamin C (ascorbic acid (AA)) for juvenile red sea bream was reported to be less than 200 mg AA kg^{-1} diet when ascorbic acid-calcium (Ca) was used in the study (Furuichi *et al.*, 1990). However, due to the unstable characteristics of AA, derivatives have been developed in Japan. L-Ascorbyl-2-phosphate magnesium (APM) has 43% the bioavailability of AA and was tested for red sea bream (Kosutarak *et al.*, 1994). Juveniles (5 g) were fed the diets containing 0, 50, 100, 200, 500 and 1000 mg APM kg^{-1} diet for 12–13 weeks. The fish fed the diet without APM showed poor growth and survival compared with those fed the diets containing APM. Among APM-fed groups, there were no significant differences in growth and survival. On the other hand, liver AA contents increased with dietary APM and the highest accumulation of AA was found in the 1000 mg APM-fed group. Furthermore, tolerance for a rising temperature was weaker in the APM-free group than in the APM-fed groups, but there was no difference in the tolerance among APM-fed groups. It is likely that at least 50 mg APM kg^{-1} diet (equivalent to 22 mg AA kg^{-1} diet) would be required for maintaining the optimal condition of red sea-bream juveniles. Appetite loss and growth retardation are typical vitamin deficiency signs. Thus,

Table 4.2. Weight gain of red sea bream at different periods when fed vitamin-sufficient and deficient diets (modified after Yone and Fuji, 1974).

Diets	Weight gain (g)	
	Day 63	Day 102
Sufficient vitamin mixture*	51	83
B_6-deficient	22	–
Choline chloride-deficient	17	–
Pantothenic acid-deficient	37	–
B_{12}-deficient	25	–
Inositol-deficient	31	–
Nicotinic acid-deficient	31	–
B_2-deficient	31	42
B_1-deficient	43	68
C-deficient	44	59
Biotin-deficient	41	63
Folic acid-deficient	47	73
ρ-Aminobenzoic acid-deficient	45	68

* Complete vitamin mixture contained the following (mg kg^{-1} diet): B_6, 40; choline chloride, 8000; pantothenic acid, 280; B_{12}, 0.09; inositol, 4000; nicotinic acid, 800; B_2, 200; B_1, 60; C, 2000; biotin, 6; folic acid, 15; ρ-aminobenzoic acid, 400.

decreased growth occurred when each vitamin was omitted from the complete vitamin mixture (Table 4.2) (Yone and Fujii, 1974).

Recently, we examined the effect of dietary AA on the quality of artificially produced juveniles using the behaviour index (Koshio *et al.*, 1997; Sakakura *et al.*, 1998; Koshio, 2001). Since red sea bream showed the behaviour called 'tilting', which is a behavioural response to stress or shock, we examined the relationship between dietary APM and tilting frequency and duration. The study demonstrated that the frequency and duration increased with increasing dietary AA. The highest peaks of these indices were found when fish were fed the diet containing 1000 mg kg^{-1} diet, but, with longer culture periods, the fish fed the diet with less than 200 mg kg^{-1} diet also showed higher frequency and longer duration compared with fish fed an AA-free diet. Since tilting behaviour can help the fish to avoid predation after restocking in the wild, we believe that, by controlling dietary vitamin C contents, it is possible to increase the survival rates and accelerate adaptation to a wild condition after restocking.

It is believed that marine fish absorb certain amounts of minerals, such as Ca, magnesium, phosphorus (P), sodium, sulphur and other minerals, from sea water to meet part of their requirements. Studies on P and Ca have been conducted on red sea bream. Dietary Ca levels did not affect the growth and body Ca or P contents, but dietary P levels did (Sakamoto and Yone, 1973). When red sea bream were fed the P-deficient diet, growth was reduced, the ash, Ca and P contents of vertebrae decreased and body lipid increased (Sakamoto and Yone, 1973;

Sakamoto and Yone, 1978b). Since the iron level in sea water and the absorption rate of iron by fish are very low, it is necessary to give a dietary supplement of iron. It was reported that the iron requirement of red sea bream is about 150 mg kg^{-1} diet (Sakamoto and Yone, 1978a).

Practical Diets

Due to complications, including labour, facilities and the high cost of live food production, the development of microdiets for larval and early juvenile stages of red sea bream has progressed in recent years. To establish the practical use of microbound diets (MBD) in mass seed production, we conducted a feeding trial using larval red sea bream (Kanazawa *et al.*, 1989). When various protein sources were combined to approximate the amino acid pattern of larval whole-body protein, using MBD together with a small amount of live food could sustain good growth and survival in 10-day-old larvae. Information on the composition of practical microdiets, crumbs and pellets for larvae and young-stage juveniles is presented in Table 4.3. Most diets for those stages have high dietary protein and energy.

Many kinds of grow-out diets for red sea bream have been developed in recent years. Compositions of commercial feeds are given in Table 4.4. Most grow-out diets contain less protein and energy compared with larval diets. Also, since red sea bream with the typical red colour have a good market value, special diets containing more carotenoids, vitamin E and vitamin C have also been developed.

There are some variations among diets used for brood-stock (Foscarini, 1988). In most cases, fresh or frozen trash fish, shrimp and squid together with compound diets were offered to spawners. However, it was found that certain nutrients affected the quality of eggs when the diets were given shortly before or during spawning. Nutrients such as protein, vitamins A and E, squid liver oil and krill oil have been reported to increase the numbers of buoyant eggs (Watanabe *et al.*, 1984a,b, 1985a,b), which have a higher hatching rate and more normal development than sinking eggs.

Feeding Practices

A typical pattern of larval and juvenile feeding schedules of red sea bream kept at 19°C is shown in Table 4.5. Formulated diets, together with live foods, are given to larval red sea bream 20 days after hatching. Table 4.6 indicates an example of a feeding schedule for juveniles (30–60 days after hatching) using formulated diets at 18–20°C.

In the grow-out stage there are two types of feeds, moist and dry feeds, which are commercially available for red sea bream in Japan. However, due to the insufficient supply and high cost of raw materials and concern about environmental issues and cost-effective operations, dry feeds will take over the market (Nakamura, 1997). One of the most important factors in aquaculture operation is

Table 4.3. Composition of commercial microdiets, crumbs and pellets for larvae and juveniles of red sea bream.

Composition	Diet A*	Diet B†	Diet C‡	Diet D§
Moisture (%)	< 10.0	< 10.0	< 10.0	–
Crude protein (%)	> 54.0	> 55.0	> 55.0	> 56.0
Crude fat (%)	> 10.0	> 10.0	> 10.0	> 9.0
Crude ash (%)	< 13.0	< 13.0	< 17.0	< 15.0
n-3 HUFA (%)	2.9	3.2	2.8	–
Vitamin C (mg %)	150	60.0	60.0	–
Gross energy (kcal g^{-1})	5.6	–	–	–
Amino acid (mg g^{-1} of wet diet)				
Arginine	28.8	36.4	33.7	–
Lysine	39.9	43.9	40.6	–
Histidine	14.1	13.6	12.8	–
Leucine	47.7	48.0	41.1	–
Isoleucine	23.8	24.3	23.2	–
Methionine	18.6	17.5	17.0	–
Valine	27.2	28.0	26.6	–
Threonine	22.9	24.7	23.8	–
Phenylalanine	25.5	27.5	25.6	–
Tryptophan	6.2	6.8	6.8	–

* < 250 or 250–400 μm particle size.
† 250–700 μm particle size.
‡ 710 to 4000 μm particle size (crumb or pellet).
§ < 360 or 400–2700 μm particle size.
Protein sources: fish-meal, krill meal, squid meal and soybean meal; carbohydrate sources: wheat and α-starch; lipid sources: fish-oils; other ingredients: wheat gluten, yeast extracts, minerals and vitamins.

Table 4.4. Composition (%) of commercial grow-out feeds for red sea bream (10–800 g).*

Composition	Diet A	Diet B	Diet C	Diet D
Crude protein	> 45 or 46	> 46 or 48	> 48	> 44
Crude fat	> 12 or 13	> 12 or 15	> 8	> 9
Crude ash	< 13 or 14	< 17	< 16	< 15
Crude fibre	< 3	< 4	< 3	< 2
Ca	–	> 2.2	> 2.3	> 2
P	–	1.4	1.6	> 1

* Protein sources: fish-meal, squid meal and krill meal; carbohydrate sources: wheat flour, α-starch; other ingredients: soybean meal, rice bran, wheat bran, fish lipid, yeast extract, minerals, vitamins and carotenoids.

understanding the environmental conditions fully. As a result, the feeding schedule should be adjusted accordingly. To determine the optimal ration size, the following are very important points which should be considered: rearing density, mesh size of the net cage, cleaning of the net, optimal water current in the cage and health condition of the fish.

The optimal ration size of dry feeds under the different water-temperature regimes is presented in Table 4.7. Red sea bream have a passive feeding behaviour and they ignore the feeds that sink to the bottom of the cage. Therefore, longer

Table 4.5. Typical pattern of a feeding schedule used for larvae and juveniles of red sea bream at 19°C.

	Days after hatching						
Feeds or diets	0	10	20	30	40	50	60
Nannochloropsis	3===============25						
Rotifers	3=======================33						
Brine shrimp			20==============40				
Formulated feeds			20==============================60				

Table 4.6. Recommended feeding practices for red sea bream, 30 to 60 days after hatching.

	Age (days after hatch)			
	30	40	50	60
Body length (mm)	12	19	26	35
Body weight (BW) (g)	0.02	0.1	0.3	0.9
Ration size (% of BW)	15	13	11	9
Feed amounts (g per 10,000 fish)	30	130	320	810
Feeding frequency (per day)	5	7	8	7

Table 4.7. Ration size (% of body weight day^{-1}) at different water temperatures for red sea bream (modified after Nakamura, 1997).

	Body weight (g)							
Water temp. (°C)	5–20	20–50	50–100	100–250	250–500	500–750	750–1000	> 1000
13–14	1.5–2.0	0.9–1.5	0.6–0.9	0.6–0.7	0.5–0.6	0.4–0.5	0.3–0.5	0.3–0.4
15–16	2.4–3.0	1.2–2.1	1.0–2.0	0.8–1.7	0.7–1.5	0.6–1.0	0.5–0.7	0.4–0.6
17–18	2.9–4.1	2.3–3.1	1.5–2.4	1.7–2.2	1.0–1.8	0.8–1.3	0.6–1.1	0.4–0.9
19–20	3.9–4.5	3.1–3.9	2.7–3.2	1.9–2.9	1.3–2.2	0.9–1.5	0.7–1.5	0.7–1.4
20–21	4.4–5.3	3.3–4.0	2.9–3.4	2.0–3.4	1.4–2.5	1.1–1.7	0.8–1.6	0.7–1.5
22–23	4.8–6.0	3.9–5.3	3.1–4.2	2.3–4.2	1.5–2.6	1.2–1.9	0.9–1.7	0.8–1.6
24–25	5.3–6.5	4.1–5.6	3.4–4.9	2.5–4.6	1.7–2.7	1.3–2.0	1.0–1.8	0.8–1.7
26–27	5.8–7.0	4.4–6.0	3.7–5.3	3.0–4.8	1.9–2.9	1.4–2.2	1.1–1.9	0.9–1.8
28	5.6–6.8	4.2–5.8	3.5–5.1	2.8–4.6	1.7–2.7	1.3–2.0	0.9–1.8	0.8–1.7

feeding duration and higher feeding frequency are necessary. Furthermore, fish should not be fed in only one location because stronger fish can take over the feeding, resulting in growth reduction of weaker fish.

References

Aoki, H., Watanabe, T., Furuichi, M. and Tsuda, H. (1997) Use of alternative protein sources as substitutes for fish meal in red sea bream diets. *Suisanzoshoku* 45, 131–139.

Aoki, H., Furuichi, M., Viyakarn, V., Yamagata, Y. and Watanabe, T. (1998) Feed protein ingredients for red sea bream. *Suisanzoshoku* 46, 121–127.

Aoki, H., Furuichi, M., Watanabe, K., Satoh, S., Yamagata, Y. and Watanabe, T. (2000) Use of low or non-fish meal diets for red sea bream. *Suisanzoshoku* 48, 65–72.

Forster, I. and Ogata, H. (1998) Lysine requirement of juvenile Japanese flounder *Paralichthys olivaceus* and juvenile red sea bream *Pagrus major*. *Aquaculture* 161, 131–142.

Foscarini, R. (1988) A review: intensive farming procedure for red sea bream (*Pagrus major*) in Japan. *Aquaculture* 72, 191–246.

Furuichi, M. and Yone, Y. (1980) Effect of dietary dextrin on the growth and feed efficiency, the chemical composition of liver and dorsal muscle, and the absorption of dietary protein and dextrin in fishes. *Bulletin of Japanese Society of Scientific Fisheries* 46, 225–229.

Furuichi, M. and Yone, Y. (1982) Availability of carbohydrate in nutrition of carp and red sea bream. *Bulletin of Japanese Society of Scientific Fisheries* 48, 945–948.

Furuichi, M., Yontanda, K. and Yone, Y. (1971) Studies on nutrition of red sea bream – V. Appropriate supply of dietary carbohydrate. *Report of Fishery Research Laboratory, Kyushu University* 1, 91–100.

Furuichi, M., Kitajima, C., Matsui, S., Yoshimatsu, T. and Tanabe, T. (1990) Vitamin C requirement of the red sea bream. In: *Abstract of Spring Meeting of the Japanese Society of Scientific Fisheries*. Japanese Society of Scientific Fisheries, Tokyo, p. 39 (in Japanese).

Kanazawa, A., Koshio, S. and Teshima, S. (1989) Growth and survival of larval red sea bream *Pagrus major* and Japanese flounder *Paralichthys olivaceus* fed microbound diets. *Journal of the World Aquaculture Society* 20, 31–37.

Koshio, S. (1999) Present status of aquaculture feed and protein replacement. *Mini Review of Data File of Fisheries Research* 8, 70–83 (in Japanese with English abstract).

Koshio, S. (2001) Critical review of the effect of vitamin C on fish behavior. In: Dabrowski, K. (ed.) *Ascorbic Acid in Aquatic Animals*. CRC Press, Boca Raton, Florida, pp. 241–253.

Koshio, S., Teshima, S. and Ishikawa, M. (1996) Nutritional evaluation of several protein sources for red sea bream. In: *Abstracts of Fall Meeting of Japanese Society of Scientific Fisheries*. Japanese Society of Scientific Fisheries, Fukuoka, Japan, p. 58 (in Japanese).

Koshio, S., Sakakura, Y., Iida, Y., Tsukamoto, K., Kida, T. and Dabrowski, K. (1997) The effect of vitamin C intake on schooling behavior of amphidromous fish, ayu (*Plecoglossus altivelis*). *Fisheries Science* 63, 619–624.

Koshio, S., Teshima, S. and Ishikawa, M. (1998) The determination of the optimal nutrient energy combination for the growth of red sea bream, *Pagrus major*, juveniles by the response surface analysis method (RSM). In: *Abstracts of Recent Advances in Finfish and Crustacean Nutrition, VIII International Symposium on Nutrition and Feeding of Fish*. GIA, Las Palmas de Gran Canaria, Spain, p. 21.

Kosutarak, P., Kanazawa, A., Teshima, S., Koshio, S. and Itoh, S. (1994) L-Ascorbyl-2-phosphate Mg as a vitamin C source for red sea bream (*Pagrus major*) juveniles.

In: Chou, L.M., Munro, A.D., Lam, T.J., Chen, T.W., Cheong, L.K.K., Ding, J.K., Hooi, K.K., Khoo, H.W., Phang, V.P.E., Shim, K.F. and Tan, C.H. (eds) *Proceedings of the Third Asian Fisheries Forum*. Asian Fisheries Society, Manila, Philippines, pp. 729–732.

Lopez-Alvarado, J. and Kanazawa, A. (1994) Effect of dietary arginine levels on growth of red sea bream larvae fed diets supplemented with crystalline amino acids. *Fisheries Science* 60, 435–439.

Nakamura, R. (1997) The grow-out feeds and the optimal feeding planning of red sea bream. *The Best Guidance of Aquaculture Feeds. Yoshoku Special Issue* 420, 122–127 (in Japanese).

Sakakura, Y., Koshio, S., Iida, Y., Tsukamoto, K., Kida, T. and Blom, J.H. (1998) Dietary vitamin C improves the quality of yellowtail (*Seriola quinqueradiata*) seedlings. *Aquaculture* 161, 427–436.

Sakamoto, S. and Yone, Y. (1973) Effect of dietary calcium/phosphorus ratio upon growth, feed efficiency and blood serum Ca and P level in red sea bream. *Bulletin of Japanese Society of Scientific Fisheries* 39, 343–348.

Sakamoto, S. and Yone, Y. (1978a) Requirement of red sea bream for dietary iron. II. *Bulletin of Japanese Society of Scientific Fisheries* 44, 223–225.

Sakamoto, S. and Yone, Y. (1978b) Effect of dietary phosphorus level on chemical composition of red sea bream. *Bulletin of Japanese Society of Scientific Fisheries* 44, 227–229.

Statistical and Information Department (2000) *Annual Statistics of Fishery and Aquaculture in 1998*. Ministry of Agriculture, Forestry and Fisheries, Tokyo, 300 pp. (in Japanese).

Sumule, O., Koshio, S., Teshima, S., Ishikawa, M. and Harada, H. (2000) Effects of the replacement of fish meal with soybean meal on growth performance and enrgy utilization of red sea bream *Pagrus major* (Temminck & Schlege) juveniles. In: *Abstracts of the Ninth International Symposium on Nutrition and Feeding in Fish*. The committee of the Ninth International Symposium on Nutrition and Feeding in Fish, Miyazaki, Japan, p. 65.

Takeuchi, T., Toyota, M., Satoh, S. and Watanabe, T. (1990) Requirement of juvenile red sea bream, *Pagrus major*, for eicosapentaenoic and docosahexaenoic acids. *Nippon Suisan Gakkaishi* 56, 1263–1269.

Takeuchi, T., Shiina, Y. and Watanabe, T. (1991) Suitable protein and lipid levels in diet for fingerlings of red sea bream *Pagrus major*. *Nippon Suisan Gakkaishi* 57, 293–299.

Takeuchi, T., Shiina, Y. and Watanabe, T. (1992) Suitable levels of n-3 highly unsaturated fatty acids in diet for fingerlings of red sea bream. *Nippon Suisan Gakkaishi* 58, 509–514.

Toyomizu, M., Akiba, Y., Horiguchi, M. and Matsumoto, T. (1982) Multiple regression and response surface analyses of the effects of dietary protein, fat and carbohydrate on the body protein and fat gains in growing chicks. *Journal of Nutrition* 112, 886–896.

Ukawa, M., Takii, K., Nakamura, M. and Kumai, H. (1994) Utilization of soybean meal for red sea bream diet. *Suisanzoshoku* 42, 335–338 (in Japanese with English abstract).

Watanabe, T., Itoh, A., Kitajima, C. and Fujita, S. (1984a) Effect of dietary protein levels on reproduction of red sea bream. *Bulletin of Japanese Society of Scientific Fisheries* 50, 1015–1022.

Watanabe, T., Ohhashi, S., Itoh, A., Kitajima, C. and Fujita, S. (1984b) Effect of nutritional composition of diets on chemical components of red sea bream broodstock and eggs produced. *Bulletin of Japanese Society of Scientific Fisheries* 50, 503–504.

Watanabe, T., Itoh, A., Satoh, S., Kitajima, C. and Fujita, S. (1985a) Effect of dietary protein levels and feeding period before spawning on chemical components of eggs produced by red sea bream broodstock. *Bulletin of Japanese Society of Scientific Fisheries* 51, 1501–1509.

Watanabe, T., Koizumi, T., Suzuki, H., Satoh, S., Takeuchi, T., Yoshida, N., Kitada, T. and Tsukashima, Y. (1985b) Improvement of quality of red sea bream eggs by feeding broodstock on a diet containing cuttlefish meal or on raw krill shortly before spawning. *Bulletin of Japanese Society of Scientific Fisheries* 51, 1511–1521.

Yone, Y. (1975) Nutritional studies of red sea bream. In: Price, K.S. Jr, Shaw, N.N. and Dobery, K.S. (eds) *Proceedings of the First International Conference on Aquaculture Nutrition*. University of Delaware, Newark, Delaware, pp. 39–64.

Yone, Y. (1976) Nutritional studies of red sea bream. *Report of Fishery Research Laboratory, Kyushu University* 3, 87–101.

Yone, Y. and Fuji, M. (1974) Studies on nutrition of red sea bream. X. Qualitative requirements for water-soluble vitamins. *Report of Fishery Research Laboratory, Kyushu University* 2, 25–32 (in Japanese with English extract).

Yone, Y., Sakamoto, S. and Furuichi, M. (1974) Studies on nutrition of red sea bream. IX. The basal diet for nutrition studies. *Report of Fishery Research Laboratory, Kyushu University* 2, 13–24.

Gilt-head Sea Bream, *Sparus aurata*

5

William Koven
National Center for Mariculture, Israel Oceanographic and Limnological Research, POB 1212, Eilat 88112, Israel

Introduction

The natural habitat of the gilt-head sea bream (*Sparus aurata*) ranges from the Mediterranean and Black Sea to the eastern Atlantic Ocean from Senegal to the UK (Kissil *et al.*, 2000a). This species is generally found in shallow lagoons along the coast but moves into deeper waters to spawn after late autumn. In many commercial rearing conditions, the larvae emerging after hatching deplete the yolk-sac after 3–4 days of endogenous feeding. At this stage, the eyes are pigmented and the mouth open, allowing the larvae to feed on rotifers (e.g. *Brachionus plicatilis*). At 12–15 days posthatching, the rapidly growing larvae are fed mostly on *Artemia* nauplii as well as rotifers through the onset of metamorphosis to the end of larval rearing (32–35 days posthatching). Prior to being fed to the larvae, rotifers and *Artemia* are routinely enriched with commercial lipid preparations to enhance their levels of essential fatty acids, which are critical for normal larval growth, development and survival. After metamorphosis, fish from 5 to 10 mg are rapidly weaned from *Artemia* to a dry high-protein (50–60%) formulated diet. At 1–3 g, the juveniles are stocked in sea cages and grown to market size (400–500 g) over 12–14 months, using high-energy extruded diets (Kissil *et al.*, 2000a). At present, sea-cage culture is more economical than land-based production systems (ponds and raceways), which need high capital investment and energy input (constant aeration and continual pumping of sea water). On the other hand, cage farming is limited to sites protected from inclement weather and may pose an environmental problem from the nutrient effluent of the fish.

Nutrient Requirements: Larval Fish

The determination of the nutrient requirements of larval sea bream has focused almost entirely on the lipids and essential fatty acids necessary to promote good growth and survival. This is primarily due to the almost exclusive use of rotifers and *Artemia* as experimental diets, which are generally limited to the manipulation of their lipid content and composition. Practical diets or microdiets, which would offer a more diverse tool for nutritional research, have not been sufficiently developed as a live-food alternative, although great strides in their application have been made in recent years.

Lipids and fatty acids

In the gilt-head sea bream, the dietary requirement for the n-3 highly unsaturated fatty acids (HUFA), primarily eicosapentaenoic acid (EPA) (20:5n-3) and docosahexaenoic acid (DHA) (22:6n-3), has been established (Koven *et al.*, 1989, 1990, 1992a,b; Mourente *et al.*, 1993; Rodríguez *et al.*, 1993). These fatty acids, in their phospholipid (PL) form, function as critical structural and physiological components of the cell membrane in most tissues (Gurr and Harwood, 1991; Sargent *et al.*, 1993a). The n-3 HUFA requirement in first-feeding sea-bream larvae ranged from 8.4 (Koven *et al.*, 1990) to 55 mg n-3 HUFA g^{-1} dry weight (DW) rotifer (Rodríguez *et al.*, 1998a). This discrepancy may be explained by the rotifer EPA : DHA ratio, which varied among studies, affecting the larval n-3 HUFA requirement. Rodríguez *et al.* (1997) found that, for a constant level of n-3 HUFA, growth was negatively correlated with the rotifer EPA : DHA ratio (0.71 : 1 to 3.6 : 1). This suggested that DHA had higher nutritional value and contributed more to growth than EPA (Mourente *et al.*, 1993; Rodríguez *et al.*, 1997).

Koven *et al.* (1992b) found an age-dependent relationship in gilt-head sea bream, where DHA was preferentially assimilated into the tissue phospholipid, phosphatidylethanolamine (PE), and was highly correlated with larval growth, while EPA in PE was poorly associated with this parameter. Unlike other tissues, neural tissue, particularly the brain and retina, is richer in PE than in the main phospholipid, phosphatidylcholine (PC). The higher biological value to larvae of DHA compared with EPA could be derived from its presence in tissues involved in visual acuity. Presumably, this would influence prey-hunting success and net energy gain (Koven *et al.*, 1992b; Sargent *et al.*, 1993b; Bell *et al.*, 1995).

Rodríguez *et al.* (1998a) concluded that 15 mg n-3 HUFA g^{-1} DW rotifer with an EPA : DHA ratio of 0.71 : 1 promoted good growth during the first 2 weeks of larval development. These results agreed well with the 17 mg (Mourente *et al.*, 1993) and 20 mg n-3 HUFA g^{-1} DW of rotifers (Salhi, 1997) with a low EPA : DHA ratio. During *Artemia* feeding, Koven *et al.* (1992a) obtained best growth in 22–36-day-old gilt-head sea-bream larvae when fed the 30 mg n-3 HUFA g^{-1} DW *Artemia*.

Unlike EPA and DHA, arachidonic acid (AA) (20:4n-6) is not a structural membrane lipid and instead is involved, as a precursor in eicosanoid synthesis, in various areas of cellular regulation, including fluid and electrolyte fluxes, the cardiovascular system, the reproductive function and the neural system (Mustafa and Srivastava, 1989). Bessonart *et al.* (1999) found that AA-supplemented

(18 mg AA g^{-1} DW) microdiets were more effective in improving survival than growth in 17–31-day-old gilt-head sea-bream larvae if provided in the presence of a low dietary EPA : DHA ratio. Similarly, Koven *et al.* (2001a) found that dietary AA (2.7 mg g^{-1} DW rotifer) fed to sea-bream larvae prior to handling stress improved survival more effectively than when fed following handling stress (4.6 mg AA g^{-1} DW *Artemia* nauplii). These findings imply the requirement of AA in early larval development, as it affects later larval and juvenile survival during the stress of metamorphosis, weaning, crowding and grading.

Practical Diets

A great deal of interest has been generated in developing an artificial larval diet or microdiet (MD) as a live-food alternative. However, in general, MDs are consumed much less than live food and are frequently sufficient only as a maintenance ration (Tandler and Kolkovski, 1991). On the other hand, the MD performance was markedly improved when co-fed with live *Artemia* nauplii (Tandler and Kolkovski, 1991; Fernández-Díaz and Yúfera, 1997; Kolkovski *et al.*, 1997a; Rosenlund *et al.*, 1997). This suggested the presence of certain nutrients in live food that elicit physiological responses in the larvae, which should be considered in larval diet development.

Kolkovski *et al.* (1997a) found that the MD ingestion rates in sea-bream larvae increased up to 120% in the presence of different concentrations of *Artemia* nauplii. These authors indicated that the free amino acids (FAA) alanine, glycine, arginine and betaine secreted by *Artemia* nauplii were responsible for stimulating food ingestion (Kolkovski *et al.*, 1997a). In a later study, the inclusion of these nutrients in an MD based on the gelatin encapsulation of liposomes improved the MD ingestion rate in 7-day-old sea-bream larvae (Koven *et al.*, 2001b).

The phospholipid PC in the diet was also shown to have a stimulatory effect on sea-bream larval feeding, while this was not observed with PE (Hadas, 1998; Koven *et al.*, 1998). Larvae fed a PC-supplemented MD had 35% higher ($P < 0.05$) ingestion rates compared with the unsupplemented MD in 21–26-day-old gilt-head larvae, although this effect diminished in 28–31-day-old larvae (Hadas, 1998; Koven *et al.*, 1998). Further studies suggested that dietary PC has a postprandial physiological influence as well, occurring in parallel or in tandem with its appetite-stimulating properties. Koven *et al.* (1993) showed a significant effect of dietary lecithin on the incorporation of labelled free fatty acid (FFA) in body neutral lipid and phospholipid (PL) in 21–45-day-old larvae. Hadas (1998) and Salhi *et al.* (1999) concluded that dietary PL contributes to lipoprotein production, thereby enhancing the efficiency of lipid transport from the enterocytes lining the digestive tract to the body tissues.

In sea-bream larvae ingesting MD supplemented with a porcine pancreatic extract (0.05%), Kolkovski *et al.* (1993) found a 30% increase in assimilation and significantly improved growth. This suggested that live food could enhance digestion by contributing enzymes to facilitate the digestion process. In contrast,

Cahu *et al.* (1995) and Moyano *et al.* (1996) found no evidence of rotifers or *Artemia* supplying proteases for digestion in various ages of marine larvae. An alternative interpretation of the contribution of live food to larval digestion is that certain factors are released from the prey upon ingestion that stimulate the conversion of indigenous zymogens to active enzymes and/or an endocrine response, which regulates digestion and nutrient assimilation. The digestive hormone bombesin influences digestion by activating the peristaltic movement of the gut and the release of hydrochloric acid (HCl) as well as increasing blood circulation to the gut wall (McDonald *et al.*, 1979). Kolkovski *et al.* (1997b), comparing the postprandial stimulation of bombesin in live food and MD, found the level of bombesin increased by 300% when *Artemia* nauplii were given as the sole food to sea-bream larvae compared with levels that were found in larvae offered only an MD. However, the nutrient factors in *Artemia* responsible for eliciting this endocrine response remain unclear.

Feeding Practices

Rotifers (e.g. *B. plicatilis*) are fed as a first food (ten rotifers ml^{-1}) to sea-bream larvae and are continually offered throughout the 32–35-day larval rearing period. Prior to being fed to larvae, rotifers previously fed baker's yeast (*Saccharomyces cerevisiae*) and/or algae (e.g. *Nannochloropsis* sp.) are enriched (500 rotifers ml^{-1}) for approximately 8 h on commercial preparations designed to increase their levels of EPA and DHA. Algae are added together with the rotifers to the rearing tanks ($\pm$ 0.2 $\times$ 10^6 algal cells ml^{-1}) to improve rotifer survival as well as imparting to the larvae unidentified metabolites that are reputed to contribute to larval health.

Instar III *Artemia* nauplii (*Artemia franciscana*), which are fed (one nauplius ml^{-1}) to larvae from approximately 18 days old to the end of larval rearing, are previously enriched (100–300 nauplii ml^{-1}) at the instar II stage (12 h posthatching from decapsulated cysts) for 18–24 h on commercial preparations, mainly to boost their levels of EPA and DHA. Apart from the essential fatty acids, other nutrients, including vitamins A, E and C and pigments, are frequently included in commercial enrichment products (Lavens and Sorgeloos, 1996).

Nutrient Requirements: Juvenile and Grow-out Fish

Proteins and amino acids

The feeding of juveniles to market weight in many commercial species represents up to 50% of the operating costs of production (Kissil *et al.*, 1997). The protein fraction of the food, which approximates 40–50% of the diet, is responsible for about 48% of the total feed price and is largely contributed by the fish-meal component ($\geq$ 44% of feed protein) (Kissil *et al.*, 1997). The highest growth rates were reported in sea bream fed 55% protein in fish-meal-based diets in 0.8–3.0 g

(Vergara *et al.*, 1996a) and 9–63 g (Santinha *et al.*, 1996) juveniles. However, fish-meal is a diminishing resource that is destined to become more expensive and scarce as resource fisheries begin to dwindle. Alternative protein sources such as rape-seed and soybean are cheaper, before processing, than quality fish-meal (kg^{-1} protein basis) (Higgs *et al.*, 1995) and are currently being evaluated.

Kissil *et al.* (2000b) fed sea bream diets that replaced 30, 60 and 100% of fish-meal with soybean-protein concentrate (SPC) or rape-seed-protein concentrate (RPC). The diet intake and weight gains were inversely related to the inclusion levels of plant proteins, possibly due to reduced palatability (e.g. phytic acid), deficiencies in essential amino acids and/or the presence of antinutritional factors. Robaina *et al.* (1995), however, found that diet intake was not significantly affected by a 10, 20 and 30% partial replacement of sardine fish-meal with soybean meal (SBM) and lupin-seed meal. On the other hand, a marked reduction in trypsin activity and protein digestibility with increasing SBM may have been attributable to dietary phytic acid. These authors concluded that lupin meal is the preferred protein alternative to SBM in diets for gilt-head sea bream and a level of up to 20% replacement of fish-meal on an equal nitrogen basis was suggested.

Kissil *et al.* (2000b) found that the efficiency of protein utilization appeared similar for diets containing increasing levels of SPC and RPC, which replaced fish-meal, except in the 100% SPC diet. They indicated that a significant depression of the protein production value in the 100% SPC diet reflected a deficiency of one or more essential amino acids, such as methionine. The poor performance of sea-bream diets based on high inclusion levels (75–100% of protein) of poultry meals was similarly attributed to deficiencies in lysine and possibly methionine (Nengas *et al.*, 1995, 1999). On the other hand, Kissil *et al.* (2000b) reported that the RPC diet, even at the 100% level, was as efficiently utilized as that of the fish-meal control, suggesting that the RPC protein quality was similar to that of fish-meal.

Robaina *et al.* (1997) reported that the growth, feed efficiency, protein efficiency ratio and protein production values were not significantly affected by the partial substitution (20, 30 and 40%) of fish-meal with corn (maize)-gluten meal (CGM) or meat and bone meal (MBM). However, delayed and increased levels of nitrogen excretion were evident in fish fed the CGM and MBM diets. This suggests an increase in deamination activity and the amount of ammonia released into the water. A similar observation was reported in SBM- and lupin-meal-containing diets, indicating that plant proteins tend to be digested more slowly and inefficiently, particularly SBM (Robaina *et al.*, 1995). In fish fed diets containing more than 20% MBM, an increase in hepatic deposition of lipids, nuclei polarization and isolated necrotic foci were observed. This may have been caused by an unfavourable ratio of saturated fatty acids and polyunsaturated fatty acids in this food component (Robaina *et al.*, 1997). These authors also found a negative correlation between dietary ash content and dietary protein digestibility in the MBM diets recommending a 30 and 20% substitution of fish-meal by CGM and MBM, respectively.

Lipids and fatty acids

Juvenile sea bream, as in larvae, have a dietary requirement for n-3 HUFA. Ibeas *et al.* (1994, 1996) found that with an EPA : DHA ratio of 1 : 2, 19 mg and 10 mg n-3 HUFA g^{-1} DW diet promoted the best growth of 42.5 g and 11.5 g sea bream, respectively. Kalogeropoulos *et al.* (1992) obtained best growth in 1 g fish fed 9 mg of EPA and DHA g^{-1} DW despite the lower EPA : DHA ratio of 1 : 1 in this study. These results suggest that, unlike in larval nutrition, in which DHA has better nutritional value than EPA, these two essential fatty acids appear to have similar nutritional input for juveniles. Ibeas *et al.* (1996) claimed that the optimum EPA : DHA ratio of 1 : 1.5 in larvae may be increasing to 1 : 1 or 2 : 1 in juveniles and adults.

Energy

Despite the growing body of information on protein sparing by dietary lipid in the gilt-head sea bream, there remain discrepancies in the results reported in the literature. Gilt-head sea bream that were fed to satiation on high-quality fish-meal diets containing 470–480 g crude protein kg^{-1} diet and 220 or 270 g lipid kg^{-1} diet exhibited markedly higher growth than fish fed similar diets containing 150 g lipid kg^{-1} diet (Caballero *et al.*, 1999). In contrast, fish fed a 497 g crude protein kg^{-1} diet containing a lower-quality fish-meal demonstrated improved growth only at 270 g lipid kg^{-1} diet (Caballero *et al.*, 1999). This suggests a possible effect of impaired essential amino acid (EAA) availability on protein sparing by dietary lipid. Vergara *et al.* (1996b) recommended a diet containing 460 g protein and 150 mg lipid kg^{-1} diet for sea-bream juveniles fed at levels approaching satiation. In these studies, a consequence of feeding close to or at satiation is an increase in body and/or liver lipid content. Company *et al.* (1999) claimed that, in order to maximize a protein-sparing effect, restricted feeding must be employed to avoid excessive fat deposition and impaired growth performance.

The main factor determining voluntary-consumption feeding in gilt-head sea bream and other fish species is the digestible energy (DE) content of the diet (Jobling and Wandsvik, 1983; Kentouri *et al.*, 1995; Paspatis and Boujard, 1996; Lupatsch *et al.*, 2001). Lupatsch *et al.* (2001) found that the efficiency of utilization (above maintenance) of daily DE in sea bream was constant at 0.50, regardless of energy intake. However, the efficiency of utilization of digestible protein (DP) varied between 0.33 to 0.60, with an optimum value of 0.47. At high dietary DP : DE ratios, the protein utilization efficiency is about 0.35, as protein is being catabolized for energy. An increase of the DE content by raising the non-protein energy fraction improved the protein efficiency by sparing protein catabolized as an energy source (Lupatsch *et al.*, 2001).

Lupatsch *et al.* (2001) calculated the energy and protein requirements for gilt-head sea bream, according to the growth potential for a specific weight and water temperature, and these are presented as a practical feeding table in Table 5.1.

Table 5.1. Recommended dietary energy and protein supply for growing *Sparus aurata* (modified from Lupatsch *et al.*, 2001).

Energy and protein requirements for different body weights						
Body weight (g per fish)	10		100		250	
Weight gain (g per fish day^{-1})*	0.25		1.00		1.82	
DE_m (kJ per fish day^{-1})†	1.22		8.25		17.66	
DE_g (kJ per fish day^{-1})‡	3.33		17.36		35.19	
DE_{m+g} (kJ per fish day^{-1})§	4.55		25.61		52.85	
DP_m (g per fish day^{-1})‖	0.034		0.172		0.326	
DP_g (g per fish day^{-1})¶	0.096		0.398		0.694	
DP_{m+g} (g per fish day^{-1})§	0.130		0.570		1.019	
Food formulation at two DE levels						
DE level of diet (MJ kg^{-1})	16	20	16	20	16	20
Diet intake (g per fish day^{-1})	0.284	0.228	1.60	1.28	3.30	2.69
DP content (g kg^{-1})	455	569	345	432	309	387
FCR	1.14	0.91	1.60	1.28	1.80	1.44
DP : DE (g MJ^{-1})	28.5	28.5	21.6	21.6	19.3	19.3

* Predicted growth for *Sparus aurata*.
† Digestible energy required for maintenance = 55.8 kJ $kg^{-0.83}$ day^{-1} (Lupatsch *et al.*, 1998).
‡ Digestible energy required for growth using energy efficiency of 0.50.
§ Digestible protein required for maintenance and growth.
‖ Digestible protein required for maintenance = 0.86 g BW $kg^{-0.70}$ day^{-1} (Lupatsch *et al.*, 1998).
¶ Digestible protein required for growth using protein efficiency of 0.47.
FCR, feed conversion ratio.

Carbohydrates

Carnivorous fish such as the gilt-head sea bream have a limited ability to use dietary carbohydrate, due to low amylolytic enzyme activity (Vergara and Jauncey, 1993). In sea bream carbohydrate digestibility was less than 77% regardless of source (Vergara and Jauncey, 1993; Lupatsch *et al.*, 1997) and increasing levels of dietary fibre further decreased the digestibility (Vergara and Jauncey, 1993). Morris and Davies (1995b) reported that lipid and carbohydrate could not be interchanged according to their metabolic energy values. Therefore, in sea-bream diets, lipids provide the majority of non-protein energy, resulting in carbohydrate accounting for less than 150 g carbohydrate kg^{-1} DW diet (Morris, 1997).

Vitamins and minerals

The requirement of gilt-head sea bream for these nutrients has concentrated on vitamins since minerals in formulated diets are generally inexpensive. Morris *et al.* (1995) characterized the pathologies of gilt-head sea-bream juveniles associated with dietary deficiencies of many of the B vitamins, including thiamine, riboflavin, pyridoxine, niacin and pantothenic acid, and demonstrated a requirement of this species for all these vitamins. Morris and Davies (1995a) determined the nicotinic acid requirement to be 63–83 mg kg^{-1} diet, while Morris and Davies (1995b) showed that a thiamine (vitamin B_1) supplement of 10 mg kg^{-1} gave significantly better growth, food conversion and apparent net protein utilization than a thiamine-deficient diet. Kissil *et al.* (1981) found that the minimum level of pyridoxine (B_6) necessary for good growth was 1.97 mg kg^{-1} dry diet. In another study, Kissil (1981) found no suppression of growth or histopathological signs in sea bream fed graded levels of biotin. However, based on the activity of pyruvate carboxylase in the liver, a dietary biotin level between 0.21 and 0.37 mg kg^{-1} DW diet appears to be required for maximum activity of this enzyme.

Ascorbic acid cannot be synthesized by most teleosts (Henrique *et al.*, 1996) and a deficiency in a number of fish species induces a variety of symptoms, including scoliosis, lordosis, haemorrhages, distorted gill filaments, fin erosion, anaemia and increased mortality (Halver, 1989). Alexis *et al.* (1997) reported many of these symptoms in gilt-head sea bream as well as granulomatous disease in the kidney and suppression of the wound-healing response.

Nutrient Requirements: Brood-stock

The gilt-head sea bream is a continuous spawner, having a short vitellogenic period (Zohar *et al.*, 1995). Over the 3–4-month spawning season, brood-stock continue to feed while a spawning female will produce a total egg biomass of 0.5–2 kg kg^{-1} body weight (BW). Consequently, the nutrient composition of the eggs is highly influenced by diet and to a much lesser extent by the endogenous body stores (Tandler *et al.*, 1995).

Proteins and amino acids

The reproductive performance of gilt-head sea bream was greatly improved when fed a diet based on a squid-meal composition (Harel *et al.*, 1992; Tandler *et al.*, 1995; Fernández-Palacios *et al.*, 1997), which is well known as a good protein and lipid source. It has been suggested that the superior performance of squid protein was related to its EAA composition, which resembles that of sea-bream egg protein (Tandler *et al.*, 1995). Tandler *et al.* (1995) reported that dietary protein influences egg quality by regulating the synthesis and selective uptake of yolk constituents. The lowest level of vitellogenin (Vg)-binding capacity was observed in oocytes from brood-stock fed diets where wheat gluten was the protein source. The supplementation of specific EAA so that dietary protein was similar to sea-bream egg protein markedly increased the level of oocyte Vg-binding sites mg^{-1} membrane protein. However, the highest level of Vg-binding sites was in oocytes from brood-stock fed squid-meal-based diets.

Lipids and fatty acids

Lipid is the main dietary constituent of brood-stock diets that directly influence the composition of the eggs and larvae (Watanabe, 1985; Mourente and Odriozola, 1990; Tandler *et al.*, 1995; Almansa *et al.*, 1999). Fernández-Palacios *et al.* (1995) reported that EPA in the eggs was more sensitive to a dietary change in n-3 HUFA than was egg DHA and that EPA as well as arachidonic acid influenced the egg fertilization rate. A low level of brood-stock dietary n-3 HUFA resulted in decreased fecundity, hatching rate and egg viability (Fernández-Palacios *et al.*, 1995; Rodríguez *et al.*, 1998b; Almansa *et al.*, 1999). It also induced a 34% decrease in larval growth and a reduction in swim-bladder inflation from 85% to 55% (Tandler *et al.*, 1995). However, excessive levels of n-3 HUFA in the brood-stock diet (31.5 g kg^{-1} DW) were associated with lower egg fecundity and yolk-sac hypertrophy in recently hatched larvae (Fernández-Palacios *et al.*, 1995). Fernández-Palacios *et al.* (1995) recommended that brood-stock diets should contain 16 g n-3 HUFA kg^{-1} DW for improved spawning performance. This agrees well with Rodríguez *et al.* (1998b), who found better egg quality from broodstock fed 18 g n-3 HUFA kg^{-1} DW diet, while Tandler *et al.* (1995) suggested that brood-stock diets must include 15 g n-3 HUFA kg^{-1} DW diet.

Practical Diets

The determination of the nutritional requirements of the gilt-head sea bream during the juvenile and grow-out stages has been a continuous research effort over the last 25 years and is summarized in Table 5.2. These studies have enabled the production of quality juveniles as well as feeds that offer faster growth, better conversion efficiency and decreased feeding costs to the farmer, allowing the industry to expand dramatically in the last few years. Today, there is a tendency to feed high-energy (up to 20% lipid) extruded diets at levels according to established feeding tables. Sea-bream production in Europe produced 57,000 tons in 1999 and is projected to reach 60,000 tons in the year 2000.

Feeding practices: brood-stock

Sea-bream culture in the Mediterranean is carried out in cages except in the Canary Islands, southern Portugal and Italy, which employ large land-based tanks or earthen ponds (Smart, 1996). Many farms use hand-feeding during grow-out to closely monitor any changes in fish appetite and health conditions. Recently, a number of farms have employed computer-controlled automatic feeders, where feeding frequency and the amount of diet can be programmed for each cage according to feeding tables and the numbers of juveniles stocked. New approaches, adopted primarily from the salmon industry, using Doppler technology and cameras placed in cages to accurately detect satiation feeding levels are currently being tested.

Table 5.2. Summary of known nutritional needs of the gilt-head sea bream during juvenile and grow-out stages (modified from Kissil *et al.*, 2000b).

Protein	• Total dietary level	% of DW diet
	Juvenile	50–60
	Grow-out	45–50
	• Amino acids (AA)	% of dietary protein
	Arginine	< 2.6
	Lysine	5.0
	Methionine + cysteine	4.0
	Tryptophan	0.6
	• Estimates of remaining AA*	
	Histidine	1.7
	Isoleucine	2.6
	Leucine	4.5
	Valine	3.0
	Phenylalanine + tyrosine	2.9
	Threonine	2.8
Lipid	• Total dietary level	% of DW diet
	Grow-out	12–24[cf]
	• (n-3) HUFA (EPA + DHA)	
	Juveniles (1–11 g)	≥ 0.9
	(12–30 g)	1
	Grow-out	1.5–2.7[cf]
Carbohydrate		% of DW diet
		20
Energy	• Daily maintenance	55.8 kJ × BW $(kg)^{-0.83}$
	• Growth requirement	23 MJ kg^{-1} live weight
Vitamins		mg kg^{-1} diet
	Pyridoxine (B_6)	3–5
	Biotin	0.37
	Nicotinic acid	63–83
	Thiamine (B_1)	> 5.0
	Riboflavin, pantothenic and ascorbic	Levels unknown
Minerals†		Commercial premixes[rv]
DP : DE ratios		28–19 g DP MJ^{-1} DE

rv, recommended values; cf, values found in commercial diets for sea bream.
* Estimates of AA based on ratios of whole-body AA to total AA.
† No information.

References

Alexis, M.N., Karanikolas, K.K. and Richards, R.H. (1997) Pathological findings owing to the lack of ascorbic acid in cultured gilthead bream (*Sparus aurata* L.). *Aquaculture* 151, 209–218.

Almansa, E., Pérez, M.J., Cejas, J.R., Badía, P., Villamandos, J.E. and Lorenzo, A. (1999) Influence of broodstock seabream (*Sparus aurata* L.) dietary fatty acids on egg quality and egg fatty acid composition throughout the spawning season. *Aquaculture* 170, 323–336.

Bell, M.V., Batty, R.S., Dick, J.R., Fretwell, K., Navarro, J.C. and Sargent, J.R. (1995) Dietary deficiency of docosahexaenoic acid impairs vision at low light intensities in juvenile herring (*Clupea harengus* L.). *Lipids* 30, 443–449.

Bessonart, M., Izquierdo, M.S., Salhi, M., Hernández-Cruz, C.M., González, M.M. and Fernández-Palacios, H. (1999) Effect of dietary arachidonic acid levels on growth and survival of gilthead seabream (*Sparus aurata* L.) larvae. *Aquaculture* 179, 265–275.

Caballero, M.J., López-Calero, G., Socorro, J., Roo, F.J., Izquierdo, M.S. and Fernández, A.J. (1999) Combined effect of lipid level and fish meal quality on liver histology of gilthead seabream (*Sparus aurata*). *Aquaculture* 179, 277–290.

Cahu, C.L., Zambonino, J.L., Le Gall, M.M. and Quazuguel, P. (1995) Early weaning of seabass: are digestive enzymes limiting? In: Lavens, P., Sorgeloos, P., Jaspers, E. and Ollevier, E. (eds) *Larvi '91 Fish and Crustacean Larviculture Symposium, 3–7 September 1995, Ghent, Belgium.* Special Publication 15, European Aquaculture Society, Ghent, Belgium, pp. 268–275.

Company, R., Calduch-Giner, J.A., Kaushik, S. and Pérez-Sánchez, J. (1999) Growth performance and adiposity in gilthead seabream (*Sparus aurata*): risks and benefits of high energy diets. *Aquaculture* 171, 279–292.

Fernández-Díaz, C. and Yúfera, M. (1997) Detecting growth in gilthead seabream, *Sparus aurata* L. larvae fed microcapsules. *Aquaculture* 153, 93–102.

Fernández-Palacios, H., Izquierdo, M.S., Robaina, L., Valencia, A., Salhi, M. and Vergara, J.M. (1995) Effect of n-3 HUFA level in broodstock diets on egg quality of gilthead seabream (*Sparus aurata* L.). *Aquaculture* 132, 325–337.

Fernández-Palacios, H., Izquierdo, M., Robaina, L., Valencia, A., Salhi, M. and Montero, D. (1997) The effect of dietary protein and lipid from squid and fish meals on egg quality of broodstock for gilthead seabream (*Sparus aurata*). *Aquaculture* 148, 233–246.

Gurr, M.I. and Harwood, J.L. (1991) *Lipid Biochemistry*. Chapman & Hall, London, England, 406 pp.

Hadas, E. (1998) The influence of dietary phospholipids on feeding rate and absorption of fatty acids in the larvae of the gilthead seabream (*Sparus aurata*). MSc thesis, Hebrew University of Jerusalem, Rehovot, Israel.

Halver, J.E. (1989) The vitamins. In: Halver, J.E. (ed.) *Fish Nutrition*, 2nd edn. Academic Press, San Diego, California, pp. 32–109.

Harel, M., Tandler, A., Kissil, G.W. and Applebaum, S. (1992) The kinetics of nutrient incorporation into body tissues of gilthead seabream, *Sparus aurata*, females and subsequent effects on egg composition and egg quality. *Israeli Journal of Aquaculture-Bamidgeh* 44, 127 [abstract].

Henrique, M.M.F., Morris, P.C. and Davies, S.J. (1996) Vitamin C status and physiological response of the gilthead seabream, *Sparus aurata* L., to stressors associated with aquaculture. *Aquaculture Research* 27, 405–412.

Higgs, D.A., Dosanjh, B.S., Prendergast, A.F., Beames, R.M., Hardy, R.W., Riley, W. and Deacon, G. (1995) Use of rapeseed/canola protein products in finfish diets. In: Lim, C.E. and Sessa, D.J. (eds) *Nutrition and Utilization Technology in Aquaculture*. AOCS Press, Champaign, Illinois, pp. 130–156.

Ibeas, C., Izquierdo, M.S. and Lorenzo, A. (1994) Effect of different levels of n-3 highly unsaturated fatty acids on growth and fatty acid composition of juvenile gilthead seabream (*Sparus aurata*). *Aquaculture* 127, 177–188.

Ibeas, C., Cejas, J., Gómez, T., Jerez, S. and Lorenzo, A. (1996) Influence of n-3 highly unsaturated fatty acid levels on juvenile gilthead seabream (*Sparus aurata*) growth and tissue fatty acid composition. *Aquaculture* 142, 221–235.

Jobling, M. and Wandsvik, A. (1983) An investigation of factors controlling food intake in Arctic charr, *Salvelinus alpinus* L. *Journal of Fish Biology* 23, 397–404.

Kalogeropoulos, N., Alexis, M.N. and Henderson, R.J. (1992) Effects of dietary soyabean and cod-liver oil levels on growth and body composition of gilthead bream (*Sparus aurata*). *Aquaculture* 104, 293–308.

Kentouri, M., Divanach, P., Geurden, I. and Anthouard, M. (1995) Evidence of adaptive behaviour in gilthead seabream (*Sparus aurata* L.) in relation to diet composition, in a self-feeding condition. *Ictyophysiologica Acta* 18, 125–169.

Kissil, G.W. (1981) Known nutritional requirements of the gilthead bream (*Sparus aurata*) in culture. In: Rosenthal, H. and Oren, O.H. (eds) *Intensive Aquaculture.* Special Publication No. 6, European Mariculture Society, Bredene, Belgium, pp. 49–55.

Kissil, G.W., Cowey, C.B., Adron, J.W. and Richards, R.H. (1981) Pyridoxine requirements of the gilthead bream, *Sparus aurata. Aquaculture* 23, 243–255.

Kissil, G.W., Lupatsch, I., Higgs, D.A. and Hardy, R.W. (1997) Preliminary evaluation of rapeseed protein concentrate as an alternative to fish meal in diets for gilthead seabream (*Sparus aurata*). *Israeli Journal of Aquaculture-Bamidgeh* 49, 135–143.

Kissil, G.W., Tandler, A., Elizur, A., Colorni, A. and Zohar, Y. (2000a) Gilthead seabream, *Sparus aurata.* In: Stickney, R.R. (ed.) *Encyclopedia of Aquaculture.* John Wiley and Sons, New York, pp. 409–416.

Kissil, G.W., Lupatsch, I., Higgs, D.A. and Hardy, R.W. (2000b) Dietary substitution of soy and rapeseed protein concentrates for fish meal and their effects on growth and nutrient utilization in gilthead seabream, *Sparus aurata* L. *Aquaculture Research*, 31, 595–602.

Kolkovski, S., Tandler, A., Kissil, G.W. and Gertler, A. (1993) The effect of dietary exogenous digestive enzymes on ingestion, assimilation, growth and survival of gilthead seabeam (*Sparus aurata*, Sparidae, Linnaeus) larvae. *Fish Physiology and Biochemistry* 12, 203–209.

Kolkovski, S., Arieli, A. and Tandler, A. (1997a) Visual and chemical cues stimulate microdiet ingestion in gilthead seabream, *Sparus aurata*, larvae. *Aquaculture International* 5, 527–536.

Kolkovski, S., Koven, W.M. and Tandler, A. (1997b) The mode of action of *Artemia* in enhancing utilization of microdiet by gilthead seabream *Sparus aurata* larvae. *Aquaculture* 155, 193–205.

Koven, W.M., Kissil, G.W. and Tandler, A. (1989) Lipid and n-3 requirement of *Sparus aurata* larvae during starvation and feeding. *Aquaculture* 79, 185–191.

Koven, W.M., Tandler, A., Kissil, G.W., Sklan, D., Friezlander, O. and Harel, M. (1990) The effect of dietary n-3 highly unsaturated fatty acids on growth, survival and swim bladder development in *Sparus aurata* larvae. *Aquaculture* 91, 131–141.

Koven, W.M., Tandler, A., Kissil, G.W. and Sklan, D. (1992a) The importance of n-3 highly unsaturated fatty acids for growth in larval *Sparus aurata* and their effect on survival, lipid composition and size distribution. *Aquaculture* 104, 91–104.

Koven, W.M., Tandler, A., Sklan, D. and Kissil, G.W. (1992b) The association of eicosapentaenoic and docosahexaenoic acids in the phospholipids of different age *Sparus aurata* larvae with growth. *Aquaculture* 116, 71–82.

Koven, W.M., Kolkovski, S., Tandler, A., Kissil, G.W. and Sklan, D. (1993) The effect of dietary lecithin and lipase as a function of age, on n-9 fatty acid incorporation in the tissue lipids of *Sparus aurata* larvae. *Fish Physiology and Biochemistry* 10, 357–364.

Koven, W.M., Parra, G., Kolkovski, S. and Tandler, A. (1998) The effect of dietary phosphatidlylcholine and its constitutent fatty acids on microdiet ingestion and fatty acid absorption rate in gilthead seabream, *Sparus aurata*, larvae. *Aquaculture Nutrition* 4, 39–45.

Koven, W., Barr, Y., Lutzky, S., Ben-Atia, I., Weiss, R., Harel, M., Behrens, P. and Tandler, A. (2001a) The effect of dietary arachidonic acid (20:4n-6) on growth, survival and resistance to handling stress in gilthead seabream (*Sparus aurata*) larvae. *Aquaculture* 193, 107–122.

Koven, W., Kolkovski, S., Hadas, E., Gamsiz, K. and Tandler, A. (2001b) Advances in the development of microdiets for gilthead seabream, *Sparus aurata*: a review. *Aquaculture* 194, 107–121.

Lavens, P. and Sorgeloos, P. (1996) *Manual on the Production and Use of Live Food for Aquaculture.* Fisheries and Technical Paper 361, FAO, Rome, Italy, 295 pp.

Lupatsch, I., Kissil, G.W., Sklan, D. and Pfeffer, E. (1997) Apparent digestibility coefficients of feed ingredients and their predictability in compound diets for gilthead seabream, *Sparus aurata* L. *Aquaculture Nutrition* 3, 81–89.

Lupatsch, I., Kissil, G.W., Sklan, D. and Pfeffer, E. (1998) Energy and protein requirements for maintenance and growth in gilthead seabream (*Sparus aurata* L.). *Aquaculture Nutrition* 4, 165–173.

Lupatsch, I., Kissil, G.W., Sklan, D. and Pfeffer, E. (2001) Effects of varying dietary protein and energy supply on growth, body composition and protein utilization in gilthead seabream (*Sparus aurata* L.). *Aquaculture Nutrition* 7, 71–80.

McDonald, T.J., Jornvale, H., Nilsson, G., Vagne, M., Ghatei, M., Bloom, S.R. and Mutt, V. (1979) Characterization of gastrin releasing peptide from porcine non-antral gastric tissue. *Biochemical and Biophysical Research Communications* 90, 227.

Morris, P.C. (1997) Nutritional needs of bass and bream. *Fish Farmer-International File*, November/December, 28–31.

Morris, P.C. and Davies, S.J. (1995a) The requirement of the gilthead seabream (*Sparus aurata* L.) for nicotinic acid. *Animal Science* 61, 437–443.

Morris, P.C. and Davies, S.J. (1995b) Thiamine supplementation of diets containing varied lipid : carbohydrate ratio given to gilthead seabream (*Sparus aurata* L.). *Animal Science* 61, 597–603.

Morris, P.C., Davies, S.J. and Lowe, D.M. (1995) Qualitative requirement for B vitamins in diets for the gilthead seabream (*Sparus aurata* L.). *Animal Science* 61, 419–426.

Mourente, G. and Odriozola, J.M. (1990) Effect of broodstock diets on lipid classes and their fatty acid composition in eggs of gilthead seabream (*Sparus aurata* L.). *Fish Physiology and Biochemistry* 8, 93–101.

Mourente, G., Rodríguez, A., Tocher, D.R. and Sargent, J.R. (1993) Effects of dietary docosahexaenoic acid (DHA, 22:6n-3) on lipid and fatty acid composition and growth in gilthead seabream (*Sparus aurata* L.) larvae during first feeding. *Aquaculture* 112, 79–98.

Moyano, F.J., Díaz, M., Alarcón, F.J. and Sarasquete, M.C. (1996) Characterization of digestive enzyme activity during larval development of gilthead seabream (*Sparus aurata*). *Fish Physiology and Biochemistry* 15, 121–130.

Mustafa, T. and Srivastava, K.C. (1989) Prostaglandins (eicosanoids) and their role in ectothermic organisms. *Advances in Comparative Environmental Physiology* 5, 157–207.

Nengas, I., Alexis, M.N., Davies, S.J. and Petichakis, G. (1995) Investigation to determine digestibility coefficients of various raw materials in diets for gilthead seabream, *Sparus aurata* L. *Aquaculture Research* 26, 185–194.

Nengas, I., Alexis, M.N. and Davies, S.J. (1999) High inclusion levels of poultry meals and related byproducts in diets for gilthead seabream *Sparus aurata* L. *Aquaculture* 179, 13–23.

Paspatis, M. and Boujard, T. (1996) A comparative study of automatic feeding and self-feeding in juvenile Atlantic salmon (*Salmo salar*) fed diets of different energy levels. *Aquaculture* 145, 245–257.

Robaina, L., Izquierdo, M.S., Moyano, F.J., Socorro, J., Vergara, J.M., Montero, D. and Fernández-Palacios, H. (1995) Soyabean and lupin seed meals as protein sources in diets for gilthead seabream (*Sparus aurata*): nutritional and histological implications. *Aquaculture* 130, 219–233.

Robaina, L., Moyano, F.J., Izquierdo, M.S., Socorro, J., Vergara, J.M. and Montero, D. (1997) Corn gluten and meat and bone meals as protein sources in diets for gilthead seabream (*Sparus aurata*): nutritional and histological implications. *Aquaculture* 157, 347–359.

Rodríguez, C., Pérez, J.A., Izquierdo, M.S., Mora, J., Lorenzo, A. and Fernández-Palacios, H. (1993) Essential fatty acid requirements of larval gilthead seabream, *Sparus aurata* (L.). *Aquaculture Fisheries Management* 24, 295–304.

Rodríguez, C., Pérez, J.A., Díaz, M., Izquierdo, M.S., Fernández-Palacios, H. and Lorenzo, A. (1997) Influence of the EPA/DHA ratio in rotifers on gilthead seabream (*Sparus aurata*) larval development. *Aquaculture* 150, 77–89.

Rodríguez, C., Pérez, J.A., Badía, P., Izquierdo, M.S., Fernández-Palacios, H. and Lorenzo Hernández, A. (1998a) The n-3 highly unsaturated fatty acid requirements of gilthead seabream (*Sparus aurata* L.) larvae when using an appropriate DHA/EPA ratio in the diet. *Aquaculture* 169, 9–23.

Rodríguez, C., Cejas, J.R., Martin, M.V., Badía, P., Samper, M. and Lorenzo, A. (1998b) Influence of n-3 highly unsaturated fatty acid deficiency on the lipid composition of broodstock gilthead seabream (*Sparus aurata* L.) and on egg quality. *Fish Physiology and Biochemistry* 18, 177–187.

Rosenlund, G., Stoss, J. and Talbot, C. (1997) Co-feeding marine fish larvae with inert and live diets. *Aquaculture* 155, 183–191.

Salhi, M (1997) Estudio de los requerimientos lipídicos de larvas de dorada (*Sparus aurata*) alimentadas con microdietas. PhD dissertation, University of Las Palmas de Gran Canaria, Canary Islands, Spain.

Salhi, M., Hernández-Cruz, C.M., Bessonart, M., Izquierdo, M.S. and Fernández-Palacios, H. (1999) Effect of different dietary polar lipid levels on gut and liver histological structure of gilthead seabream (*Sparus aurata*) larvae. *Aquaculture* 179, 253–263.

Santinha, P.J.M., Gomes, E.F.S. and Coimbra, J.O. (1996) Effects of protein level of the diet on digestibility and growth of gilthead seabream, *Sparus aurata* L. *Aquaculture Nutrition* 2, 81–87.

Sargent, J.R., Bell, J.G., Bell, M.V., Henderson, R.J. and Tocher, D.R. (1993a) The metabolism of phospholipids and polyunsaturated fatty acids in fish. In: Lahlou, B. and Vitello, P. (eds) *Aquaculture: Fundamental and Applied Research.* Coastal and Estuarine Studies 43, American Geophysical Union, Washington, DC, pp. 103–124.

Sargent, J.R., Bell, M.V. and Tocher, D.R. (1993b) Docosahexaenoic acid and development of brain and retina in marine fish. In: Drevon, C.A., Baksaas, I. and Krokan, H.E. (eds) *Omega–2 Fatty Acids: Metabolism and Biological Effects.* Birkhauser, Basle, Switzerland, pp. 139–149.

Smart, G. (1996) Farming bass and bream. *Fish Farmer-International File*, July–August, 29–31.

Tandler, A. and Kolkovski, S. (1991) Rates of ingestion and digestibility as limiting factors in the successful use of microdiets in *Sparus aurata* larval rearing. In: Lavens, P., Sorgeloos, P., Jaspers, E. and Ollevier, E. (eds) *Larvi '91 Fish and Crustacean Larviculture Symposium, 3–7 September 1995, Ghent, Belgium*. Special Publication 15, European Aquaculture Society, Ghent, Belgium, pp. 169–171.

Tandler, A., Harel, M., Koven, W.M. and Kolkovski, S. (1995) Broodstock and larvae nutrition in gilthead seabream *Sparus aurata* – new findings on its mode of involvement in improving growth, survival and swimbladder inflation. *Israeli Journal of Aquaculture-Bamidgeh* 47, 95–111.

Vergara, J.M. and Jauncey, K. (1993) Studies on the use of dietary energy by gilthead seabream (*Sparus aurata* L.) juveniles. In: Kaushik, S.J. and Luquet, P. (eds) *Fish Nutrition in Practice. Proceedings of the 4th International Symposium on Fish Nutrition and Feeding*. INRA Editions, Paris, France, pp. 453–458.

Vergara, J.M., Fernández-Palacios, H., Robaina, L., Jauncey, K., De La Higuera, M. and Izquierdo, M. (1996a) The effects of varying dietary protein level on the growth, feed efficiency, protein utilization and body composition of gilthead seabream fry. *Fisheries Science* 62, 620–623.

Vergara, J.M., Robaina, L., Izquierdo, M.S. and De la Higuera, M. (1996b) Protein sparing effect of lipids in diets for fingerlings of gilthead seabream. *Fisheries Science* 62, 624–628.

Watanabe, T. (1985) Importance of the study of broodstock nutrition for further development of aquaculture. In: Cowey, C.B., Mackie, A.M. and Bell, J.G. (eds) *Nutrition and Feeding in Fish*. Academic Press, London, UK, pp. 395–414.

Zohar, Y., Harel, M., Hassin, S. and Tandler, A. (1995) Gilthead Seabream. In: Bromage, N.R. and Roberts, R.J. (eds) *Broodstock Management and Egg and Larval Quality*. Blackwell Science, Oxford, UK, pp. 94–117.

Atlantic Salmon, *Salmo salar*

6

Trond Storebakken
AKVAFORSK, Institute of Aquaculture Research AS, N-1432 Sunndalsøra, Norway

Introduction

Farming of Atlantic salmon (*Salmo salar*) began in Norway in the late 1960s. During the 1980s and 1990s the production spread, mainly to other north-western European countries and to Chile. In 1980 annual world production was less than 10,000 t and this increased to 850,000 t whole fish produced in 2000. It is projected to approach 2 million t in 2010.

The growth rate of the fish in the farms is continuously improving, due to genetic selection, improved diets and feeding and improved management. The improvement in growth ascribed to genetic selection can be as much as 15–20% per generation if the salmon are from a combined family–phenotype selection programme (Gjedrem, 1983). Recent Canadian results have shown that transgenic salmon, which express growth hormone receptor activity in muscle tissues, have a growth rate 2.6–2.9 times that of the control population, mainly because of increased feed intake (Cook *et al.*, 2000a). Selection or transgenic techniques may also influence the composition of the growth, result in improved feed conversion (Thodesen *et al.*, 1999; Cook *et al.*, 2000a) and change the rate of metabolism (Thodesen *et al.*, 1999; Cook *et al.*, 2000b,c). Economical feed conversions in Norwegian salmon farming have changed from more than 2 kg feed dry matter (DM) kg^{-1} gain in 1980 to approximately 1.15 kg kg^{-1} at the end of the 1990s. In salmon the biological feed conversion can vary from less than 0.6 to more than 1.3 kg intake kg^{-1} gain, and the proportion of protein and energy in diet partitioned into gain can vary from 25 to 60%, depending on the formulation of the diet and the size of the fish. Thus, requirement figures must be based on the requirements of the animal, related to growth, nutrient partitioning and the specific physiological processes in which the nutrient is required.

Requirements differ throughout the life cycle. This is especially important in salmon, which undergo physiological changes during smoltification and transfer from a life in fresh water to salt water. In addition to developing the ability to maintain the ion balance in the hyperosmotic salt water, the salmon also undergoes changes in connection with preferred substrates for energy metabolism. Furthermore, the composition of the growth becomes more fatty as the salmon grows larger. Sexual maturation is another life stage that may call for attention in respect of diet formulation. Initially, growth is stimulated by an increase in the levels of sex steroids. Later in the maturation process, nutrients are directed into gonads rather than somatic growth.

The freshwater stage of the salmon is characterized by a moderately fatty composition of the growth (Shearer *et al.*, 1994). During smoltification and immediately after transfer to salt water, the body fat content drops rapidly by around 0.1 kg of body weight. Thereafter, the growth becomes increasingly fatty as the salmon grows. During sexual maturation, gonadal development has priority, with a subsequent reduction in body fat. The growth rates are also much higher in juvenile than in on-growing salmon, when related to body weight (Austreng *et al.*, 1987), while reduced growth is often seen during smoltification.

In the 1970s, moist diets, consisting of minced trash fish, binder meal with micronutrients and canthaxanthin, crustacean by-products and fish-oil, were the 'only feed which salmon in saltwater would accept'. Steam-pelleted, crumbled diets were only used in the freshwater stage. The process of developing diets especially for Atlantic salmon was largely started in 1972, with the work of Erland Austreng at the Department of Poultry and Fur Animal Science at the Agricultural University of Norway. Among the most significant initial findings was that increasing dietary lipid levels from 8–10%, a level commonly accepted at that time, up to 18% in steam-pelleted diets increased growth and survival (Austreng, 1976). The dietary protein level and the cost of gain decreased correspondingly. In the middle of the 1970s, the quality of ingredients in the dry feeds was so high that the salmon also accepted it in salt water. Since then, dietary lipid concentrations have increased and protein concentrations decreased, facilitated by the introduction of extruded diets in the 1980s and vacuum-coating techniques in the 1990s.

Nutrient Requirements

Protein and amino acids

Salmon digest protein efficiently, and more than half of the protein is hydrolysed and absorbed in the pyloric region (Krogdahl *et al.*, 1999). Digestion and absorption of protein are highly different from one feed ingredient to another, or even within feed ingredients processed by the same method. This is illustrated in Table 6.1. Salmon utilizes dietary protein from high-energy diets more efficiently for growth than from low-energy diets. Protein retention values higher than 50% are not uncommon (Table 6.2).

Requirements for only three essential amino acids – lysine, arginine and methionine – have been published for salmon (Table 6.3). The estimates of arginine requirements in post-smolts differ by more than 25%. The main differences in the two experiments were that Lall *et al.* (1994) obtained faster growth

Table 6.1. Apparent amino acid and crude protein digestibilities in various protein-rich feed ingredients.

	Capelin meal*†‡	Herring meal*†§	Herring meal‡‖	Menhaden meal‡‖	Anchovy meal*‡‖	Canola meal*	Soybean meal*	Wheat gluten¶	Bacterial meal**
Crude protein	81–87	74–83	81–85	78	77	74	71	100	82
Amino acids									
Ala	89–94	87–88	86–92	88	81	80	74	96	85
Arg	90–96	89–92	91–96	91	82	85	77	98	92
Asp	84–86	68–79	83–84	70	79	74	67	94	82
Cys	72	76				79		100	52
Glu	89–94	84–88	86–91	86	82	88	75	99	86
Gly	78–85	75–80	81–82	72	82	75	63	98	80
His	88–96	84–90	75–90	85	76	85	77	98	81
Ile	87–95	84–91	85–93	91	82	75	67	98	83
Leu	91–95	87–91	86–93	89	82	77	70	99	84
Lys	88–95	82–88	86–91	84	81	77	67	95	92
Met	89	86				89	71	98	83
Phe	88–94	84–90	84–92	89	80	78	70	99	77
Pro	86–89	84–80	83–89	84	74	81	77	100	84
Ser	90–94	84–86	90–90	86	81	77	73	99	80
Thr	88–93	79–87	81–89	85	81	75	62	97	82
Trp	84–85	53–93	84–88	73	81		45	98	
Tyr	89–97	83–82	89–97	86	87	76	68	99	74
Val	89–95	84–90	85–93	90	82	73	67	98	85

* Anderson *et al.* (1992), in salt water.
† Norse LT-94. Low-temperature dried capelin meal with high hygienic quality of the fish.
‡ Anderson *et al.* (1995), in fresh water.
§ Flame-dried.
‖ Steam-dried.
¶ Vital wheat gluten. Storebakken *et al.* (2000a), in salt water.
** BioProtein. Skrede *et al.* (1998), in salt water.
Ala, alanine; Arg, arginine; Asp, aspartic acid; Cys, cystine; Glu, glutamic acid; Gly, glycine; His, histidine; Ile, isoleucine; Leu, leucine; Lys, lysine; Met, methionine; Phe, phenylalanine; Pro, proline; Ser, serine; Thr, threonine; Trp, tryptophan; Tyr, tyrosine; Val, valine.

and more efficient feed conversion than those of the salmon in the experiment by Berge *et al.* (1997), thus giving a more reliable result. Since requirement estimates are still lacking for most indispensable amino acids for salmon, dietary digestible amino acid profiles may be based on whole-body amino acid composition (Wilson and Cowey, 1985) or on amino acid requirements for other salmonids.

It is common practice to include 50–55% crude protein in diets for juvenile salmon and to reduce the dietary protein content to 40–45% in grower diets, provided the protein-rich feed ingredients are of high quality.

Lipids and fatty acids

The need for high fat content in the diet for Atlantic salmon reflects the body composition of the fish. The high dietary lipid level stresses the need to use

Table 6.2. Partitioning of dietary nitrogen into growth and excretion in salmon fed diets with fish-meal as the main protein source.

Life stage	Water	Fish weight (kg)	DP : DE (mg kJ^{-1})	N retention (%)	Faecal N loss (%)	Other N losses (%)	Reference
Parr	Fresh	0.08	16.7–27.9	55–47	8–6	37–46	Grisdale-Helland and Helland, 1997
During smoltification	Fresh	0.07	19.5–27.1	51–43	6–6	43–51	
Post-smolt	Salt	0.10		56–57	8–8	36–34	Helland and Grisdale-Helland, 1998a
Post-smolt	Salt	0.80	18.6	44	12	44	Thodesen *et al.*, 1999

DP, digestible protein; DE, digestible energy; N, nitrogen.

Table 6.3. Known essential L-amino acid requirements for salmon.

Amino acid	g kg^{-1} DM*	g kg^{-1} N × 6.25*	Fish size	Reference
Arginine	16	41	Post-smolt	Lall *et al.*, 1994
	21.2–21.6	50–51	Post-smolt	Berge *et al.*, 1997
Lysine	19.9	39.8	Fingerling	Anderson *et al.*, 1993
	16–18		Post-smolt	Berge *et al.*, 1998
Methionine	11†	24†	Fingerling	Rollin *et al.*, 1994

* Estimated by broken-line regression of growth on dietary amino acid concentration.
† Dietary cysteine level not specified.
DM, dry matter; N, nitrogen.

high-quality oils and to know the effects of the lipid source on growth and salmon health, as well as product quality. Lipids are sources of energy, essential fatty acids, eicosanoids and components of the cell membrane (phospholipids) and they assist in the uptake of lipid-soluble nutrients.

Salmon must have oil with a low melting-point, as saturated fats are poorly digested. The pyloric caeca and the proximal intestine are the main sites of fat digestion and absorption but some absorption also occurs in the distal portion of the intestine (Krogdahl *et al.*, 1999). The digestibility of fish-oil from fish-meal-based extruded diets ranges from 90 to 95% in salmon (Aksnes, 1995; Storebakken *et al.*, 1998a, 2000a). Most vegetable oils are digested at a similar or higher rate in rainbow trout (Austreng *et al.*, 1979).

Atlantic salmon, like other fish, are unable to synthesize fatty acids of the n-3 and n-6 families. These fatty acids must be provided in the diet. The main symptoms of essential fatty acid deficiency in salmon, in addition to reduced growth and increased mortality, is a decrease in the essential fatty acids in the blood and liver phospholipids and a subsequent increase in 20:3n-9 (Ruyter *et al.*, 2000a). The precursor fatty acids, 18:3n-3 (linolenic acid) and 18:2n-6 (linoleic acid) are desaturated and elongated into the longer polyunsaturated fatty acids (PUFA), such as 20:5n-3 (eicosapentaenoic acid (EPA)) and 22:6n-3 (docosahescaenoic acid (DHA)), typical 'marine' fatty acids, and 20:4n-6 (arachidonic acid) by the same enzyme systems.

In juvenile salmon fed semipurified diets with graded amounts of methyl esters of either 18:2n-6, 18:3n-3 or a 1 : 1 mixture of 20:5n-3 and 22:6n-3, increasing levels up to 1% of dietary DM improved growth and survival. Fry fed the mixture of fatty acids performed better than fish fed 18:3n-3 alone (Ruyter *et al.*, 2000b). The requirement for n-3 fatty acids is approximately 1% of dry feed for juveniles if the fatty acids are supplemented as 20:5n-3 and 22:6n-3. No requirement has been established for n-6 fatty acids. A requirement by larger salmon for essential fatty acids has not been published.

Fish-oils and fish-lipid residue of fish-meal are the two main sources of fat in salmon feed, but fluctuations in the fish-oil supply and an anticipated shortage of fish-oil during the first half of the 1990s made it increasingly necessary to substitute at least part of the fish-oil with vegetable oils. The effects of different combinations of fish-oils (high in n-3 PUFA) and vegetable oils (rich in 18:2n-6, 20:4n-6 and eventually 18:3n-3) on the growth and health of salmon have been evaluated. Generally, growth reductions or increased mortalities as a result of substituting half or more of the fish-oils with vegetable oils have not been observed with salmon, as long as the quantitative requirement for n-3 fatty acids is met (Thomassen and Røsjø, 1989; Dosanjh *et al.*, 1998).

The fatty acid composition of the salmon fillet largely reflects that of the dietary oil, with some enrichment of n-3 PUFA (Hardy *et al.*, 1987; Thomassen and Røsjø, 1989; Polvi and Ackmann, 1992; Dosanjh *et al.*, 1998). This may also affect the quality of the farmed salmon. However, Hardy *et al.* (1987) did not obtain any significant differences in flavour or texture traits when a sensory panel evaluated salmon gaining from 36 to 52% weight on low-fat (18–20%) diets. The dietary lipids tested were herring or menhaden fish-oils, soybean oil and tallow. However, when 0.25 kg salmon were fed low-fat diets consisting of capelin fish-oil, high- or low-erucic acid rape-seed oil or soybean oil, the fish fed either of the rape-seed oils had lower 'salmon odour' than the group fed the fish-oil (Thomassen and Røsjø, 1989). Furthermore, the fish fed the soybean oil had significantly less 'salmon taste' than those fed the low-erucic acid rape-seed oil.

During the last two decades, the understanding of the term 'high-energy' salmon grower diet has shifted from 25–30% fat to 35–40% fat. Johnsen *et al.* (1993) and Hillestad *et al.* (1998) did not find any significant effects on growth by increasing dietary fat concentrations from 23 to 30% in diets for salmon with a start weight of 0.2 and 0.3 kg, respectively. Einen and Roem (1997) observed a reduction in growth for the highest fat content when gradually increasing dietary fat levels from 26 to 39%, and suggested that digestible protein : digestible energy (DP : DE) ratios around 19 g MJ^{-1} are optimal for salmon weighing from 1 to 2.5 kg, decreasing to 16–17 g MJ^{-1} for fish weighing from 2.5 to 5 kg. In support of these findings, Refstie *et al.* (2000) obtained an increase in growth by increasing dietary fat from 32 to 39%, with a corresponding decrease in DP : DE from 21 to 17 g MJ^{-1}, in salmon with a final weight of 2.7–2.8 kg.

Dietary fat content also influences the chemical composition of the fish. The proportion of fat in the viscera increases significantly with increasing dietary fat content (Einen and Roem, 1997; Hillestad *et al.*, 1998; Refstie *et al.*, 2000). The fat content of the fillets is less influenced by dietary fat content than that of the viscera, even with an increase in dietary fat content from 20 to 40% (Einen and Skrede, 1998), but increased fillet fat deposition has been observed in salmon fed diets with increased dietary fat content from 32 to 39% (Refstie *et al.*, 2000). Smoked fillets from salmon fed diets with a range of dietary fat concentrations from 26 to 39% were also presented to an expert panel, and the only significant difference found was a slightly elevation in fattiness in the fillets with the highest fat content. Other traits, such as hardness, juiciness and various measures for odour, flavour and colour, were not significantly affected by dietary fat level (Einen and Skrede, 1998). Similarly, no significant differences were found when salmon fed diets with 32 or 39% fat were subjected to sensory evaluation (Bjerkeng *et al.*, 1998). Thus, the effects on the quality of the edible parts of the salmon of increasing the dietary fat content from 20 to 35–40% seems to have been small.

The 'high-energy' grower diets do not just affect growth rates and body composition, but also have positive effects on feed conversion and release of nutrients from the fish farm. Johnsen *et al.* (1993) found that the nitrogen load to the environment was reduced by 35% kg^{-1} salmon produced by increasing the dietary fat content from 22 to 30%. This is ascribed both to an overall more efficient feed conversion and to a reduction in deamination of protein because of reductions in the protein intake per kg gain. The same authors also calculated a parallel reduction in organic-matter load to the environment by 22% and a reduction of phosphorus load by 22%.

Carbohydrates

Atlantic salmon have no specific requirement for dietary carbohydrates, since they can synthesize sufficient amounts of glucose by gluconeogenesis. There are mainly two types of carbohydrates in commercial salmon diets: starches and non-starch polysaccharides (NSP). Starches are added to the diet mainly for binding and expansion during extrusion. NSP come from vegetable feed ingredients such as grains or legumes.

The starches must be gelatinized to improve availability, but salmon still have a limited ability to hydrolyse gelatinized starch. One reason may be that salmon have low, if any, detectable activity of α-amylase in the intestine (Sørensen, 1995). Thus, the amylose chain is not hydrolysed into oligo- and disaccharides before they are subjected to the disaccharidases. Intestinal brush-border disaccharidases are active in salmon, and maltase has the highest activity. Most of the disaccharidase activity is found in the pyloric caeca and the proximal part of the intestine, which is also the main site of starch hydrolysis (Krogdahl *et al.*, 1999).

Starch digestibility in salmon appears to be higher in fresh or brackish water than in salt water. Figure 6.1 illustrates that starch digestibility is reduced with increasing dietary concentrations, and that this reduced digestibility of starch is accompanied by reduced digestibility of fat. This necessitates extra precaution to avoid excessive use of starch in salmon feeds. Atlantic salmon also has a poor ability to regulate blood glucose when the dietary carbohydrate load is excessive (Hemre and Hansen, 1998). However, the salmon regulates blood glucose fairly efficiently when the load is more moderate (Hemre *et al.*, 1996). Small salmon regulate blood glucose less efficiently than larger fish (Waagbø *et al.*, 1993), and the ability to regulate is higher at optimal water temperature than at low temperature (Hemre *et al.*, 1995b).

A small inclusion of starch in the diet is beneficial, since a protein-sparing effect of starch has been observed in salmon in the freshwater stage (Hemre *et al.*, 1995a; Grisdale-Helland and Helland, 1997). With larger salmon in salt water, replacement of protein with starch did not result in protein-sparing (Helland and Grisdale-Helland, 1998a). Currently, it is common practice to use 6–15% gelatinized starch in salmon diets, depending on fish size and feed processing. This level of inclusion does not impair salmon growth (Hemre *et al.*, 1995a; Aksnes,

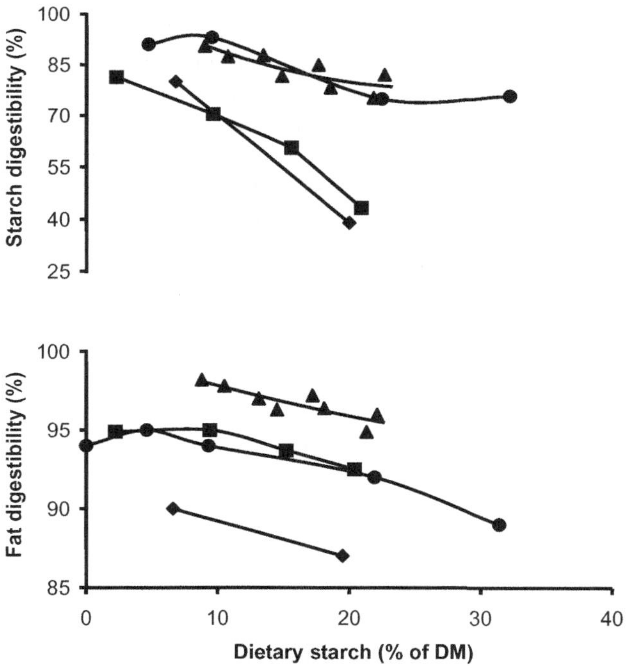

Fig. 6.1. Apparent digestibility of starch and fat in salmon fed diets with different dietary starch concentrations. ■, Aksnes (1995), 0.6 kg salmon in salt water; ●, Hemre *et al.* (1995a), 80 g post-smolt in brackish water (20 g l^{-1}); ▲, Grisdale-Helland and Helland (1997), 80 g fingerlings in fresh water; ◆, Krogdahl *et al.* (1999), 1.9 kg salmon in salt water; DM, dry matter.

1995). A further increase in dietary starch reduces growth and feed conversion (Helland and Grisdale-Helland, 1998a).

Insoluble NSP, such as cellulose, mainly act as fillers in the stomach and intestine and do not affect uptake of nutrients (Mundheim and Opstvedt, 1990). Soluble NSP, such as mixed β-glucans and arabinoxylans in grains, and pectic and acidic polysaccharides in legumes, increase the viscosity of the digesta and the water content of the faeces and reduce digestibility of fat- and lipid-soluble components. Refstie (2000) has estimated that, if fish-meal were substituted with 30% soybean meal in a high-energy (40% fat) salmon diet, 35 g less fat would be absorbed per kg feed intake, mainly due to soluble NSP in the soybean.

Energy

Lipid, protein and carbohydrates are converted into energy. The energy utilization is improved by reducing dietary protein and increasing dietary fat, thereby decreasing the DP : DE ratio. The energy expenditures in faecal losses normally account for 10% or less if the quality of the feed ingredients is high and dietary carbohydrate is kept at a tolerable level (Grisdale-Helland and Helland, 1997; Helland and Grisdale-Helland, 1998a,b). However, in post-smolts newly transferred to sea water, faecal energy losses were elevated due to low fat digestibility (Helland and Grisdale-Helland, 1998a). Ammonia and urea represent a minor loss in energy, and the heat increment accounted for only 9% of gross energy intake of post-smolts newly released to the sea (Helland and Grisdale-Helland, 1998b). Maintenance and activity can account for a considerable amount of the energy budget. Activity inflicted by stress and too high a water current in tanks may represent unnecessary energy expenditure.

In a salmon that is genetically selected against early sexual maturation and against excessive adipose deposition, the following DP : DE ratios may be considered optimal:

Fingerlings:	23 g MJ^{-1}
Smoltification:	20 g MJ^{-1}
Grower:	
Up to 2.5 kg	19 g MJ^{-1}
Over 2.5 kg	16–17 g MJ^{-1}

Vitamins

In this chapter, only mechanisms and symptoms confirmed in experiments with Atlantic salmon will be described. Among the lipid-soluble vitamins (A, D, E, K), vitamins A and E have received most attention in salmon. Vitamin C is the water-soluble vitamin that has been subject to most investigation in salmon, while requirements are established for only a few of the B vitamins.

Lipid-soluble vitamins

It is common practice to add a supplement of 2000 to 2500 IU of vitamin A to salmon feeds. However, fish-meal and fish-oil normally contain considerable amounts of vitamin A, and unsupplemented practical diets have been reported to contain more than 70,000 IU vitamin A kg^{-1} (Grisdale-Helland *et al.*, 1991). In view of this, the justification for adding a supplement of vitamin A to salmon diets based on fish-meal (full meal from pelagic, fatty species) may be questioned. In addition to reduced vitamin A stores in the liver, salmon parr respond to diets depleted in vitamin A (1300 IU kg^{-1}) by a reduction in both humoral and cell-mediated immune responses (Thompson *et al.*, 1994). Vitamin A is toxic when fed in high doses, but even juvenile salmon can tolerate diets with more than 75,000 IU vitamin A kg^{-1} without pronounced reductions in growth or increased mortality (Grisdale-Helland *et al.*, 1991). Vitamin A and astaxanthin are closely related in salmon, since astaxanthin is provitamin A and is capable of supporting the need for vitamin A, even in juvenile salmon (Christiansen *et al.*, 1995). The quantitative requirement for vitamin A has not been established in Atlantic salmon.

It is common practice to add a supplement of between 1500 and 2500 IU of vitamin D_3 to salmon feeds. Also, commonly used feed ingredients, such as fish-meal and fish-oil, are rich in this vitamin. No differences in growth, survival or plasma calcium level were observed when salmon fingerlings were fed three different levels (0.04, 2.21 and 26.7 mg kg^{-1}) of vitamin D_3 for 11 weeks (Horvli *et al.*, 1998). However, tissue levels of vitamin D_3 were elevated in accordance with the increase in dietary dose. Plasma concentration of 25-OH D_3 was significantly elevated in the fish fed the highest dose of D_3, while the concentration of 1,25-di-OH D_2 decreased. The requirement for vitamin D has not been established in Atlantic salmon.

The main function of vitamin E is as a lipid-soluble antioxidant. Vitamin E is also important for a normal reproductive result in warm-blooded animals. α-Tocopherol has a considerably higher vitamin E effect than the β, δ and γ forms. Only α-tocopherol is abundant in marine fish, and fish-oils normally contain approximately 300 mg kg^{-1}.

Major symptoms of vitamin E deficiency in addition to reduced growth and increased mortality are: skeletal muscle degenerations due to lacking antioxidant protection of the cell membranes, anaemia and increased erythrocyte fragility, ceroid accumulation in the liver and a response to handling characterized by transitory fainting with interruption in swimming (Poston *et al.*, 1976; Lall *et al.*, 1988; Hamre *et al.*, 1994). The immune response is only moderately affected by vitamin E deficiency, and the response of vitamin E-deficient salmon fingerlings to bacterial infection has been inconsistent (Lall *et al.*, 1988; Hardie *et al.*, 1990). In juvenile salmon (from first feeding) the minimum requirement of α-tocopheryl acetate has been estimated as 60 mg kg^{-1} dry feed (Hamre and Lie, 1995a). This requirement figure is higher than that found in other salmonids. Hamre *et al.* (1997) suggested that this was influenced by a low tissue ascorbic acid level in the fish, limiting regeneration of oxidized tocopherol. In salmon fingerlings, Lall *et al.*

(1988) found that the requirement for vitamin E was about 30 mg α-tocopheryl acetate kg^{-1} dry feed. Vitamin E supplementation cannot be seen independently from the quantity and quality of the dietary oil. Vitamin E deteriorates rapidly when oils become oxidized (Koshio *et al.*, 1994). The level of dietary n-3 PUFA seems to be of minor importance for the accumulation of vitamin E in salmon tissues (Hamre and Lie, 1995b), but there are examples showing that the beneficial effects of high-n-3 PUFA fish-oils on fish growth and health at low sea temperatures are only seen when the dietary level of vitamin E is high (Waagbø *et al.*, 1993). Furthermore, a combination of high dietary vitamin E concentration and a high-n-3 PUFA fish-oil has been shown to facilitate efficient carotenoid pigmentation (Christiansen *et al.*, 1993). Tocopherols are not only antioxidants *in vivo* but also protect the salmon fillet against oxidation during storage (Sigurgisladottir *et al.*, 1994; Hamre *et al.*, 1998; Parazo *et al.*, 1998). In order to improve the oxidative stability of the fish during storage and to ensure efficient pigmentation, it is not uncommon to supplement high-energy diets for salmon growers with several times the vitamin E requirement, as high as 150 mg kg^{-1} or more.

There are three forms of vitamin K, K_1 (phytomenadione, synthesized from plants), K_2 (menaquinone, synthesized by bacteria) and K_3 (menadione, the form commonly obtained by chemical synthesis). The effect and toxicity of the three forms differ among various animal species. It is common practice to add approximately 0.04 g vitamin K_3 kg^{-1} feed. In juvenile salmon fed diets with a high concentration of vitamin A, growth was inferior when K_3 was fed compared with K_1 (Grisdale-Helland *et al.*, 1991). This questions the present use of K_3 in salmon feeds. Menadione is unstable during extrusion of fish feeds (Marchetti *et al.*, 1999). The requirement for vitamin K in salmon is not known.

Water-soluble vitamins

The information on most water-soluble vitamins for Atlantic salmon is limited, and only two requirement figures have been established for B vitamins.

Pyridoxine deficiency in salmon fingerlings is characterized by increased mortality, 'shock syndrome' behavioural changes, degenerative changes in kidneys, ovary and liver, paucity of the thyroid colloid and hyperplasia of haematopoietic tissue in the kidney (Herman, 1985). In addition, Albrektsen *et al.* (1994a) showed that deficient fingerlings had reduced lipid content in the muscle, reduced n-3 PUFA in liver phosphatidylethanolamine and phosphatidylcholine and reduced activity of aspartate aminotransferase in the liver. The requirement for pyridoxine in juvenile salmon is covered by adding 5 mg kg^{-1} to a practical fish-meal-based diet (Albrektsen *et al.*, 1994a).

In biotin-deficient fry, growth is decreased and mortality increased. In addition, increased liver glycogen and reduced activity of pyruvate carboxylase in the liver, hypertrophy and hyperplasia of the gill tissue and extensive fusions of the secondary gill lamellae have been observed in deficient salmon. The requirement for biotin is covered by 0.3 mg biotin kg^{-1} feed for juvenile salmon (Mæland *et al.*, 1998).

The requirement for inositol is no higher than 300 mg kg^{-1} of feed for juvenile salmon (Waagbø *et al.*, 1998). The requirement should be covered by the inositol content of the feed ingredients in most practical diets. Waagbø *et al.* (1998), however, recommended a supplementation of starter diets with 200 mg kg^{-1} to compensate for fluctuations in inositol content in the ingredients, leaching loss and the eventual increased requirement in the fry.

Based on growth, survival and body composition in an experiment where salmon fingerlings were fed diets with different amounts of choline chloride and soya lecithin, Poston (1991) suggested that the choline requirement in salmon fingerlings is between 430 and 1300 mg kg^{-1}. Hung *et al.* (1997) did not observe improved growth or changes in whole-body fat and protein in fingerlings when supplementing the diets with more than 880 mg choline kg^{-1}. This indicates that the higher range of the requirement estimate is 880 mg kg^{-1} or less.

Ascorbic acid is a dietary essential in Atlantic salmon (Mæland and Waagbø, 1998). Crystalline ascorbic acid is unstable, especially during extrusion, even with coated vitamin C preparations (Gadient and Fenster, 1994). Thus, more stable forms of vitamin C, such as ascorbyl-2-sulphate, ascorbyl-2-monophosphate or polyphosphate, are currently used in today's salmon feeds. Ascorbyl-2-sulphate does not seem equivalent to ascorbic acid as a dietary vitamin C source in Atlantic salmon (Sandnes *et al.*, 1990), but ascorbyl-2-monophosphate has high vitamin C activity in salmon (Sandnes and Waagbø, 1991). Mixtures of ascorbyl mono-, di- and triphosphates or mono- and diphosphates showed similar, high stability during feed production and storage, and are equally bioavailable (Roem and Oines, 1993). Lall *et al.* (1989) found that giving a supplement of 50 mg of ascorbic acid kg^{-1} was sufficient for growth and feed conversion for juvenile salmon. The absence of vitamin C caused deficiency symptoms such as lethargy, scoliosis, lordosis, broken back and anaemia. Sandnes *et al.* (1992) estimated the need for a supplement of vitamin C in a fish-meal-based practical diet to 10–12 mg ascorbic acid equivalents kg^{-1} in salmon from first feeding to 23 g of weight.

Minerals

The requirement for essential minerals varies with the life-cycle stage and growth rate, while the need for supplementation of the feeds varies with the mineral composition of the water and diet composition. This is due to the salmon's ability to take up minerals from the water and physiological changes in connection with smoltification and sexual maturation. There is also significant genetic variation in the apparent absorption of several minerals (potassium (K), calcium (Ca), magnesium (Mg), zinc (Zn) and iron (Fe)) (Thodesen *et al.*, 2001), indicating a potential to select salmon genetically for improved utilization of essential minerals. There is often a poor relationship between the mineral profile of the diet and the amounts of various elements that are available to the fish. In general, fish-meal-based diets contain higher amounts of available macrominerals than vegetable-protein-based diets, while the adequacy of many of the microminerals remains to be determined.

Calcium is abundant in both hard fresh water and salt water. Salmon actively take up calcium, unless the ammonium ion concentration is too high (Vinogradov *et al.*, 1987). Thus, salmon can partially or fully satisfy their requirement for calcium by uptake from hard fresh water or salt water (Flik *et al.*, 1995). In contrast to what is often found in warm-blooded animals, excess dietary calcium does not inhibit the uptake of phosphorus (P) in Atlantic salmon (Vielma and Lall, 1998). The requirement for calcium has not been established in Atlantic salmon.

For salmon in fresh water, it is important to supplement the diets to cover the phosphorus requirement, but not in excess, since phosphorus is the first limiting factor for plant growth in fresh water. Practical experience has shown that the combination of regulatory limitations for phosphorus supplementation to the feed and the use of feed ingredients with low phosphorus availability makes phosphorus one of the most limiting essential nutrients in salmon production. The availability of phosphorus to the salmon is highly variable depending on the form in which it is fed. For example phytic acid phosphorus in vegetable-feed ingredients has low availability to salmon, while some inorganic phosphorus salts are almost 100% available (Nordrum *et al.*, 1997; Storebakken *et al.*, 1998b). Typically the availability of phosphorus from fish-bone is half of that from sodium phosphate, while the availability of phosphorus from mono- and dibasic calcium phosphates is intermediate between the sodium salt and fish-bone (Nordrum *et al.*, 1997). Previous studies indicate that the phosphorus requirement is around 6 g total P kg^{-1} dietary DM for juvenile salmon in fresh water (Ketola, 1975) and for salmon in salt water (Lall and Bishop, 1977). More recently, Åsgård and Shearer (1997) found that the requirement was 10–11 g total P kg^{-1} DM or 9 g digestible P kg^{-1} DM for rapidly growing juvenile salmon.

Salmon take up magnesium from the water. In addition to deficiency symptoms generally found in fish, magnesium deficiency results in insufficient osmoregulation in salmon smolts. Whole-body, serum and vertebrae magnesium concentrations are better indicators of subclinical deficiency than growth (El-Mowafi and Maage, 1998). The requirement for dietary magnesium remains to be established for salmon in soft fresh water. With salmon kept in fresh water supplemented with some salt water (54 mg Mg l^{-1}), El-Mowafi and Maage (1998) found that 196 mg Mg kg^{-1} resulted in a lower magnesium content in serum, vertebrae and the whole body of salmon parr than diets with 325 mg Mg kg^{-1} or more.

Manganese (Mn) can also be taken up from the water. Experiments with salmon parr in water with a salinity of 2.5 p.p.t. indicate that a total concentration of 15 mg total dietary Mn kg^{-1} was necessary to maintain a normal level of manganese in vertebrae and whole fish (Lorenzen *et al.*, 1996). More recent studies (Maage *et al.*, 2000) have shown that growth and mortality in salmon fry were not affected over a range of dietary manganese concentrations exceeding 1.1 mg kg^{-1}. However, 7 mg total Mn kg^{-1} was required to maintain the normal activity of superoxide dismutase in the liver, and 7.5–10 mg kg^{-1} was needed for normal mineralization. Whole-body Mn levels in salmon parr are also reduced by excessive supplementation of iron to the diet (Andersen *et al.*, 1996).

It is advisable to include salt in smoltification diets, as this increases Na–K adenosine triphosphatase activity (Basulto, 1976). Requirements for sodium and potassium and their nutritional interactions have not been published with respect to Atlantic salmon.

The availability of iron depends on the dietary source; haem iron is about twice as available as Fe from iron sulphate, while elemental iron (Andersen *et al.*, 1997) and ferric oxide (Fe_2O_3) (Maage and Sveier, 1998) are not available to salmon. High dietary doses of iron are metabolically stressful to the salmon, and they do not seem to adequately downregulate iron uptake when the body stores are high (Kvingedal *et al.*, 1996). Thus, precautions should be taken to avoid overdosing of iron in diets for salmon. There are discrepancies concerning the iron requirement of salmon. Andersen *et al.* (1996) estimated the dietary iron requirement to be between 60 and 100 mg kg^{-1} in salmon parr. However, when re-evaluating the same data, Shearer (2000) suggested that the dietary Fe requirement in juvenile salmon may be as high as 200 mg kg^{-1}.

In studies dealing with copper (Cu) toxicity in salmon parr, growth declined with increasing dietary supplementation above 35 mg Cu kg^{-1}, while there was no increase in mortality when the diets were supplemented up to 1750 mg Cu kg^{-1} (Berntssen *et al.*, 1999). No copper requirement has been established in salmon. However, no differences in growth or haematology were observed when parr were fed a basal diet with 3.5 mg kg^{-1}, or supplemented with from 5 to 100 mg Cu kg^{-1} from copper sulphate ($CuSO_4$). The only significant difference in tissue Cu concentration was that the salmon supplemented with 5 mg kg^{-1} had a slightly elevated liver Cu concentration (Lorenzen *et al.*, 1994).

The availability of zinc is highly affected by dietary composition. For example the availability of Zn is strongly reduced by phytic acid from vegetable-feed ingredients (Storebakken *et al.*, 1998b). Other protein ingredients, such as wheat gluten, have been shown to enhance zinc absorption in salmon (Storebakken *et al.*, 2000a), while the dietary oil source has no effect on the zinc status of the fish (Maage and Waagbø, 1990). Increased dietary ash content may also result in a reduced uptake of zinc (Shearer *et al.*, 1992). The Zn requirement during first feeding is high, and deficiency symptoms, such as retarded growth, high condition factor and elevated iron levels, appear rapidly. Supplementation of 37–57 mg Zn kg^{-1} in the feed was needed to avoid deficiency symptoms, while as much as 57–97 mg kg^{-1} was needed to sustain whole-body Zn levels in fast-growing fry (Maage *et al.*, 1993). In larger (40 g) fingerlings, 17 mg Zn kg^{-1} was sufficient to avoid deficiency symptoms, while 67 mg Zn kg^{-1} was needed to maintain whole-body and serum Zn concentrations within a normal range (Maage and Julshamn, 1993).

Selenium (Se) is important for the glutathione peroxidase system. Deficiency symptoms in juvenile salmon other than reduced tissue selenium levels are: reductions in growth and packed blood-cell volume, reduced vitamin E concentrations in liver, blood and brain, lost integrity of the endoplasmic reticulum in pancreatic tissue, reduction in glutathione peroxidase activity in the liver and plasma, and increased hepatic glutathione-*S*-transferase, plasma pyruvate kinase and kidney glutathione levels (Poston *et al.*, 1976; Bell *et al.*, 1987). The availability of Se varies among dietary sources. In smolts Se-methionine was efficiently digested, while the availability of Se from Na_2SeO_3, Se-cystine and fish-meal was lower. However, selenium from Na_2SeO_3 and Se-cystine is much more efficiently incorporated into plasma glutathione peroxidase than Se from the other two sources (Bell and Cowey, 1989). The requirement for selenium has not been established in salmon.

Carotenoid pigments

Astaxanthin (3, 3′-dihydroxy-β, β-carotene-4, 4′-dione) or canthaxanthin (β, β-carotene-4, 4′-dione) are added to the diet of salmon mainly to give a pink colour to the flesh. Astaxanthin is the major carotenoid in wild Atlantic salmon (Khare *et al.*, 1973), is more efficiently utilized for flesh pigmentation (Storebakken *et al.*, 1985) and gives a redder hue to the flesh colour than canthaxanthin (Skrede and Storebakken, 1986). Thus, astaxanthin is the most commonly used carotenoid in salmon production today. Astaxanthin is metabolized by the salmon (Schiedt, 1998). The first step in reductive metabolism of astaxanthin metabolism is formation of idoxanthin (3, 3′, 4′-trihydroxy-β, β-carotene-4, 4-one). Idoxanthin is found in the blood and accumulated in the flesh of Atlantic salmon. Accumulation of idoxanthin may be high in the flesh of small salmon but decreases as the fish grows larger. Astaxanthin is taken up from the intestine as free astaxanthin. It is not unusual for the salmon to lose as much as 40–60% of the dietary astaxanthin unabsorbed in faeces. The rate of astaxanthin deposition is low in Atlantic salmon compared with other salmonid species (Storebakken *et al.*, 1986). The salmon rarely accumulates more than 10–12 mg astaxanthin kg^{-1} flesh, and 10–15% of dietary astaxanthin is normally recovered. Astaxanthin is not only a donor of pigment in salmon, as Christiansen *et al.* (1994, 1995) found that dietary astaxanthin was essential for growth and survival in salmon fry during first feeding. Based on their results, they recommended that first-feeding diets should contain a minimum of 5 mg astaxanthin kg^{-1}. The capacity to deposit carotenoids develops as the salmon grows, and the utilization of dietary carotenoid first starts getting efficient when the fish has reached a certain weight, around 1 kg. Generally, the response to increasing the dietary astaxanthin concentration beyond 50–60 mg kg^{-1} is marginal (March and MacMillan, 1996). The carotenoid uptake in salmon is facilitated by dietary lipid, since carotenoid uptake and pigmentation increase with increasing dietary fat content (Torrissen *et al.*, 1994; Bjerkeng *et al.*, 1998).

Astaxanthin represents a major proportion of the feed cost – more than 15% if the dietary astaxanthin concentration is 50–60 mg kg^{-1}. Thus, it is of crucial importance for the salmon farmer to use a pigmentation regime that results in satisfactory pigmentation at minimum cost. The target carotenoid concentration in the salmon fillet depends on the preference in the market. Generally, the demands for pigmentation are satisfied with 6–8 mg total carotenoid kg^{-1} flesh. As a consequence of the low capacity to utilize dietary carotenoid and the long time required to reach satisfactory pigmentation, salmon should receive pigmented feed throughout the salt-water period in order to ensure a satisfactory pigmentation. One pigmentation strategy that has proved useful is to start with a moderate astaxanthin concentration (25–35 mg kg^{-1}) in the feed once the smolts are transferred to the sea pens. The dietary concentration is then increased to 45–60 mg kg^{-1} when the salmon grows and the capacity to take up dietary astaxanthin is improved, at approximately 1 kg. Thereafter, sampling at least every 2 months must carefully monitor pigmentation to determine if the dietary carotenoid concentration may be reduced or increased. An alternative strategy, which has also been practised, is to start out with a high dietary astaxanthin concentration (50–60 mg kg^{-1}) in order to saturate the tissues with astaxanthin. At 1.5 to 2 kg body weight the carotenoid concentration is reduced to 35–45 mg kg^{-1}, and the concentration is maintained at this level until slaughter.

Practical Diets

Early findings (Austreng, 1977) in experiments replacing fish-meal with graded amounts of fish-oil and maize meal to parr with a starting weight of 1 g indicate that the optimal crude protein content with respect to growth and survival is around 44% in pelleted feeds. Increasing the dietary fat of a 60% protein diet to 12.6 and 16.0% caused a linear increase in growth and the number of fish becoming smolts after 15 months (Austreng, 1976). These findings resulted in commercially produced salmon diets containing 45–50% crude protein and 17–18% fat. With the introduction of extruded feeds came the possibility to further increase the dietary fat content, and salmon at the end of the freshwater stage grow rapidly on a diet containing 30% fat, 55% crude protein and 10.5% starch (Grisdale-Helland and Helland, 1997). The growth rates of salmon during smoltification appear to be little affected by dietary macronutrient composition, but both carcass and visceral fat contents decrease with decreasing dietary fat content (Helland and Grisdale-Helland, 1998b).

Currently, most diets for Atlantic salmon are produced by extrusion. Salmon feeds should sink. A range of compositions for practical salmon diets is presented in Table 6.4. Whole steam-dried fish-meal is the predominating protein source. Defatted or full-fat soybean is used to a limited extent in grower diets, but not in diets for juveniles due to inherent antinutrients (Storebakken *et al.*, 2000b). Wheat gluten is used mainly for feed technical reasons, while maize-gluten meals are commonly used as a supplementary source of protein. At the end of the 1990s it became increasingly common to substitute part of the fish-oil with soybean or canola oil, mainly because of the limited availability of fish-oil in the market.

Table 6.4. Typical ranges for formulation and chemical composition of diets for different size classes of Atlantic salmon.

	Starter	Fingerling	Smolt	Grower	Finishing	Brood stock
Ingredients (%)						
Fish-meal	35–60	35–60	35–60	25–50	25–50	25–50
Soy products	0	0–5	0–5	0–15	0–15	0–15
Gluten products	0–15	0–15	0–15	5–20	5–20	5–20
Cereal grains	8–15	8–15	8–15	10–18	10–18	10–18
Oils	10–15	10–20	10–20	20–30	10–30	10–30
Others*	3–5	3–5	3–5	3–5	3–5	3–5
Composition of DM (%)						
Crude protein	50–60	50–60	50–55	35–55	35–55	50–60
Crude fat	18–25	18–30	18–30	30–40	20–40	20–355
Starch	6–12	6–12	6–12	7–15	7–15	7–15

* Vitamin and micromineral premixes, macrominerals, pellet binders, carotenoids, other additives.
DM, dry matter.

Feeding Practice

It is of crucial importance for both the growth, feed conversion and health of the salmon, and the economy of the fish farmer that the fish receives the right amounts of feed at correct feeding frequency. The feeding should also be carried out in a manner that enables the salmon to eat all that is fed. The optimal particle diameter for juvenile salmon is 2.5% of 'fork-length' (Wànkowski, 1979). Similar experiments have not been published for salmon in salt water, but practical experience indicates that this value is also useful for larger fish.

When feeding salmon, it is important to spread the feed so that dominant individuals are prevented from driving other fish away from where the feed is dropped. This is achieved by utilizing the water current when automatic feeders are used in tanks and spreading the feed before it reaches the water in pens. Feeding in pens is carried out by several methods, such as manual shovelling or the use of air- or water-driven distributors. Optimal feeding frequency reflects the energy requirement of the fish and the rate of passage through the gastrointestinal tract. Thus, optimal feeding frequency depends on the fish size and water temperature: most frequent feeding is required for small fish and when the water temperature is optimal for growth. Starter diets float for some time, and frequent feeding results in continuous access to feed during this stage. Even though rapid growth can be obtained by less frequent feeding, it is common practice to feed juvenile salmon at least every half-hour. It is common practice also to feed salmon in marine pens every 0.5–2 h during daylight when using automatic feeders. However, maximum growth can be obtained by three to four meals a day in summer if sufficient time is used when feeding fish in each pen. It is important that sufficiently large amounts are fed at once, so that competition over the feed is avoided.

Salmon should be fed according to appetite for rapid growth and efficient feed conversion. Underfeeding should be avoided, especially for juvenile fish, as hunger-induced aggression results in excessive fin erosion and bimodality in growth (Storebakken and Austreng, 1987). Overfeeding is not necessary for rapid growth; it negatively affects the economy and may also pollute the fish farm. The correct amount to feed in a day mainly depends on fish size and water temperature, and it is possible to control feeding of juvenile salmon successfully based on these two variables and known feed efficiency (Austreng *et al.*, 1987; Storebakken and Austreng, 1987). One efficient method of predicting growth is to calculate growth factors, corrected for fish size and water temperature (based on Iwama and Tautz, 1981):

$$\text{Growth factor (GF)} = 1000 \times (W_t^{1/3} - W_0^{1/3})/dd$$

where W_t and W_0 are fish weights at days 0 and t, and dd is the product of average temperature and feeding days. W_t can be calculated as $(W_0^{1/3} + (\text{GF}/1000) \times dd)^3$. Thus, the estimated feed requirement is $(W_t - W_0)/\text{FE}$, where FE is feed efficiency (kg gain kg^{-1} fed). The GF is not independent of fish size, and typical values for salmon are (Einen and Mørkøre, 1997):

		GF
First feeding	(0.15–1 g)	1.0 (0.6)
Parr	(1–10 g)	1.5 (0.9–1.4)
Parr	(> 10 g)	2.0 (0.6–1.9)
Salmon in salt water		3.5 (2.5–2.7)

The values were relevant for Norwegian fish farms in the mid-1990s, while the values in parentheses are calculated from Austreng *et al.* (1987). The appetite of the fish and the optimum daily ration of the salmon, however, also depend on factors such as genetic selection, time of the year, sexual maturation and composition of the diets. It is still not feasible to incorporate such information into sufficiently accurate feeding management models. Appetite feeding based on observations of feeding activity in the surface of deep net pens will result in underfeeding. Thus, it is increasingly common to control automatic feeding based on detection of uneaten feed in the bottom of the pen, by the use of a collector attached to detectors. Once a certain amount of uneaten feed particles is detected, feeding is reduced or delayed. In spite of the labour costs involved, it is common to combine computer-controlled automatic feeding with manual feeding, mainly to check that the salmon is behaving normally in the pens.

References

Aksnes, A. (1995) Growth, feed efficiency and slaughter quality of salmon, *Salmo salar* L., given feeds with different ratios of carbohydrate and protein. *Aquaculture Nutrition* 1, 241–248.

Albrektsen, S., Hagve, T.A. and Lie, Ø. (1994a) The effect of dietary vitamin B_6 on tissue fat content and lipid composition in livers and gills of Atlantic salmon (*Salmo salar*). *Comparative Biochemistry and Physiology* 109A, 403–411.

Albrektsen, S., Waagbø, R., Lie, Ø. and Sandnes, K. (1994b). Contents and organ distribution of vitamin B_6 in Atlantic salmon (*Salmo salar*) and turbot (*Psetta maxima*) during the reproductive cycle. *Comparative Biochemistry and Physiology* 109A, 705–712.

Andersen, F., Maage, A. and Juelshamn, K. (1996) An estimation of dietary iron requirement of Atlantic salmon, *Salmo salar* L., parr. *Aquaculture Nutrition* 2, 41–47.

Andersen, F., Lorentzen, M., Waagbø, R. and Maage, A. (1997) Bioavailability and interactions with other micronutrients of three dietary iron sources in Atlantic salmon, *Salmo salar*, smolts. *Aquaculture Nutrition* 3, 239–246.

Anderson, J.S., Lall, S.P., Anderson, D.M. and Chandrasoma, J. (1992) Apparent and true availability of amino acids from common feed ingredients for Atlantic salmon (*Salmo salar*) reared in sea water. *Aquaculture* 108, 111–124.

Anderson, J.S., Lall, S.P., Anderson, D.M. and McNiven, M. (1993) Quantitative dietary lysine requirement of Atlantic salmon (*Salmo salar*) fingerlings. *Canadian Journal of Fisheries and Aquatic Science* 50, 316–322.

Anderson, J.S., Lall, S.P., Anderson, D.M. and McNiven, M.A. (1995) Availability of amino acids from various fish meals fed to Atlantic salmon (*Salmo salar*). *Aquaculture* 138, 291–301.

Åsgård, T. and Shearer, K.D. (1997) Dietary phosphorus requirement of juvenile Atlantic salmon, *Salmo salar* L. *Aquaculture Nutrition* 3, 17–23.

Austreng, E. (1976) Fett og protein i fôr til laksefisk. I. Fettinhold i tørrfôr til laksunger (*Salmo salar* L.). *Meldinger fra Norges Landbrukshøgskole* 55(5), 16 pp.

Austreng, E. (1977) Fett og protein i fôr til laksefisk. IV. Proteininnhold i tørrfôr til laksunger (*Salmo salar* L.). *Meldinger fra Norges Landbrukshøgskole* 56(19), 10 pp.

Austreng, E., Skrede, A. and Eldegard, Å. (1979) Effect of dietary fat source on the digestibility of fat and fatty acids in rainbow trout and mink. *Acta Agricultura Scandinavia* 29, 119–126.

Austreng, E., Storebakken, T. and Åsgård, T. (1987) Growth rate estimates for cultured Atlantic salmon and rainbow trout. *Aquaculture* 60, 157–160.

Basulto, S. (1976) Induced seawater tolerance in connection with inorganic salts in the feeding of Atlantic salmon (*Salmo salar* L.). *Aquaculture* 8, 45–55.

Bell, J.G. and Cowey, C.B. (1989) Digestibility and bioavailability of dietary selenium from fishmeal, selenite, selenomethionine and selenocystine in Atlantic salmon (*Salmo salar*). *Aquaculture* 81, 61–68.

Bell, J.G., Cowey, C.B., Adron, J.W. and Pirie, B.J.S. (1987) Some effects of selenium deficiency on enzyme activities and indices of tissue peroxidation in Atlantic salmon parr (*Salmo salar*). *Aquaculture* 65, 43–54.

Berge, G.E., Lied, E. and Sveier, H. (1997) Nutrition of Atlantic salmon (*Salmo salar*): the requirement and metabolism of arginine. *Comparative Biochemistry and Physiology* 117A, 501–509.

Berge, G.E., Sveier, H. and Lied, E. (1998) Nutrition of Atlantic salmon (*Salmo salar*): the requirement and metabolic effect of lysine. *Comparative Biochemistry and Physiology* 120A, 477–485.

Berntssen, M.H.G., Lundbye, A.K. and Maage, A. (1999) Effects of copper concentration on growth, feed utilisation and nutritional status of Atlantic salmon (*Salmo salar*) fry. *Aquaculture* 174, 167–181.

Bjerkeng, B., Refstie, S., Fjalestad, K.T., Storebakken, T., Rødbotten, M. and Roem, A.J. (1998) Quality parameters of the flesh of Atlantic salmon (*Salmo salar*) as affected by dietary fat content and full-fat soybean meal as a partial substitute for fish meal in the diet. *Aquaculture* 157, 297–309.

Christiansen, R., Waagbø, R. and Torrissen, O.J. (1993) Effects of polyunsaturated fatty acids and vitamin E on flesh pigmentation in Atlantic salmon (*Salmo salar*). In: *Fish Nutrition in Practice: Biarritz (France), 24–27 June 1991*. Les Colloques, no. 61, INRA, Paris, pp. 333–343.

Christiansen, R., Lie, Ø. and Torrissen, O.J. (1994) Effect of astaxanthin and vitamin A on growth and survival during first feeding of Atlantic salmon, *Salmo salar* L. *Aquaculture and Fisheries Management* 25, 903–914.

Christiansen, R., Lie, Ø. and Torrissen, O.J. (1995) Growth and survival of Atlantic salmon, *Salmo salar* L., fed different dietary levels of astaxanthin: first-feeding fry. *Aquaculture Nutrition* 1, 189–198.

Cook, J.T., McNiven, M.A., Richardson, G.F. and Sutterlin, A.M. (2000a) Growth rate, body composition and feed digestibility/conversion of growth-enhanced transgenic Atlantic salmon (*Salmo salar*). *Aquaculture* 188, 15–32.

Cook, J.T., McNiven, M.A. and Sutterlin, A.M. (2000b) Metabolic rate of pre-smolt growth enhanced Atlantic salmon (*Salmo salar*). *Aquaculture* 188, 33–45.

Cook, J.T., Sutterlin, A.M. and McNiven, M.A. (2000c) Effect of food deprivation on oxygen consumption and body composition of growth-enhanced transgenic Atlantic salmon (*Salmo salar*). *Aquaculture* 188, 47–63.

Dosanjh, B.S., Higgs, D.A., McKenzie, D.J., Randall, D.J., Eales, J.G., Rowshandeli, N., Rowshandeli, M. and Deacon, G. (1998) Influence of dietary blends of menhaden oil and canola oil on growth, muscle lipid composition, and thyroidal status of Atlantic salmon (*Salmo salar*) in sea water. *Fish Biochemistry and Physiology* 19, 123–134.

Einen, O. and Mørkøre, T. (1997) *Fôringslære for akvakultur*. Landbruksforlaget, Oslo, Norway, 235 pp.

Einen, O. and Roem, A.J. (1997) Dietary protein/energy ratios for Atlantic salmon in relation to fish size, growth, feed utilization and slaughter quality. *Aquaculture Nutrition* 3, 115–126.

Einen, O. and Skrede, G. (1998) Quality characteristics in raw and smoked fillets of Atlantic salmon, *Salmo salar*, fed high-energy diets. *Aquaculture Nutrition* 4, 99–108.

El-Mowafi, A.F.A. and Maage, A. (1998) Magnesium requirement of Atlantic salmon (*Salmo salar* L.) parr in seawater-treated fresh water. *Aquaculture Nutrition* 4, 31–38.

Flik, G., van der Velden, J.A. and Wendelaar Bonga, E. (1995) Calcium transport processes in fishes. In: Wood, C.M. and Shuttleworth, T.J. (eds) *Cellular and Molecular Approaches to Fish Ionic Regulation. Fish Physiology*, Vol. 14, Academic Press, San Diego, pp. 317–342.

Gadient, M. and Fenster, R. (1994) Stability of ascorbic acid and other vitamins in extruded fish feeds. *Aquaculture* 124, 207–211.

Gjedrem, T. (1983) Genetic variation in quantitative traits and selective breeding in fish and shellfish. *Aquaculture* 33, 51–72.

Grisdale-Helland, B. and Helland, S.J. (1997) Replacement of protein by fat and carbohydrate in diets for Atlantic salmon (*Salmo salar*) at the end of the freshwater stage. *Aquaculture* 152, 167–180.

Grisdale-Helland, B., Helland, S.J. and Åsgård, T. (1991) Problems associated with the present use of menadione sodium bisulfite and vitamin A in diets for Atlantic salmon. *Aquaculture* 92, 351–358.

Hamre, K. and Lie, Ø. (1995a) Minimum requirement of vitamin E for Atlantic salmon, *Salmo salar* L., at first feeding. *Aquaculture Research* 26, 175–184.

Hamre, K. and Lie, Ø. (1995b) α-Tocopherol levels in different organs of Atlantic salmon (*Salmo salar* L.) – Effect of smoltification, dietary levels of n-3 polyunsaturated fatty acids and vitamin E. *Comparative Biochemistry and Physiology* 111A, 547–554.

Hamre, K., Hjeldnes, B., Kryvi, H., Sandberg, S., Lorentzen, M. and Lie, Ø. (1994) Decreased concentration of hemoglobin, accumulation of lipid oxidation products and unchanged skeletal muscle in Atlantic salmon (*Salmo salar*) fed low dietary vitamin E. *Fish Physiology and Biochemistry* 12, 421–429.

Hamre, K., Waagbø, R., Berge, R.K. and Lie, Ø. (1997) Vitamin C and E interact in juvenile Atlantic salmon (*Salmo salar*, L). *Free Radical Biology and Medicine* 22, 137–149.

Hamre, K., Berge, R.K. and Lie, Ø. (1998) Oxidative stability of Atlantic salmon (*Salmo salar*, L.) fillet enriched in α-, γ-, and δ-tocopherol through dietary supplementation. *Food Chemistry* 62, 173–178.

Hardie, L.J., Fletcher, T.C. and Secombes, C.J. (1990) The effect of vitamin E on the immune response of the Atlantic salmon (*Salmo salar* L.). *Aquaculture* 87, 1–13.

Hardy, R.W., Scott, T.M. and Harrell, L.W. (1987) Replacement of herring oil with menhaden oil, soybean oil, or tallow in the diets of Atlantic salmon raised in marine net-pens. *Aquaculture* 65, 267–277.

Helland, S.J. and Grisdale-Helland, B. (1998a) The influence of dietary carbohydrate and protein levels on energy and nitrogen utilization of Atlantic salmon in seawater. In: McCracken, K.J., Unsworth, E.F. and Wylie, A.R.G. (eds) *Energy Metabolism of Farmed Animals*. CAB International, Wallingford, UK, pp. 391–394.

Helland, S.J. and Grisdale-Helland, B. (1998b) The influence of replacing fish meal in the diet with fish oil on growth, feed utilization and body composition of Atlantic salmon (*Salmo salar*) during the smoltification period. *Aquaculture* 162, 1–10.

Hemre, G.I. and Hansen, T. (1998) Utilization of different dietary starch sources and tolerance to glucose loading in Atlantic salmon (*Salmo salar*) during parr–smolt transformation. *Aquaculture* 161, 145–157.

Hemre, G.I., Sandnes, K., Lie, Ø., Torrissen, O.J. and Waagbø, R. (1995a) Carbohydrate nutrition in Atlantic salmon, *Salmo salar* L.: growth and feed utilization. *Aquaculture Research* 26, 149–154.

Hemre, G.I., Torrissen, O., Krogdahl, A. and Lie, Ø. (1995b) Glucose tolerance in Atlantic salmon, *Salmo salar* L., dependence on adaptation to dietary starch and water temperature. *Aquaculture Nutrition* 1, 69–75.

Hemre, G.I., Waagbø, R., Hjeltnes, B. and Aksnes, A. (1996) Effect of gelatinized wheat and maize starch in diets for large Atlantic salmon (*Salmo salar* L.) on glycogen retention, plasma glucose and fish health. *Aquaculture Nutrition* 2, 33–39.

Herman, R.L. (1985) Histopathology associated with pyridoxine deficiency in Atlantic salmon (*Salmo salar* L.). *Aquaculture* 46, 173–177.

Hillestad, M., Johnsen, F., Austreng, E. and Åsgård, T. (1998) Long-term effects of dietary fat level and feeding rate on growth, feed utilization and carcass quality of Atlantic salmon. *Aquaculture Nutrition* 4, 89–97.

Horvli, O., Lie, Ø. and Aksnes, L. (1998) Tissue distribution of vitamin D_3 in Atlantic salmon, *Salmo salar*: effect of dietary level. *Aquaculture Nutrition* 4, 127–131.

Hung, S.S.O., Berge, G.M. and Storebakken, T. (1997) Growth and digestibility effects of soya lecithin and choline chloride on juvenile Atlantic salmon. *Aquaculture Nutrition* 3, 141–144.

Iwama, G.K. and Tautz, A.F. (1981) A simple growth model for salmonids in hatcheries. *Canadian Journal of Fisheries and Aquatic Science* 38, 649–656.

Johnsen, F., Hillestad, M. and Austreng, E. (1993) High energy diets for Atlantic salmon: effects on pollution. In: *Fish Nutrition in Practice. Biarritz, France, 24–27 June 1991*. Les Colloques, no. 61, INRA, Paris, pp. 391–401.

Ketola, H.G. (1975) Requirements of Atlantic salmon for dietary phosphorus. *Transactions of the American Fisheries Society* 104, 548–551.

Khare, A., Moss, G.P., Weedon, B.C.L. and Matthews, A. (1973) Identification of astaxanthin in scottish salmon. *Comparative Biochemistry and Physiology* 45B, 971–973.

Koshio, S., Ackman, R.G. and Lall, S.P. (1994) Effects of oxidized herring and canola oils in diets on growth, survival, and flavor of Atlantic salmon, *Salmo salar*. *Journal of Agricultural and Food Chemistry* 42, 1164–1169.

Krogdahl, Å., Nordrum, S., Sørensen, M. and Røsjø, C. (1999) Effects of diet composition on apparent nutrient absorption along the intestinal tract and of subsequent fasting on mucosal disaccharidase activities and plasma nutrient concentrations in Atlantic salmon *Salmo salar* L. *Aquaculture Nutrition* 5, 121–133.

Kvingedal, A.M., Dehli, A. and Rørvik, K.-A. (1996) Transferrin mRNA in relation to liver iron storage in farmed Atlantic salmon (*Salmo salar*). *Fish Physiology and Biochemistry* 15, 317–322.

Lall, S.P. and Bishop, F.J. (1977) Studies on mineral and protein utilization by Atlantic salmon grown in sea water. *Technical Report Fisheries and Marine Services, Research and Development* 688, 1–16.

Lall, S.P., Olivier, G., Hines, J.A. and Ferguson, H.W. (1988) The role of vitamin E in nutrition and immune response in Atlantic salmon (*Salmo salar*). Proceedings of the 11th Atlantic Regional Fish Health workshop. *Bulletin of the Aquaculture Association of Canada* 88(2), 76–78.

Lall, S.P., Olivier, G., Weerakoon, D.E.M. and Hines, J.A. (1989) The effect of vitamin C defieciency and excess on immune response in Atlantic salmon (*Salmo salar*). In: *Proceedings of the Third International Symposium on Feeding and Nutrition in Fish.* Toba, Japan, pp. 427–441.

Lall, S.P., Kaushik, S.J., Le Bail, P.Y., Keith, R., Anderson, J.S. and Plisetskaya, E. (1994) Quantitative arginine requirement of Atlantic salmon (*Salmo salar*) reared in sea water. *Aquaculture* 124, 13–25.

Lorenzen, M., Maage, A. and Julshamn, K. (1994) Effects of dietary selenite or selenomethionine on tissue selenium levels of Atlantic salmon (*Salmo salar*). *Aquaculture* 121, 359–367.

Lorenzen, M., Maage, A. and Juelshamn, K. (1996) Manganese supplementation of a practical, fish meal based diet for Atlantic salmon parr. *Aquaculture Nutrition* 2, 121–125.

Maage, A. and Julshamn, K. (1993) Assessment of zinc status in juvenile Atlantic salmon (*Salmo salar*) by measurement of whole body and tissue levels of zinc. *Aquaculture* 117, 179–191.

Maage, A. and Sveier, H. (1998) Addition of dietary iron(III) oxide does not increase iron status of growing Atlantic salmon. *Aquaculture International* 6, 249–252.

Maage, A., Lorentzen, M., Bjornevik, M. and Julshamn, K. (1993) Zinc requirement of Atlantic salmon (*Salmo salar*) fry during start feeding. In: *Fish Nutrition in Practice: Biarritz, France, 24–27 June 1991*. Les Colloques, no. 61, INRA, Paris, pp. 873–882.

Maage, A., Lygren, B. and El-Mowafi, A.F.A. (2000) Manganese requirement of Atlantic salmon (*Salmo salar*) fry. *Fisheries Science* 66, 1–8.

Maage, R. and Waagbø, R. (1990) Zinc and selenium in tissues of young Atlantic salmon *Salmo salar* fed diets containing different lipid sources at two levels of vitamin E. *Fiskeridirektoratets Skrifter, Serie Ernœring* 3(2), 21–29.

Mæland, A. and Waagbø, R. (1998) Examination of the qualitative ability of some cold water marine teleosts to synthesise ascorbic acid. *Comparative Biochemistry and Physiology* 121A, 249–255.

Mæland, A., Waagbø, R., Sandnes, K. and Hjeldnes, B. (1998) Biotin in practical fish-meal based diet for Atlantic salmon *Salmo salar* L. fry. *Aquaculture Nutrition* 4, 241–247.

Marcetti, M., Tossani, N., Marchetti, S. and Bauce, G. (1999) Stability of crystalline and coated vitamins during manufacture and storage of fish feeds. *Aquaculture Nutrition* 5, 115–120.

March, B.E and MacMillan, C. (1996) Muscle and plasma concentrations of astaxanthin in rainbow trout, Chinook salmon and Atlantic salmon in response to different dietary levels of astaxanthin. *Progressive Fish Culturist* 58, 178–186.

Mundheim, H. and Opstvedt, J. (1990) Effects of dietary level of protein and fiber on apparent protein digestibility in the rainbow trout (*Oncorhynchus mykiss*) and salmon (*Salmo salar*) and comparison of protein digestibility in mink (*Mustea vison*), rainbow trout and salmon. In: Takeda, M. and Watanabe, T. (eds) *The Current Status of Fish Nutrition in Aquaculture. Proceedings of the 3rd International Symposium on Feeding and Nutrition of Fish.* 28 August–1 september 1998. Toba, Japan, pp. 195–200.

Nordrum, S., Åsgård, T., Shearer, K.D. and Arnessen, P. (1997) Availability of phosphorus in fish bone meal and inorganic salts to Atlantic salmon (*Salmo salar*) as determined by retention. *Aquaculture* 157, 51–61.

Parazo, M.P.M., Lall, S.P., Castell, J.D. and Ackman, R.G. (1998) Distribution of α- and γ-tocopherols in Atlantic salmon (*Salmo salar*) tissues. *Lipids* 33, 697–704.

Polvi, S.M. and Ackman, R.G. (1992) Atlantic salmon (*Salmo salar*) muscle lipids and their response to alternative dietary fatty acid sources. *Journal of Agricultural and Food Chemistry* 40, 1001–1007.

Poston, H.A. (1991) Response of juvenile Atlantic salmon fry to feedgrade lecithin and choline. *Progressive Fish Culturist* 53, 224–228.

Poston, H.A., Combs, J.R. and Lebovitz, L. (1976) Vitamin E and selenium interrelations in the diet of Atlantic salmon (*Salmo salar*): gross, histological and biochemical deficiency signs. *Journal of Nutrition* 106, 892–904.

Refstie, S. (2000) Evaluation of soybean meal as a fish meal substitute in feeds for Atlantic salmon and rainbow trout. DSc thesis, Agricultural University of Norway, Ås, Norway.

Refstie, S., Storebakken, T., Baeverfjord, G. and Roem, A. (2000) Long-term protein and lipid growth of Atlantic salmon (*Salmo salar*) fed diets with partial replacement of fish meal by soy protein products at high lipid levels. *Aquaculture* 193, 91–106.

Roem, A.J. and Oines, S. (1993) The utilization of two commercial vitamin C polyphosphates by Atlantic salmon. In: *Fish Nutrition in Practice. Biarritz, France, 24–27 June 1991*. Les Colloques, no 61, INRA, Paris, pp. 199–202.

Rollin, X., Hidalgo, Y., Valdez, M., Teller, E. and Vanbelle, M. (1994) Quantitative methionine requirement of Atlantic salmon (*Salmo salar*) juveniles. *Aquaculture* 124, 61.

Ruyter, B., Røsjø, C., Einen, O. and Thomassen, M.S. (2000a) Essential fatty acids in Atlantic salmon: time course of changes in fatty acid composition of liver, blood and carcass induced by a diet deficient in n-3 and n-6 fatty acids. *Aquaculture Nutrition* 6, 109–117.

Ruyter, B., Røsjø, C., Einen, O. and Thomassen, M.S. (2000b) Essential fatty acids in Atlantic salmon: effects of increasing dietary doses of n-6 and n-3 fatty acids on growth, survival and fatty acid composition of liver, blood and carcass. *Aquaculture Nutrition* 6, 119–127.

Sandnes, K. and Waagbø, R. (1991) Enzymatic hydrolysis of ascorbate-2-monophosphate and ascorbate-2-sulphate *in vitro*, and bioactivity of ascorbate-2-monophosphate in Atlantic salmon (*Salmo salar*). *Fiskeridirektoratets Skrifter, Serie Ernæring* 4, 33–39.

Sandnes, K., Hansen, T., Killie, J.E.A. and Waagbø, R. (1990) Ascorbate-2-sulfate as a dietary vitamin C source for Atlantic salmon (*Salmo salar*): 1. Growth, bioactivity, haematology and humoral immune response. *Fish Physiology and Biochemistry* 8, 419–427.

Sandnes, K., Torrissen, O. and Waagbø, R. (1992) The minimum dietary requirement of vitamin C in Atlantic salmon (*Salmo salar*) fry using Ca ascorbate-2-monophosphate as dietary source. *Fish Physiology and Biochemistry* 10, 315–319.

Schiedt, K. (1998) Absorption and metabolism of carotenoids in birds, fish and crustaceans. In: Britton, G., Liaaen-Jensen, S. and Pfander, H. (eds) *Carotenoids*. Vol. 3: *Biosynthesis and Metabolism*. Birkenhäuser, Basle, Switzerland, pp. 285–358.

Shearer, K.D. (2000) Experimental design, statistical analysis and modeling of dietary nutrient requirements in studies with fish: a critical review. *Aquaculture Nutrition* 6, 91–102.

Shearer, K.D., Maage, A., Opstvedt, J. and Mundheim, H. (1992) Effects of high-ash diets on growth, feed efficiency, and zinc status of juvenile Atlantic salmon (*Salmo salar*). *Aquaculture* 106, 345–355.

Shearer, K.D., Åsgård, T., Andorsdòttir, G. and Aas, G.H. (1994) Whole body elemental and proximate composition of Atlantic salmon (*Salmo salar*) during the life cycle. *Journal of Fish Biology* 44, 785–797.

Sigurgisladottir, S., Parrish, C.C., Lall, S.P. and Ackmann, R.G. (1994) Effects of feeding natural tocopherols and astaxanthin on Atlantic salmon (*Salmo salar*) fillet quality. *Food Research International* 27, 23–32.

Skrede, A., Berge, G.M., Storebakken, T., Herstad, O., Aarstad, K.G. and Sundstøl, F. (1998) Digestibility of bacterial protein grown on natural gas in mink, pigs, chicken and Atlantic salmon. *Animal Feed Science and Technology* 76, 103–116.

Skrede, G. and Storebakken, T. (1986) Characteristics of color in raw, baked and smoked wild and pen-reared Atlantic salmon. *Journal of Food Science* 51, 804–808.

Sørensen, M. (1995) Lokalisering av fordøyelse og absorpsjon av stivelse i tarm hos laks (Localisation of digestion and absorption of starch in salmon intestine). MSc thesis, Agricultural University of Norway, Ås, Norway (in Norwegian and English).

Storebakken, T. and Austreng, E. (1987) Ration level for salmonids. I. Growth, survival, body composition and feed conversion in Atlantic salmon fry and fingerlings. *Aquaculture* 60, 189–206.

Storebakken, T., Foss, P., Austreng, E. and Liaaen-Jensen, S. (1985) Carotenoids in diets for salmonids. II. Epimerization studies in Atlantic salmon. *Aquaculture* 44, 259–269.

Storebakken, T., Foss, P., Huse, I., Wandsvik, A. and Berg-Lea, T. (1986) Carotenoids in diets for salmonids. III. Utilization of canthaxanthin from dry and wet diets by Atlantic salmon, rainbow trout and sea trout. *Aquaculture* 51, 245–255.

Storebakken, T., Kvien, I.S., Shearer, K.D., Grisdale-Helland, B., Helland, S.J. and Berge, G.M. (1998a) The apparent digestibility of diets containing fish meal, soybean meal or bacterial meal fed to Atlantic salmon (*Salmo salar*): evaluation of different faecal collection methods. *Aquaculture* 169, 195–210.

Storebakken, T., Shearer, K.D. and Roem, A.J. (1998b) Availability of protein, phosphorus and other elements in fish meal, soy-protein concentrate and phytase-treated soy-protein-concentrate-based diets to Atlantic salmon, *Salmo salar*. *Aquaculture* 161, 365–379.

Storebakken, T., Shearer, K.D., Baeverfjord, G., Nielsen, B.G., Åsgård, T., Scott, T. and De Laporte, A. (2000a) Digestibility of macronutrients, energy and amino acids, absorption of elements and absence of intestinal enteritis in Atlantic salmon, *Salmo salar*, fed diets with wheat gluten. *Aquaculture* 184, 115–132.

Storebakken, T., Refstie, S. and Ruyter, B. (2000b) Soy products as fat and protein sources in fish feeds for intensive aquaculture. In: Drackley, J.K. (ed.) *Soy in Animal Nutrition.* Federation of Animal Science Societies, pp. 127–170.

Thodesen, J., Grisdale-Helland, B., Helland, S. and Gjerde, B. (1999) Feed intake, growth and feed utilization of offspring from wild and selected Atlantic salmon (*Salmo salar*). *Aquaculture* 180, 237–246.

Thodesen, J., Storebakken, T., Shearer, K.D., Rye, M., Bjerkeng, B. and Gjerde, B. (2001) Genetic variation in mineral absorption of large Atlantic salmon reared in seawater. *Aquaculture* 194, 263–271.

Thomassen, M.S. and Røsjø, C. (1989) Different fats in feed for salmon: influence on sensory parameters, growth rate and fatty acids in muscle and heart. *Aquaculture* 79, 129–135.

Thompson, I., Fletcher, T.C., Houlihan, D.F. and Secombes, C.J. (1994) The effect of dietary vitamin A on the immumocompetence of Atlantic salmon (*Salmo salar* L.). *Fish Physiology and Biochemistry* 12, 513–523.

Torrissen, K.R., Lied, E. and Espe, M. (1994) Differences in digestion and absorption of dietary protein in Atlantic salmon (*Salmo salar*) with genetically different trypsin isosymes. *Journal of Fish Biology* 45, 1087–1104.

Vielma, J. and Lall, S.P. (1998) Phosphorus utilization by Atlantic salmon (*Salmo salar*) reared in freshwater is not influenced by higher dietary calcium uptake. *Aquaculture* 160, 117–128.

Vinogradov, G.A., Komov, V.T. and Klerman, A.K. (1987) Ionic regulation in Atlantic salmon *Salmo salar* at different stages of ontogenesis. *Journal of Ichthyology* 27, 852–857.

Waagbø, R., Sandnes, K., Jørgensen, J., Engstad, R., Glette, J. and Lie, Ø. (1993) Health aspects of dietary lipid source and vitamin E in Atlantic salmon (*Salmo salar*). II. Spleen and erythrocyte phospholipid fatty acid composition, nonspecific immunity and disease resistance. *Fiskeridirektoratets Skrifter Serie Ernæring* 6, 63–80.

Waagbø, R., Sandnes, K. and Lie, Ø. (1998) Effects of inositol supplementation on growth, chemical composition and blood chemistry in Atlantic salmon, *Salmo salar* L., fry. *Aquaculture Nutrition* 4, 53–59.

Wànkowski, J.W.J. (1979) Morphological limitations, prey size selectivity, and growth response of juvenile Atlantic salmon, *Salmo salar*. *Journal of Fish Biology* 14, 89–100.

Wilson, R.P. and Cowey, C.B. (1985) Amino acid composition of whole body tissue of rainbow trout and Atlantic salmon. *Aquaculture* 48, 373–376.

Atlantic Halibut, *Hippoglossus hippoglossus*

7

Barbara Grisdale-Helland[1] and Ståle J. Helland[1,2]
[1]AKVAFORSK (Institute of Aquaculture Research AS), N-6600 Sunndalsøra, Norway; [2]Department of Animal Science, Agricultural University of Norway, N-1432 Ås-NLH, Norway

Introduction

In 1999, about 400 metric tons (t) of farmed Atlantic halibut with a value of 36 million Norwegian crowns (NOK) were processed in Norway, up from 275 t in 1998 and 113 t in 1997. Halibut larvae have a long yolk-sac stage requiring carefully controlled temperature, salinity, light and microbial conditions and a low disturbance level (Mangor-Jensen *et al.*, 1998). Feeding of larvae involves the use of live organisms, followed by gradual replacement with dry diets. Juveniles and grow-out fish are fed dry pellets in tanks on land or in sea pens. The fish are normally processed at a weight of about 5 kg after a production time of about 37–44 months, depending on the techniques used (Engelsen, 1998). Females may reach this weight 6 months before males (Engelsen, 1998). Currently, larvae are produced from brood stock originally captured in the wild and from first-generation domesticated fish.

Nutrient Requirements

Protein and amino acids

During the yolk-sac stage and the initial period of exogenous feeding, free and then later protein-derived amino acids are the major sources of energy for Atlantic halibut larvae (Rønnestad, 1993; Rønnestad and Naas, 1993; Finn *et al.*, 1995). The major amino acids in the yolk proteins are alanine and leucine, whereas aspartic acid and glutamic acid predominate in the proteins of the larval body and in juveniles (Finn *et al.*, 1995; Kim and Lall, 2000).

The optimal protein level in diets for Atlantic halibut appears to be dependent on the size of the fish. Hjertnes and Opstvedt (1989) reported a linear increase in the growth rate of halibut grown from 7 to about 180 g, in response to increasing the dietary protein level from 41% to 62% on a dry matter (DM) basis and decreasing the carbohydrate level from 27% to 3%. Similarly, we found a linear response in the growth of halibut from 60 to 125 g when the dietary protein level was increased from 41% to 56% and the carbohydrate level was decreased from 24% to 7% (Helland and Grisdale-Helland, unpublished). In contrast, Hjertnes *et al.* (1993) found no effect when the dietary protein level was increased from 40% to 55%, concomitant with a decrease in dietary carbohydrate from 24% to 8%, on halibut grown from 34 to about 300 g. However, the authors reported a significant increase in growth when the dietary protein level was further increased to 60% and the carbohydrate level was reduced to 3%. Reducing the protein level of this diet from 60% to 55% and increasing dietary fat from 26% to 30% resulted in significantly lower growth. Halibut grown from about 150 g to 550 g benefited from an increase in the dietary protein level from 41% to 51% (dietary carbohydrate reduced from 27% to 15%), but higher dietary protein levels, obtained through reductions in either carbohydrate or fat, did not result in increased growth of this size of fish (Aksnes *et al.*, 1996; Helland and Grisdale-Helland, 1998). In the weight range of 0.6–1.5 kg, the growth of halibut was not improved by increasing the dietary protein level from 37% to 50% (DM basis) with a reduction in dietary fat content (Berge and Storebakken, 1991) or from 48% to 54% (replacing either dietary fat or carbohydrate) (Grisdale-Helland and Helland, 1998).

The digestibility by halibut of protein in dry diets based on low-temperature-produced fish-meal is in the range of 82–86% (Grisdale-Helland and Helland, 1998; Berge *et al.*, 1999) and that in moist diets has been found to be 75–88% (Berge and Storebakken, 1991; Berge *et al.*, 1991; Haugen, 1999). Protein digestibility, as measured in mink, has been reported to influence the amount of feed consumed by halibut to reach satiation (Aksnes and Mundheim, 1997).

Few alternative protein sources to fish-meal have been tested in halibut diets. The replacement of fish-meal with soybean-protein concentrate (SPC) (44% of total dietary protein) and a methionine supplement in the diet of 600–900 g halibut had no effect on nitrogen digestibility, but resulted in slightly lower feed efficiency, probably due to the fibre content in SPC (Berge *et al.*, 1999). The fish compensated for the lower overall nutritive value of the SPC diet by increasing feed intake and there was no effect on growth. Replacement of fish-meal with full-fat soybean meal (30% of dietary protein) had no effect on feed intake, growth or feed utilization (B. Grisdale-Helland, S.J. Helland, G. Baeverfjord and G.M. Berge, unpublished data). Wheat gluten is also an acceptable replacement for fish-meal in diets for Atlantic halibut, but may require supplementation of lysine when levels as high as 30% of diet are used (Helland *et al.*, 2000). A lysine requirement of 19–21 g kg^{-1} weight gain (2.9–3.5% of dietary DM) for 60–125 g halibut fed a 60% protein diet is suggested from these results. This is in the upper range of what has been reported for other species (NRC, 1993).

Halibut in the 50–300 g weight range fed high-energy, fish-meal-based diets consume about 400 g protein and 4200 kcal gross energy (GE) kg^{-1} gain (Table 7.1). Larger fish consume about the same amount of protein and 4700 kcal GE kg^{-1} gain. The retention of protein in halibut fed high-energy diets containing 47–60% protein is generally high (Table 7.2). Isotope studies of

Table 7.1. Protein and energy consumed by Atlantic halibut per kg gain*.

Initial weight (g)	Final weight (g)	Dietary protein content (g kg^{-1})	Dietary energy content (kcal kg^{-1})	Protein consumed†	Energy consumed‡	Reference
5	539	630	5115	536	4326	A
5	539	550	6095	436	4828	A
60	125	560	5688	383	3896	B
60	125	596	6071	359	3657	B
72	100	540	5497	501	5091	C§
140	260	511	5712	331	3705	D
170	315	484	5903	320	3920	E
630	864	472	6262	366	4852	F
750	860	509	5688	422	4708	G§
1020	1260	479	5832	410 (338)	4995 (4254)	H
2420	2700	535	5378	418	4183	I

* Experiments in which feed intake was measured. If not specified, data were calculated from available information. When several diets were tested in the same experiment, that with the lowest protein level giving highest gain was chosen. When gross energy was not measured, it was calculated using the following factors (kcal kg^{-1}): protein 5400, fat 9300, carbohydrate 4100.

† g kg^{-1} gain (digestible protein in parentheses).

‡ kcal gross energy kg^{-1} gain (digestible energy in parentheses).

A, Aksnes *et al.*, 1996; B, S.J. Helland and B. Grisdale-Helland, unpublished; C, Nortvedt and Tuene, 1995; D, Helland and Grisdale-Helland, 1998; E, B. Grisdale-Helland, S.J. Helland, G. Baeverfjord and G.M. Berge, unpublished; F, Berge *et al.*, 1999; G, Haugen, 1999; H, Grisdale-Helland and Helland, 1998; I, Helland *et al.*, 1997.

§ Data from four centrepoint tanks.

protein synthesis rates also suggest that this species uses protein efficiently for growth (Fraser *et al.*, 1998).

Energy

It is generally known that fish eat to meet their energy need and therefore, although bulk may restrain consumption, fish should be able to compensate for a lower energy-density diet with increased intake. Results from a study with 1 kg halibut fed diets differing in dietary levels of carbohydrate and protein show, however, that compensation for low digestible-energy (DE) content by increased intake does not always occur, even when bulk is apparently not a limiting factor (Grisdale-Helland and Helland, 1998).

The fasting metabolic rate for halibut has been estimated to be 1.4 (Jonassen *et al.*, 2000) to 2.7 g oxygen (O_2) kg$^{-0.78}$ day^{-1} (Davenport *et al.*, 1990) (calculated using the weight coefficient 0.78 determined for plaice (*Pleuronectes platessa*) and flounder (*Platichthys flesus*) by Fonds *et al.* (1992)). The routine metabolic

Table 7.2. Retention of consumed protein and energy in whole body of Atlantic halibut (%).

Initial weight (g)	Final weight (g)	Dietary protein content (g kg^{-1})	Dietary energy content (kcal kg^{-1})	Protein retained (%)	Energy retained (%)	Reference*
60	125	560	5688	43	47	B
60	125	596	6071	48	49	B
72	100	540	5497	33		C†
140	260	511	5712	46	58	D
170	315	484	5903	52	55	B
630	864	472	6262	47	57	F

* See Table 7.1 for references.
† Data from four centrepoint tanks.

rate has been estimated to be in the range 1.3–3.6 g O_2 $kg^{-0.78}$ day^{-1} (Davenport *et al.*, 1990; Hallaråker *et al.*, 1995; Simensen, 1999; Jonassen *et al.*, 2000; B. Grisdale-Helland, S.J. Helland, G. Baeverfjord and G.M. Berge, unpublished). The highest of these metabolic-rate estimates for both fed and fasted fish are from Davenport *et al.* (1990) and may be overestimates because the fish, which were collected in the wild, were noted to react poorly to handling.

Lipids and fatty acids

It is widely accepted that docosahexaenoic acid (DHA), 22:6n-3, and eicosapentaenoic acid (EPA), 20:5n-3, are essential fatty acids in diets for marine fish. Data from other marine fish species suggests that the requirement for n-3 highly unsaturated fatty acids (HUFA) is 1–2% of the diet for larvae and 0.5–1% for larger fish (reviewed by Sargent *et al.*, 1995; Harel and Place, 1998). The lack of Δ-5-desaturase activity in marine fish, preventing conversion of linoleic acid, 18:2n-6, to arachidonic acid (AA), 20:4n-6, which is essential for the production of eicosanoids, implies the essentiality of providing AA in larvae diets (Sargent *et al.*, 1997, 1999). McEvoy *et al.* (1998) noted, though, that normal pigmentation in halibut larvae was negatively correlated with the AA concentration in the brain. The best pigmentation occurred in the larvae fed *Artemia* enriched with oil containing EPA : AA ratios above 4 : 1 and no correlation between the DHA : EPA ratio in the oil and pigmentation was found (McEvoy *et al.*, 1998). In contrast, Shields *et al.* (1999a) suggested that the poorer pigmentation of halibut larvae fed *Artemia* compared with those fed copepods (*Eurytemora velox*) was related to a relative deficiency of DHA in the *Artemia* and was not related to the levels or ratio of EPA and AA. The copepods also contained a higher ratio of phospholipids to triacylglycerols (TG) than the *Artemia*. As reported by Shields *et al.* (1999a), phospholipids may be more easily digested than TG and may also be required in the diet of fish larvae because of low *de novo* synthesis.

In early trials with grow-out Atlantic halibut, the digestibility of fat was found to be in the 78–94% range (Berge and Storebakken, 1991; Berge *et al.*, 1991), whereas more recent studies show consistently higher digestibility (95–99%) (Grisdale-Helland and Helland, 1998; Berge *et al.*, 1999; Haugen, 1999).

Carbohydrates

Gawlicka *et al.* (2000) noted that the amylase activity level was low in halibut larvae before metamorphosis compared with that found for other marine fish larvae. The amylase activity in metamorphic larvae was considerably higher than that found in earlier stages (Gawlicka *et al.*, 2000).

The digestibility of carbohydrate in 1 kg Atlantic halibut is reduced when the dietary level is increased (Grisdale-Helland and Helland, 1998; Grisdale-Helland and Helland, unpublished; Helland and Grisdale-Helland, unpublished results; Fig. 7.1). Brudeseth (1996) found that, of the disaccharidases, only maltase activity in halibut was affected by the dietary carbohydrate level. Comparing Atlantic halibut and Atlantic salmon, Brudeseth (1996) found that these species exhibited similarities in both the total capacity of the disaccharidases and the enzymatic capacity of similar gut segments.

Metabolic adaptation to increasing dietary carbohydrate has been shown. Garcia-Riera and Hemre (1996) reported that halibut fed diets containing 5% or 8% carbohydrate showed rapid removal of glucose from the blood and deposition in the liver compared with fish fed a diet containing 2% carbohydrate. The latter fish had the highest plasma glucose levels 16 h after a peritoneal injection of glucose, and exhibited greater retention of glucose in muscle, heart, gills, kidney and the gastrointestine.

Vitamins

Before the development of dry diets, it was common to feed large halibut whole or parts of fish, such as herring, or moist diets. This is still being practised with brood stock. A preliminary study by Goff and Lall (1989) indicated that the liver content of vitamins A, E and C and total flavin could be increased in Atlantic halibut through vitamin supplementation of these types of diets.

The content of vitamin A in halibut larvae during the initial first-feeding period appear to be affected by the type of live food used. Rønnestad *et al.* (1998) observed that larvae fed only *Artemia* enriched with DHA Selco (INVE, Belgium) contained lower levels of all-trans retinal and all-trans retinol and had a higher incidence of malpigmentation than larvae initially fed copepods (mainly *Temora longicornis*). High contents of lutein and astaxanthin were found in *Temora*, whereas cryptoxanthin/cantaxanthin and an unknown retinoid compound were found in *Artemia*.

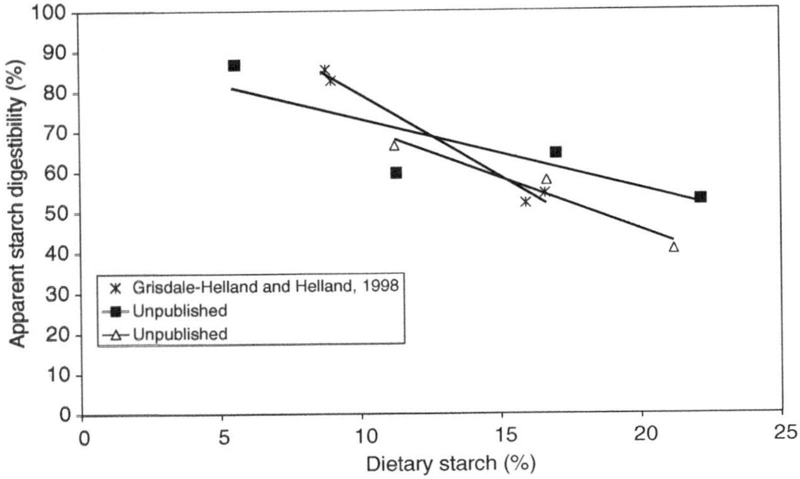

Fig. 7.1. Relationship between dietary starch concentration and apparent starch digestibility in Atlantic halibut fed diets containing extruded wheat.

Halibut juveniles do not exhibit L-gulonolactone oxidase activity, which is involved in the synthesis of ascorbic acid (Mæland and Waagbø, 1998). Halibut larvae (10 mg) fed zooplankton (mainly *T. longicornis* and containing 750 mg ascorbic acid kg^{-1} dry weight (DW)) were observed to maintain stable tissue concentrations of ascorbic acid, suggesting that the requirement for this vitamin is met at this level (Rønnestad *et al.*, 1999a). Young halibut (500–1000 mg) weaned on to formulated diets showed increasing whole-body ascorbic acid levels in response to increasing dietary ascorbate polyphosphate (300–3000 mg ascorbic acid equivalents kg^{-1} diet), although retention of the vitamin was estimated to be very low (Mæland *et al.*, 1999). The dietary level of ascorbic acid has not been shown to affect growth or mortality of young halibut (Mæland and Waagbø, 1998; Mæland *et al.*, 1999).

Vitamin B_6 in halibut larvae feeding on zooplankton containing 4–5 mg B_6 kg^{-1} DW stabilized during development at about 5 mg kg^{-1} DW (Rønnestad *et al.*, 1997). The decline in the vitamin B_6 content in whole larvae per g growth, from the yolk-sac stage until first feeding, was 3.5 mg, or about 17 nmol (Rønnestad *et al.*, 1997), similar to the requirement exhibited by many different species of animals (15 nmol g^{-1} weight gain) (Coburn, 1994).

Practical Diets

In semi-intensive systems, zooplankton is still necessary for the production of acceptable halibut juveniles (Berg, 1997). As mentioned above, feeding with marine copepods generally improves pigmentation (Shields *et al.*, 1999a) and also influences the vitamin A status, eye rod/cone ratio and eye migration of halibut larvae (Rønnestad *et al.*, 1998; Shields *et al.*, 1999a,b). However, because of the limited supply of copepods, studies are continuing on enrichment of *Artemia* for use as halibut larval diets (Olsen *et al.*, 1999; Shields *et al.*, 1999b).

Inclusion of free amino acids in first-feeding diets for halibut is difficult because of leaching. Inclusion in liposomes may be possible, but only small particles can be made. Alternatively, either liposomes may be included in microdiets fed to larvae or free amino acids may be used to enrich *Artemia* (Rønnestad *et al.*, 1999b).

Some commercial diets are produced specifically for halibut larvae and grow-out fish. The protein content in these diets (DM basis) range from 62–68% for particles less than 1 mm to 48–67% for 8–20 mm pellets. Concomitant with the decrease in protein content with increasing pellet size is an increase in fat from 12–20% to 16–27%.

Feeding Practices

The production technology used in halibut farming was initially based on experience from turbot and cod, whereas the nutrition and feeding of juvenile and grow-out halibut have had their base mostly in salmon production. In Norway, most of the larvae are produced semi-intensively and fed zooplankton and *Artemia* nauplii, whereas the rest are produced in intensive systems (Olsen *et al.*, 1999), as is the case in Scotland (Shields *et al.*, 1999b) and Iceland. After first feeding, the majority of halibut in Norway are raised on land in tanks with circular water flow (with or without shelves) (Engelsen, 1998). Octagonal concrete tanks and raceways are also in use. Salmon cages, modified to give a stiff bottom, also have good potential for use with halibut, but although they may be cheaper to establish than a land operation, they require greater production and technical control (Engelsen, 1998). Automatic feeders, supplemented with hand-feeding, are normally used in halibut production.

Little research has been done on optimal feeding practices for halibut. Feed intake of halibut is influenced by pellet size and/or mixture of pellet sizes (Helland *et al.*, 1997). In a pilot trial, growth of 390–600 g halibut was found to be significantly better when the fish were fed in excess daily or every second day compared with every third day. Feed intake measurements showed no effect of feeding interval on feed efficiency and it was concluded that this size of halibut could be fed every second day (Helland *et al.*, 1993). Our results from a trial with 23 g halibut fed from one to 72 times per day revealed only small differences in growth and feed efficiency (Grisdale-Helland and Helland, unpublished).

References

Aksnes, A. and Mundheim, H. (1997) The impact of raw material freshness and processing temperature for fish meal on growth, feed efficiency and chemical composition of Atlantic halibut (*Hippoglossus hippoglossus*). *Aquaculture* 149, 87–106.

Aksnes, A., Hjertnes, T. and Opstvedt, J. (1996) Effect of dietary protein level on growth and carcass composition in Atlantic halibut (*Hippoglossus hippoglossus* L.). *Aquaculture* 145, 225–233.

Berg, L. (1997) Commercial feasibility of semi-intensive larviculture of Atlantic halibut (*Hippoglossus hippoglossus*). *Aquaculture* 155, 333–340.

Berge, G.M. and Storebakken, T. (1991) Effect of dietary fat level on weight gain, digestibility, and fillet composition of Atlantic halibut. *Aquaculture* 99, 331–338.

Berge, G.M., Krogdahl, Å., Strømsnes, Ø., Grønseth, F.A., Myhre, P. and Austreng, E. (1991) Digestibility determination in Atlantic halibut (*Hippoglossus hippoglossus*). *Fiskeridirektoratets Skrifter, Serie Ernæring* 4, 117–125.

Berge, G.M., Grisdale-Helland, B. and Helland, S.J. (1999) Soy protein concentrate in diets for Atlantic halibut (*Hippoglossus hippoglossus*). *Aquaculture* 178, 139–148.

Brudeseth, L.K. (1996) Hydrolyse av karbohydrater i tarmmucosa hos atlantisk laks (*Salmo salar*), kveite (*Hippoglossus hippoglossus*) og mink (*Mustela vison*). Effekt av variasjon i fôrets innhold av stivelse og antinæringsstoff fra soya. (Hydrolysis of carbohydrates in the intestinal mucosa of Atlantic salmon (*Salmo salar*), halibut (*Hippoglossus hippoglossus*) and mink (*Mustela vison*). Effect of variation in dietary content of starch and antinutrients from soya.) Cand. Scient. thesis, University of Oslo, Oslo, Norway (in Norwegian).

Coburn, S.P. (1994) A critical review of minimal vitamin B_6 requirements for growth in various species with a proposed method of calculation. *Vitamins and Hormones – Advances in Research and Applications* 48, 259–300.

Davenport, J., Kjørsvik, E. and Haug, T. (1990) Appetite, gut transit, oxygen uptake and nitrogen excretion in captive Atlantic halibut, *Hippoglossus hippoglossus* L. and lemon sole, *Microstomus kitt* (Walbaum). *Aquaculture* 90, 267–277.

Engelsen, R. (1998) *Matfiskoppdrett av kveite. En økonomisk betraktning.* (Production of grow-out halibut. An economic consideration.) Report to the Norwegian Research Council, Rolf Engelsen AS, Bergen, 74 pp. (in Norwegian).

Finn, R.N., Rønnestad, I. and Fyhn, H.J. (1995) Respiration, nitrogen and energy metabolism of developing yolk-sac larvae of Atlantic halibut (*Hippoglossus hippoglossus*). *Comparative Biochemistry and Physiology* 111A, 647–671.

Fonds, M., Cronie, R., Vethaak, A.D. and van der Puyl, P. (1992) Metabolism, food consumption and growth of plaice (*Pleuronectes platessa*) and flounder (*Platichthys flesus*) in relation to fish size and temperature. *Netherlands Journal of Sea Research* 29, 127–143.

Fraser, K.P.P., Lyndon, A.R. and Houlihan, D.F. (1998) Protein synthesis and growth in juvenile Atlantic halibut, *Hippoglossus hippoglossus* (L.): application of ^{15}N stable isotope tracer. *Aquaculture Research* 29, 289–298.

Garcia-Riera, M.P. and Hemre, G.-I. (1996) Organ responses to ^{14}C-glucose injection in Atlantic halibut, *Hippoglossus hippoglossus* (L.), acclimated to diet of varying carbohydrate content. *Aquaculture Research* 27, 565–571.

Gawlicka, A., Parent, B., Horn, M.H., Ross, N., Opstad, I. and Torrissen, O.J. (2000) Activity of digestive enzymes in yolk-sac larvae of Atlantic halibut (*Hippoglossus hippoglossus*): indication of readiness for first feeding. *Aquaculture* 184, 303–314.

Goff, G.P. and Lall, S.P. (1989) An initial examination of the nutrition and growth of Atlantic halibut (*Hippoglossus hippoglossus*) fed whole herring with a vitamin supplement. *Bulletin of the Aquaculture Association of Canada* 89(3), 56–58.

Grisdale-Helland, B. and Helland, S.J. (1998) Macronutrient utilization by Atlantic halibut (*Hippoglossus hippoglossus*): diet digestibility and growth of 1 kg fish. *Aquaculture* 166, 57–65.

Hallaråker, H., Folkvord, A. and Stefansson, S.O. (1995) Growth of juvenile halibut (*Hippoglossus hippoglossus*) related to temperature, day length and feeding regime. *Netherlands Journal of Sea Research* 34, 139–147.

Harel, M. and Place, A.R. (1998) The nutritional quality of live feeds for larval fish. *Bulletin of the Aquaculture Association of Canada* 98(4), 6–11.

Haugen, T. (1999) Growth, feed conversion efficiency and digestion in Atlantic halibut: shorttime effects of different levels of temperature and fat/protein ratio in feed. Cand. Scient. thesis, University of Bergen, Bergen, Norway.

Helland, S.J. and Grisdale-Helland, B. (1998) Growth, feed utilization and body composition of juvenile Atlantic halibut (*Hippoglossus hippoglossus*) fed diets differing in the ratio between the macronutrients. *Aquaculture* 166, 49–56.

Helland, S.J., Grisdale-Helland, B. and Holm, J. (1993) Feeding regimes and compensatory growth of halibut. In: *6th International Symposium on Fish Nutrition and Feeding.* Hobart, Australia, 4–7 October, 1993. Abstract.

Helland, S.J., Grisdale-Helland, B. and Berge, G.M. (1997) Feed intake and growth of Atlantic halibut (*Hippoglossus hippoglossus*) fed combinations of pellet sizes. *Aquaculture* 156, 1–8.

Helland, S.J., Grisdale-Helland, B., Berge, G.M. and Bekkevold, K.R. (2000) Replacement of fish meal with wheat gluten in diets for Atlantic halibut: implications for lysine supplementation. In: *Ninth International Symposium on Nutrition and Feeding in Fish.* Miyazaki, Japan, 21–25 May, 2000, p. 68.

Hjertnes, T. and Opstvedt, J. (1989) Effects of dietary protein levels on growth in juvenile halibut (*Hippoglossus hippoglossus* L.). In: *Proceedings of the Third International Symposium on Feeding and Nutrition in Fish.* 28 Aug–1 Sept, 1989, Toba, Japan, pp. 189–193.

Hjertnes, T., Gulbrandsen, K.E., Johnsen, F. and Opstvedt, J. (1993) Effect of dietary protein, carbohydrate and fat levels in dry feed for juvenile halibut (*Hippoglossus hippoglossus* L.). In: Kaushik, S.J. and Luquet, P. (eds) *Fish Nutrition in Practice.* Les Colloques, no. 61, INRA, Paris, pp. 493–496.

Jonassen, T.M., Imsland, A.K., Kadowaki, S. and Stefansson, S.O. (2000) Interaction of temperature and photoperiod on growth of Atlantic halibut *Hippoglossus hippoglossus.* L. *Aquaculture Research* 31, 219–227.

Kim, J.-D. and Lall, S.P. (2000) Amino acid composition of whole body tissue of Atlantic halibut (*Hippoglossus hippoglossus*), yellowtail flounder (*Pleuronectes ferruginea*) and Japanese flounder (*Paralichthys olivaceus*). *Aquaculture* 187, 367–373.

McEvoy, L.A., Estevez, A., Bell, J.G., Shields, R.J., Gara, B. and Sargent, J.R. (1998) Influence of dietary levels of eicosapentaenoic and arachidonic acids on the pigmentation success of turbot (*Scophthalmus maximus* L.) and halibut (*Hippoglossus hippoglossus* L.). *Bulletin of the Aquaculture Association of Canada* 98(4), 17–20.

Mæland, A. and Waagbø, R. (1998) Examination of the qualitative ability of some cold water marine teleosts to synthesise ascorbic acid. *Comparative Biochemistry and Physiology Part A* 121, 249–255.

Mæland, A., Rosenlund, G., Stoss, J. and Waagbø, R. (1999) Weaning of Atlantic halibut *Hippoglossus hippoglossus* L. using formulated diets with various levels of ascorbic acid. *Aquaculture Nutrition* 5, 211–219.

Mangor-Jensen, A., Harboe, T., Shields, R.J., Gara, B. and Naas, K.E. (1998) Atlantic halibut, *Hippoglossus hippoglossus* L., larvae cultivation literature, including a bibliography. *Aquaculture Research* 29, 857–886.

Nortvedt, R. and Tuene, S. (1995) Multivariate evaluation of feed for Atlantic halibut. *Chemometrics and Intelligent Laboratory Systems* 29, 271–282.

NRC (National Research Council) (1993) *Nutrient Requirements of Fish.* National Academy Press, Washington, DC, 114 pp.

Olsen, Y., Evjemo, J.O. and Olsen, A. (1999) Status of the cultivation technology for production of Atlantic halibut (*Hippoglossus hippoglossus*) juveniles in Norway/Europe. *Aquaculture* 176, 3–13.

Rønnestad, I. (1993) No efflux of free amino acids from yolk-sac larvae of Atlantic halibut (*Hippoglossus hippoglossus* L.). *Journal of Experimental Marine Biology and Ecology* 167, 39–45.

Rønnestad, I. and Naas, K.E. (1993) Routine metabolism in Atlantic halibut at first feeding – a first step towards an energetic model. In: Walther, B.T. and Fyhn, H.J. (eds) *Physiology and Biochemistry of Marine Fish Larval Development.* University of Bergen, Bergen, Norway, pp. 279–284.

Rønnestad, I., Lie, Ø. and Waagbø, R. (1997) Vitamin B_6 in Atlantic halibut, *Hippoglossus hippoglossus* – endogenous utilization and retention in larvae fed natural zooplankton. *Aquaculture* 157, 337–345.

Rønnestad, I., Helland, S. and Lie, Ø. (1998) Feeding *Artemia* to larvae of Atlantic halibut (*Hippoglossus hippoglossus* L.) results in lower larval vitamin A content compared with feeding copepods. *Aquaculture* 165, 159–164.

Rønnestad, I., Hamre, K., Lie, Ø. and Waagbø, R. (1999a) Ascorbic acid and α-tocopherol levels before and after exogenous feeding. *Journal of Fish Biology* 55, 720–731.

Rønnestad, I., Thorsen, A. and Finn, R.N. (1999b) Fish larval nutrition: a review of recent advances in the roles of amino acids. *Aquaculture* 177, 201–216.

Sargent, J.R., Bell, J.G., Bell, M.V., Henderson, R.J. and Tocher, D.R. (1995) Requirement criteria for essential fatty acids. *Journal of Applied Ichthyology* 11, 183–198.

Sargent, J.R., McEvoy, L.A. and Bell, J.G. (1997) Requirements, presentation and sources of polyunsaturated fatty acids in marine fish larval feeds. *Aquaculture* 155, 117–127.

Sargent, J.R., McEvoy, L., Estevez, A., Bell, G., Bell, M., Henderson, J. and Tocher, D. (1999) Lipid nutrition of marine fish during early development: current status and future directions. *Aquaculture* 179, 217–229.

Shields, R.J., Bell, J.G., Luizi, F.S., Gara, B., Bromage, N.R. and Sargent, J.R. (1999a) Natural copepods are superior to enriched *Artemia* nauplii as feed for halibut larvae (*Hippolgossus hippoglossus*) in terms of survival, pigmentation and retinal morphology: relation to dietary essential fatty acids. *Journal of Nutrition* 129, 1186–1194.

Shields, R.J., Gara, B. and Gillespie, M.J.S. (1999b) A UK perspective on intensive hatchery rearing methods for Atlantic halibut (*Hippoglossus hippoglossus*). *Aquaculture* 176, 15–25.

Simensen, L.M. (1999) Growth, activity and oxygen consumption in juvenile halibut (*Hippoglossus hippoglossus* L.) reared under four different photoperiods. Cand. Scient. thesis, University of Bergen, Bergen, Norway.

Japanese Flounder, *Paralichthys olivaceus*

8

Kotaro Kikuchi[1] and Toshio Takeuchi[2]
[1]Abiko Research Laboratory, Central Research Institute of Electric Power Industry, 1646 Abiko, Abiko, Chiba 270-1194, Japan;
[2]Department of Aquatic Biosciences, Tokyo University of Fisheries, Konan, Minato, Tokyo 108-8477, Japan

Introduction

Seedling production has been conducted with various flat-fish species for stock enhancement in the coastal zone of Japan. However, Japanese flounder (*hirame* in Japanese), *Paralichthys olivaceus* (left-eye flounder), is the only flat-fish species that has been commercially produced. Aquaculture of Japanese flounder started in the mid-1970s and commercial production became extensive in the early 1980s with the development of seedling production and farming techniques. The production has increased gradually during the past two decades, was 8583 t in 1997, nearly equal to that of the wild catch, and ranked fourth among marine cultured finfish in Japan next to yellowtail, *Seriola quinqueradiata*, red sea bream, *Pagrus major*, and coho salmon, *Oncorhynchus kisutch* (Fig. 8.1). Flounders are highly valued and their market price is two to three times higher than that of yellowtail and red sea bream. Locally available trash fish, such as sand lance or

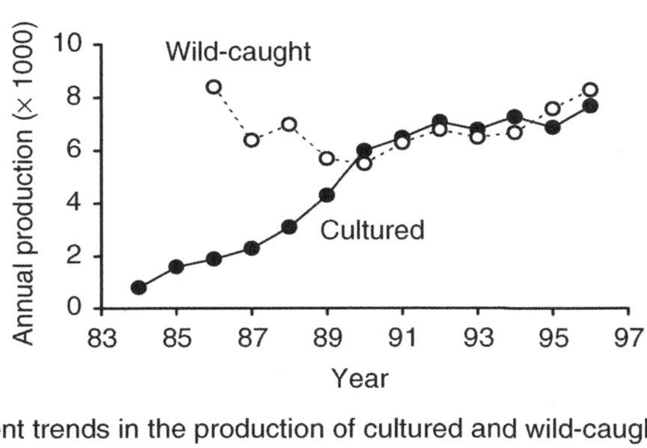

Fig. 8.1. Recent trends in the production of cultured and wild-caught Japanese flounder.

sardines, had been used for grow-out of Japanese flounder. However, the use of dry and moist pellets has increased recently due to the shortage of these trash fish. This chapter provides up-to-date information on the nutritional availability of dietary protein, lipids and carbohydrates, for grow-out of Japanese flounder. Vitamin and mineral requirements are summarized by Takeuchi (1998).

Nutrient Requirements

Protein and amino acids

Similar to other flat-fish and finfish species (Cowey *et al.*, 1970), ten amino acids – arginine, histidine, isoleucine, leucine, lysine, methionine, phenylalanine, threonine, tryptophan and valine – are essential for Japanese flounder (Kanazawa, 1990). Amino acid requirements examined with 3 g fish (Forster and Ogata, 1998) are within the range found for other fresh and marine fish species (Takeuchi, 1998). Growth improvement with dietary taurine, a free amino acid, was demonstrated and research on its biochemical function has been conducted recently (Takeuchi, 1998)

Because of the low availability of dietary carbohydrate and lipid as an energy source, growth of Japanese flounder depends highly on the protein. The optimum ratio of digestible energy to digestible crude protein for juvenile Japanese flounder is considered to be 8.6 kcal g^{-1} (Sato, 1998), which is lower than that for rainbow trout, *Oncorhynchus mykiss* (13.0) (Takeuchi *et al.*, 1978), and carp, *Cyprinus carpio* (9.7–11.6) (Takeuchi *et al.*, 1979).

With declining catches of sardines, the main source of fish-meal for aquaculture diets in Japan, alternative ingredients for dietary fish-meal have been the focus of attention for the last decade. Previous studies indicate that 20–40% of fish-meal protein can be replaced by feather meal (Kikuchi *et al.*, 1994b), 50% by defatted soybean meal (SBM) (Kikuchi *et al.*, 1994b), 20% by meat and bone-meal (MBM) (Kikuchi *et al.*, 1997), 60% by meat meal (Sato and Kikuchi, 1997) and 40% by corn (maize)-gluten meal (CGM) (Fig. 8.2; Kikuchi, 1999a) in the diet of juvenile Japanese flounder, and 20% by malt-protein flour for fingerlings

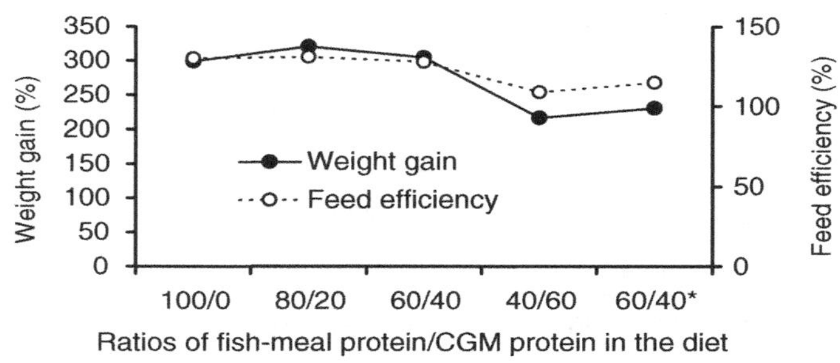

Fig. 8.2. Weight gain and feed efficiency of Japanese flounder fed diets with different ratios of fish-meal and corn (maize)-gluten meal (CGM) for 8 weeks (Kikuchi *et al.*, 1999a). *Crystalline amino acids were not supplied.

(Yamamoto *et al.*, 1995). However, most of these protein sources require supplementation of essential amino acids, such as methionine and lysine, to achieve growth comparable to that of the control diet (fish-meal as sole protein source) (Fig. 8.2).

Apparent protein digestibility of meat meal and SBM were comparable to that of fish-meal; however, that of feather meal, MBM, CGM and malt-protein flour was lower (Sato, 1998; Yamamoto *et al.*, 1998). All essential amino acids of CGM were poorly available for Japanese flounder (Yamamoto *et al.*, 1998). A combination of SBM (30% in the diet) and blood meal (10%) successfully replaced 47% of fish-meal protein in the diet without amino acid supplements (Kikuchi, 1999b). Furthermore, inclusion of blue mussel meat (5% dry matter) for an equal weight of SBM in this diet improved the growth of flounder markedly, mostly due to increased feed consumption. Stimulation of feeding with the mussel meat was demonstrated with a diet in which 3% fish-meal protein was replaced by mussel protein (Kikuchi, 1998). Blue mussel meat is an effective protein source that can replace more than 60% of fish-meal protein in the diet with incremental increases in growth and feed utilization and without supplemental amino acids (Kikuchi and Sakaguchi, 1997). Thus, a considerable amount of fish-meal protein can be replaced by several alternative protein sources in the diet of Japanese flounder, as summarized in Table 8.1. However, the results were obtained from a 6–8 week feeding trial with fish of less than 10 g initial body weight. Long-term rearing is required to determine the practical potential of these ingredients.

Lipids and fatty acids

Information on essential fatty acid (EFA) requirements of Japanese flounder is limited to those of the larval stage feeding an *Artemia* (Takeuchi, 1997). Growth

Table 8.1. Replacement levels of alternative protein sources for fish-meal in the diet of juvenile Japanese flounder.

Protein source	Replacement of dietary fish-meal (% protein)	Crude protein of test diets (%)	Amino acid supplements	Reference
Defatted soybean meal	50	49–51	Necessary	Kikuchi *et al.* (1994b)
Maize-gluten meal	40	54–57	Necessary	Kikuchi (1999a)
Silkworm pupa meal	20	45–48	Not necessary	K. Kikuchi (unpublished)
Feather meal	40	53–56	Necessary	Kikuchi *et al.* (1994a)
Meat meal	60	52–56	Not clear	Sato and Kikuchi (1997)
Meat and bone-meal	20	48–50	Not clear	Kikuchi *et al.* (1997)
Blue mussel meat	> 60	44–48	Not necessary	Kikuchi and Sakaguchi (1997)
Cuttlefish meal	100	51	Not necessary	Kikuchi *et al.* (1993)

of larval flounder was improved with an increasing dietary n-3 highly unsaturated fatty acid (HUFA) level in *Artemia* nauplii. However, no difference was found between docosahexaenoic acid (DHA) and eicosapentaenoic acid (EPA) (Furuita *et al.*, 1999). Positive effects of DHA enrichment on growth of larvae was also reported with microparticulate diets (Kanazawa *et al.*, 1994). Although arachidonic acid (AA) is considered to be a dietary essential for juvenile turbot (Takeuchi, 1997), the importance of dietary AA for Japanese flounder has not yet been clarified (Kanazawa *et al.*, 1994; Furuita *et al.*, 1998). Takeuchi (1998) recommended that dietary requirements of n-3 HUFA in Japanese flounder are 1.1–1.4% for fish larger than 20 g body weight, and 3% or more with at least 3% of DHA for larvae. Dietary supplement of EPA and DHA is believed to affect brain development of larval Japanese flounder (Furuita *et al.*, 1998).

Dietary supplementation of soybean and hen's-egg lecithin (1% in the diet) also accelerates the growth of 8 g initial body-weight flounder and the effect was superior for soybean lecithin (Kanazawa, 1993). Among phosphatidylcholine (PC), phospatidylethanolamine and phospatidylinositol fractions of soybean lecithin, only PC showed a positive effect on the growth of larval flounder (Kanazawa, 1993). Takeuchi (1998) recommended that the dietary supplement of PC should be more than 1% for larval Japanese flounder.

Availability of dietary lipids as an energy source for Japanese flounder has been examined recently. Juvenile flounder of 4–5 g initial body weight were fed 18 experimental diets with six protein levels (40, 45, 50, 55, 60 and 65%) and three lipid levels each (pollack liver oil, 10 to 28%) twice daily to apparent satiation for 8 weeks (Sato, 1998). The growth of fish depended mostly on dietary protein level and increasing level of dietary lipid did not produce positive effects at all protein levels (Fig. 8.3). The protein efficiency ratio (PER) was statistically

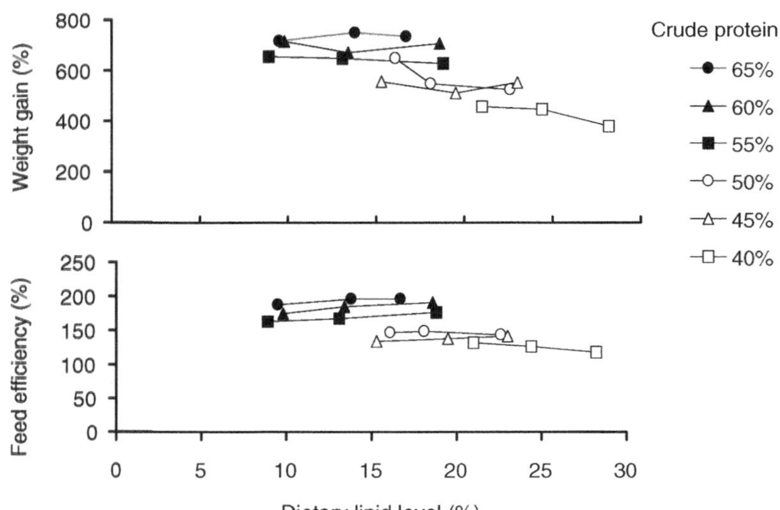

Fig. 8.3. Effect of dietary protein and lipid levels on weight gain and feed efficiency of Japanese flounder (Sato, 1998).

similar among dietary treatments regardless of dietary protein and lipid levels. Different results were obtained from more recent research in our laboratory with fish of 55 g and 245 g initial body weight. The PER of fish fed the diet containing the highest lipid level (20.3%) was significantly higher than that of fish fed the diet with the lowest lipid content (9.8%) in both feeding trials. However, a large quantity of dietary lipids may cause adverse effects on the health condition of the cultured fish by increasing the blood triglyceride level, liver weight and crude lipid content of the liver and muscle (Sato, 1998). Although more study is required to determine the optimum lipid level in the diet, based on the existing experimental findings, it is unlikely that dietary lipids serve as an effective energy source for the growth of Japanese flounder, as was shown in Atlantic salmon, *Salmo salar*, rainbow trout, *O. mykiss*, and yellowtail, *S. quinqueradiata*.

Carbohydrates

Although carbohydrate is one of the most important energy sources for domesticated farm animals, most fish species have a limited ability to utilize carbohydrates. The availability of dietary glucose, maltose, dextrin and potato starch was examined at a 26% inclusion level with juvenile Japanese flounder (Kikuchi *et al.*, 1998). Growth and feed efficiency (FE) were higher in fish fed dextrin and potato-starch diets, and tended to decrease with decreasing molecular weight of dietary carbohydrates. Dietary inclusion of glucose and maltose resulted in a marked increase of blood sugar levels immediately after feeding (Kikuchi *et al.*, 1998), as was shown in other fish species (Furuichi and Yone, 1982). The availability of dietary gelatinized (α) and raw (β) potato starch was compared. However, the difference has not been clarified. Use of carboxymethylcellulose as a binder is not recommended because it inhibited the protease activity of the flounder (Yamamoto and Akiyama, 1995).

The potential utilization of dietary carbohydrates as an energy source was examined with flounder of 3.5 g initial body weight and five diets containing different ratios of fish-meal to potato starch (Fig. 8.4) (Kikuchi *et al.*, 1998). Although the PER was not affected by dietary composition, the FE decreased with increasing potato starch in the diet. Furthermore, a significant growth reduction

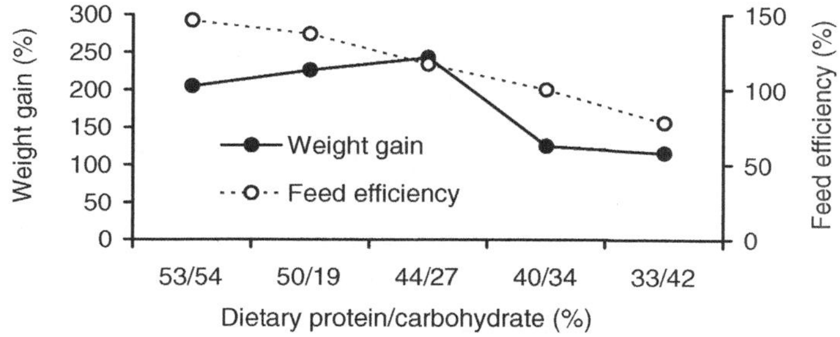

Fig. 8.4. Weight gain and feed efficiency of Japanese flounder fed diets with different ratios of protein to carbohydrate for 6 weeks (Kikuchi *et al.*, 1998).

was found in fish fed diets with less than 40% crude protein, with a concomitant decrease of protein utilization. Adverse effects of increasing carbohydrates (wheat starch) in the diet on growth and feed utilization are reported in other feeding experiments with fish of 50 g and 310–360 g initial body weight and the effect was more serious in larger fish (Kikuchi *et al.*, 1992). Based on these results, it is considered that Japanese flounder utilize dietary carbohydrates poorly as an energy source and there are no protein-sparing effects of dietary carbohydrate.

Practical Diets and Feeding Practices

Raw fish and moist pellets have been mainly used for grow-out of Japanese flounder. More than five companies produce commercial dry pellets for grow-out, although the total production is not reported. Commercial diets have protein levels ranging from 48 to 56% and lipid levels from 6 to 12% – a high-protein and low-lipid diet (Table 8.2). The protein level tends to increase and the lipid decrease with the growth of fish. Commercial diets contain 75–78% of animal protein, fish and krill meals, 12–13% of wheat and potato starch and maize powder and 10–12% of other constituents. The latter include vitamin and mineral mixtures, pollack liver oil, beer yeast and soybean lecithin. To our knowledge, plant-protein sources as alternatives to fish-meal have not been used for the commercial diet yet. Both pelleted and extruded diets are available for the culture; however, there is a strong tendency to use extruded diets.

Eight to ten pellet sizes ranging from 1 to 18 mm in diameter are used, depending on fish size. Fish are fed diets by hand one to five times a day, each to satiation, at almost all farms. Table 8.3 shows the recommended feeding rate and

Table 8.2. Proximate composition of commercial dry pellets for grow-out of Japanese flounder (%).

Producer	Crude protein*	Crude lipid*	Crude fibre	Ash
A	> 48–52	> 8–10	< 2	< 15
B	> 47–50	> 8–10	< 2.5	< 15
C	> 50–56	> 6–12	< 4	< 17

* Minimum values guaranteed.

Table 8.3. Recommended pellet size, feeding rate and frequency per day of dry pellets for the culture of Japanese flounder.

Producer	February	April	June	August	October	December	February
Temperature (°C)*	15.2	17.7	21.5	27.2	22.7	18.0	14.9
Fish size (g)*	6.4	36.0	94.0	231.0	452.0	686.0	835.0
Pellet size (mm)	2–3	4–5	5–6	8–9	10–11	12–13	16–18
Feeding rate (%)	5.8	1.3	2	1.8	1.3	0.4	0.2
Frequency	5	3	3	3	2	1–2	1–2

* Average temperature and estimated growth.

feeding frequency per day with the growth of fish. The feeding rate increases with the increase of water temperature up to 25°C and decreases thereafter. The fish grows from juvenile to 900 g in 14 months with commercial diets.

References

Cowey, C.B., Adron, J.W. and Blair, A. (1970) Studies on the nutrition of marine flatfish. The essential amino acid requirements of plaice and sole. *Journal of the Marine Biological Association of the United Kingdom* 50, 87–95.

Forster, I. and Ogata, H.Y. (1998) Lysine requirement of juvenile Japanese flounder *Paralichthys olivaceus* and juvenile red sea bream *Pagrus major*. *Aquaculture* 161, 131–142.

Furuichi, M. and Yone, Y. (1982) Availability of carbohydrate in nutrition of carp and red sea bream. *Bulletin of Japanese Society of Scientific Fisheries* 48, 945–948.

Furuita, H., Takeuchi, T. and Uematsu, K. (1998) Effect of eicosapentaenoic and docosahexaenoic acids on growth, survival and brain development of larval Japanese flounder (*Paralichthys olivaceus*). *Aquaculture* 161, 269–279.

Furuita, H., Konishi, K. and Takeuchi, T. (1999) Effect of different levels of eicosapentaenoic and docosahexaenoic acids in *Artemia* nauplii on growth, survival and salinity tolerance of larvae of the Japanese flounder, *Paralichthys olivaceus*. *Aquaculture* 170, 59–69.

Kanazawa, A. (1990) Nutrition and feed in *hirame* culture. *Suisanzoshoku* 38, 398–399 (in Japanese).

Kanazawa, A. (1993) Essential phospholipids of fish and crustaceans. In: Kaushik, S.J. and Luquet, P. (eds) *Fish Nutrition in Practice*. INRA, Paris, pp. 519–530.

Kanazawa, A., Matsumoto, H., Teshima, S., Koshio, S. and Osada, T. (1994) *Abstracts for the Meeting of the Japanese Society of Fisheries Science*, 1–4 October, Mie. 74 pp. (in Japanese).

Kikuchi, K. (1998) Blue mussels in the diet of juvenile Japanese flounder. *UJNR Technical Report* 26, 269–274.

Kikuchi, K. (1999a) Partial replacement of fish meal with corn gluten meal in diets for Japanese flounder *Paralichthys olivaceus*. *Journal of the World Aquaculture Society* 30, 357–363.

Kikuchi, K. (1999b) Use of defatted soybean meal as a substitute for fish meal in diets of Japanese flounder (*Paralichthys olivaceus*). *Aquaculture* 179, 3–11.

Kikuchi, K. and Sakaguchi, I. (1997) Blue mussel as an ingredient in the diet of juvenile Japanese flounder. *Fisheries Science* 63, 837–838.

Kikuchi, K., Honda, H. and Kiyono, M. (1992) Effect of dietary protein level on growth and body composition of Japanese flounder, *Paralichthys olivaceus*. *Suisanzoshoku* 40, 335–340 (in Japanese, with English abstract).

Kikuchi, K., Honda, H., Kiyono, M. and Miyazono, I. (1993) Total replacement of fish meal with other protein sources in the diet of Japanese flounder, *Paralichthys olivaceus*. *Suisanzoshoku* 41, 345–351.

Kikuchi, K., Furuta, T. and Honda, H. (1994a) Utilization of feather meal as a protein source in the diet of juvenile Japanese flounder. *Fisheries Science* 60, 203–306.

Kikuchi, K., Furuta, T. and Honda, H. (1994b) Utilization of soybean meal as a protein source in the diet of juvenile Japanese flounder, *Paralichthys olivaceus*. *Suisanzoshoku* 42, 601–604.

Kikuchi, K., Sato, T., Furuta, T., Sakaguchi, I. and Deguchi, Y. (1997) Utilization of meat and bone meal as a protein source in the diet of juvenile Japanese flounder. *Fisheries Science* 63, 29–32.

Kikuchi, K., Sato, T. and Deguchi, Y. (1998) Effect of dietary carbohydrates on the growth of juvenile Japanese flounder. *Suisanzoshoku* 46, 541–546 (in Japanese, with English abstract).

Sato, T. (1998) Development of formulated feeds for juvenile Japanese flounder. PhD thesis, Tokyo University of Fisheries, Tokyo, Japan (in Japanese).

Sato, T. and Kikuchi, K. (1997) Meat meal as a protein source in the diet of juvenile Japanese flounder. *Fisheries Science* 63, 877–880.

Takeuchi, T. (1997) Essential fatty acids requirements of aquatic animals with emphasis on fish larvae and fingerlings. *Reviews in Fisheries Science* 5, 1–25.

Takeuchi, T. (1998) Nutritional requirements of larval and juvenile Japanese flounder. In: *Text Book of Basic and Theoretical Course in Sea Farming – XII*. Fisheries Agency and Japan Sea Farming Association, Tokyo, pp. 1–23 (in Japanese).

Takeuchi, T., Yokoyama, M., Watanabe, T. and Ogino, C. (1978) Optimum ratio of dietary energy to protein for rainbow trout. *Bulletin of Japanese Society of Scientific Fisheries* 44, 729–732 (in Japanese, with English abstract).

Takeuchi, T., Watanabe, T. and Ogino, C. (1979) Optimum ratio of dietary energy to protein for carp. *Bulletin of Japanese Society of Scientific Fisheries* 45, 983–987 (in Japanese, with English abstract).

Yamamoto, T. and Akiyama, T. (1995) Effect of carboxymethylcellulose, α-starch and wheat gluten incorporated in diets as binders on growth, feed efficiency, and digestive energy activity of fingerling Japanese flounder. *Fisheries Science* 61, 309–313.

Yamamoto, T., Unuma, T. and Akiyama, T. (1995) Utilization of malt protein flour in fingerling Japanese flounder diets. *Bulletin of National Research Institute of Aquaculture* 24, 33–42.

Yamamoto, T., Unuma, T. and Akiyama, T. (1998) Apparent availabilities of amino acids from several protein sources for fingerling Japanese flounder. *Bulletin of National Research Institute of Aquaculture* 27, 27–35.

North American Flounders

9

Harry V. Daniels[1] and Margie L. Gallagher[2]
[1]*Department of Zoology, North Carolina State University, 207 Research Station Road, Plymouth, NC 27962, USA;* [2]*Institute for Coastal and Marine Resources, East Carolina University, 125 Ragsdale, Greenville, NC 27858-4353, USA*

Introduction

Flounder culture in North America is in its infancy. Meaningful research on native flounders only began during the early 1990s. This research has focused on three flat-fish with potential for aquaculture in North America: summer flounder (*Paralichthys dentatus*), southern flounder (*Paralichthys lethostigma*) and winter flounder (*Pseudopleuronectes americanus*) (Daniels *et al.*, 1996; Rivara and Bengtson, 1999; Howell and Litvak, 2000). These fish are found primarily on the Atlantic coast of the USA and have different primary natural ranges that overlap slightly at their extremes (Fig. 9.1). Some fish of each species can be found beyond these ranges but they are found in greatest abundance within the primary natural ranges. Winter flounder are found north of Cape Cod to Labrador, Canada (Buckley, 1989). Summer flounder are found from Cape Cod to Cape Hatteras (Rogers and van den Avyle, 1983). Southern flounder are found from Cape Hatteras along the coast to Florida and along the northern coastline of the Gulf of Mexico (Reagan and Wingo, 1985). Winter flounder are right-eyed flounder belonging to the family Pleuronectidae. Summer and southern flounder are left-eyed flounder belonging to the family Bothidae. In the wild, winter flounder rarely exceed 2.25 kg, while summer and southern flounders grow to a size of about 12–15 kg.

The different natural ranges and the corresponding differences in environmental conditions will probably have a profound effect on the environmental requirements for the culture of these flounders. However, because these fish have only recently been investigated for their aquaculture potential, culture techniques specific to each fish have not yet been established. Instead, much of the technology used during the hatchery and grow-out phases for one fish is also used, with few modifications, for the other two flat-fish. But culture conditions such as light intensity, water depth and stocking density appear to have a

Fig. 9.1. Primary natural ranges for three North American flounders.

greater effect on the success of flounder culture than for other marine finfish. For example, researchers have commonly observed that flounder stocked at low densities often do not feed actively and may lose weight or eventually die. Therefore, stocking density is a fundamental variable with greater potential importance for the success of a grow-out operation than the nutritional value of the feed. Because culture conditions have such an effect on feeding, and feeding is an integral part of nutrition, a general description of common culture practices is provided to familiarize the reader with flounder culture. As specific culture techniques are developed for each of the North American flounders, future modifications of nutritional requirements may be necessary to accommodate changes in feeding behaviour and improved fish growth brought about by better culture conditions.

Larviculture techniques for flounder are similar to those used for other marine finfish. North American flounders spawn small eggs (0.7–1.1 mm in diameter) during the late autumn to early spring. As with most marine fish, yolk-sac larvae have small mouth openings and require live feed at first feeding. Therefore, appropriately sized live feeds must be cultured and maintained in a flounder hatchery. Hatchery managers prefer rotifers (*Brachionus plicatilis*) as first food for larvae because they are easy to culture, swim slowly and are readily eaten by larval fish. Although nutritionally very poor, rotifers can easily be enriched on emulsified oils (Rainuzzo *et al.*, 1994). Based on the size requirements, rotifers can be filtered to select for the appropriate size for the fish. Southern and summer flounder are capable of consuming rotifers within a size range of 50–150 μm at first feeding. Winter flounder eat slightly smaller-sized rotifers because of their smaller mouths. Larval flounder consume rotifers for the first 2–3 weeks and then begin consuming *Artemia* nauplii. Postmetamorphic flounder are weaned on to dry feeds at around 50–60 days posthatch and remain on these extruded diets throughout the grow-out cycle. Most foodfish are cultured in shallow, indoor or covered tanks with either recirculating or flow-through water maintained at an exchange rate of 100–200% h^{-1}.

Nutrient Requirements

Protein and amino acids

Although the nutritional requirements of other flat-fish, such as Japanese flounder (*Paralichthys olivaceus*) have been extensively investigated, very little is known about the nutritional requirements of North American flat-fish (Bengtson, 1991; Howell and Litvak, 2000). A few studies have been done on southern and summer flounder (Bengtson, 1991; Daniels and Gallagher, 2000) but there is little information available on the nutritional requirements of winter flounder (Hebb *et al.*, 1997).

One of the unique aspects of the flounder nutrition work described here is that several studies have sought to link physiological and biochemical assessments to nutrition in addition to the more traditional measurements of growth and survival. For example, Bisbal and Bengtson (1991) used biochemical methods to assess the overall nutritional status of summer flounder. They reported that under normal conditions the ribonucleic : deoxyribonucleic acid (RNA : DNA) ratio rose as the larval fish advanced from 16 days posthatch to 60 days. But, when the animals were starved, the RNA : DNA ratio declined markedly from the basal level during the starvation period.

Generally, fish have higher protein requirements on a gram protein kJ^{-1} (or $kcal^{-1}$) basis compared with land animals. This phenomenon is due in part to the fact that fish do not have to maintain a constant body temperature and they exert less energy to maintain position and move in water (Lovell, 1989). This effect is apparently exacerbated in flat-fish because of their sedentary nature.

Although studies investigating protein in diets for various species of juvenile flat-fish are variable and often contradictory, it is clear that protein requirements are higher for flat-fish than for other finfish. Guillaume *et al.* (1991) reported that the protein requirement of turbot (*Scophthalmus maximus*), sole (*Solea solea*) and plaice (*Pleuronectes platessa*) was between 57 and 60% of the diet. Aksnes *et al.* (1996) predicted a 74% protein requirement for halibut (*Hippoglossus hippoglossus*). Danielssen and Hjertnes (1991) found that turbot from the same egg batch grew best at protein levels in excess of 45%. However, significant growth differences ($P < 0.001$) of comparable magnitude were noted between egg batches from different parents when the dietary protein was the same, indicating that parentage is at least as important as dietary protein.

Recent work by Daniels and Gallagher (2000) indicated that growth of summer flounder was significantly better with 56% protein in the diet compared with 50, 45, 40 and 35% protein. In addition, red blood-cell counts were significantly higher in fish fed 52% or 56% protein, and haematocrits were highly correlated ($P < 0.01$) with dietary protein. Juvenile southern flounder fed isocaloric diets showed significantly higher growth and lower feed conversion ratios ($P < 0.05$) when fed diets containing at least 48%. These fish grew 65–125% larger than those fed less than 48% protein. Based on these results, a dietary protein level of 48% appears to be sufficient for optimum growth of juvenile southern flounder. Winter flounder also appear to have a minimum protein requirement of 40–45% for optimum growth of juvenile fish (Hebb *et al.*, 1997). In this study, highest specific growth rate, feed conversion and protein efficiency were obtained with a diet containing 45% protein and 10% lipid. Therefore, North American flounders, as a group, have a demonstrated requirement for high levels of dietary protein.

As with many fish, the major source of protein for studies with flat-fish has been fish-meal. Although there are currently no reports on the replacement of fish-meal in diets for southern or summer flounder, Kikuchi and co-workers have demonstrated that Japanese flounder, a congener of the southern and summer flounder, can effectively use soybean meal (Kikuchi, 1999a,b; Kikuchi *et al.*, 1994a), feather meal (Kikuchi *et al.*, 1994b) and meat and bone-meal (Kikuchi *et al.*, 1997) in diets; however, none of these sources can totally replace fish-meal (Kikuchi *et al.*, 1993).

Lipids and fatty acids

Lipid requirements of larval flat-fish have received considerable attention, because of the roles that essential fatty acids play in both early survival and development (Kanazawa, 1997; Venizelos and Benetti, 1999). Poor nutrition during the rotifer-feeding stage has also been linked to incomplete or abnormal pigmentation of juvenile fish. Specifically, insufficient intake of docosahexaenoic acid (DHA), 22:6n-3, during the early larval stages is a major cause of pigmentation problems. Minimum requirements for DHA have not been established for North American flounders. Hatchery managers typically use commercially available enrichment products with a high DHA content to ensure adequate nutrition. Abnormal pigmentation does not appear to affect flesh quality but, instead, is a marketing problem because of the unusual appearance of the whole fish. Because the market value of fillets is considerably lower than that of whole fresh fish, abnormally pigmented flounder represent a loss of potential income for the producer. Optimum larval nutrition reduces abnormal pigmentation to below 10–20% of the population. The residual amount of abnormal pigmentation observed, even with high DHA levels, has led some researchers to conclude that other factors, such as genetics, also contribute to abnormal pigmentation (Kanazawa, 1997). Anecdotal evidence of abnormally pigmented flounder caught from the wild suggests that abnormal pigmentation occurs with some frequency in natural populations. It is likely that these fish are more susceptible to predation because of their inability to adequately camouflage themselves in the sand or mud.

Bisbal and Bengtson (1991) found that fatty acid requirements for larval summer flounder during metamorphosis could be met by brine shrimp (*Artemia*) enriched with eicosapentaenoic acid (EPA), 20:5n-3, and DHA or with EPA alone. Larvae that did not receive either EPA and DHA or EPA alone had a high incidence of albinism and incomplete eye migration. Additionally, Rainuzzo *et al.* (1997) demonstrated that the type of algae on which rotifers were raised could influence the number of flounder larvae completing metamorphosis, but that enrichment with highly unsaturated fatty acids (HUFA) could overcome this problem. These results are similar to those reported for winter flounder fed commercially available diets containing different levels of DHA (Blair *et al.*, 1998).

Alves *et al.* (1999) demonstrated that the addition of algae (*Tetraselmis suecica* and *Isochrysis galbana*) increased survival of summer-flounder larvae. The authors suggested that improved survival was not due to the direct nutritional value of the algae to the larvae, but rather to a number of possible indirect factors, such as maintenance of the nutritional value of rotifers to the larvae, increased feeding due to turbidity or reduction in bacterial load. Devresse *et al.* (1994) reported improved pigmentation in Japanese flounder (*P. olivaceus*) and European turbot (*S. maximus*) larvae fed rotifers and *Artemia* enriched with DHA. These authors reported that the DHA : EPA ratio seemed to be the most important factor in proper pigmentation. McEvoy *et al.* (1998) found that live foods enriched with EPA enhanced normal pigmentation in turbot and halibut (*H. hippoglossus*) but that enrichment with arachidonic acid, 20:4n-6, led to malpigmentation.

Recent work in our laboratory using enriched rotifers and *Artemia* demonstrated that southern-flounder larvae require n-3 HUFA for good growth and survival. Larvae fed rotifers lacking n-3 HUFA at first feeding did not survive to 26 days posthatch and did not develop past the 'B' stage (van Maaren and Daniels, 2000). In addition, EPA appeared to accumulate in the tissues. The addition of n-6 fatty acids from maize oil did not improve growth, nor did the addition of arachidonic acid. However, the role of arachidonic acid in these fish was unclear since unenriched rotifers contained significant amounts of the fatty acid. HUFA are also required for juvenile flounder. Whalen *et al.* (1999) reported that yellowtail flounder (*Pleuronectes ferrugineus*) required approximately 2.5% HUFA in the diet. Deficient diets produced fish with fatty livers and inferior growth.

Practical Diets and Feeding Practices

At present, practical diet formulations for flounder should contain 45–50% protein, 10–15% lipid and less than 10% ash, with a caloric density of no greater than 3.0 kcal g^{-1}, which is somewhat less in this sedentary flat-fish compared with other species (Table 9.1). Information on brood-stock diets does not exist. Most producers and researchers feed live mullet or frozen fish, shrimp and squid to brood-stock and supplement these feeds with pelleted food when possible. Daily feeding rates should be adjusted when using semimoist food or raw fish (Table 9.2). Flounder in a grow-out system are fed several times per day, but brood-stock are usually fed only once daily or on alternate days.

Although Kikuchi and co-workers (Kikuchi *et al.*, 1994a,b, 1997; Kikuchi, 1999a,b) have demonstrated the suitability of replacing fish-meal with various alternative sources of protein (i.e. defatted soybean meal, maize-gluten meal, meat and bone-meal) for Japanese flounder, the ability of these ingredients to meet the needs of southern, summer and winter flounder has not been investigated. Therefore, it is prudent to feed fish-meal-based feeds at this time.

Practical methods for weaning hatchery-reared flounder from a diet of live feed (*Artemia*) on to dry formulated diets have not been well established (Daniels and Hodson, 1999). The weaning success of flat-fish is unpredictable.

Table 9.1. General feed formulation for North American flounder.

Ingredient	% of dry weight
Fish-meal	50–60
Ground winter wheat	20–30
Menhaden oil	5–6
Vitamin mix	1.5
Mineral mix	1.5
Ascorbate phosphate	0.5
Choline chloride	0.2
Proximate composition	
Protein	45–50
Lipid	10–15
Ash	10

Table 9.2. Feeding rates for various-sized flounder fed different feeds.

Fish size		Daily feeding rate (%)*		
Length (cm)	Weight (g)	Raw fish	Moist pellet	Dry pellet
4–5	1–2	60	24	15
10–15	8–32	20	8	5
20–25	90–180	10–15	4–6	2–4
> 25	> 150	3–5	1–4	1

* Weight of feed/weight of fish.

Survival rates during weaning are highly variable and can range from 45% to 90% between batches (Lee and Litvak, 1996a,b). Several factors appear to affect weaning success: fish age, feeding behaviour, duration of the change-over period when both live food and formulated diets are fed and the properties of the formulated diets, such as colour, texture and buoyancy. As fish grow and consume more food, the cost of producing brine shrimp becomes prohibitive. To reduce production costs, fish should be weaned as early as possible.

Artemia are a high-cost food for larval flounder but are necessary for good growth and development. Techniques for weaning premetamorphic fish to dry feeds have met with limited success. As a result, most weaning programmes are started when fish have completed metamorphosis and have reached an age of 50–55 days posthatch. Many commercial weaning programmes employ a gradual reduction of live food until dry diets are the sole feed. This gradual weaning is usually completed in 2–3 weeks. Despite the prolonged weaning period, the mortality of fish that refuse dry feed or from cannibalism between cohorts can reach 20–30%. Frequent grading of fish is often employed to minimize mortality from cannibalism (Ward and Bengtson, 1999), but this practice is stressful to fish and can lead to some mortality from excessive handling.

Once fish are weaned on to dry diets, stocking densities must be maintained at a sufficient level to encourage active feeding (Table 9.3). Young fish should be fed to satiation up to six times per day, while fish larger than 10 g can be fed four times per day. Flounder readily accept dry diets so semimoist feeds are not needed for grow-out. Contrary to popular belief, flounder will come to the surface to feed on floating (extruded) pellets. The use of floating feed offers several advantages to the culturist. Feeding activity can be more easily observed than with sinking feeds and uneaten pellets are quickly removed from the tank either with a net or in the overflow of the tank water. Flounder generally only eat sinking feeds as they are in the process of falling to the bottom. Pellets that reach the tank bottom are normally not consumed and may get trapped under the fish, making it more difficult to remove uneaten feed from the tank. Therefore, flounder diets for

Table 9.3. Example of stocking densities used to maintain active feeding and optimum growth of three North American flounders.*

	Stocking densities	
Body weight (g)	ind. m^{-2}	kg m^2
1.5	1600	2.4
10	400	4
50	200	10
100	150	15
250	100	25
500	60	30

* Summer flounder (*Paralichthys dentatus*); southern flounder (*Paralichthys lethostigma*); winter flounder (*Pseudopleuronectes americanus*).

grow-out should be extruded so that they float on the water surface or sink very slowly to the tank bottom.

References

Aksnes, A., Hjertnes, T. and Opstvedt, J. (1996) Effect of dietary protein level on growth and carcass composition in Atlantic halibut (*Hippoglossus hippoglossus* L.). *Aquaculture* 145, 225–223.

Alves, D., Specker, J.L. and Bengtson, D.A. (1999) Investigations into the causes of early larval mortality in cultured summer flounder (*Paralichthys dentatus* L.). *Aquaculture* 176, 155–172.

Bengtson, D.A. (1991) Aquaculture of summer flounder (*Paralichthys dentatus*), status of knowledge, current research and future research priorities. *Aquaculture* 176, 39–49.

Bisbal, G.A. and Bengtson, D.A. (1991) Effect of dietary (n-3) HUFA enrichment on survival and growth of summer flounder, *Paralichthys dentatus*, larvae. In: Lavens, P., Sorgeloos, P., Japsers, E. and Ollevier, F. (eds) *LARVI 91-Fish and Crustacean Larviculture Symposium*. Special Publication No. 15, European Aquaculture Society, Ghent, Belgium, pp. 56–58.

Blair, T., Batt, J., Melanson, R., Kirk, S. and Castell, J. (1998) Evaluation of several commercial enrichment media for enhancing the nutritional value of rotifers fed to winter flounder (*Pleuronectes americanus*). *Bulletin of the Aquaculture Association of Canada* 98(4), 35–37.

Buckley, J. (1989) *Species Profiles: Life Histories and Environmental Requirements of Coastal Fishes and Invertebrates (North Atlantic) – Winter Flounder*. US Fish Wildlife Service Biological Report 82(11.87), Washington DC, US Army Corps of Engineers, TR EL-82-4, Vicksburg MS, 12 pp.

Daniels, H.V. and Gallagher, M.L. (2000) Effect of dietary protein level on growth and blood parameters in summer flounder, *Paralichthys dentatus*. *Journal of Applied Aquaculture* 10, 45–52.

Daniels, H.V. and Hodson, R.G. (1999) Weaning success of southern flounder juveniles: effect of changeover period and diet type on growth and survival. *North American Journal of Aquaculture* 61, 47–51.

Daniels, H.V., Berlinsky, D.L., Hodson, R.G. and Sullivan, C.V. (1996) Effects of stocking density, salinity and light intensity on growth and survival of southern flounder *Paralichthys lethostigma* larvae. *Journal of the World Aquaculture Society* 27, 153–159.

Danielssen, D.S. and Hjertnes, T. (1991) Effect of dietary protein levels in diets for turbot (*Scophthalmus maximus* L.) to market size. In: Kaushik, S.J. and Luguet, P. (eds) *Fish Nutrition in Practice*. INRA Editions, Versailles, France, pp. 89–96.

Devresse, B., Leger, P., Sorgeloos, P., Murata, O., Nasu, T., Ikeda, S., Rainuzzo, J.R., Reitan, K.I., Kjorsvik, E. and Olsen, Y. (1994) Improvement of flat fish pigmentation through the use of DHA-enriched rotifers and *Artemia*. *Aquaculture* 124, 287–288 [abstract].

Guillaume, J., Coustans, M., Metailler, R., Person-Le Ruyet, J. and Robin, J. (1991) Flatfish, turbot, sole and plaice. In: Wilson, R.P. (ed.) *Handbook of Nutrient Requirements of Finfish*. CRC Press, Boca Raton, Florida, pp. 77–82.

Hebb, C.D., Castell, J.D., Anderson, D.M. and Batt, J. (1997) Nutritional studies on growth and protein utilization during the juvenile stage of winter flounder (*Pleuronectes americanus*). *Bulletin of the Aquaculture Association of Canada* 97(2), 45–47.

Howell, W.H. and Litvak, M.K. (2000) Winter flounder culture. In: Stickney, R. (ed.) *Encyclopedia of Aquaculture*. John Wiley and Sons, New York, pp. 998–1005.

Kanazawa, A. (1997) Importance of dietary lipids in flatfish. In: Howell, W.H., Keller, B.J., Park, P.K., McVey, J.P., Takayanagi, K. and Uekita, Y. (eds) *Nutrition and Technical Development of Aquaculture. Proceedings of the Twenty-sixth US–Japan Aquaculture Symposium. 16–18 September 1997, Durham, New Hampshire.* UJNR Technical Report No. 26, UNH Sea Grant, Durham, New Hampshire, USA, pp. 181–186.

Kikuchi, K. (1999a) Partial replacement of fish meal with corn gluten meal in diets for Japanese flounder *Paralichthys olivaceus. Journal of the World Aquaculture Society* 30, 357–363.

Kikuchi, K. (1999b) Use of defatted soybean meal as a substitute for fish meal in diets of Japanese flounder (*Paralichthys olivaceus*). *Aquaculture* 179, 3–11.

Kikuchi, K., Honda, H., Kiyono, M. and Miyazono, I. (1993) Total replacement of fish meal with other protein sources in the diet of Japanese flounder, *Paralichthys olivaceus. Suisanzoshoku* 41, 345–351.

Kikuchi, K., Furuta, T. and Honda, H. (1994a) Utilization of soybean meal as a protein source in the diet of juvenile Japanese flounder *Paralichthys olivaceus. Suisanzoshoku* 42, 601–604.

Kikuchi, K., Furuta, T. and Honda, H. (1994b) Utilization of feather meal as a protein source in the diet of juvenile Japanese flounder. *Fisheries Science* 60, 203–206.

Kikuchi, K., Sato, T., Furuta, T., Sakaguchi, I. and Deguchi, Y. (1997) Use of meat and bone meal as a protein source in the diet of juvenile Japanese flounder. *Fisheries Science* 63, 29–32.

Lee, G.W.Y. and Litvak, M.K. (1996a) Weaning of wild young-of-the-year winter flounder (*Pleuronectes americanus*: Walbaum) on a dry diet: effects on growth, survival and feed efficiency ratios. *Journal of the World Aquaculture Society* 27, 30–39.

Lee, G.W.Y. and Litvak, M.K. (1996b) Weaning of metamorphosed winter flounder (*Pleuronectes americanus*) reared in the laboratory: comparison of two commercial artificial diets on growth, survival and conversion efficiency. *Aquaculture* 144, 251–263.

Lovell, R.T. (1989) *Nutrition and Feeding of Fish.* Van Nostrand Reinhold, New York, 260 pp.

McEvoy, L., Estevez, A., Bell, J.G., Shields, R.J., Gara, B. and Sargent, J.R. (1998) Influence of dietary levels of eicosapentaenoic and arachidonic acids on the pigmentation success of turbot (*Scophthalmus maximus* L.) and halibut (*Hippoglossus hippoglossus* L.). *Bulletin of the Aquaculture Association of Canada* 98, 17–20.

Rainuzzo, J.R., Reitan, K.I., Jorgensen, J. and Olsen, Y. (1994) Lipid composition in turbot larvae fed live feed cultured by emulsions of different lipid classes. *Comparative Biochemistry and Physiology* 107A, 699–710.

Rainuzzo, J.R., Reitan, K.I. and Olsen, Y. (1997) The significance of lipid at early stages of marine fish: a review. *Aquaculture* 155, 103–115.

Reagan, R.E. Jr and Wingo, W.M. (1985) *Species Profiles: Life Histories and Environmental Requirements of Coastal Fishes and Invertebrates (Gulf of Mexico) – Southern Flounder.* Washington DC, Fish and Wildlife Service/OBS-82/11.30, US Fish and Wildlife Service, TR EL-82-4.9, US Army Corps of Engineers, Vicksburg MS, 14 pp.

Rivara, G. and Bengtson, D.A. (1999) Summer flounder culture in the northeast: update on recent research and industry status. *Journal of Shellfish Research* 18, 276–285.

Rogers, S.G. and van den Avyle, M.J. (1983) *Species Profiles: Life Histories and Environmental Requirements of Coastal Fishes and Invertebrates (South Atlantic) – Summer Flounder.* Fish and Wildlife Service/OBS-82/11.15, US Fish and Wildlife Service, Washington DC, TR EL-82-4, US Army Corps of Engineers, Vicksburg, Mississippi,, 14 pp.

van Maaren, C.C. and Daniels, H.V. (2000) A practical guide to the morphological development of southern flounder, *Paralichthys lethostigma*, from hatch through metamorphosis. *Journal of Applied Aquaculture* 10(2), 1–9.

Venizelos, A. and Benetti, D.D. (1999) Pigment abnormalities in flatfish. *Aquaculture* 176, 181–188.

Ward, F.A. and Bengtson, D.A. (1999) Partitioning of fish and diet selection as methods for the reduction of cannibalism in *Paralichthys dentatus* larviculture. *Journal of the World Aquaculture Society* 30, 302–310.

Whalen, K.S., Brown, J.A., Parrish, C.C., Lall, S.P. and Goddard, J.S. (1999) Effect of dietary n-3 HUFA on growth and body composition of juvenile yellowtail flounder (*Pleuronectes ferrugineus*). *Bulletin of the Aquaculture Association of Canada* 98, 21–22.

Yellowtail, *Seriola quinqueradiata*

10

Toshiro Masumoto
Laboratory of Fish Nutrition, Kochi University, Nankoku, Kochi, 783-8502, Japan

Introduction

Yellowtail (*Seriola quinqueradiata*) is one of the economically important finfish for aquaculture in Japan. Commercial production of yellowtail began in the 1950s, and production has expanded rapidly since the 1960s. The total production of yellowtail in Japan was about 140,000 tons in 1997, representing about 70% of the total production of aquaculture marine finfish in Japan. Fish are normally cultured in floating net cages in coastal zones. Since artificial seed production has not been used for yellowtail, fingerlings of yellowtail, known as *mojyako* in Japanese, are caught in the wild while they are schooling along coastal currents with drifting objects (e.g. algae). Those fingerlings (< 10 g) are caught in spring and sold to growers when the fish has grown to 50–100 g in body weight. The duration of the growth period depends on the desired market size of the fish. Most yellowtail growers target a market size of about 2–5 kg, while some growers even raise the fish to a weight of 7–8 kg. A typical growth pattern and target size for the market are shown in Fig. 10.1.

Commercial yellowtail culture is done in sea cages made of nylon netting or metal. The size and number of cages vary. A relatively small-scale production site may have five cages of 10 m × 10 m × 8 m in size, while a relatively larger-scale production site often has more than 20 cages of 18 m × 22 m × 8 m or 12 m × 12 m × 12 m in size (H. Matsumoto, Kochi, 2000, personal communication). The optimum water temperature for yellowtail to grow is 20–29°C. Food consumption of yellowtail is drastically reduced at water temperatures below 17°C, particularly when a dry diet is fed. Optimum water salinity for yellowtail is 29.8 to 36.3 p.p.t. (with a chloride concentration of 16.5–20.1%) (Inoue, 1983). A dissolved oxygen (DO) level of more than 5.7 mg l^{-1} is desirable for good growth. Growth reduction and abnormal behaviour have been reported when the DO level becomes lower than 4.3–5.7 and 2.9–4.3 mg l^{-1}, respectively. The stocking density of yellowtail depends on cage-site conditions, such as temperature, DO, mesh size and water exchange rate in the cage. The

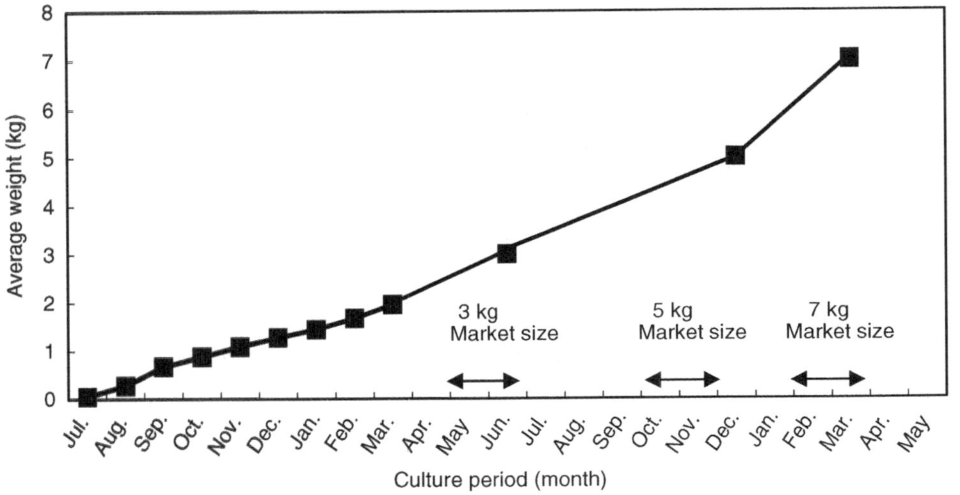

Fig. 10.1. Typical growth curve of farmed yellowtail.

recommended densities for various growth stages are 6.9, 5.6, 8.0 and 7.0 kg m^{-3} for individual fish body weight of 0.05, 0.2, 0.8 and 3.5 kg, respectively. However, in practical situations, the fish are often kept at densities of more than twice these densities.

Nutrient Requirements

Protein and amino acids

The protein requirement of fish is affected by several factors, such as growth rate, nutritional quality of the ingredients (protein digestibility and amino acid composition) and digestible energy content of the diet. Providing the optimum dietary energy levels for animals is important because a diet deficient in energy will result in the use of dietary protein for energy rather than for protein synthesis. On the other hand, excess energy in the diet may result in decreased nutrient intake by the animal or excessive fat deposition in the body. Therefore, a balanced energy-to-protein ratio (E/P) is important. Table 10.1 shows E/P values for yellowtail derived from reported optimum growth rates under various temperature and growth-stage conditions. These values indicate that a crude protein-to-gross energy ratio (CP/GE) of 20–26 g kJ^{-1} ensures good growth, with a feed conversion ratio (FCR) of approximately 1.0 for yellowtail at a high temperature, while the value tends to be lower (about 20) in a cooler water temperature. Quantitative requirements of essential amino acid requirements for methionine, lysine, arginine and histidine are given in Table 10.2. These values were obtained by feeding graded levels of the respective crystalline amino acids in

Table 10.1. Estimated values of crude protein : gross energy (CP : GE) ratio based on the best growth from published results (modified from Talbot *et al.*, 1999b). Energy equivalents (kJ g^{-1} of protein: 23.4, lipid: 39.8 and nitrogen-free extract: 17.2 were used (Cho *et al.*, 1982).

Reference	Feed type	Water temp.* (°C)	Rearing period (days)	Start–end BW (g)	SGR† (% bw day^{-1})	FCR†	PER§	CP (% in diet)	CL (% in diet)	GE (mJ kg^{-1})	CP : GE (g mJ^{-1})
Start feeding before summer											
Takeuchi *et al.* (1992)	EP	19–22	30	1.8–16.4	7.4	0.87	2.3	53.6	15.6	20.7	25.9
Takeuchi *et al.* (1992)	EP	20–24	35	3.7–35.9	6.5	1.03	1.9	55.2	10.3	18.6	29.7
Masumoto *et al.* (1997)	SMP	23–26	40	12–120	5.6	1.08	1.8	52.0	16.0	22.0	23.7
Watanabe *et al.* (1998)	EP	24	45	13.1–158	5.5	0.89	2.4	50.4	24.6	24.0	20.9
Sakamoto *et al.* (1995)	EP	25	36	40–179	4.2	1.00	2.2	54.0	20.4	23.1	23.5
Viyakarn *et al.* (1992)	EP	25	36	39–179		1.00	2.0	54.0	22.0	23.5	23
Start feeding in summer											
Shimeno *et al.* (1980)	SMP	26–29	30	106–290	3.4	1.42	1.3	52.6	15.2	21.5	24.5
Shimeno *et al.* (1985)	SMP	26–29	30	88–214	3	1.36	1.3	57.1	16.2	21.6	26.4
Takeda *et al.* (1975)	SMP	25–29	28	65–201	4	1.19	1.3	71.0	8.0	20.0	25
Start feeding after summer											
Watanabe *et al.* (1998)	EP	27–22	64	130–341	1.5	1.41	1.5	50.4	24.6	24.0	20.9
Sakamoto *et al.* (1995)	EP	25–15	99	425–1240	1.4	1.60	2.2	50.7	26.0	24.3	21.1
Viyakarn *et al.* (1992)	EP	26–21	112	365–759	0.7	2.63	0.77	54.0	20.0	23.1	23.4
Watanabe *et al.* (1992)	EP	25–28–18	97	160–688	1.5	1.40	1.6	49.5	24.5	23.6	21
Watanabe *et al.* (1992)	EP	21–28–20	182	1190–3260	0.6	2.13	1.1	48.3	29.4	24.8	19.5

* Order of values indicate direction of water temperature change during the experiment.
† PER = ((Ln end wt – Ln start wt)/days duration of experiment) × 100.
‡ g feed fed/g wt gain.
§ g wt gain/g protein fed.
EP, extruded type pellets; SMP, single moist pellets made from mash, oil and water; PER, protein efficiency ratio; CL, crude lipid; SGR, specific growth rate.

a test diet containing a mixture of intact protein sources such as fish-meal and soy protein. The amino acid profile of the test diet was similar to that of fish muscle. Signs of gross deficiency were observed in fish fed a diet with the lowest level of methionine (0.57% of the diet^{-1}). These fish exhibited bilateral cataracts within 20 days after the start of the feeding study. To meet the amino acid requirement, it is important to know the protein and amino acid availability from protein ingredients for yellowtail. The values of protein digestibility and amino acid availability are presented in Tables 10.3 and 10.4.

Energy

It is generally believed that fish, like other animals, control their feed intake in order to meet their energy intakes in tune with their growth rates (Kaushik and Médale, 1994). Thus, energy requirement is an essential factor in the formulation and feeding of feed for efficient fish production. Masumoto *et al.* (1997) determined the energy requirement of juvenile yellowtail (23–26°C, initial body weight (BW) 12 g and final BW 120 g) for maintenance and maximum gain. The estimated maintenance energy level based on body composition analysis for fasting fish was 134 kJ kg^{-1} BW day^{-1}, and the level for zero growth was estimated from the results of feeding experiments to be 214 kJ kg^{-1} BW day^{-1}. The energy

Table 10.2. Quantitative amino acid requirements of yellowtail. Requirement values are expressed as per cent of diet. In parentheses, the numerators are requirements as per cent of dietary protein and the denominators are per cent of crude protein in the diet.

Amino acid	Requirement	Reference
Arginine	1.43–1.63 (3.40–3.88/42)	Ruchimat *et al.* (1998)
Histidine	0.65–0.85 (1.49–1.95/43)	Ruchimat (1998)
Lysine	1.78 (4.13/43)	Ruchimat *et al.* (1997b)
Methionine	1.11 (2.56/43)	Ruchimat *et al.* (1997a)

Table 10.3. Apparent crude protein (CP) and gross energy (GE) digestibility (%) for feedstuffs commonly used in yellowtail diets* (T. Masumoto, unpublished).

	Apparent digestibility (%) of	
	CP	GE
Fish-meal (moist pellets)	89	ND
Fish-meal (extruded pellets)	84	83
Maize-gluten meal	37	33
Meat bone-meal	80	82
Meat meal	97	99
Poultry feather meal	68	70
Soybean meal	93	62

* Measured for more than 20 fish.
ND, not determined.

requirement for maximum growth based on the relationship between growth rate and energy intake was estimated to be 772 kJ kg^{-1} BW day^{-1}. These values are higher than those for carp, *Cyprinus carpio*, or rainbow trout, *Oncorhynchus mykiss*. For example, the maintenance energy requirement is 40–60 kJ kg^{-1} BW day^{-1} for carp (Ohta and Watanabe, 1996b) and about 90 kJ kg^{-1} BW day^{-1} for rainbow trout (Akiyama and Nose, 1980; Ohta and Watanabe, 1996a). The energy requirement for maximal growth of rainbow trout is 335–377 kJ kg^{-1} BW day^{-1} (Ohta and Watanabe, 1996a).

The energy requirements for maximal growth at various water temperatures and growth stages can be estimated by recalculating growth performance results from available published reports (Table 10.5). Expressed in terms of per unit of weight gain, the requirement would be about 2000, 3000 and 4000–5000 kJ 100 g^{-1} for fish weighing < 200, 200–300 and > 500 g, respectively. Younger fish require less energy per unit weight gain than do older fish. Expressed in terms of daily requirement, the values are about 700–1000, 800–1000 and 300–500 kJ kg^{-1} BW day^{-1}, respectively. These data suggest that the energy requirement is more affected by water temperature than by the growth stage of fish. The energy requirement of fish should be met by the digestible energy content in the diet. Therefore, measurements of the digestible energy of each

Table 10.4. Apparent amino acid and protein digestibility (%) of some dietary ingredients (Masumoto *et al.*, 1996). Values are mean (± SD) of two replications.

Amino acids	Brown fish-meal	Casein	Meat meal	Soybean-protein conc.	Maize-gluten meal	Full-fat soybean meal
Alanine	89.7 ± 1.3	94.5 ± 1.1	86.0 ± 0.3	84.9 ± 2.1	47.3 ± 9.8	77.3 ± 5.4
Arginine	92.5 ± 2.1	96.0 ± 0.6	82.2 ± 1.5	89.9 ± 3.9	47.6 ± 9.4	85.4 ± 5.8
Aspartic acid	89.3 ± 3.3	95.5 ± 0.8	79.2 ± 1.1	90.3 ± 1.1	44.1 ± 9.3	82.0 ± 6.6
Cysteine	90.3 ± 2.5	96.2 ± 0.6	43.8 ± 8.5*	87.2 ± 9.9	47.1 ± 10.5	85.0 ± 10.5
Glutamic acid	91.9 ± 3.0	98.2 ± 0.2	81.6 ± 1.0	92.0 ± 2.1	48.8 ± 9.6	86.4 ± 5.6
Glycine	92.0 ± 1.5	91.6 ± 1.0*	89.8 ± 0.4*	81.8 ± 4.6	42.4 ± 8.2	74.8 ± 7.6
Histidine	93.0 ± 2.9	98.3 ± 0.3	86.0 ± 0.4	92.5 ± 0.1	50.8 ± 8.8	53.0 ± 19.7*
Isoleucine	90.2 ± 3.3	97.3 ± 0.4	75.9 ± 0.9	87.9 ± 0.9	45.0 ± 9.7	79.3 ± 6.3
Leucine	90.7 ± 3.1	97.7 ± 0.5	77.5 ± 1.2	86.9 ± 1.8	46.5 ± 10.4	78.0 ± 7.1
Lysine	93.1 ± 2.9	98.0 ± 0.4	85.0 ± 0.6	91.2 ± 1.4	47.6 ± 8.6	83.4 ± 6.2
Methionine	92.2 ± 3.2	98.1 ± 0.4	83.8 ± 0.5	86.8 ± 1.0	50.2 ± 9.8	76.0 ± 7.3
Phenylalanine	88.8 ± 3.1	97.4 ± 0.6	78.4 ± 1.0	88.9 ± 0.2	47.1 ± 10.4	79.4 ± 7.3
Proline	69.9 ± 7.0*	98.8 ± 0.2	87.0 ± 0.5	88.9 ± 2.4	51.1 ± 9.9	82.6 ± 7.3
Serine	89.6 ± 2.6	96.8 ± 0.3	73.8 ± 1.7	86.1 ± 4.3	46.0 ± 10.0	79.0 ± 7.4
Threonine	88.9 ± 2.8	95.3 ± 0.4	73.8 ± 1.9	83.0 ± 2.6	43.4 ± 8.0	74.8 ± 5.8
Tyrosine	90.1 ± 3.4	98.3 ± 0.3	76.3 ± 1.4	89.1 ± 3.7	50.6 ± 9.5	82.0 ± 7.3
Valine	85.7 ± 2.7	95.2 ± 0.1	72.3 ± 1.4*	79.7 ± 0.6	40.0 ± 6.5	69.1 ± 5.1
Average	89.3	96.7	78.4	87.5	46.8	78.1
APD	88.7 ± 0.0	95.4 ± 0.6	80.3 ± 0.0	87.3 ± 2.3	49.7 ± 9.1	83.2 ± 2.4

* Significantly different ($P < 0.05$) from APD within the same column.
APD, apparent protein digestibility.

Table 10.5. Requirements of CP and GE values estimated from published best growth results.

Reference	Water temp. (°C)	DFR (%)	FCR	Req. for 100 g BW gain* CP (g)	Req. for 100 g BW gain* GE (kJ)	Req. for 1 kg BW fish per day† CP (g)	Req. for 1 kg BW fish per day† GE (kJ)
Start feeding before summer							
Takeuchi *et al.* (1992)	19.0–22.4	4.4	0.87	47	1801	24	911
Takeuchi *et al.* (1992)	20.2–24.1	4.5	1.03	57	1916	25	837
Masumoto *et al.* (1997)	23–26	4.3	1.08	56	2376	22	946
Watanabe *et al.* (1998)	24	3.3	0.89	45	2136	17	792
Sakamoto *et al.* (1995)	25	3.6	1.00	54	2310	19	832
Viyakarn *et al.* (1992)	25	3.5	1.00	54	2350	19	823
Start feeding in summer							
Shimeno *et al.* (1980)	26–29	4.6	1.42	75	3053	24	989
Shimeno *et al.* (1985)	26–29	4.1	1.36	78	2938	23	886
Takeda *et al.* (1975)	25–29	4.8	1.19	84	2380	34	960
Start feeding after summer							
Watanabe *et al.* (1998)	27.3–21.8	2.8	1.41	71	3384	14	672
Sakamoto *et al.* (1995)	25–15	2	1.6	81	3888	10	486
Viyakarn *et al.* (1992)	26–20.6	2.8	2.63	142	6075	15	647
Watanabe *et al.* (1992)	25–28–18	2.4	1.40	69	3304	12	566
Watanabe *et al.* (1992)	21–28–20	1.4	2.13	103	5282	7	347

* 100 g × FCR × CP (g) or GE (kJ) content values given in Table 10.1.
† 1000 g × DFR × CP (g) or GE (kJ)/1000 content values given in Table 10.1.
DFR, daily feeding rate.

ingredient are essential for establishing an adequate feeding rate for fish. Digestible energy values for common ingredients used in yellowtail culture are given in Table 10.3. The value for maize-gluten meal is much lower than that reported for other fish. The apparent digestibility for GE for the soybean meal is lower than that for CP, suggesting that there are undigestible carbohydrates for yellowtail in soybean meal.

Lipids and fatty acids

Yellowtail require the n-3 rather than the n-6 series of fatty acids (Furukawa *et al.*, 1966; Tsukahara *et al.*, 1967). Moreover, unlike rainbow trout or other freshwater fish species, yellowtail are unable to use 18:3 (n-3) as an essential fatty acid (EFA) and require n-3 highly unsaturated fatty acids (HUFA) with 20 or more carbon atoms (Yone, 1978). A comparison of the effect of various fish-oil sources on the growth of yellowtail (40 g) showed that oil high in n-3 HUFA (squid liver oil, sardine oil and skip-jack oil) had better nutritional value than that of oil containing low levels of n-3 HUFA (pollack liver oil and herring oil) (Deshimaru *et al.*, 1982a). The optimum dietary lipid level for young yellowtail (45–80 g) was 9% when squid oil containing 27% n-3 HUFA was used and 15% when pollack liver oil containing 13% n-3 HUFA was used (Deshimaru *et al.*, 1982b). Based on these results, the n-3 HUFA requirement for yellowtail is estimated to be approximately 2% of the diet (Deshimaru *et al.*, 1982b). It appears that docosahexaenoic acid (DHA), 22:6n-3, is physiologically more important than eicosapentaenoic acid (EPA), 20:5n-3, at the larval stage of yellowtail. When yellowtail larvae were fed either EPA- or DHA-fortified rotifers, both groups of fish showed equally good survival. However, larvae fed EPA-enriched rotifer showed significantly poor results for activity tests compared with larvae fed DHA-enriched rotifer (Furuita *et al.*, 1996). They suggested that fortification with DHA is necessary for the live food for yellowtail because live foods, such as rotifers or *Artemia*, do not contain enough DHA to meet the EFA requirement of larval stage.

Carbohydrates

The optimum carbohydrate level for yellowtail was investigated by feeding a fish-meal-based diet supplemented with gelatinized starch at levels of 0, 10, 20 or 40% (Shimeno *et al.*, 1979). The growth rate, feed efficiency and energy retention, as well as body fat and glycogen content, were higher in fish fed diets with 10% and 20% starch levels than in fish fed diets with 0% and 40% starch levels. The protein digestibility of fish fed a diet with a starch level of 40% was lower than that of fish fed a diet with 20% starch (56% vs. 78%). Furuichi *et al.* (1986) also reported that the growth rate of fish fed a diet with 20% α-starch was similar to that of fish fed a diet with 10% starch, although the feed efficiency was significantly lower in the former group of fish. These results suggested that the optimum gelatinized starch level is less than 20%.

The effects of various carbohydrate sources on the growth rate of yellowtail were also tested. At the 15% inclusion level, the growth of fish fed starch was better than that of fish fed fructose (Shimeno *et al.*, 1979). Yellowtail cannot utilize dietary glucose well. In fact, the digestibility of glucose is higher than that of gelatinized starch, although the growth performance of fish fed glucose is inferior to that of fish fed starch at a 10% dietary level (Furuichi *et al.*, 1986). Carnivorous fish, such as yellowtail, cannot utilize as much dietary carbohydrate as omnivorous fish, such as common carp. This lower ability of carbohydrate utilization is related to the lower amylase activity, carbohydrate metabolism enzyme activity and insulin response. Among these factors, the lower response of insulin secretion seems to be the most responsible. The insulin secretion response is substantially lower and slower in yellowtail than in other fish, such as common carp or red sea bream (*Pagrus major*), as indicated in Fig. 10.2 (Furuichi and Yone, 1981, 1982a,b). It was reported that the plasma insulin level in yellowtail

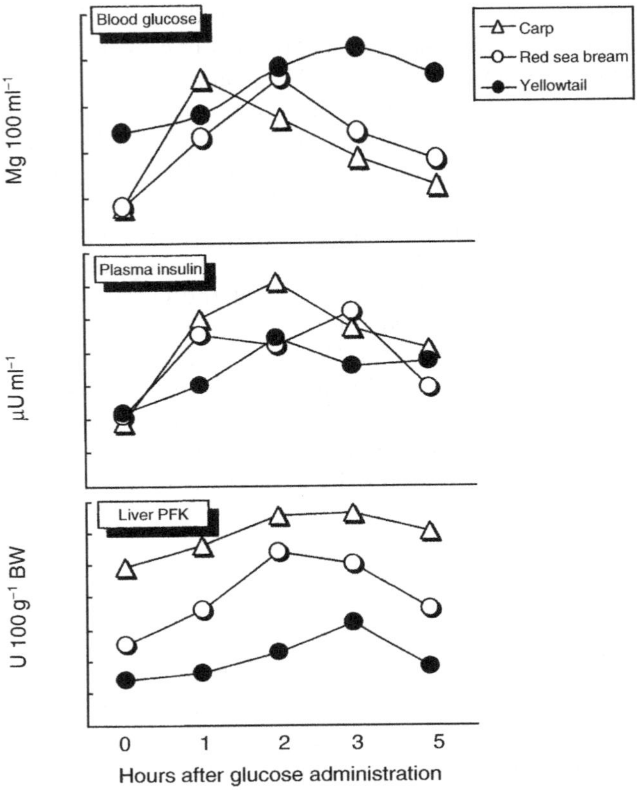

Fig. 10.2. Changes in blood glucose, plasma insulin levels and liver phosphofructokinase (PFK) activity of fish after orally administered glucose (Furuichi and Yone, 1981, 1982a,b).

reached a maximum at about 2 h and that the activity of carbohydrate metabolism enzyme activity became high at 2–3 h after administration of glucose. Therefore, most of the dietary carbohydrate, which is absorbed before the rise in enzyme activities, is excreted without being utilized. This suggests that poor utilization of carbohydrate by yellowtail is due to carbohydrate metabolism rather than lower absorption of carbohydrate.

Vitamins and minerals

The quantitative requirements of the yellowtail have been determined by feeding purified casein-gelatin diets containing graded levels of vitamins. The requirement value can vary depending on the criteria used – e.g. maximum weight gain, feed conversion efficiency or tissue vitamin concentration. Table 10.6 shows vitamin requirement values based on maximum growth rate and liver concentration

Table 10.6. Vitamin requirements of yellowtail determined based on growth and maximum liver storage expressed as dietary concentration (mg kg^{-1} dry diet) and daily requirement (mg kg^{-1} BW day^{-1}) (Hosokawa, 1999).

	Requirement			
	(mg kg^{-1} dry diet)		(mg kg^{-1} BW day^{-1})	
Vitamin	Growth	Storage	Growth	Storage
Thiamine hydrochloride	1.2	11.2	0.4	3.6
Riboflavin	2.9	11.0	1.0	3.6
Pyridoxine hydrochloride	2.5	11.7	0.8	3.8
Nicotinic acid	12.0	–	3.9	–
Pantothenic acid	13.5	35.9	4.5	11.5
Inositol	190	423.0	53.0	116
Biotin	0.22	0.67	0.06	0.19
Folic acid	0.8	1.2	0.24	0.39
Choline chloride	2100.0	2920.0	615.0	856
Cyanocobalamin	0.053	–	0.017	–
Ascorbic acid	122.0	–	34.0	34.0
Retinol (as acetate)	5.68	–	5100	–
α-Tocopherol	119.0	–	35.0	35.0

(Hosokawa, 1999). Vitamin deficiency signs are similar to those reported for other fish species. However, the onset of deficiency is quicker than that in most other species, indicating the fast-growing nature of yellowtail fingerlings (Table 10.7).

The interaction between macronutrients and vitamins was studied. When the dietary protein level was increased from 59% to 79%, the pyridoxine requirement increased by 1.4-fold. On the other hand, the α-tocopherol requirement increased from 35 to 93 mg kg^{-1} and 160 mg kg^{-1} as the dietary lipid level increased from 8% to 15% and 25%, respectively (Hosokawa, 1999).

The effects of megadoses of vitamins E and C on immune responses in yellowtail fingerlings were studied. When the fish were fed diets containing two to five times or 16–50 times the required amount of dietary vitamin E or vitamin C, respectively, phagocytic activity and antibody titre in fish fed those diets increased compared with those in fish fed diets with the required levels of vitamin E or C. As a result, fish fed megadoses of these two vitamins increased their resistance to Pseudotuberculosis and streptococcicosis (Hosokawa, 1999).

There have been only a few studies on the mineral requirements for yellowtail. Hosokawa *et al.* (1985) examined the effects on the growth of yellowtail of casein-based semipurified diets supplemented with various levels of mineral mixture (Halver, 1957). The best growth was obtained from the diet containing an 8% mineral level. Fish fed a diet without the mineral mixture showed an initial loss of appetite, followed by dark coloration, scoliosis, anaemia and death, but these signs were improved by feeding a recovery diet containing

Table 10.7. Vitamin deficiency and number of days to develop deficiency signs in yellowtail (Hosokawa, 1999). Common vitamin deficiency signs such as anorexia, reduced growth or increased mortality are not included.

Vitamin	Deficiency signs	Days
Thiamine	Dark coloration, congestion in fins and opercula	24
Riboflavin	Congestion in fins and eyes, dark coloration, cloudy lens	30
Pyridoxine	Epileptiform fins, convulsions, ataxia	14
Nicotinic acid	Skin lesions, haemorrhage in body surface, loss of caudal fin	20
Pantothenic acid	Blue coloration, clubbed gills	10
Inositol	Dark coloration, ataxia, skin lesions	20
Biotin	Haemorrhage in fins, opercula and liver	24
Folic acid	Congestion in fins and opercula, dark coloration, macrocytic anaemia	24
Choline	Dark coloration, poor feed conversion	3
Cyanocobalamin	Congestion in fins and opercula, haemorrhagic liver	25
Ascorbic acid	Scoliosis, dark coloration, ataxia, haemorrhage in body surface, hypochromic anaemia	20
Vitamin A	Dark coloration, ataxia, haemorrhage in fins and eyes	19
Calciferol	None	–
α-Tocopherol	Dark coloration, ataxia, haemorrhage in fins and opercula	28
Menadione	None	–

the mineral mixture (Makino, 1990). In omission tests, some of the deficiency signs described above were observed in fish fed diets devoid of iron and phosphorus, indicating the essentiality of these two minerals (Makino, 1990). The minimum levels of dietary iron and phosphorus were estimated to be about 60–160 mg kg^{-1} and 6.7 g kg^{-1} dry diet, respectively. These values are in agreement with the values reported for red sea bream and Atlantic salmon (*Salmo salar*). It has also been reported that iron-proteinate (a commercial compound binding iron to soybean hydrolysate) was superior to iron citrate or iron sulphate as the iron source (Kubota *et al.*, 1983; Kuwabara *et al.*, 1983).

When yellowtail were fed diets devoid of sodium (Na), potassium (K) and chloride (Cl), they exhibited no apparent deficiency signs. Thus, the yellowtail's requirements for these minerals may be met with absorption of these minerals from sea water by drinking. Fish fed diets devoid of calcium and magnesium showed slightly reduced growth rates and feed efficiency, which may be an indication of the essentiality of these minerals in the diet. The suitable mineral-mixture level was tested in a fish-meal-based diet and it was found to be 1% (Ikeda, 1976). However, the necessity of iron and phosphorus supplementation in a fish-meal-based diet has been reported.

Special diets

It has been suggested that krill meal, especially the astaxanthin in the meal, is an effective component in the reproductive performance of red sea bream brood-stock. Verakunpiriya *et al.* (1997) added a supplement of synthetic astaxanthin instead of krill meal to a practical diet of dry pellets and examined the reproductive performance of yellowtail brood-stock. They found that the supplementation of astaxanthin in dry pellets was effective for brood-stock yellowtail and the optimum level appears to be 30 p.p.m., but egg quality was reduced at levels above 30 p.p.m.

Feeding Practices

Yellowtail culture had been expanded due to massive catches of low-cost fish used as food, such as sand-lance and sardine. However, in recent years, the catch of these fish has decreased and their cost has therefore increased. This has forced many yellowtail producers to change from feeding fish to the use of a formulated diet. As shown in Fig. 10.3, formulated diet production has increased and the production of dry diet (mainly extruded pellets) is now almost the same as that of mash (mainly used for moist pellets). Feeding extruded pellets for the first year of

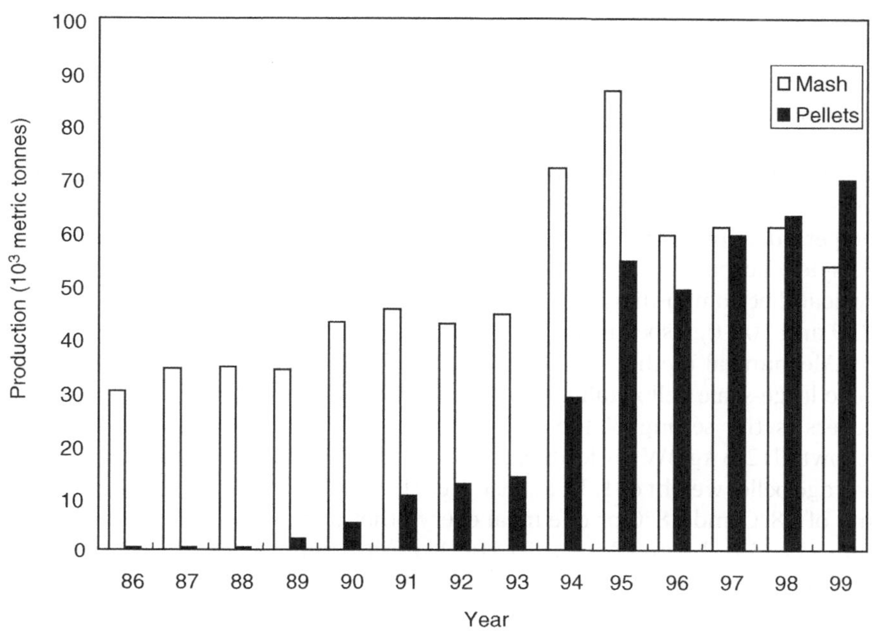

Fig. 10.3. Diet production (mash and pellets) for yellowtail (source: Japan Fish Feed Association, Tokyo, Japan).

Table 10.8. Typical diet selection pattern for yellowtail production (from Isobe and Matsushita, 1994).

	1st year				2nd year			
	Spring	Summer	Autumn	Winter	Spring	Summer	Autumn	Winter
Pattern	Apr.–June	Jul.–Sep.	Oct.–Dec.	Jan.– Mar.	Apr.–June	Jul.–Sep.	Oct.–Dec.	Jan.–Mar.
1	EP	FF						
2	EP	OMP*	FF					
3	EP	OMP					FF	
4	EP			OMP			FF	
5	EP			OMP	EP		FF	
6	EP						FF	

* Wet portion of OMP is frozen feed fish. The inclusion level of frozen fish depends on fish farmer and water temperature. The frozen fish are partially thawed, ground and combined with a dry mash.
EP, extruded pellets; FF, frozen fish (chopped sardine or sand-lance); OMP, Oregon-type moist pellets.

yellowtail culture during the growing season (high water temperature) is popular. However, the use of fish or moist pellets is still common when water temperatures are reduced (< 15°C) because yellowtail particularly dislike a dry-type diet at such a low temperature. In yellowtail culture, the type of diet used depends on economic factors and culture conditions (for example, winter water temperature) of each fish farm (Table 10.8). It was estimated that formulated diets now account for about 40% of the total food used for yellowtail production. Typical diet composition and recommended feeding rates are shown in Tables 10.9 and 10.10, respectively.

Mechanical methods of delivering diet have recently been introduced into some large-scale yellowtail farms. One important factor in the use of automatic feeders is the setting of the delivery rate. Feed delivery rate was tested for yellowtail (2.6 kg BW) by feeding an extruded pellet (11 mm in diameter, with an average pellet weight of 1.31 g). Fish were fed one meal daily at a water temperature of 28°C and 18°C or one meal every other day at 18°C; the diet-delivering rates were 3.6, 3.4 and 3.9 kg diet 1000 kg^{-1} fish min^{-1}, respectively (Talbot *et al*, 1999a,b). This means that the yellowtail consume seven to eight pellets per minute. Moreover, under these conditions, the yellowtail ate more than 80% of the total feed amount within 10 min. This feeding activity is similar to the feeding activity of salmon and trout.

Table 10.9. Example of ingredients and nutrient composition of extruded-type pellets for yellowtail (Sakamoto, 1994).

	%
Ingredients	
Fish-meal	55
Krill meal	10
Wheat flour	7
Potato starch	3
Wheat gluten	3
Mineral mix	2
Vitamin mix	2
Feed oil	18
Nutrient content	
Crude protein	47
Crude lipid	24
Crude starch	12
Crude ash	10
Moisture	8

Table 10.10. Suggested feeding rate (% of BW) for yellowtail (Sakamoto Feeds Co., Ltd).

	Water temperature (°C)														
BW (g)	14	15	16	17	18	19	20	21	22	23	24	25	26	27	28
10					8.3	8.8	9.3	9.6	9.8	10	10.2				
20					7.6	7.9	8.3	8.5	8.8	9	9.3				
30					6.3	6.7	7	7.5	7.7	7.8	8.1	8.3			
40				4.9	5.3	5.8	6.1	6.4	6.6	6.8	7.1	7.2	7.4		
50				4	4.3	5	5.3	5.4	5.6	5.8	6	6.3	6.6	6.9	
100				3	3.1	3.5	3.8	4	4.1	4.3	4.5	4.8	5.1	5.4	5.6
200			2.2	2.3	2.6	2.7	2.8	2.9	3.3	3.5	3.6	3.9	4.1	4.4	4.6
300			2	2.1	2.4	2.5	2.6	2.8	3.2	3.3	3.3	3.4	3.7	3.8	4.1
400			1.8	1.9	2.2	2.3	2.4	2.5	2.8	2.9	3.1	3.3	3.4	3.6	3.8
500	1.1	1.3	1.5	1.7	2	2.1	2.3	2.3	2.5	2.6	2.8	3	3	3.1	3.3
700	1.1	1.3	1.4	1.6	1.7	1.8	2	2.1	2.2	2.2	2.3	2.5	2.6	2.6	2.7
1000	1	1.2	1.2	1.3	1.4	1.5	1.5	1.6	1.6	1.7	1.7	1.8	1.9	1.9	1.9
1250	1	1.2	1.2	1.3	1.4	1.5	1.5	1.5	1.6	1.6	1.7	1.8	1.8	1.9	1.9
1500	0.9	1.1	1.1	1.2	1.3	1.4	1.4	1.5	1.5	1.6	1.6	1.7	1.7	1.8	1.9
1750	0.9	1	1.1	1.1	1.2	1.3	1.4	1.4	1.5	1.5	1.6	1.6	1.7	1.8	1.8
2000	0.8	0.9	1	1.1	1.2	1.2	1.3	1.4	1.4	1.5	1.5	1.6	1.6	1.7	1.8
2500	0.8	0.9	1	1	1.1	1.2	1.3	1.4	1.4	1.5	1.5	1.6	1.6	1.7	1.7
3000	0.7	0.8	0.9	0.9	1	1	1.1	1.2	1.3	1.4	1.5	1.5	1.6	1.7	1.6
3500	0.6	0.7	0.8	0.9	1	1	1.1	1.2	1.2	1.3	1.4	1.4	1.5	1.6	1.5
4000	0.6	0.7	0.8	0.8	0.9	0.9	1	1	1.1	1.2	1.2	1.3	1.4	1.5	1.4
4500	0.5	0.7	0.7	0.8	0.9	0.9	1	1	1.1	1.1	1.2	1.3	1.3	1.4	1.4
5000	0.5	0.6	0.7	0.8	0.8	0.8	0.9	1	1	1.1	1.1	1.2	1.2	1.3	1.3
5500	0.4	0.6	0.6	0.7	0.8	0.8	0.8	0.9	1	1	1.1	1.1	1.1	1.2	1.2
6000	0.4	0.5	0.6	0.6	0.7	0.8	0.8	0.8	0.9	0.9	1	1	1.1	1.1	1.1

References

Akiyama, T. and Nose, T. (1980) Changes in body weight, condition factor and body composition of fingerling salmon with various sizes during starvation. *Bulletin of National Research Institute of Aquaculture* 1, 71–78 (in Japanese).

Cho, C.Y., Slinger, S.J. and Bayley, H.S. (1982) Bioenergetics of salmonid fishes: energy intake, expenditure and productivity. *Comparative Biochemistry and Physiology* 73B, 25–41.

Deshimaru, O., Kuroki, K. and Yone, Y. (1982a) Nutritive values of various oils for yellowtail. *Bulletin of Japanese Society of Scientific Fisheries* 48, 1155–1157 (in Japanese).

Deshimaru, O., Kuroki, K. and Yone, Y. (1982b) Suitable levels of lipids and ursodeoxycholic acid in diet for yellowtail. *Bulletin of Japanese Society of Scientific Fisheries* 48, 1265–1270 (in Japanese).

Furuichi, M. and Yone, Y. (1981) Change of blood sugar and plasma insulin levels of fishes in glucose tolerance test. *Bulletin of Japanese Society of Scientific Fisheries* 47, 761–764.

Furuichi, M. and Yone, Y. (1982a) Changes in activities of hepatic enzymes related to carbohydrate metabolism of fishes in glucose and insulin-glucose. *Bulletin of Japanese Society of Scientific Fisheries* 48, 463–466.

Furuichi, M. and Yone, Y. (1982b) Effect of insulin on blood sugar levels of fishes. *Bulletin of Japanese Society of Scientific Fisheries* 48, 1289–1291.

Furuichi, M., Taira, H. and Yone, Y. (1986) Availability of carbohydrate in nutrition of yellowtail. *Nippon Suisan Gakkaishi* 52, 99–102.

Furuita, H., Takeuchi, T., Watanabe, T., Fujimoto, H., Sekiya, S. and Imaizumi, K. (1996) Requirement of larval yellowtail for eicosapentaenoic acid, docosahexaenoic acid, and n-3 highly unsaturated fatty acid. *Fisheries Science* 62, 372–379.

Furukawa, A., Tsukahara, H. and Funae, K. (1966) Studies on feed for fish V. Results of the small floating net culture test to establish the artificial diet as complete yellowtail foods. *Bulletin of Naikai Regional Fisheries Research Laboratory* 23, 45–56.

Halver, J.E. (1957) Nutrition of salmonid fishes – III. Water soluble vitamin requirements of chinook salmon. *Journal of Nutrition* 62, 225–243.

Hosokawa, H. (1999) Studies on the vitamin nutrition in yellowtail. PhD thesis, United Graduate School of Agricultural Sciences Ehime University, Ehime, Japan (in Japanese).

Hosokawa, H., Takeda, M., Tachi, T. and Hayami, H. (1985) Study on a basal diet for yellowtail – II. Mineral mixture. In: *Abstracts of Oral Presentation at the Autumn Meeting of Nippon Suisan Gakkai.* Nippon Suisan Gakkai, Tokyo, Japan, p. 54 (in Japanese).

Ikeda, Y. (1976) Diagnostic studies on haematological and chemical constituents of blood in cultured yellowtail. PhD thesis, Kyoto University, Kyoto, Japan (in Japanese).

Inoue, H. (1983) Environmental condition and its control on yellowtail culture. *Fisheries Research* 2, 89–91 (in Japanese).

Isobe, K. and Matsushita, I. (1994) Optimum feeding plan for yellowtail at grow-out stage. *Yosyoku* 8, 60–63 (in Japanese).

Kaushik, S.J. and Médale, F. (1994) Energy requirements, utilization and dietary supply to salmonids. *Aquaculture* 124, 81–97.

Kubota, S., Kamiya, N., Takeda, M., Shimeno, S. and Kuwabara, H. (1983) Pathological and histological study on young yellowtail fed various ferrous compounds. In: *Abstracts of Oral Presentation at the Meeting of Nippon Gyobyo Gakkai.* Nippon Gyobyo Gakkai, Tokyo, Japan, p. 6 (in Japanese).

Kuwabara, H., Takeda, M., Shimeno, S., Hosokawa, H., Kubota, S., Kamiya, N. and Kajiyama, H. (1983) Effect of adding ferrous compounds to yellowtail diets. In: *Abstracts of Oral Presentation at the Meeting of Nippon Suisangakkai.* Nippon Suisangakkai, Fukuyama, Japan, p. 43 (in Japanese).

Makino, H. (1990) Study on mineral requirements in yellowtail. MSc thesis, Kochi University, Nankoku, Japan (in Japanese).

Masumoto, T., Ruchimat, T., Ito, Y., Hosokawa, H. and Shimeno, S. (1996) Amino acid availability values for several protein sources for yellowtail (*Seriola quinqueradiata*). *Aquaculture* 146, 109–119.

Masumoto, T., Ruchimat, T., Hosokawa, H. and Shimeno, S. (1997) Energy requirement and energy partitioning in juvenile yellowtail. *Bulletin of Marine Science and Fisheries, Kochi University* 17, 79–87 (in Japanese).

Ohta, M. and Watanabe, T. (1996a) Energy requirement for maintenance of body weight and activity, and for maximum growth in rainbow trout. *Fisheries Science* 62, 737–744.

Ohta, M. and Watanabe, T. (1996b) Dietary energy budgets in carp. *Fisheries Science* 62, 745–753.

Ruchimat, T. (1998) Studies on amino acid requirement of yellowtail (*Seriola quinqueradiata*). PhD thesis, United Graduate School of Agricultural Sciences Ehime University, Ehime, Japan.

Ruchimat, T., Masumoto, T., Hosokawa, H. and Shimeno, S. (1997a) Quantitative methionine requirement of yellowtail (*Seriola quinqueradiata*). *Aquaculture* 150, 113–122.

Ruchimat, T., Masumoto, T., Hosokawa, H., Itoh, Y. and Shimeno, S. (1997b) Quantitative lysine requirement of yellowtail (*Seriola quinqueradiata*). *Aquaculture* 158, 331–339.

Ruchimat, T., Masumoto, T., Itoh, Y. and Shimeno, S. (1998) Quantitative arginine requirement of juvenile yellowtail *Seriola quinqueradiata. Fisheries Science* 64, 348–349.

Sakamoto, H. (1994) Development of dry pellet for mariculture. In: Watanabe, T. (ed.) *Use of Alternative Protein Sources in Aquaculture.* Koseisya-Koseikaku, Tokyo, pp. 23–34. (in Japanese).

Sakamoto, H., Watanabe, T. and Takeuchi, T. (1995) Optimum levels of protein and lipid in a newly developed soft-dry pellet for yellowtail, *Seriola quinqueradiata. Suisanzosyoku* 43, 345–352 (in Japanese).

Shimeno, S., Hosokawa, H. and Takeda, M. (1979) The importance of carbohydrate in the diet of a carnivorous fish. In: Halver, J.E. and Tiews, K. (eds) *Finfish Nutrition and Fishfeed Technology*, Vol. 1. Heenemann, Berlin, pp. 127–143.

Shimeno, S., Hosokawa, H., Takeda, M. and Kajiyama, H. (1980) Effects of calories to protein ratios in formulated diet on the growth, feed conversion and body composition of young yellowtail. *Bulletin of Japanese Society of Scientific Fisheries* 46, 1083–1087 (in Japanese).

Shimeno, S., Hosokawa, H., Takeda, M., Kajiyama, H. and Kaisho, T. (1985) Effect of dietary lipid and carbohydrate on growth, feed conversion and body composition in young yellowtail. *Bulletin of Japanese Society of Scientific Fisheries* 51, 1893–1898.

Takeda, M., Shimeno, S., Hosokawa, H., Kajiyama, H. and Kaisyo, T. (1975) The effect of dietary calorie–protein ratio on the growth, feed conversion and body composition of young yellowtail. *Bulletin of the Japanese Society of Scientific Fisheries* 41, 443–447 (in Japanese).

Takeuchi, T., Shiina, Y., Watanabe, T., Sekiya, S. and Imaizumi, K. (1992) Suitable protein and lipid levels for fingerlings of yellowtail. *Nippon Suisan Gakkaishi* 58, 1333–1339.

Talbot, C., Corneillie, S. and Korsøen, Ø. (1999a) Pattern of feed intake in four species of fish under commercial farming conditions: implications for feeding management. *Aquaculture Research* 30, 509–518.

Talbot, C., Garcia-Gómez, A., De-la-Gándara, F. and Muraccioli, P. (1999b) Food intake, growth, and body composition in Mediterranean yellowtail (*Seriola dumerilli*) fed isonitrogenous diets containing different lipid levels. In: *Recent Advances in Mediterranean Aquaculture Finfish Species Diversification, Proceedings of the Seminar of the CIHEAM Network TECAM, Zaragoza, Spain, 24–28 May 1999.*

Tsukahara, H., Furukawa, A. and Funae, K. (1967) Studies on feed for fish VIII. The effects of dietary fat on the growth of yellowtail (*Seriola quinqueradiata* Temminak et Schegel). *Bulletin of Naikai Regional Fisheries Research Laboratory* 24, 29–50.

Verakunpiriya, V., Mushiake, K., Kawano, K. and Watanabe, T. (1997) Supplemental effect of astaxanthin in broodstock diets on the quality of yellowtail eggs. *Fisheries Science* 63, 816–823.

Viyakarn, V., Watanabe, T., Aoki, H., Tsuda, H., Sakamoto, H., Okamoto, N., Iso, N., Satoh, S. and Takeuchi, T. (1992) Use of soybean meal as a substitute for fish meal in a newly developed soft-dry pellet for yellowtail. *Nippon Suisan Gakkaishi* 58, 1991–2000.

Watanabe, T., Viyakarn, V., Kimura, H., Ogawa, K., Okamoto, N. and Iso, N. (1992) Utilization of soybean meal as a protein source in a newly developed soft-dry pellet for yellowtail. *Nippon Suisan Gakkaishi* 58, 1761–1773.

Watanabe, T., Aoki, H., Shimamoto, K., Hadzuma, M., Maita, M., Yamagata, Y., Kiron, V. and Satoh, S. (1998) A trial to culture yellowtail with non-fishmeal diets. *Fisheires Sience* 64, 505–512.

Yone, Y. (1978) Essential fatty acids and lipid requirements of marine fish. In: Yone, Y. (ed.) *Dietary Lipids in Aquaculture.* Koseisya-Koseikaku, Tokyo, pp. 43–59 (in Japanese).

11 Red Drum, *Sciaenops ocellatus*

Delbert M. Gatlin III
Department of Wildlife and Fisheries Sciences and Faculty of Nutrition, Texas A&M University System, College Station, TX 77843-2258, USA

Introduction

The red drum (*Sciaenops ocellatus*), which is also commonly called redfish or channel bass, is a euryhaline sciaenid that is native to the Gulf of Mexico and Atlantic Ocean. This fish historically supported commercial and recreational fisheries for many decades. However, commercial overfishing in the Gulf of Mexico resulted in the state of Texas prohibiting the sale of native red drum in 1981 (Matlock, 1990), and other harvest restrictions were imposed by various state and federal regulatory agencies in the early 1980s. As a result of these actions, research efforts related to the aquacultural production of this species for enhancement of wild populations as well as for food production were accelerated.

A variety of systems have been used to culture red drum. Juvenile fish are commonly raised in fertilized ponds prior to release into the wild for stock enhancement. Red drum for food markets have been reared to 1–2 kg in various culture systems, including earthen ponds, cages, net pens and raceways operating in flow-through or recirculating modes. Under favourable environmental conditions, red drum can be grown from a weight of 1 g to approximately 1 kg in 8–12 months (Sandifer *et al.*, 1993).

Nutrition of red drum is one area of research that has received considerable attention because of the influence nutrition has on fish growth and health as well as cost of production in intensive aquaculture. Nutritious and cost-effective diets have been developed to support the aquacultural production of red drum based on their established requirements for some of the most critical nutrients and energy. Dietary requirements for other nutrients continue to be investigated in addition to evaluation of diet formulations and feeding practices. This chapter will summarize information concerning nutritional requirements of the red drum, as well as discussing various aspects related to diet formulation and feeding strategies employed in its aquaculture. Nutrition and feeding of red drum during and after larval development will be considered separately, as these stages differ considerably in the fish's life history.

Nutrition of Red-drum Larvae

Techniques for controlled spawning of red drum in captivity were developed in the 1970s and have allowed for the production of large quantities of eggs and larvae with limited effort (Arnold, 1988). Mature broodfish, typically ranging from 10 to 25 kg, are held in captivity and commonly fed combinations of shrimp, squid or fish at a rate of approximately 2.5% of body weight per day, every other day. These fish are conditioned to spawn by subjecting them to seasonal variations in temperature and photoperiod, which are commonly compressed into cycles of 120 or 150 days. The fish typically begin spawning when autumn conditions are simulated and, once spawning is initiated, environmental conditions can be adjusted slightly so that the fish continue to spawn for several months. These broodfish typically produce several hundred thousand eggs that measure less than 1 mm in diameter. The buoyant eggs, which in nature are carried by surface currents into brackish estuarine nurseries, are commonly removed from the brood tank and concentrated with the aid of a skimming device at the water's surface. The eggs hatch approximately 24 h after being spawned, and larvae derive nourishment from their yolk-sac for at least another 48–72 h. After they begin exogenous feeding, the larvae primarily consume zooplankton, such as rotifers and copepods, until reaching a size of approximately 50 mm in length (Holt, 1990).

Under aquacultural conditions, the type of zooplankton provided will depend on the larval-rearing system employed. If larval red drum are to be reared under controlled conditions in tanks, the marine rotifer (*Brachionus plicatilis*) is a preferred first food (Wohlschlag *et al.*, 1990). Stock monocultures of unicellular algae, such as *Tetraselmis chuii* or *Isochrysis* sp., are typically used to support the growth of rotifers (Treece and Wohlschlag, 1990a). Algae and rotifers have been successfully propagated separately in continuous culture as well as together in batch culture. Rotifers have also been grown with yeast and fish-oil emulsion with or without algae (Wohlschlag *et al.*, 1990). Typically, rotifers are provided for larval red drum cultured in tanks at a density of 3–5 organisms ml^{-1} to ensure an adequate number of prey (Holt *et al.*, 1981). This density of rotifers is generally maintained for 7–10 days, after which the larval red drum should be large enough to consume newly hatched brine shrimp (*Artemia* sp.) nauplii, which are provided at a density of 0.5–2 organisms ml^{-1} (Treece and Wohlschlag, 1990b). Procedures involved in hatching *Artemia* cysts and preparing the nauplii for feeding to larval fish are well established (Sorgeloos and Personne, 1975). Enriching rotifers and *Artemia* nauplii with highly unsaturated fatty acids (HUFA) by culturing them with fish-oil emulsion has proved beneficial in the rearing of red-drum larvae (Craig *et al.*, 1994). During the time in which larvae are fed rotifers and *Artemia* nauplii, inert microparticulate diets may also be provided.

However, commercial larval diets have not generally been able to completely replace live food organisms for larval red drum cultured intensively.

As red-drum larvae make the transition to juveniles, they may be gradually converted to an artificial dry diet. After 10–14 days of being fed *Artemia*, a high-protein starter crumble for salmonids is typically introduced to the fish, and they can normally be weaned to the artificial diet in 3–4 days. *Artemia* should continue to be fed until all fish are consuming the artificial diet. Cannibalism may be a problem in rearing red drum in intensive tank culture and can be minimized by maintaining a uniform size distribution.

Another way of providing zooplankton for larval red drum is to stock the fish in fertilized, brackish-water ponds when the water temperature is approximately 22°C or higher. This approach generally requires less labour and is commonly employed at facilities that produce juvenile fish for stock enhancement as well as at commercial production facilities. There are various organic and inorganic fertilizers that have been used to promote the growth of phytoplankton and zooplankton in larval-rearing ponds. Inorganic nutrients may include nitrogen in the form of urea and phosphorus as phosphoric acid. One of the most common organic fertilizers used in ponds for red-drum larvae is cottonseed meal. A variety of fertilization regimes, which are initiated approximately 3 weeks before stocking larvae, have been successfully used in growing the fish to juvenile size in ponds (Porter and Maciorowski, 1984). The preferred zooplankton for larval red drum in brackish-water ponds includes rotifers and copepods. Approximately 2 weeks after stocking the larval red drum, a high-protein, finely ground starter diet is typically distributed in the pond so that the fish will gradually become accustomed to eating a prepared diet. Any uneaten diet also may serve as organic fertilizer. Under good environmental conditions, red-drum larvae may grow to 37.5 mm in 30 days (Holt, 1990).

Nutrient Requirements of Red-drum Juveniles

In nature, juvenile red drum of 60–100 mm consume small benthic invertebrates along with fish. Shrimp, crabs and fish constitute the major food items for red drum larger than 100 mm, and they consume more fish relative to crustaceans as they grow (Bass and Avault, 1975). Under aquacultural conditions, red drum readily consume artificial diets. Studies to determine the red drum's requirements for specific nutrients were initiated in the early 1980s (Robinson, 1988) and continue to date. The following sections present specific information concerning the dietary requirements of red drum for various nutrients and energy.

Protein and amino acids

The minimum dietary requirement of red drum for protein consisting of a balanced mixture of amino acids has been addressed in several studies because satisfying this requirement is necessary to ensure adequate growth and health of the fish, while providing excessive levels of this most expensive component is generally uneconomic. Most of these studies have been conducted with young, rapidly growing fish to determine their minimum dietary protein requirement for maximum weight gain. Requirement values have generally ranged from 35 to 45% of diet (Daniels and Robinson, 1986; Serrano *et al.*, 1992). Most recently, the metabolic protein requirements of red drum for maintenance and maximum growth were established at approximately 2 and 20 g digestible protein (DP) kg^{-1} body weight day^{-1}, respectively (McGoogan and Gatlin, 1998). Dietary protein requirements may vary due to a number of factors, including fish size, water temperature, protein quality, feed allowance and non-protein energy (NRC, 1993).

In addition to supplying amino acids for protein synthesis, dietary protein may also be catabolized for energy. Carnivorous fish species, in particular, appear to be very proficient at using dietary protein for energy due to the efficient way in which ammonia from deaminated protein is excreted via the gills with limited energy expenditure (NRC, 1993). A digestible energy (DE) level of approximately 15 kJ g^{-1} diet (or 35–45 kJ DE g^{-1} protein) has been determined to be adequate for maximum weight gain and desirable body composition of red drum (Daniels and Robinson, 1986; Serrano *et al.*, 1992). Digestible energy requirements of red drum for maintenance and maximum growth were determined to be approximately 90 and 900 kJ DE kg^{-1} body weight day^{-1}, respectively (McGoogan and Gatlin, 1998).

Dietary requirements of red drum for some of the indispensable amino acids have also been determined. Development of a suitable test diet to quantify amino acid requirements of red drum was required because diets composed of purified ingredients, such as casein and gelatin, were not readily accepted by this species (Moon and Gatlin, 1989). Diets consisting of a limited quantity of fish-muscle protein together with crystalline amino acids to simulate the amino acid pattern of red-drum muscle protein have readily allowed quantification of this species' amino acid requirements. An alternative but less successful approach with red drum has been the use of diets containing intact unbalanced proteins supplemented with crystalline amino acids (Brown *et al.*, 1988; Craig and Gatlin, 1992).

Requirements for total sulphur amino acids (methionine plus cystine) and lysine are typically the most critical to quantify because the levels of these amino acids in feedstuffs are usually most limiting relative to the amounts required by fish. The total sulphur amino acid requirement of red drum was determined to be 3.0% of dietary protein (Moon and Gatlin, 1991). This sulphur amino acid requirement appeared to be more limiting than the lysine requirement, which was quantified to be approximately 4.4% of dietary protein (Brown *et al.*, 1988; Craig and Gatlin, 1992). The threonine requirement of red drum has also been quantified, at 2.3% of dietary protein (Boren and Gatlin, 1995). Established relationships between patterns of indispensable amino acids in muscle tissue and levels required in the diet may allow other amino acid requirements of red drum to be estimated after quantifying the requirement for only one or two of the most limiting amino acids (Wilson and Poe, 1985; Moon and Gatlin, 1991).

Lipids and fatty acids

The red drum, like most other carnivorous species, has been shown to efficiently utilize dietary lipid for energy (Williams and Robinson, 1988; Ellis and Reigh, 1991; Serrano *et al.*, 1992). A total of 7 and 11% menhaden oil in diets containing 40% crude protein have produced maximum weight gain of red drum. Lipid included as high as 28% of diet did not enhance growth of red drum relative to diets containing lower levels but did increase lipid deposition in the body (McGoogan and Gatlin, 1999). Marine oils containing HUFA of the linolenic acid (n-3) family are needed to satisfy the essential fatty acid requirements of red drum (Lochmann and Gatlin, 1993a). Juvenile red drum were determined to require eicosapentaenoic acid (20:5n-3) and docosahexaenoic acid (22:6n-3) at approximately 10% of dietary lipid (Lochmann and Gatlin, 1993a,b). The red drum also appeared to have a very limited ability to elongate and desaturate shorter-chain fatty acids (Lochmann and Gatlin, 1993a).

Carbohydrates

Fish do not have a specific dietary requirement for carbohydrates, but the presence of these compounds in diets may provide a rather inexpensive source of energy. However, the ability of fish to utilize dietary carbohydrate for energy varies considerably, with most carnivorous species having the most limited ability (NRC, 1993). Although the red drum is a carnivorous fish in nature, it has not generally been shown to be adversely affected by high levels of soluble carbohydrate in the diet, although lipid is used preferentially to carbohydrate. In two different studies, red drum generally had better weight gain and feed efficiency when fed diets containing more lipid and less carbohydrate relative to other diets estimated to be isoenergetic (Ellis and Reigh, 1991; Serrano *et al.*, 1992). Results from other studies have indicated that red drum was not adversely affected by as much as 35% soluble carbohydrate in the diet, although in some instances weight gain and feed efficiency were slightly reduced (Daniels and Robinson, 1986; Davis and Robinson, 1987; Williams and Robinson, 1988).

Fibrous carbohydrates, such as cellulose and hemicellulose, are essentially indigestible by fish and thus do not make a positive contribution to their nutrition. Typically, the level of crude fibre in red-drum diets is restricted to less than 7% to limit the amount of undigested material entering the culture system.

Vitamins and minerals

There are 15 vitamins established as being essential for terrestrial animals as well as for many of the fish species that have been extensively studied (NRC, 1993). Currently there is specific information on the quantitative dietary requirements of red drum for only choline (Craig and Gatlin, 1996) and ascorbic acid (Aguirre and Gatlin, 1999). Typically, a vitamin premix based on vitamin requirements of other fish species is added to commercial diets for red drum to provide adequate levels of each vitamin independent of levels in ingredients. This supplement will also provide a margin of safety for losses associated with diet processing and storage.

It has been established with several fish species that they require the same minerals as terrestrial animals for tissue formation and other metabolic functions, including osmoregulation (NRC, 1993). However, some water-borne minerals may contribute to satisfying the metabolic requirements of fish and interact with dietary requirements.

Of the macrominerals, phosphorus has received considerable attention because its presence in water and utilization by fish are generally limited. The total phosphorus requirement of red drum has been determined to be 0.86% of diet (Davis and Robinson, 1987). However, the availability of phosphorus from feedstuffs may vary considerably, as noted with red drum (Gaylord and Gatlin, 1996); thus, supplementing diets on the basis of available phosphorus is important. In addition to phosphorus supplementation, inclusion of 2% sodium chloride and/or 2% potassium chloride in practical diets has been shown to have positive effects on growth of red drum in fresh water and brackish water (6 p.p.t.), but no such effects were observed in full-strength artificial sea water (Gatlin *et al.*, 1992). The beneficial effects of adding a supplement of salt to diets for red drum in fresh water and brackish water appeared to be due to a metabolic need for these ions in hypotonic environments. Dietary requirements of red drum for other macrominerals have not been investigated.

Of the microminerals, zinc (Zn) has been demonstrated in some fish species to be most important to supplement in diets due to its low level in practical feedstuffs and/or interactions with other dietary components that may reduce its bioavailability (NRC, 1993). At this time, only the dietary zinc requirement of red drum has been determined (Gatlin *et al.*, 1991). A minimum requirement of 20–25 mg Zn kg^{-1} diet was quantified using semipurified diets, although higher levels are generally supplemented. Other micromineral requirements have not been determined for red drum at this time. However, an inexpensive trace-mineral premix is typically added to most nutritionally complete diets to ensure adequacy (NRC, 1993).

Practical Diets

Costs associated with diets and feeding generally constitute the largest variable expense in intensive production of red drum, as with other fish species; therefore, formulation of cost-effective diets can significantly influence the profitability of the aquacultural enterprise. In addition, red drum under intensive production schemes must derive essentially all nutrients from prepared diets; therefore, proper nutrition is critical to maintaining efficient growth and optimal health of the fish.

The accuracy of diet formulation can be improved if information about the digestibility or availability of various nutrients from feedstuffs is known. Such information has been obtained for red drum for most of the commonly used feedstuffs (Gaylord and Gatlin, 1996; McGoogan and Reigh, 1996). Feedstuffs of marine origin, including various fish-meals, have been most effective in diet formulations for carnivorous species such as red drum because they are generally quite palatable and high in protein, lipid and DE. However, these feedstuffs are usually quite expensive and may substantially increase the cost of diet formulations. Other feedstuffs from animal by-products have been used to replace fish-meal with some success (Moon and Gatlin, 1994; Meilahn *et al.*, 1996); however, their quality can be variable. Several protein feedstuffs of plant origin, such as soybean meal and cottonseed meal, are less expensive but have had limited use in diet formulations because they may be deficient in at least one indispensable amino acid and are usually less palatable to carnivorous species (Reigh and Ellis, 1992; Davis *et al.*, 1995). However, in one study, relatively high dietary levels of soybean meal could be included without reducing growth or feed intake of red drum if a minimum of 10% of dietary protein was provided by fish-meal (McGoogan and Gatlin, 1997).

A variety of by-products from grains such as maize and wheat have been used in fish diets to supply available carbohydrate for energy and improve pellet stability. Relatively high levels of these feedstuffs can be included in red-drum diets that are manufactured by extrusion processing, because this species is not adversely affected by soluble carbohydrate (Ellis and Reigh, 1991; Serrano *et al.*, 1992).

Lipids of marine origin are commonly included in diets of red drum to provide essential fatty acids and improve palatability. Marine lipids containing high levels of n-3 HUFA have also been shown to confer increased tolerance of red drum to cold water temperatures due to incorporation of these fatty acids into membrane lipids (Craig *et al.*, 1995). Alternative lipid sources, including saturated and relatively unsaturated animal and vegetable lipids, have also been shown to be well utilized by red drum, as long as the fish's requirement for n-3 HUFA is met (Craig and Gatlin, 1995; Tucker *et al.*, 1997). Due to this species' tendency to deposit large amounts of lipid in the liver and visceral cavity, alternative lipids, such as medium-chain triglycerides, have also been investigated (Craig and Gatlin, 1995; Davis *et al.*, 1999). These compounds generally reduce lipid deposition but do not support rapid fish growth. Another type of dietary lipid that has been shown to enhance the growth of very young red drum is phospholipid in the form of soybean lecithin (Craig and Gatlin, 1997). However, it is not likely that the cost of phospholipid addition to diets composed of practical feedstuffs is warranted once the fish are larger than 50 g. Commercially available vitamin and mineral premixes are also commonly included in diets for red drum, although more specific information about requirements for these nutrients is needed.

A variety of production diets have been evaluated with red drum in controlled feeding trials (McGoogan and Gatlin, 1997; Thoman *et al.*, 1999). Results of these studies generally indicate that a crude protein level of 40% or greater is needed to support the most rapid growth of red drum. Providing the proper amount of dietary energy relative to protein (and other nutrients) is also critical for ensuring adequate nutrient intake, since the fish's intake may be regulated by energy (NRC, 1993). However, a recent study with red drum indicated that diet intake was more directly influenced by body size than by dietary energy density (Thoman *et al.*, 1999). A balanced dietary energy-to-protein ratio is also important in order to maximize use of dietary protein for protein synthesis rather than for catabolism for energy. Optimizing the dietary energy-to-protein ratio was able to reduce ammonia excretion of red drum (McGoogan and Gatlin, 1999, 2000).

Table 11.1 shows a practical diet formulation that meets all known requirements of red drum. Research to determine the cost-effectiveness of different diet formulations for red drum has been rather limited to date but deserves further consideration.

Juvenile red drum are typically fed crumbled pellets until they reach a size at which extruded pellets can be readily consumed. Production diets are often produced by extrusion processing because the resulting pellets have high water stability and low density, which may allow them to float. This characteristic assists the aquaculturist in monitoring the fish's intake, especially in large culture systems. A general guide relating diet particle size to the size of red drum is presented in Table 11.2.

Table 11.1. A model diet formulation containing 40% crude protein and 15.1 kJ digestible energy g^{-1} for growing red drum from juvenile to adult size (from Gaylord and Gatlin, 1996).

Ingredient	g kg^{-1} as fed
Menhaden fish-meal	32.3
Soybean meal (48% protein)	26.0
Meat and bone-meal	10.0
Wheat grain	26.0
Menhaden fish-oil	5.0
Mineral premix*	0.1
Vitamin premix*	0.1
Dicalcium phosphate	0.5

* Similar to those used for other warm-water fish species.

Table 11.2. Suggested diet particle sizes for red drum of various sizes.

Fish weight (g)	Diet particle size (mm)
0.25–0.5	0.25
0.6–1.0	0.50
1.1–2.0	1.5
2.25–4.0	2.0
5.0–20	2.5
25–50	3.0
55–150	3.9
175–300	4.7
325–500	6.25
> 500	9.4

Feeding Practices

Appropriate feeding schedules and practices must be employed in the aquacultural production of red drum, as with other species, to ensure that maximum benefit can be derived from prepared diets. It is generally desirable to provide red drum with as much diet as they will consume on a regular basis. However, excessive feeding should normally be avoided because it not only wastes expensive diet but may also cause the water quality to deteriorate.

Feeding schedules for red drum, like other species, are influenced by fish size and water temperature. In general, smaller fish consume more feed when expressed as a percentage of body weight than do larger fish. Red drum weighing less than 5 g typically consume prepared diets at over 7% of body weight day^{-1}. In contrast, 50 g red drum consumed diet at approximately 5% of body weight day^{-1} and fish over 250 g consumed less than 2% of body weight day^{-1} (Thoman *et al.*, 1999). Smaller fish should also be fed more frequently than larger fish due to their higher metabolic rates. In addition, water temperature may also significantly influence feed intake, with reduced consumption occurring above and below the optimal temperature range. Specific feeding schedules for red drum have been empirically derived. The means by which diet is administered to red drum is largely dictated by the design and size of the culture systems.

References

Aguirre, P. and Gatlin, D.M., III (1999) Dietary vitamin C requirement of red drum (*Sciaenops ocellatus*). *Aquaculture Nutrition* 5, 247–249.

Arnold, C.R. (1988) Controlled year-round spawning of red drum *Sciaenops ocellatus* in captivity. In: Arnold, C.R., Holt, G.J. and Thomas, P. (eds) *Red Drum Aquaculture. Proceedings of a Symposium on the Culture of Red Drum and Other Warm Water Fishes.* Contributions in Marine Science Supplement to Vol. 30, Marine Science Institute, Port Aransas, Texas, pp. 65–70.

Bass, R.J. and Avault, J.W. Jr (1975) Food habits, length–weight relationships, condition factor, and growth of juvenile red drum, *Sciaenops ocellata*, in Louisiana. *Transactions of the American Fisheries Society* 104, 35–45.

Boren, R.S. and Gatlin, D.M., III (1995) Dietary threonine requirement of juvenile red drum (*Sciaenops ocellatus*). *Journal of the World Aquaculture Society* 26, 279–283.

Brown, P.B., Davis, D.A. and Robinson, E.H. (1988) An estimate of the dietary lysine requirement of juvenile red drum *Sciaenops ocellatus*. *Journal of the World Aquaculture Society* 19, 109–112.

Craig, S.R. and Gatlin, D.M., III (1992) Dietary lysine requirement of juvenile red drum *Sciaenops ocellatus*. *Journal of the World Aquaculture Society* 23, 133–137.

Craig, S.R. and Gatlin, D.M., III (1995) Coconut oil and beef tallow, but not tricaprylin, can replace menhaden oil in the diet of red drum (*Sciaenops ocellatus*) without adversely affecting growth and fatty acid composition. *Journal of Nutrition* 125, 3041–3048.

Craig, S.R. and Gatlin, D.M., III (1996) Dietary choline requirement of juvenile red drum (*Sciaenops ocellatus*). *Journal of Nutrition* 126, 1696–1700.

Craig, S.R. and Gatlin, D.M., III (1997) Growth and body composition of juvenile red drum (*Sciaenops ocellatus*) fed diets containing lecithin and supplemental choline. *Aquaculture* 151, 259–267.

Craig, S.R., Arnold, C.R. and Holt, G.J. (1994) The effects of enriching live foods with highly unsaturated fatty acids on the growth and fatty acid composition of larval red drum *Sciaenops ocellatus*. *Journal of the World Aquaculture Society* 25, 424–431.

Craig, S.R., Neill, W.H. and Gatlin, D.M., III (1995) Effects of dietary lipid and environmental salinity on growth, body composition, and cold tolerance of juvenile red drum (*Sciaenops ocellatus*). *Fish Physiology and Biochemistry* 14, 49–61.

Daniels, W.H. and Robinson, E.H. (1986) Protein and energy requirements of juvenile red drum (*Sciaenops ocellatus*). *Aquaculture* 53, 243–252.

Davis, D.A. and Robinson, E.H. (1987) Dietary phosphorus requirement of juvenile red drum *Sciaenops ocellatus*. *Journal of the World Aquaculture Society* 18, 129–136.

Davis, D.A., Jirsa, D. and Arnold, C.R. (1995) Evaluation of soybean proteins as replacements for menhaden fish meal in practical diets for the red drum *Sciaenops ocellatus*. *Journal of the World Aquaculture Society* 26, 48–58.

Davis, D.A., Lazo, J.P. and Arnold, C.R. (1999) Response of juvenile red drum (*Sciaenops ocellatus*) to practical diets supplemented with medium chain triglycerides. *Fish Physiology and Biochemistry* 21, 235–247.

Ellis, S.C. and Reigh, R.C. (1991) Effects of dietary lipid and carbohydrate levels on growth and body composition of juvenile red drum, *Sciaenops ocellatus*. *Aquaculture* 97, 383–394.

Gatlin, D.M., III, O'Connell, J.P. and Scarpa, J. (1991) Dietary zinc requirement of the red drum, *Sciaenops ocellatus*. *Aquaculture* 92, 259–265.

Gatlin, D.M., MacKenzie, D.S., Craig, S.R. and Neill, W.H. (1992) Effects of dietary sodium chloride on red drum juveniles in waters of various salinities. *Progressive Fish-Culturist* 54, 220–227.

Gaylord, T.G. and Gatlin, D.M., III (1996) Determination of digestibility coefficients of various feedstuffs for red drum (*Sciaenops ocellatus*). *Aquaculture* 139, 303–314.

Holt, J. (1990) Growth and development of red drum eggs and larvae. In: Chamberlain, G.W., Miget, R.J. and Haby, M.G. (eds) *Red Drum Aquaculture*. Texas A&M University Sea Grant College Program, Galveston, Texas, pp. 46–50.

Holt, J., Godbout, R. and Arnold, C.R. (1981) Effects of temperature and salinity on egg hatching and larvae survival of red drum *Sciaenops ocellata*. *Fishery Bulletin* 79, 569–573.

Lochmann, R.T. and Gatlin, D.M., III (1993a) Evaluation of different types and levels of triglycerides, singly and in combination with different levels of *n*-3 highly unsaturated fatty acid ethyl esters in diets of juvenile red drum, *Sciaenops ocellatus*. *Aquaculture* 114, 113–130.

Lochmann, R.T. and Gatlin, D.M., III (1993b) Essential fatty acid requirement of juvenile red drum (*Sciaenops ocellatus*). *Fish Physiology and Biochemistry* 12, 221–235.

McGoogan, B.B. and Gatlin, D.M., III (1997) Effects of replacing fish meal with soybean meal in diets for red drum *Sciaenops ocellatus* and potential for palatability enhancement. *Journal of the World Aquaculture Society* 28, 374–385.

McGoogan, B.B. and Gatlin, D.M., III (1998) Metabolic requirements of red drum, *Sciaenops ocellatus*, for protein and energy based on weight gain and body composition. *Journal of Nutrition* 128, 123–129.

McGoogan, B.B. and Gatlin, D.M., III (1999) Dietary manipulations affecting growth and nitrogenous waste production of red drum, *Sciaenops ocellatus*. I. Effects of dietary protein and energy levels. *Aquaculture* 178, 333–348.

McGoogan, B.B. and Gatlin, D.M., III (2000) Dietary manipulations affecting growth and nitrogenous waste production of red drum, *Sciaenops ocellatus*. II. Effects of energy level and nutrient density at various feeding rates. *Aquaculture* 182, 271–285.

McGoogan, B.B. and Reigh, R.C. (1996) Apparent digestibility of selected ingredients in red drum (*Sciaenops ocellatus*) diets. *Aquaculture* 141, 233–244.

Matlock, G.C. (1990) The life history of red drum. In: Chamberlain, G.W., Miget, R.J. and Haby, M.G. (eds) *Red Drum Aquaculture*. Texas A&M University Sea Grant College Program, Galveston, Texas, pp. 1–21.

Meilahn, C.W., Davis, D.A. and Arnold, C.R. (1996) Effects of commercial fish meal analogue and menhaden fish meal on growth of red drum fed isonitrogenous diets. *Progressive Fish-Culturist* 58, 111–116.

Moon, H.Y. and Gatlin, D.M., III (1989) Amino acid nutrition of the red drum (*Sciaenops ocellatus*): determination of limiting amino acids and development of a suitable amino acid test diet. In: *Proceedings of the Third International Symposium on Feeding and Nutrition in Fish*. Toba, Japan. Japan Translation Center Ltd, Tokyo, pp. 201–208.

Moon, H.Y. and Gatlin, D.M., III (1991) Total sulfur amino acid requirement of juvenile red drum, *Sciaenops ocellatus*. *Aquaculture* 95, 97–106.

Moon, H.Y.L. and Gatlin, D.M., III (1994) Effects of dietary animal proteins on growth and body composition of the red drum (*Sciaenops ocellatus*). *Aquaculture* 120, 327–340.

NRC (National Research Council) (1993) *Nutrient Requirements of Fish*. National Academy Press, Washington, DC, 114 pp.

Porter, C.W. and Maciorowski, A.F. (1984) Spotted seatrout fingerling production in saltwater ponds. *Journal of the World Mariculture Society* 15, 222–232.

Reigh, R.C. and Ellis, S.C. (1992) Effects of dietary soybean and fish-protein ratios on growth and body composition of red drum (*Sciaenops ocellatus*) fed isonitrogenous diets. *Aquaculture* 104, 279–292.

Robinson, E.H. (1988) Nutritional requirements of red drum: a review. In: Arnold, C.R., Holt, G.J. and Thomas, P. (eds) *Red Drum Aquaculture. Proceedings of a Symposium on the Culture of Red Drum and Other Warm Water Fishes*. Contributions in Marine Science Supplement to Vol. 30, Marine Science Institute, Port Aransas, Texas, pp. 11–20.

Sandifer, P.A., Hopkins, J.S., Stokes, A.D. and Smiley, R.D. (1993) Experimental pond grow-out of red drum, *Sciaenops ocellatus*, in South Carolina. *Aquaculture* 118, 217–228.

Serrano, J.A., Nematipour, G.R. and Gatlin, D.M., III (1992) Dietary protein requirement of the red drum (*Sciaenops ocellatus*) and relative use of dietary carbohydrate and lipid. *Aquaculture* 101, 283–291.

Sorgeloos, P. and Personne, G. (1975) Technological improvements for the cultivation of invertebrates as food for fishes and crustaceans. II. Hatching and culturing of brine shrimp, *Artemia salina* L. *Aquaculture* 6, 303–317.

Thoman, E.S., Davis, D.A. and Arnold, C.R. (1999) Evaluation of growout diets with varying protein and energy levels for red drum (*Sciaenops ocellatus*). *Aquaculture* 176, 343–353.

Treece, G.D. and Wohlschlag, N. (1990a) Raising food organisms for intensive larval culture: I. Algae. In: Chamberlain, G.W., Miget, R.J. and Haby, M.G. (eds) *Red Drum Aquaculture*. Texas A&M University Sea Grant College Program, Galveston, Texas, pp. 57–65.

Treece, G.D. and Wohlschlag, N. (1990b) Raising food organisms for intensive larval culture: III. Artemia. In: Chamberlain, G.W., Miget, R.J. and Haby, M.G. (eds) *Red Drum Aquaculture*. Texas A&M University Sea Grant College Program, Galveston, Texas, pp. 71–77.

Tucker, J.W. Jr, Lellis, W.A., Vermeer, G.K., Roberts, D.E. Jr and Woodward, P.N. (1997) The effects of experimental starter diets with different levels of soybean or menhaden oil on red drum (*Sciaenops ocellatus*). *Aquaculture* 149, 323–339.

Williams, C.D. and Robinson, E.H. (1988) Response of red drum to various dietary levels of menhaden oil. *Aquaculture* 70, 107–120.

Wilson, R.P. and Poe, W.E. (1985) Relationship of whole body and egg essential amino acid patterns to amino acid requirement patterns in channel catfish, *Ictalurus punctatus*. *Comparative Biochemistry and Physiology* 80B, 385–388.

Wohlschlag, N.S., Maotang, L. and Arnold, C.R. (1990) Raising food organisms for intensive larval culture: II. Rotifers. In: Chamberlain, G.W., Miget, R.J. and Haby, M.G. (eds) *Red Drum Aquaculture*. Texas A&M University Sea Grant College Program, Galveston, Texas, pp. 66–70.

Southern Bluefin Tuna, *Thunnus maccoyii* 12

Brett Glencross, Chris Carter, John Gunn, Robert van Barneveld, Kirsten Rough and Steven Clarke
FRDC-CRC, Southern Bluefin Tuna Subprogram, PO Box 120, Henley Beach, SA 5022, Australia

Introduction

The southern bluefin tuna (SBT) (*Thunnus maccoyii*, Castelnau) is one of two species of 'bluefin' tunas and the key tuna species currently being farmed in Australia. Reported at sizes of up to 220 cm in length and 200 kg in weight, the SBT is one of the largest members of the Scombridae (mackerels and tunas) (Fig. 12.1). Its close relative, the northern bluefin tuna (NBT) (*Thunnus thynnus*) is the only species in the family that grows to a larger size (Kailola *et al.*, 1993). The NBT is the species that predominates in the developing tuna aquaculture industries in the Mediterranean, North America and Japan.

Tunas are unusual among teleosts because they are warm-bodied (Carey and Teal, 1969). They are able to maintain body temperatures above ambient water temperature through the use of counter-current heat exchangers. Bluefin tunas have the most advanced heat-exchange systems of all the tunas, and are able to thermally insulate a number of organs and tissues (e.g. brain, viscera and muscles) from ambient water and from other tissues (Stevens and McLeese, 1984). SBT are high-level to apex predators even as juveniles, and have very broad dietary ranges (Young *et al.*, 1997).

The SBT aquaculture industry runs seasonally. Each year, beginning in the Australian summer (December–February), juvenile fish of approximately 20 kg (2–4 years of age) are caught by purse-seine nets in the Great Australian Bight, off the southern Australian coastline. The fish are then transferred to specially designed towing pontoons and slowly towed back to Boston Bay near Port Lincoln in South Australia. During the next 4–8 months, through autumn, winter and into spring, the tuna are primarily fed baitfish to improve both their condition and their biomass. By October, most of the fish have been harvested and the industry begins preparations for the return of the fish to the Great Australian Bight during the following summer.

Fig. 12.1. Southern bluefin tuna on the market floor at Tsukiji market in Japan.

The product is sold almost exclusively to the Japanese sashimi markets, the largest and most influential of which is Tsukiji market in Tokyo (Fig. 12.2). Other key markets for farmed tuna include Osaka, Nagoya and Yokohama, though most regional markets will service some quantities of farmed tuna (Glencross, 1999).

Diet development for SBT has been particularly challenging for many reasons. The costs of research with this species are extremely high, having to maintain operations of numerous sea-cages, boats, crews and associated infrastructure. Each of the 1000 SBT used annually for research have been worth close to US$600. The annual cost of diet for research is approximately US$60,000–100,000. Working with SBT presents many problems other than the financial costs. The sheer size of the fish involved (usually around 1 m in length and 20 kg in weight) and the nature of its physical requirements demand that experiments be conducted in sea cages and that handling be kept to an absolute minimum. For similar reasons, a large number of replicates to minimize experimental variance and assist the design of statistically sound experiments has been limited. Because of these difficulties several alternative strategies have been developed to examine some growth parameters, such as levels of tissue and blood nucleic acids and protein, and *in vitro* digestibility studies (Carter *et al.*, 1998, 1999). Although these methods hold much promise, to date they have not been able to replace growth trials as the key form of performance assessment.

Fig. 12.2. Tsukiji wholesale market floor – fresh chilled product, October 1998.

Nutrient Requirements

The difficulties associated with conducting structured and highly replicated trials with SBT have prevented the determination of nutritional requirements through standard dose–response techniques (Glencross *et al.*, 1999a). As a consequence relative requirements have been determined based on the modelling of the performance of SBT fed various practical diet formulations. This approach can provide useful insights that can be applied in the early development of artificial feeds within the constraints of a restricted research environment. To date, it has contributed to the rapid development of a moist feed that supports performance equivalent to that obtained with SBT fed baitfish under research conditions (i.e. small pontoons, low stock density and controlled feeding).

Protein and amino acids

No specific studies examining protein utilization by SBT have been attempted to date. Despite this, several alternative and/or indirect methods of measuring protein utilization by and amino acid requirements (Table 12.1) of SBT have been examined (van Barneveld *et al.*, 1997; Carter *et al.*, 1998, 1999; Glencross *et al.*, 1999a).

Considerable effort has been made to develop both *in vitro* assessment techniques and surrogate assessment methods (Carter *et al.*, 1998, 1999; Bransden *et al.*, 1999; Table 12.2). Results from these studies have proved useful in the refinement of moist-diet development for SBT (Table 12.3). More recent studies have examined growth rate as a function of protein intake, taking into consideration the effect of water temperature and dietary energy (Glencross *et al.*, 1999b). These interpretations suggest that optimal daily protein intake for growth is no greater than 10 g day^{-1} kg^{-1} of biomass.

The capacity for protein synthesis (mg ribonucleic acid (mg (RNA) g^{-1} protein) provides a key indicator of protein metabolism and a value of 8 has been measured in SBT for approximate maintenance (Carter *et al.*, 1998). This value is at least four times higher than that predicted for a fish based on the weight, temperature and feed intake of the SBT (Houlihan *et al.*, 1995) but very similar to that of an unfed mammal (Milward *et al.*, 1973). A consequence would be significantly higher energy demands associated with protein turnover in SBT than in other fishes, even at zero growth. This suggests a close relationship between protein metabolism and energy requirements in SBT.

Table 12.1. Essential amino acid composition of key muscle types (g kg^{-1} on a dry-matter basis) of SBT and a proposed ideal amino acid balance, relative to lysine (data from van Barneveld *et al.*, 1997).

Amino acid	White muscle (g kg^{-1})	Red muscle (g kg^{-1})	Ideal balance (%)
Methionine	10.9	11.0	35
Threonine	17.2	18.2	56
Valine	19.2	20.4	63
Isoleucine	17.1	18.7	57
Leucine	27.6	29.3	90
Phenylalanine	14.9	15.1	47
Lysine	30.7	32.5	100
Histidine	30.3	30.5	96
Arginine	22.3	23.9	73

Table 12.2. *In vitro* digestibility of key feed ingredients by SBT digesta enzymes (data derived from Carter *et al.*, 1999).

Ingredient (nitrogen %)	Digestible protein	Digestible organic matter
Fish-meal – Tasmanian (11.1%)	81.7 ± 2.01%	74.7 ± 0.45%
Fish-meal – Chilean (9.8%)	65.8 ± 0.65%	54.7 ± 1.66%
Pilchards (*Sardinops sagax*) (12.2%)	82.3 ± 1.53%	78.3 ± 2.73%
Squid meal (10.6%)	36.7 ± 1.08%	27.2 ± 1.74%
Casein (13.3%)	98.3 ± 0.19%	97.8 ± 2.03%
Wheat gluten (11.5%)	94.7 ± 2.39%	71.4 ± 1.90%
Soybean meal (6.9%)	45.1 ± 3.48%	49.2 ± 3.19%
Wheat flour (1.8%)	88.3 ± 1.02%	58.31 ± 0.98%

The indispensable amino acid requirements of SBT have also been estimated based on the composition of both red and white muscle (Table 12.1). From this an ideal amino acid balance has been estimated (van Barneveld *et al.*, 1997). Similar to other fish species lysine has been estimated as the key limiting amino acid. As a consequence, diets have generally been formulated to consider this limitation within the scope of key ingredients used.

Energy

Based on the non-parametric analysis of data from several moist diets and baitfish diets fed under various feeding regimes during the 1998 season trials, a response surface model was developed. This model suggested that both protein and energy were key response factors in the diet of SBT, but that dietary energy had a greater influence than dietary protein (Glencross *et al.*, 1999b).

Modelling of the nutritional value of feed with respect to the effects of dietary energy on feed conversion suggested that dietary fat levels were a key factor in influencing the energetic value of the diets to SBT. This aspect will be discussed further in the following section on lipids.

The apparent enormous dietary energy requirement of SBT is consistent with what is known of their general physiological requirement to thermoregulate. The energetic processes required to generate metabolic heat in an environment that is constantly against a thermal gradient is likely to place an enormous energetic demand on these fish (Carey *et al.*, 1984).

Table 12.3. Fatty acid composition of SBT muscle lipid (data derived from Nichols *et al.*, 1998).

Fatty acid	% of total fatty acids
16:1n-7	0.5
18:0	11.1
18:1n-9	7.8
18:2n-6	0.9
18:3n-3	0.1
20:1n-9	0.2
20:4n-6	3.0
20:5n-3	4.8
22:1n-11	0.0
22:6n-3	44.1
SFA	34.2
MUFA	11.5
PUFA	54.3
n-6	4.4
n-3	50.0

PUFA, polyunsaturated fatty acids; SFA, saturated fatty acids; MUFA, mono-unsaturated fatty acids.

The decline in diet intake throughout a production season and the slowing of growth has provided a key indicator to levels of dietary intake required for maintenance. At the maintenance level, where growth has ceased, an average of about 150 kJ of gross energy (GE) was consumed per kilogram of fish per day. In most cases, this was under conditions where ambient water temperatures were less than 15°C. Based on the data from Glencross *et al.* (2002), the energetic requirement for 1 kg of gain, at a temperature of about 16°C was 56 MJ of GE. This was about half of that reported from earlier studies (Glencross *et al.*, 1999b), though still close to twice that estimated for Atlantic salmon, *Salmo salar* (NRC, 1993). This variability highlights the dramatic improvements made in diet performance and nutritional assessment over a few years.

Similar to the recent studies examining growth rates as a function of average protein intake has been the evaluation of GE intakes as a function of growth. As would be anticipated, GE intakes were consistent with protein intake assessments (Fig. 12.3). The plateauing of growth in this case was observed with a daily energy intake of about 300 kJ kg^{-1} day^{-1}. As with the previous interpretations of protein intake it should be noted that there was no compensation for variation in ambient water temperatures.

Lipids and fatty acids

Lipids

Dietary lipids play an important part in the nutrition of SBT, primarily because of their apparent enormous need for energy. To date this has been demonstrated by the relationship between the fat content of a range of moist pellets and baitfish, and the food conversion ratio (FCR) of these feeds on a dry-matter basis (Fig. 12.4). This interpretation suggests that the nutritional performance of the feed can be improved by increasing the dietary fat content (Glencross *et al.*, 1999b).

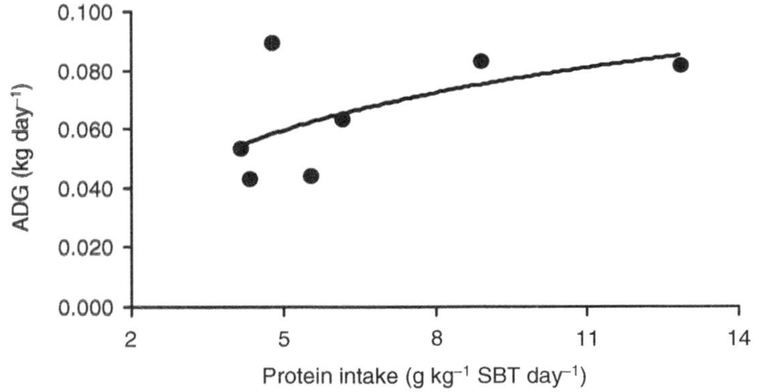

Fig. 12.3. Average daily gain (ADG) as a function of daily dietary protein intake (g kg^{-1} SBT day^{-1}). (Data derived from Glencross *et al.*, 2002.)

Fatty acids

Examination of the fatty acid profile of lipid from flesh of wild-caught SBT (Table 12.3) reveals very high levels of the long-chained, n-3 polyunsaturated fatty acid (PUFA) docosahexaenoic acid (DHA) (22:6n-3) (Nichols *et al.*, 1998). Levels of eicosapentaenoic acid (EPA) (20:5n-3) are notably low, as are other long-chained PUFAs. The high level of DHA and a high n-3 : n-6 ratio suggest that SBT may have high requirements for the n-3 series of fatty acids, notably DHA.

Carbohydrates

Characterization of the digestive enzyme capabilities of SBT has identified some α-amylase activity, though not enough to be considered as viable for large-scale carbohydrate digestion (van Barneveld *et al.*, 1997). As a consequence of this finding, starch has been used as a filler/binder in some experimental diets.

Vitamins and minerals

No studies have yet examined the vitamin or mineral requirements for SBT. Supplementation of vitamins to baitfish used as feed has been undertaken by some sectors of the industry, despite no conclusive proof as to its value. Vitamin fortification has been attempted by either coating the baitfish in a vitamin-containing gel or by direct injection of the baitfish with a vitamin premix. Anecdotal evidence suggests that vitamin fortification is only of value when poorer-quality feeds (e.g. unfresh and rancid fish) were used and/or the SBT

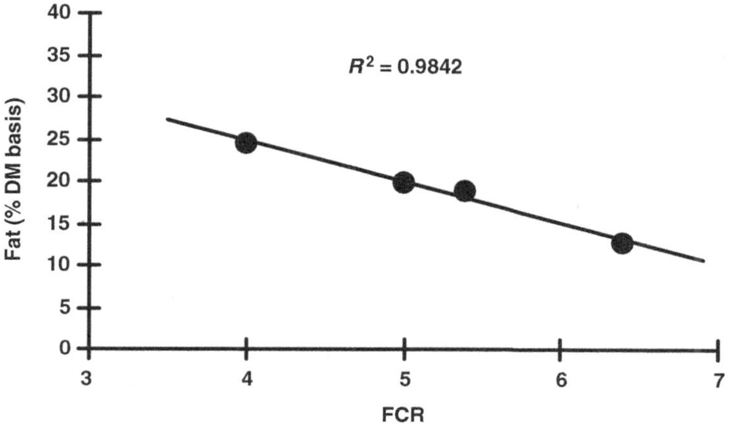

Fig. 12.4. Food conversion ratio (FCR) response to varying levels of fat (dry-matter (DM) fat) in various diets, modelled on a dry-matter consumption basis (data derived from Glencross *et al.*, 1999a).

were in poor condition. Vitamin premixes have been based primarily on those formulated for salmonids (van Barneveld *et al.*, 1997).

Supplementation of baitfish feeding with a vitamin-enriched moist pellet has been attempted by both the R&D and industry sectors as a pilot initiative (Glencross *et al.*, 1998). This practice showed considerable improvements in survival of SBT, though more studies are needed. The use of vitamin-fortified diets has also been shown in pilot studies to be beneficial in fortifying the α-tocopherol content of the flesh.

Practical Diets

The key practicalities of diets for the SBT aquaculture industry have been dictated by the need for large quantities of readily consumed, nutritious diet. Baitfish of various species have proved to be an ideal food, though the risks associated with using baitfish diets are many (Clarke *et al.*, 1997). To date, satisfactory manufactured diets for commercial use are available, but need further improvement. Though considerable progress has been made in recent years with the development of specialized moist pellets, the practicalities of using these feeds have not seen them gain widespread acceptance. Development of better bound and more attractive moist pellets and/or development of semi-dry or soft-dry pellets, primarily through extrusion, appears to hold the key to future progress in diet development for this species.

Currently the SBT aquaculture industry exclusively uses baitfish as the prime diet source. The actual species fed to the juvenile SBT varies throughout the year and from year to year, depending on availability of local and international baitfish stocks. Primarily, the pilchard (*Sardinops sagax*) has been the feed of choice. Origins, quality and composition of this feed, however, have varied considerably; they range from local Australian stocks to those from California and/or Japan. On some farms, Atlantic herring (*Clupea harengus*) and small blue mackerel (*Scomber scombrus*) have also been fed. Some farmers have used whole squid as a supplement. Some vitamin supplements have also been used.

To facilitate the weaning process of the juvenile SBT, a moist (400 g kg^{-1} water content) pellet has been fed to the fish during the initial phases of the production cycle in midsummer, when the fish feed extremely vigorously. The moist pellets typically have a water content in excess of 30% (Table 12.4). These have been primarily made through the addition of a formulated meal to ground baitfish to form a moist, doughlike mash, which is screw-pressed to make pellets. Under research conditions, growth performance of SBT fed moist diets is similar to that of fed baitfish (Fig. 12.5).

Most dry diets trialled to date have been hard-extruded products. While the proximate (relative protein and fat contents) composition of these feeds has not differed dramatically from that of the other feeds (Table 12.4), progress with them has been slow. The development of dry manufactured diets has been limited, primarily due to the problems encountered in weaning the juvenile SBT on to a dry artificial diet. Though weaning on to moist diets has been

Table 12.4. Proximate composition (protein, fat and moisture) of key experimental diets. Values are means ± SEM (n = 3).

	Pilchards	Dry pellets	Moist pellets
Protein (%)	17.7 ± 1.8	46.4 ± 1.1	35.9 ± 1.1
Fat (%)	7.4 ± 2.6	13.1 ± 0.6	12.2 ± 1.0
Water (%)	68.9 ± 2.1	6.0 ± 0.3	38.4 ± 0.8

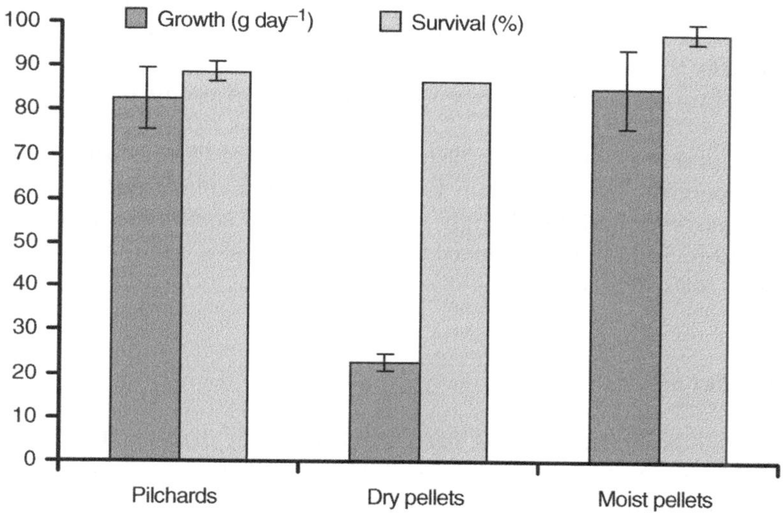

Fig. 12.5. Performance of juvenile SBT fed one of three generic feed types (Glencross *et al.*, 1999a).

particularly successful, similar levels of success have not yet been achieved with dry pellets.

Feeding Practices

The feeding practices associated with SBT aquaculture have developed primarily to accommodate the tuna's huge appetite and minimize the labour intensity required. The key feeding practices vary depending on the type of food being fed.

Feeding with baitfish

Despite their generalist diet in the wild, SBT exhibit a high degree of selectivity when feeding in captivity. Preferred food appears to be clupeoids, the pilchard (*Sardinops sagax*) and other related species. Squid and other species of fish have been used with varying success. If SBT have been fed pilchards when first acclimatized into the farms, it has been very difficult to wean them on to other food types (including moist pellets).

Baitfish are fed to SBT either after thawing or in frozen blocks. Feeding of thawed baitfish is carried out by hand, requiring the shoveling of thawed baitfish into the sea cages. Some farmers have used pumps to move baitfish from boat holds to the cages via pipes and hoses. In contrast, frozen-block feeding involves placing 30 kg frozen blocks of baitfish into floating feed cages within each sea cage. As the blocks of baitfish thaw, the thawed baitfish fall through the mesh of the feed cages, where they are then consumed by the SBT. Frozen-block feeding gained considerable popularity in the late 1990s as a means of feeding almost continuously through the day, while minimizing labour inputs. Many operators currently use a combination of frozen-block and thawed feeding to ensure optimal FCR.

Feed intakes of SBT fed baitfish vary seasonally. During periods of high water temperatures (> 20°C), daily feed intake has been observed to be as high as 15% of their body weight, though 10–12% is more common. At lower water temperatures, feed intake is reduced to a level as low as 2% of their body weight (Glencross *et al.*, 2001).

Feeding with pellet diets

To date, all pellet feeding has been carried out by hand, although recent developments have examined the use of mechanized feeders. Typically this has been carried out twice daily, reducing to a single feeding during the later stages of the growing season, in response to slower feeding activity (Glencross *et al.*, 2002). Hand-feeding of moist pellets has been carried out in a similar fashion to that of baitfish, usually by shovelling chilled (not frozen) pellets to the feeding fish. Feeding has usually been regulated by visual assessment of feeding activity. Since 1998, underwater surveillance cameras have also been used to assist in determining the feeding activity and estimating feed losses on the floor of the sea cages (Glencross *et al.*, 2002). Total diet intake of pellets has always been less than that of baitfish, though it is greater on a dry-matter basis (Glencross *et al.*, 1999b).

Feeding frequency

Research over the last 3 years has indicated that feeding SBT 7 days a week does not result in significantly higher growth rates or better condition factors than a feeding strategy in which the fish are fed on 5 or 6 days a week. The industry now uses a variety of feeding strategies, with most farmers feeding 5–6 days a week during the summer growing season and less during winter.

By tracking the variations in visceral temperature using archival tags (Gunn *et al.*, 1994), it has been possible to record over periods of weeks to months when SBT have eaten in the farms (Fig. 12.6). Archival tag data have been used to compare the feeding patterns of fish fed under different feeding strategies to answer the key question of whether fish eat at every opportunity and, if not, is this related to how often they are fed? In 1999, the feeding frequency of fish fed to satiation, twice daily, 4, 5 and 6 days per week was compared, once in summer and again in winter. Table 12.5 shows the proportion of days on which food was offered and feeding took place. These show that, during summer, there was very little difference between treatments in the frequency of feeding. However, in winter, there is clear evidence that the fish that were fed only 4 days per week were feeding more frequently than those fed 5 and 6 days. During the summer, fish fed 5 and 6 days a week ate more and as a result they gained more weight than those fed 4 days per week. During the winter, fish fed 4 days per week fed more often and ate more at each feeding than those fed 5 and 6 days per week. As

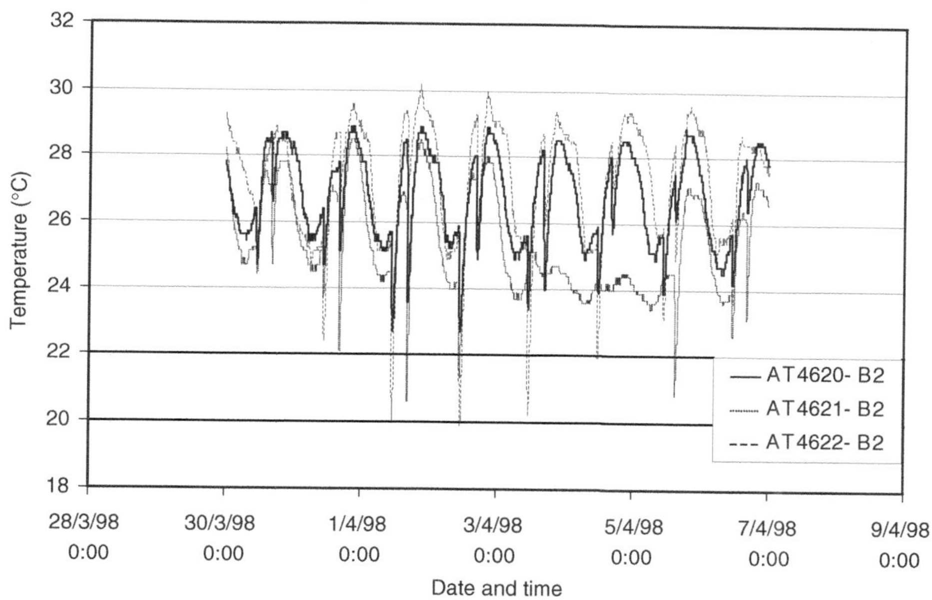

Fig. 12.6. Variation in visceral temperature in three SBT fed twice each day, 7 days per week. Note the differences in the visceral temperature and the similarity in the timing of feeding in each fish as indicated by the sharp rise in visceral temperature.

Table 12.5. Proportion of days on which a fish was offered food and a feeding event was recorded in the archival tag data.

	4 days week^{-1}	5 days week^{-1}	6 days week^{-1}
Summer	95.7%	95.3%	95.6%
Winter	78.3%	55.6%	50.60
Number of fish	13	11	9
(number of days fed summer/winter)	(46/49)	58/59)	(70/69)

a result, they either grew faster or held their weight, whereas the fish fed 5 and 6 days per week lost weight. By the end of the two experiments, there were no significant differences in the overall growth rates of fish in the three treatments.

Acknowledgements

We acknowledge the support of the Australian Fisheries Research and Development Corporation (FRDC), the Cooperative Research Centre for Aquaculture (CRC), the Tuna Boat Owners Association of South Australia, PIVOT Ltd, Ridley AgriProducts, the South Australian Research and Development Institute and the Commonwealth Scientific and Industrial Research Organization Division of Marine Research.

References

Bransden, M.P., Carter, C.G. and Nowak, B.F. (1999) Atlantic salmon (*Salmo salar*) as a surrogate species for the development of an artificial diet for the southern bluefin tuna (*Thunnus maccoyii*). In: *Proceedings of the World Aquaculture Society*. World Aquaculture Society, Sydney, Australia, p. 99.

Carey, F.G. and Teal, J.M. (1969) Regulation of body temperature by the bluefin tuna. *Comparative Biochemistry and Physiology* 28, 205–213.

Carey, F.G., Teal, J.M. and Stevens, E.D. (1984) Bluefin tuna warm their viscera during digestion. *Journal of Experimental Biology* 109, 1–20.

Carter, C.G., Seeto, G.S., Smart, A., van Barneveld, R.J. and Clarke, S.M. (1998) Correlates of growth in farmed southern bluefin tuna, *Thunnus maccoyii* (Castelnau). *Aquaculture* 161, 157–170.

Carter, C.G., Bransden, M.P., van Barneveld, R.J. and Clarke, S.M. (1999) Alternative methods for nutrition research on the southern bluefin tuna, *Thunnus maccoyii* (Castelnau): *in vitro* digestibility. *Aquaculture* 179, 57–70.

Clarke, S., Smart, A., van Barneveld, R. and Carter, C. (1997) The development and optimisation of manufactured feeds for farmed southern bluefin tuna. *Austasia Aquaculture*, 11(3), 59–62.

Glencross, B.D. (1999) *Tuna Market and Product Quality Aspects, Japan October 1998*. SARDI Research Report Series No. 45, SARDI, Adelaide, South Australia, Australia, 44 pp.

Glencross, B.D., van Barneveld, R.J., Foster, C.K., Bayly, G.B. and Clarke, S.M. (1998) Supplementing the bait-fish diet of farmed southern bluefin tuna (*Thunnus maccoyii*) with a semi-moist pellet. In: *Proceedings of the Australian Marine Science Association Annual Meeting*. SARDI, Adelaide, South Australia, Australia, p. 19.

Glencross, B.D., van Barneveld, R.J., Carter, C.G. and Clarke, S.M. (1999a) On the path to a manufactured feed for farmed southern bluefin tuna. *World Aquaculture Magazine* 30(3), 42–46.

Glencross, B.D., van Barneveld, R.J., Carter, C.G. and Clarke, S.M. (1999b) Factors influencing feed intake and feed conversion in farmed southern bluefin tuna (*Thunnus maccoyii*). In: *Proceedings of the World Aquaculture Society, 26 April–2 May 1999, Sydney, Australia*. NSW Fisheries, Sydney, Australia, p. 287.

Glencross, B.D., Clarke, S.M., Buchanan, J.G., Carter, C.G. and van Barneveld, R.J. (2002) Temporal growth of farmed, juvenile southern bluefin tuna, *Thunnus maccoyii* (Castelnau), fed a moist-pellet diet. *Journal of the World Aquaculture Society.*

Gunn, J.S., Polacheck, T., Davis, T.L.O., Sherlock, M. and Betlehem, A. (1994) The development and use of archival tags for studying the migration, behaviour and physiology of Southern Bluefin Tuna, with an assessment of the potential for transfer of the technology to groundfish research. In: *Proceedings of ICES Mini Symposium on Fish Migration*, p. 21.

Houlihan, D.F., McCarthy, I.D., Carter, C.G. and Martin, P. (1995) Protein turnover and amino acid flux in fish larvae. *Proceedings of ICES Marine Science Symposium* 201, 87–99.

Kailola, P.J., Williams, M.J., Stewart, P.C., Reichelt, R.E., McNee, A. and Grieve, C. (1993) *Australian Fisheries Resources.* Bureau of Resource Sciences, Canberra, Australian Capital Territory, Australia, 422 pp.

Milward, D.J., Garlick, P.J., James, W.P.T., Nnanyelugo, D.O. and Ryatt, J.S. (1973) Relationships between protein synthesis and RNA content in skeletal muscle. *Nature* 241, 204–205.

Nichols, P.D., Virtue, P., Mooney, B.D., Elliot, N.G. and Yearsley, G.K. (1998) *Seafood the Good Food: the Oil (Fat) Content and Composition of Australian Commercial Fishes, Shellfishes and Crustaceans.* CSIRO Marine Research, Hobart, Austrlia.

NRC (National Research Council) (1993) *Nutrient Requirements of Fish.* National Academy Press, Washington, DC, 114 pp.

Stevens, E.D. and McLeese, J.M. (1984) Why bluefin tuna have warm tummies: temperature effects on trypsin and chymotripsin. *American Journal of Physiology (Regulatory Integrative Physiology)* 246(15), R487–R494.

van Barneveld, R.J., Smart, A., Clarke, S.M., Carter, C.G., Davis, B.J., Tivey, D.R. and Brooker, J.D. (1997) Nutritional management of sea-caged southern bluefin tuna (*Thunnus maccoyii*). In: Corbett, J.L., Choct, M., Nolan, J.V. and Rowe, J.B. (eds) *Recent Advances in Animal Nutrition in Australia '97.* Department of Animal Science, University of New England, Armidale, Australia, pp. 88–97.

Young, J.W., Lamb, T.D., Le, D, Bradford, R.W. and Whitelaw, A.W. (1997) Feeding ecology and interannual variation in diet of southern bluefin tuna, *Thunnus maccoyii*, in relation to coastal and oceanic waters off eastern Tasmania, Australia. *Environmental Biology of Fishes* 50, 275–291.

Milkfish, *Chanos chanos* 13

Chhorn Lim,[1] Ilda G. Borlongan[2] and Felicitas P. Pascual[3]
[1]USDA-ARS, MSA, Aquatic Animal Health Research Laboratory, PO Box 952, Auburn, AL 36830, USA; [2]Aquaculture Department, Southeast Asian Fisheries Development Center, Tigbauan, Iloilo, the Philippines; [3]Aquaculture Specialists, Inc., Nanga, Guimbal, Iloilo, the Philippines

Introduction

Milkfish (*Chanos chanos* Forsskal), the only species known in the family of Chanidae, is an important food fish in the Indo-Pacific regions (Chen, 1976). In some countries, milkfish fingerlings are also used as bait for the tuna fishing industry. This species is best suited for culture in the tropics because of its fast growth, efficient use of natural foods, herbivorous food habit, propensity to consume a variety of supplemental feeds, resistance to disease and handling and tolerance to a wide range of environmental conditions. Milkfish are euryhaline and can thrive in waters of 0 to 150 p.p.t. salinity (Crear, 1980).

Milkfish is cultured on a large scale only in the Philippines, Indonesia and Taiwan. Small-scale or experimental production is being practised in a few other Asian countries, such as Thailand, Malaysia, Vietnam and Sri Lanka, and in Hawaii, Guam and Kiribati. Milkfish farming is believed to have begun in Indonesia some 700 years ago (Ronquillo, 1975) and was introduced into the Philippines and Taiwan more than 300 years ago (Ling, 1977). In 1983, milkfish was the single most important species produced through aquaculture in these countries, using more than 500,000 ha of brackish-water and freshwater areas to produce over 365,000 t (Lee and Banno, 1990). In 1990, the cultured area decreased to 410,000 ha but the production increased to about 460,000 t (Lee, 1995; Bagarinao, 1999). In Taiwan, approximately 12,850 ha of ponds were used for milkfish production in 1990 and yielded an average of 7.0 t ha^{-1}. The Phillippines produced 210,900 t of milkfish from 156,177 ha of brackish-water ponds and 19,000 t from about 10,000 ha of freshwater pens, at an average yield of 1.3–1.9 t ha^{-1}. In 1990, Indonesia devoted the largest pond area (231,000 ha) for milkfish production but the average yield was only 0.6 t ha^{-1}.

The discrepancy between the yield in these countries is mainly attributed to differences in culture technology, skill and management inputs, such as stocking rate, size, pest and predator control, fertilization, feed and feeding and water management. For example, the multiple-size stocking method used in the Philippines yielded an annual production of 2.2–2.7 t ha^{-1} (Pamplona and Mateo, 1985). The deep-water culture practised in Taiwan with the use of high stocking rates and formulated diets produced 8–10 t ha^{-1} $year^{-1}$ (Liao and Chen, 1986). This chapter reviews information on nutrient requirements, feeds and feeding practices of milkfish.

Nutrient Requirements

Protein and amino acids

Milkfish, like other fish, do not have an absolute protein requirement *per se*, but require a balanced mixture of essential and non-essential amino acids. The minimum amount of dietary protein required for the optimum growth, good feed efficiency and survival of 40 mg milkfish fry has been reported to be about 40% (Lim *et al.*, 1979). However, it has been reported that the protein requirements of fish vary with species, size or age, protein quality, dietary level of energy, water quality, presence of natural food and feeding and culture management (NRC, 1983).

Milkfish require the same ten essential amino acids (arginine, histidine, isoleucine, leucine, lysine, methionine, phenylalanine, threonine, tryptophan and valine) as other species (Borlongan and Coloso, 1993). The quantitative requirements of juvenile milkfish for these essential amino acids, determined by breakpoint analysis of the growth curve, are presented in Table 13.1. It has been

Table 13.1. Essential amino acid requirement of juvenile milkfish (from Borlongan and Coloso, 1993).

Amino acid	Requirement (% of dietary protein)*
Arginine	5.2
Histidine	2.0
Isoleucine	4.0
Leucine	5.1
Lysine	4.0
Methionine†	2.5
Phenylalanine‡	4.2 or 2.8
Threonine	4.5
Tryptophan	0.6
Valine	3.6

* The basal diets contained 40–45% protein.

† In the presence of cystine at 0.75% of dietary protein.

‡ In the presence of tyrosine at 1.0% and 2.6% of dietary protein, respectively.

suggested that arginine, leucine and lysine may be the first limiting amino acids since they occurred at high concentrations in the amino acid pattern of protein from the whole body of milkfish juveniles (Coloso *et al.*, 1988). More recent studies, however, showed that the requirement levels of these three essential amino acids and those of tryptophan and valine were lower than those found in milkfish body proteins, but those of the other essential amino acids were similar to the values of the tissue proteins (Borlongan and Coloso, 1993).

The non-essential amino acids can be adequately synthesized by fish but their presence in the diets has a sparing effect in that it reduces the need for fish to synthesize them. Two special examples of sparing action are the conversion of phenylalanine to tyrosine and methionine to cystine. These non-essential amino acids can only be synthesized from the essential amino acid precursor (NRC, 1993). Milkfish actually have a requirement for aromatic amino acids (5.22% of dietary protein), which can be met by either phenylalanine alone or a proper mixture of phenylalanine and tyrosine. In the presence of tyrosine at 1.0% of dietary protein, the phenylalanine requirement was estimated to be 4.22%. The phenylalanine requirement decreased to 2.80% of dietary protein when the tyrosine level was increased to 2.66%. Dietary tyrosine was found to replace up to 46% of the total aromatic amino acid requirement for milkfish (Borlongan, 1992a; Borlongan and Coloso, 1993). Since most practical diets contain adequate levels of phenylalanine and tyrosine, the sum of these two amino acids normally exceeds the dietary needs of most fish (NRC, 1993). A similar relationship exists between sulphur-containing amino acids (methionine and cystine). The presence of cystine in the diet will reduce some of the requirement for methionine (NRC, 1993). The methionine requirement of milkfish was estimated to be 2.7% of protein at 0.28% dietary cystine. In the presence of 1.48% cystine, the methionine requirement was reduced to 1.50% of dietary protein (Sastrillo, 1980).

Milkfish appear to have the ability to utilize crystalline amino acids. Supplementation of 0.5% L-tryptophan to a vitamin-free casein diet provided better growth of young milkfish than a diet containing a combination of casein and gelatin (Lee and Liao, 1976). Chiu *et al.* (1986) reported that supplementation of 2.8% lysine hydrochloride to a maize-gluten meal-based diet significantly improved the growth and feed efficiency of milkfish fry.

Milkfish use proteins of animal origin better than plant proteins. Among animal proteins, fish meal and meat and bone-meal have higher nutritive value than shrimp-head meal. Among plant proteins, soybean meal was superior to copra and *Leucena leucocephala* leaf meals (Samsi, 1979). Borlongan and Coloso (1994) showed that leaf meals from swamp cabbage (kangkong, *Ipomea reptans*), sweet potato (kamote, *Ipomea batatas*), ipil-ipil (*L. leucocephala*) and cassava (kamoteng kahoy, *Manihot esculanta*) can be used to replace 15% of fish-meal protein in a basal diet containing 53% Peruvian fish-meal and 15% soybean meal without affecting the growth performance of milkfish. The true protein digestibility of casein, gelatin, fish-meal, soybean meal and *L. leucocephala* leaf meal by three sizes of milkfish held in fresh water or sea water is given in Table 13.2. Gelatin had the highest digestibility regardless of fish size and water salinity.

Table 13.2. Apparent protein digestibility of some feedstuffs by three sizes of milkfish in fresh water (F) and sea water (S) (from Ferraris *et al.*, 1986).

		Apparent protein digestibility (%)		
Feed ingredient	Water	2 g	60 g	165 g
Casein	F	58	83	87
	S	73	49	65
Gelatin	F	94	94	94
	S	96	98	97
Fish-meal	F	45	65	73
	S	71	62	71
Soybean meal (defatted)	F	53	69	94
	S	74	54	58
L. leucocephala leaf meal	F	47	41	42
	S	61	31	−10

The digestibility of other feedstuffs increased with fish size and tended to be lower for fish held in sea water than in fresh water. *Leucena leucocephala* leaf meal was the least digestible (Ferraris *et al.*, 1986).

Energy

Providing the optimum energy in the diet is important because a deficiency in non-protein energy means that part of the protein will be used to meet the energy requirement. Except for a study of Pascual (1984), no studies have been conducted to determine the energy requirements of milkfish. For milkfish juveniles weighing 0.5–0.8 g, poor growth was observed in fish fed diets containing below 2500 kcal of metabolizable energy kg^{-1} diet. However, increasing dietary energy levels beyond 3500 kcal kg^{-1} did not further improve weight gain. A dietary protein-to-energy ratio of 118–127 mg $kcal^{-1}$ was suggested. Lim *et al.* (1979) obtained maximum growth of milkfish fry (40 mg) with a 40% protein diet having 2740 kcal of digestible energy kg^{-1} diet. In a study to evaluate different supplemental feeds for milkfish in brackish-water ponds, Sumagaysay *et al.* (1991) reported that a diet containing 27.4% protein and 4236 kcal of gross energy kg^{-1} resulted in a maximum increase in profit. In the absence of clear-cut information on energy requirement, data available for tilapia and common carp are suggested.

Lipids and fatty acids

Lipids are important sources of highly digestible energy and are the only source of essential fatty acids needed by fish for normal growth and development. Milkfish fed lipid-free or 7% lauric acid (LA) diets grew significantly less than those fed diets containing 6% LA plus 1% linoleic (18:2n-6), or 0.5% linoleic and 0.5% linolenic (18:3n-3) acids. The diet supplemented with 1% 18:3n-3 provided the highest weight gain but was not significantly different from those fed the diet with 1% 18:2n-6. Based on these results, it appears that milkfish have a dietary requirement for both linoleic (n-6) and linolenic (n-3) acids (Bautista and de la Cruz, 1988). Borlongan (1992b), however, showed that the n-3 rather than the n-6 series fatty acids were essential for the growth and survival of milkfish juveniles, because supplementation of 2% 18:3n-6 to the tristearin-based diets did not enhance milkfish growth as effectively as diets supplemented with n-3 series fatty acids. Signs of essential fatty acid deficiency in milkfish were growth depression, fin erosion, darkening of the upper pigmented body surface, high mortality, increased levels of monoenoic acids, decreased levels of polyunsaturated fatty acids and liver abnormalities, such as lipid infiltration in the blood-vessels and cellular swelling (Bautista and de la Cruz, 1988; Borlongan, 1992b). However, juvenile milkfish grew well and did not exhibit essential fatty acid deficiency signs when fed diets containing at least 1% 18:3n-3 and/or 1.0% of the long-chain polyunsaturated fatty acids (PUFA) 20:5n-3 and 22:6n-3 (Borlongan, 1992b).

Because of the presence of significant amounts of PUFA in the livers of milkfish, despite their absence in the natural food, milkfish may have the ability to bioconvert short-chain n-6 and n-3 unsaturated fatty acids into long-chain n-6 and n-3 PUFA (Benitez and Gorricita, 1983). Kanazawa (1985), however, observed that milkfish grew better on a diet containing 20:4n-6 than on a diet containing 18:2n-6. The growth-promoting effect of 20:5n-3 and 22:6n-3 has also been observed to be superior to that of 18:3n-3 (Borlongan, 1992b). Thus, the ability of milkfish to bioconvert 18:2n-6 and 18:3n-3 to longer-chain PUFA, as has been suggested by Benitez and Gorriceta (1983), may be insufficient to meet their physiological needs.

Milkfish do not tolerate as high a level of dietary lipid as do salmonids. In studies using cod-liver oil (Alava and de la Cruz, 1983) and the combination of a 1 : 1 ratio of cod-liver oil and maize oil (Camacho and Bien, 1983), a lipid level of 7–10% has been reported as optimum for milkfish fingerlings. This level was sufficient to maintain liver structural and cellular integrity. Lipid levels below 7% resulted in decreased granulation and loss of nuclei of liver cells. Dietary lipid levels exceeding 10% caused minor disruption of hepatocytes from the formation of large lipid vacuoles, the loss of hepatic cord with development of fibrous tissues and the occurrence of pyknotic nuclei (Alava and de la Cruz, 1983).

Carbohydrates

Milkfish, like other finfish, do not have a specific requirement for carbohydrates. However, carbohydrates are always included in fish diets because they are the least expensive source of energy, function as a pellet binder and serve as precursors for the formation of various metabolic intermediates essential for growth. Carbohydrates have also been shown to have a sparing effect on the utilization of dietary protein in several aquaculture species. Milkfish can probably utilize carbohydrates as an energy source very efficiently, as do tilapia and common carp. Purified diets containing up to 35% dextrin have been used successfully in various nutritional studies. Practical diets in Taiwan currently used for pond feeding contain 45% or more of total carbohydrate.

Vitamins and minerals

The importance of vitamins and minerals in milkfish diets has been recognized as these nutrients have normally been included in experimental or production diets. Except for the study of Minoso *et al.* (1999), which demonstrated the dietary essentiality of phosphorus and iron, no information is available on milkfish vitamin and mineral requirements. However, milkfish probably require the same vitamins and minerals as do other aquaculture species. Various vitamin and mineral mixes designed for cold-water and warm-water fishes have been used by different workers in milkfish nutrition research with satisfactory results. Thus, in the absence of information on these subjects, vitamin and mineral allowances established for other tropical species, such as tilapia and common carp, are recommended.

Practical Diets

Newly hatched milkfish fry use their yolk as a source of nutrients. As soon as the yolk-sac is absorbed, they begin to swim up and search for food. Live foods such as rotifers (*Brachionus* sp.) are used as the first food up to day 14. Newly hatched brine shrimp (*Artemia*) nauplii are commonly used from day 15 to harvest (usually day 21). However, wild-caught milkfish fry or 15-day-old hatchery-produced milkfish larvae have been reared successfully to adult stages with artificial diets. Before the development of compounded diets, during 1- to 2-week holding periods, wild-caught milkfish fry were fed mashed boiled egg-yolks, good-quality rice bran or wheat flour. Such agricultural products or by-products as rice bran, leaf meal, breadcrumbs or soybean meal are occasionally used for fish grown in nursery or grow-out ponds when there is insufficient growth or depletion of natural foods.

In recent years, efforts have been made to develop prepared diets for milkfish using available information on milkfish nutrient requirements and information derived from other species. Microbound larval diets using carrageenan as a binder were developed. These diets have been used successfully for milkfish larvae in combination with rotifers from day 2 to 14 and solely from day 15 to 21 as a substitute for *Artemia* (Borlongan *et al.*, 2000). Diets for fry have been formulated and used for fish grown in fresh water (Santiago *et al.*, 1983, 1989) and sea water (Alava and Lim, 1988). An example of a practical diet formula for milkfish fry is given in Table 13.3.

During the grow-out phase, milkfish are regarded as a herbivore. Compounded diets containing 23%–27% crude protein have been used successfully for deep-water pond culture of milkfish in Taiwan (Liao and Chen, 1986). Since the milkfish is an efficient feeder and feeds at the bottom of the food chain, natural pond foods make a valuable contribution to its nutrient requirements. Thus, these diets are assumed to be sufficient for satisfactory milkfish growth. A model formula of a practical pond diet for milkfish is presented in Table 13.4.

Adult milkfish, known in the Philippines as *sabalo*, are considered to be omnivores. However, little is known about the diets and feeding of milkfish brood-stock. Milkfish that matured and spawned naturally in floating cages at the SEAFDEC (Southeast Asian Fisheries Development Center) Aquaculture Department in the Philippines were fed with shrimp pellets containing 42%

Table 13.3. Model formula of a practical diet (40% protein) for milkfish fry (modified after Lim, 1991).

Ingredient	Per cent in diet
Fish-meal	30.0
Soybean meal	20.0
Squid meal or shrimp meal	15.0
Wheat flour	25.45
Marine fish-oil	8.0
Vitamin mix*	1.0
Trace mineral mix†	0.5
Antioxidant‡	0.05

* Complete vitamin mix for warm-water fish (NRC, 1993).
† Trace mineral mix should provide the following minerals (mg kg^{-1} diet): manganese, 25: zinc, 80; iron, 60; copper, 5; iodine, 5; cobalt, 0.05; and selenium, 0.04.
‡ Butylated hydroxyanisole or butylated hydroxytoluene.

Table 13.4. Model formula of a practical pond feed (27% protein) for milkfish (modified after Lim, 1991).

Ingredient	Per cent in diet
Fish-meal, anchovy or menhaden	10.0
Soybean meal, 48% protein	34.5
Grains or grain by-products	48.0
Pellet binder*	3.0
Oil	2.0
Dicalcium phosphate	1.5
Vitamin mix†	0.5
Trace mineral mix‡	0.5

* Pellet binder may be hemicellulose or lignin sulphonate.
† Vitamin mix for supplemental diets for warm-water fishes (NRC, 1993).
‡ The same trace mineral mix as that used in practical diets for milkfish fry (Table 13.3).

protein (Lacanilao and Marte, 1980). At present, a 36% protein and 8% lipid with 1% lecithin diet has been developed for milkfish brood-stock. Fed to breeders at the onset of the spawning season, this diet was found to improve the reproductive performance of milkfish (FDS, 1994). At Tung Hsing Hatchery, Pingtung, Taiwan, pond-reared brood-stock were fed with a variety of diets, including rice bran, wheat meal, soybean meal and eel diet (Lin, 1985). At Tungkang Marine Laboratory, Taiwan, a diet consisting of 70% eel diet, 14.75% wheat-germ meal, 14.75% soybean meal and 0.5% vitamin E was used during the spawning season (Chang *et al.*, 1993). Purina trout chow or laboratory-prepared diets containing 32–46% protein have been used successfully at the Oceanic Institute in Hawaii (Kelley and Lee, 1986). In Japara, Indonesia, a 36% protein diet, in which 50% of the protein was supplied by fish-meal, has been used successfully for maturing brood-stock reared in tanks (Poernomo *et al.*, 1985). Thus, in the absence of information on milkfish brood-stock nutrient requirements, the diets for brood-stock should contain a high level of good-quality protein. Also, marine fish-oil rich in PUFA should be included. In addition, the diets should be fortified with extra vitamins and trace minerals. An example of a milkfish brood-stock diet is given in Table 13.5.

Milkfish accept a variety of diets, in meal form and in moist, sinking or floating pellets. Crude feedstuffs are offered in meal form, whereas the compounded diets are usually processed into sinking pellets. Milkfish can use meal-form diets effectively. However, compounded diets should be pelleted to minimize dissolution and separation of nutrients and subsequent waste. Crude diet sources may be uneconomical when pelleted for pond feeding.

Pelleted diets must have desirable physical characteristics, especially water stability and pellet size. The diets must remain water-stable long enough to minimize nutrient loss and diet wastage. Hard and durable pellets are necessary when diets are to be crumbled for feeding smaller fish. The most common pellet

Table 13.5. Model formula of a brood-stock diet (36% protein) for milkfish (modified after Lim, 1991).

Ingredient	Per cent in diet
Fish-meal, anchovy or menhaden	25.0
Soybean meal	34.0
Wheat flour	15.0
Rice bran	20.9
Pellet binder*	3.0
Fish-oil	2.5
Dicalcium phosphate	1.0
Vitamin mix†	1.0
Trace mineral mix‡	0.5

* Pellet binder may be hemicellulose or lignin sulphonate.
† Complete vitamin mix for warm-water fishes (NRC, 1993).
‡ The same trace mineral mix as that used in practical diets for milkfish fry (Table 13.3).

size used for feeding milkfish to marketable size (400 to 500 g) is approximately 4–5 mm in diameter and 6–8 mm long. Diets in meal or crumbled forms of different particle sizes are used for fry and fingerlings.

Feeding Practices

Feeding rates for milkfish are affected by size, water quality (such as temperature, salinity and dissolved oxygen), feeding frequency, and nutrient density of the diets, especially energy content. As with other fish, the feed consumption rate of milkfish is inversely related to fish size. For example, with a diet containing 40% protein and 3450 kcal of metabolizable energy (ME) kg^{-1}, a daily feeding rate of 20% of the biomass is optimum for 7.7 mg milkfish fry reared under laboratory conditions (Lim, 1978). For fish averaging 0.60 g, feeding at 9% of the body weight resulted in a 130% increase in weight gain over the 5% feeding rate (Chiu *et al.*, 1986). In pond environments where natural food is present, milkfish grown to marketable size are fed with commercial pellets containing 23–27% protein at a daily rate of 3–4% of body weight (Benitez, 1984).

Milkfish, like most other species, benefit from multiple daily feedings. The growth and feed efficiency of 0.6 g fingerlings fed at 5% or 9% of body weight increased by about 20% when the feeding frequency was increased from four to eight times daily (Chiu *et al.*, 1986). Under pond conditions, milkfish are normally fed two to three times daily (Liao and Chen, 1986). Diets are offered to fish by hand or by automatic feeders. The latter are commonly used in Taiwan and in intensive culture in the Philippines. Automatic feeders equipped with one or two pipes extending toward the pond are installed on the dikes. The feeders contain devices that can be adjusted to deliver measured quantities of diets at given time intervals. The pond bottom of the feeding area is checked 1 h after feeding for the purpose of adjusting diet allowances.

References

Alava, V.R. and De la Cruz, M.C. (1983) Quantitative dietary fat requirement of *Chanos chanos* fingerlings in a controlled environment. Paper presented at the Second International Milkfish Aquaculture Conference, Iloilo, Philippines, 4–8 October 1983.

Alava, V.R. and Lim, C. (1988) Artificial diets for milkfish, *Chanos chanos* (Forsskal), fry reared in sea water. *Aquaculture* 71, 339–346.

Bagarinao, T. (1999) *Ecology and Farming of Milkfish*. SEAFDEC Aquaculture Department, Tigbauan, Iloilo, the Philippines, 171pp.

Bautista, M.N. and de la Cruz, M.C. (1988) Linoleic (ω6) and linolenic (ω3) acids in the diet of fingerling milkfish (*Chanos chanos* Forsskal). *Aquaculture* 71, 347–358.

Benitez, L.V. (1984) Milkfish nutrition. In: Juario, J.V., Ferraris, R.P. and Benitez, L.V. (eds) *Advances in Milkfish Biology and Culture*. Island Publishing House, Manila, pp. 133–143.

Benitez, L.V. and Gorricita, I.R. (1983) Lipid composition of milkfish grown in ponds by traditional aquaculture. In: Cho, C.Y., Cowey, C.B. and Watanabe, T. (eds) *Finfish Nutrition in Asia – Methodological Approaches to Research and Development.* IDRC-233e, Ottawa, Canada, pp. 145–151.

Borlongan, I.G. (1992a) Dietary requirement of milkfish (*Chanos chanos* Forsskal) juveniles for total aromatic amino acids. *Aquaculture* 102, 309–317.

Borlongan, I.G. (1992b) The essential fatty acid requirement of milkfish (*Chanos chanos* Forsskal). *Fish Physiology and Biochemistry* 9, 401–407.

Borlongan, I.G. and Coloso, R.M. (1993) Requirements of juvenile milkfish (*Chanos chanos* Forsskal) for essential amino acids. *Journal of Nutrition* 123, 125–132.

Borlongan, I.G. and Coloso, R.M. (1994) Leaf meals as protein sources for milkfish, *Chanos chanos* (Forsskal). In: De Silva, S.S. (ed.) *Fish Nutrition Research in Asia, Proceedings of the Fifth Asian Fish Nutrition Workshop.* Asian Fisheries Society Special Publication No. 9, Asian Fisheries Society, Manila, the Philippines, pp. 63–68.

Borlongan, I.G., Marte, C.L. and Nocillado, J.N. (2000) Development of larval diet for milkfish (*Chanos chanos*). *Journal of Applied Ichthyology* 16, 68–72.

Camacho, A.S. and Bien, N. (1983) Studies on the nutrient requirement of milkfish *Chanos chanos* (Forsskal). Paper presented at the Technical Symposium on Aquaculture, University of the Philippines in the Visayas, Iloilo, the Philippines, 19 February 1983.

Chang, S.L., Su, M.S. and Liao, I.C. (1993) Milfish fry production in Taiwan. In: Lee, C.S., Su, M.S. and Liao, I.C. (eds) *Fish Hatchery in Asia, Proceeding of the Finfish Hatchery in Asia '91.* Oceanic Institute, Honolulu, Hawaii, pp. 157–171.

Chen, T.P. (1976) *Aquaculture Practice in Taiwan.* Fishing News Books, Farnham, Surrey, England, 162 pp.

Chiu, Y.N., Sumagaysay, N.S. and Sastrillo, M.A.S. (1986) Effect of feeding frequency and feeding rate on the growth and feed efficiency of milkfish *Chanos chanos* Forsskal juveniles. *Asian Fisheries Science* 1, 27–31.

Coloso, R.M., Benitez, L.V. and Tiro, L.B. (1988) The effect of dietary protein-energy levels on growth and metabolism of milkfish (*Chanos chanos* Forsskal). *Comparative Biochemistry and Physiology* 89A, 11–17.

Crear, D. (1980) Observations on the reproductive state of milkfish populations *Chanos chanos* from hypersaline ponds on Christmas Island (Pacific Ocean). In: Avault, J.W. Jr (ed.) *Proceedings World Mariculture Society*, Vol. 21. Louisiana State University, Baton Rouge, Louisiana, pp. 548–556.

FDS (Feed Development Section) (1994) *Feed and Feeding of Milkfish, Nile Tilapia, Asian Sea Bass, and Tiger Prawn.* Aquaculture Extension Manual No. 21, SEAFDEC Aquaculture Department, Tigbauan, Iloilo, the Philippines, 97 pp.

Ferraris, R.P., Catacutan, M.R., Mabelin, R.L. and Jazul, A.P. (1986) Digestibility in milkfish, *Chanos chanos* (Forsskal): effects of protein source, fish size and salinity. *Aquaculture* 59, 93–105.

Kanazawa, A. (1985) Nutritional factors in fish reproduction. In: Lee, C.S. and Liao, I.C. (eds) *Reproduction and Culture of Milkfish.* Oceanic Institute, Waimanalo, Hawaii, pp. 115–125.

Kelley, C. and Lee, C.S. (1986) Artificial propagation. In: Lee, C.S., Gordon, M.S. and Watanabe, W.O. (eds) *Aquaculture of Milkfish (*Chanos chanos*): State of the Art.* Oceanic Institute, Waimanalo, Hawaii, pp. 83–116.

Lacanilao, F.L. and Marte, C.L. (1980) Sexual maturation of milkfish in floating cages. *Asian Aquaculture* 3, 4–6.

Lee, C.S. (1995) *Aquaculture of Milkfish (*Chanos chanos*).* Oceanic Institute, Hawaii, 141 pp.

Lee, C.S. and Banno, J. (1990) Milkfish culture and production in Southeast Asia – present and future. In: *Proceedings, Regional Workshops on Milkfish Culture and Production in Southeast Asia – Present and Future*. Tarawa, Kiribati, pp. 43–65.

Lee, D.L. and Liao, I.C. (1976) A preliminary study on the purified test diet for young milkfish, *Chanos chanos*. In: *Proceedings, International Milkfish Workshop Conference*. Tigbauan, Iloilo, the Philippines, pp. 104–120.

Liao, I.C. and Chen, T.I. (1986) Milkfish culture methods in Southeast Asia. In: Lee, C.S., Gordon, M.S. and Watanabe, W.O. (eds) *Aquaculture of Milkfish (*Chanos chanos*): State of the Art*. Oceanic Institute, Honolulu, Hawaii, pp. 209–238.

Lim, C. (1978) Effect of feeding rate on the survival and growth of milkfish (*Chanos chanos*) fry in a controlled environment. In: *SEAFDEC Aquaculture Department Quarterly Research Report II (4)*. Tigbauan, Iloilo, the Philippines, pp. 17–19.

Lim, C. (1991) Milkfish, *Chanos chanos*. In: Wilson, R.P. (ed.) *Handbook of Nutrient Requirements of Finfish*. CRC Press, Boca Raton, Florida, pp. 97–104.

Lim, C., Sukhawongs, S. and Pascual, F.P. (1979) A preliminary study on the protein requirements of *Chanos chanos* (Forsskal) fry in a control environment. *Aquaculture* 17, 195–201.

Lin, L.T. (1985) My experience in artificial propagation of milkfish – studies on natural spawning of pond reared broodstock. In: Lee, C.S. and Liao, I.C. (eds) *Reproduction and Culture of Milkfish*. Oceanic Institute, Waimanalo, Hawaii, pp. 185–203.

Ling, S.W. (1977) *Aquaculture in Southeast Asia – a Historical Overview*. University of Washington Press, Seattle, Washington, 108 pp.

Minoso, M.G., Borlongan, I.G. and Satoh, S. (1999). Essentiality of phosphorus, magnesium, iron, zinc and manganese in milkfish diets. *Fisheries Science* 65, 721–725.

NRC (National Research Council) (1993) *Nutrient Requirements of Warmwater Fishes and Shellfishes*. National Academy Press, Washington, DC, 102 pp.

NRC (National Research Council) (1993) *Nutrient Requirements of Fish*. National Academy Press, Washington, DC, 114 pp.

Pamplona, S.D. and Mateo, R.T. (1985) Milkfish farming in the Philippines. In: Lee, C.S. and Liao, I.C. (eds) *Reproduction and Culture of Milkfish*. Ocean Institute, Honolulu, Hawaii, pp. 141–163.

Pascual, F.P. (1984) The energy-protein requirement of *Chanos chanos* fingerlings. Poster paper presented during the International Symposium on Feeding and Nutrition in Fish, 10–13 July 1984, University of Aberdeen, UK.

Poernomo, A., Lim, C., Vanstone, W.E., Daulay, T. and Anindiastuti (1985) Maturation of captive milkfish (*Chanos chanos*) in tanks. In: Lim, C. (ed.) *Fish Nutrition and Feed Technology Research in Indonesia*. Research Institute for Freshwater Fisheries, Central Research Institute for Fisheries, Agency for Agricultural Research and Development, Jakarta, Indonesia, pp. 78–103.

Ronquillo, I.A. (1975) Biological studies on bangos (*Chanos chanos*). *Philippine Journal of Fisheries* 9, 18–37.

Samsi, S. (1979) Effects of various protein sources on the growth and survival rates of milkfish (*Chanos chanos* Forsskal) fingerlings in a controlled enviroment. MSc thesis, University of the Philippines in the Visayas, Iloilo, the Philippines.

Santiago, C.B., Banes-Aldaba, M. and Sungalia, E.T. (1983) Effect of artificial diets on growth and survival of milkfish fry in fresh water. *Aquaculture* 34, 247–252.

Santiago, C.B., Pantastico, J.B., Baldia, S.F. and Reyes, O.S. (1989) Milkfish (*Chanos chanos*) fingerling production in freshwater ponds with the use of natural and artificial feeds. *Aquaculture* 77, 307–318.

Sastrillo, M.A. (1980) Methionine requirement of milkfish (*Chanos chanos* Forsskal) juveniles. MSc thesis, University of the Philippines in the Visayas, Iloilo, the Philippines, 38 pp.

Sumagaysay, N.S., Marquez, F.E. and Chiu-Chern, Y.N. (1991) Evaluation of different feeds for milkfish (*Chanos chanos*) reared in brackishwater ponds. *Aquaculture* 93, 177–189.

14 Rainbow Trout, *Oncorhynchus mykiss*

Ronald W. Hardy
Hagerman Fish Culture Experiment Station, University of Idaho, 3059F National Fish Hatchery Road, Hagerman, ID 83332, USA

Introduction

Rainbow trout are classified as *Oncorhynchus mykiss*, and as such belong to the same genus as Pacific salmon, and to the family Salmonidae, which includes Atlantic salmon (*Salmo salar*), various trout (*Salvelinus* sp.), Arctic char (*Salvelinus alpinus*), Arctic grayling (*Thymallus arcticus*) and whitefish (*Coregonus* sp.). Rainbow trout are native to areas around the North Pacific Ocean, from southern California through Alaska, the Aleutians and the western Pacific areas of the Kamchatka Peninsula and Okhotska Sea drainages. Rainbow trout primarily inhabit fresh water, but in the eastern and western North Pacific anadromous stocks are found. These stocks follow a life cycle similar to that of Pacific salmon, in that they spend a part of their life in the ocean, but return to lakes and rivers for spawning and the fry and juvenile stages of their life history. Most strains of rainbow trout can adapt to life in sea water, once they reach the post-juvenile stage (*c.* 75–100 g), through a gradual increase in salinity of their rearing water. This is one of the qualities of rainbow trout that has led to their prominence as a farmed species of fish.

Rainbow trout have been cultured for hundreds of years, and are the most widely farmed trout in the world. Rainbow trout can tolerate a wide range of water temperatures and other environmental variables, such as water quality, but they require highly oxygenated water and thrive in water temperatures of 13–18°C. They are a highly valued foodfish, and can be grown to have pigmented (red) or non-pigmented (white) flesh, depending upon their diet. In North America, Britain, Denmark, France and Italy, most trout farming occurs in fresh water, using flow-through water-supply systems. In Chile and Scandinavian countries, rainbow trout are grown in marine cages after an initial phase in fresh water. Diet formulations for rainbow trout grown in marine cages differ somewhat from those for trout grown in fresh water (Table 14.1). Global production of rainbow trout was 358,456 t in 1995, second only to Atlantic salmon among salmonids (Tacon, 1998). France, Chile, Denmark and Italy

Table 14.1. Generalized practical feed formulations used for rainbow trout reared in sea water and in fresh water.

	Sea-water diet (g kg^{-1})	Freshwater diet (g kg^{-1})
Ingredient	550	400
Fish-meal		80
Poultry by-product meal		0–50
Blood meal	50	50–100
Soybean meal	144	120–250
Wheat grain and by-products	10	10
Vitamin premix	1	1
Trace-mineral premix	4	4
Choline chloride (60%)	1	1
Ascorbic acid	240	120–210
Fish-oil	0.1	–
Carophyll Pink®		
Proximate composition	8%	8%
Moisture	43%	45%
Crude protein	28%	18–26%
Crude fat		

accounted for 48% of global production in 1995. The USA produced 25,777 t in 1997, about 7% of global production. Of this, 67% was produced in the state of Idaho, with North Carolina, California and Pennsylvania accounting for a majority of the remaining US production.

Rainbow trout live in lakes, streams and rivers, consuming zooplankton as fry, followed by insects, crustaceans and other fish as they grow. Spawning occurs in spring associated with rising water temperatures, although considerable variability is found, with coastal rainbow trout spawning in late December. Females deposit anywhere from 500 to 2500 large eggs (50–150 mg egg^{-1}) in nests dug in gravel, and males fertilize the eggs as they are deposited. The time required for fertilized eggs to develop and hatch depends upon water temperature. At 4.5°C rainbow-trout eggs require 80 days to hatch, at 10°C, 31 days and at 15°C, 19 days (Leitritz and Lewis, 1980). Eggs are extremely sensitive to handling and shock from 2 days postfertilization until the blastophore is completely closed, 9 days at 10°C. Once the eggs become pigmented (about 16 days at 10°C), the period of sensitivity is over and the eggs can be handled until just before hatching. At hatching, fry are attached to their large egg-yolk. These fry are called yolk-sac fry, or alevins, and they burrow into spaces within gravel, where they continue to develop and grow, utilizing their yolk-sac for all necessary energy and nutrients to grow. When the yolk-sac is nearly gone and has been surrounded by skin on the ventral side of the fish, the fry are said to be 'buttoned up'. The time needed for alevins to reach this stage depends on water temperature, but at 10°C is approximately 20 days and at 15°C, 10 days or less from hatching. The fry are then ready to feed, and emerge from the gravel seeking food on the water's surface. At this point they are said to be 'swim-up' fry. The entire sequence from spawning to emergence from the gravel is timed such that the fish emerge when natural food is abundant in spring. Since streams differ in water temperature and food abundance throughout the geographical range of rainbow trout, local populations are adapted to local conditions and spawning and fry emergence are timed appropriately.

Rainbow-trout growth rates depend on water temperature and food abundance, and wild fish generally reach maturity at 3–4 years of age. Most spawning trout are first spawners, but a small proportion of spawners, mainly females, survive to spawn again. Growth and maturation in rainbow trout are indeterminate, meaning that there is no set rate or age. Rather, environmental factors determine growth and maturation, with fish in cold, harsh environments generally living longer than those in warmer, benign environments. Maximum size is variable, with 17–23 kg rainbow trout sometimes being captured in Kooteney Lake, British Columbia. These fish would be 5–6 years old (Behnke, 1992). However, rainbow trout in streams typically weigh 100 g at 1 year of age and 300–450 g after 3 years.

Livingston Stone first described rainbow-trout farming in the USA in a paper in 1872 (Stickney, 1996). Stone stated that fish were typically reared for 4 years to reach market size, and typical ex-farm sale prices ranged from $0.50 to $1.25 per lb. The McCloud River Hatchery in California began producing rainbow-trout eggs in 1881, and shipped eggs throughout the country to state and federal agencies and to private individuals interested in undertaking rainbow-trout rearing (Stickney, 1996). Rainbow-trout culture became a farming business in the early 1900s, with a third of farms being fee-fishing operations, at least until the early 1950s. In Idaho, the first commercial trout farm was started in 1909, near Twin Falls. This area contains many suitable trout-farm sites supplied with abundant, constant temperature (14.5°C) spring water from the eastern Snake River aquifer, over 23 m^3 s^{-1} in flow (Brannon and Klontz, 1989). Trout farming expanded greatly in the early 1950s, supported in part by the development of pelleted feeds. Prior to this, farms produced their own feed, using combinations of the by-products from slaughterhouses and dry meals, formed into moist pellets.

Several characteristics of rainbow trout contributed to the expansion of its culture. First, rainbow trout are easy to spawn, and their fry are large at first feeding compared with most other freshwater fish. Thus, they can be fed prepared feed from first feeding. Secondly, they grow rapidly and are in demand as a foodfish. Thirdly, as mentioned earlier, they tolerate a wide range of water temperatures and there are numerous water sources in temperate regions in which they can be grown. A final characteristic of rainbow trout that contributed to their success as a farmed fish was that their spawning time could be manipulated by selection and by photoperiod adjustment to make eggs available year-round. This allowed farms to supply rainbow trout to markets throughout the year, increasing the availability of fresh fish. More recently, egg suppliers have developed all-female stocks, thus eliminating the problem of precocious maturation among males, which reduced their market value and lowered returns to farmers. Triploid rainbow trout are also available to growers, but the advantages of triploid trout, e.g. the fish are sterile, somatic growth is not interrupted by gonadal growth, are not realized until the fish are over 500 g. Thus, in the USA, triploid trout are not widely used in commercial rainbow-trout farming.

Farming systems for rainbow trout are similar throughout the world. Fish are raised in flowing water in earthen or concrete raceways, with stocking densities depending upon water flow and water quality, e.g. temperature and dissolved oxygen content. Water flow is usually gravity-driven. Usually, raceways are arranged in series, with water flowing from one to another with a small drop between the raceways. Upper raceways are typically stocked at higher densities than lower raceways as water quality decreases along a series of raceways. In Idaho, typical rearing densities are 1.8 kg l^{-1} min^{-1} water flow (15 lb per g.p.m.) in raceways receiving first-use water, and up to 9.6 kg l^{-1} min^{-1} when all raceways in a series are combined (Brannon and Klontz, 1989). In Italy, long earthen raceways, sometimes more than 1000 m long, resembling wide, shallow streams, are used to raise trout.

Eggs hatch indoors in hatching jars or trays supplied with upwelling water. Hatching success is typically 95%. When fish hatch, they have a large yolk that sustains them until first feeding. Feed is provided almost continuously during the first 7–10 days of feeding. As the fish grow, feeding frequency is reduced and they are transferred to larger troughs or circular tanks. When fish reach 15–20 g, they are stocked in outdoor raceways, where they remain until harvest – in the USA at about 500–600 g after 6–7 months of rearing. In Europe, fish are harvested at a larger size, typically 1–1.5 kg, while fish reared in marine net cages are harvested at 2.5–3.5 kg.

The lack of a suitable test diet to which vitamins could be added or deleted limited progress on establishing the essentiality of vitamins to trout. What was lacking was a vitamin-free diet that, when supplemented with all known vitamins, would support normal growth of trout without the development of deficiencies. A close approximation was reported in 1947 (McLaren *et al.*, 1947a), but, although this development was a major advancement, long-term feeding trials with diets lacking several vitamins did not result in clinical deficiencies, demonstrating that the diet was not completely vitamin-free (McLaren *et al.*, 1947b). A refinement was reported within a few years, and the essentiality of pantothenic acid, folic acid and inositol was demonstrated (Wolf, 1951). However, no need was demonstrated for ascorbic acid, biotin, vitamin B_{12}, nicotinic acid or vitamin E, indicating that the diet contained sufficient amounts of these vitamins to sustain the fish. The problem was overcome by further refinement (Halver and Coates, 1957), allowing the qualitative and quantitative requirements of vitamins to be determined in fish (Halver 1957a, 1989). This diet, called H-440, was subsequently modified to test for the amino acid requirements of salmonids (Halver, 1957b). Through the next few decades, new information on nutritional requirements led to further refinement of the semipurified test diet for salmonids, and to new formulations designed specifically to be deficient in certain essential nutrients (Cho and Cowey, 1991). Most importantly, the new semipurified diets support growth rates that are only slightly lower than those seen in fish fed practical diets, a far cry from the growth rates supported by earlier semipurified diets (Hardy, 1991). This point is important, because the manifestation of clinical deficiencies of essential nutrients depends in part on the growth rates of fish and subsequent depletion of body stores. Formulations for several semipurified research diets for trout are shown in Table 14.2.

Table 14.2. Semipurified diet formulations for salmonids (H-440, Oregon test diet (OTD) and Guelph test diet for trout).

Ingredient	H-440	OTD	Guelph
Vitamin-free casein	38.0	49.5	40.0
Gelatin	12.0	8.7	4.0
Dextrin	28.0	15.6	9.0
Starch	–	–	11.5
Alpha-cellulose	–	7.7	3.0
Carboxymethylcellulose	–	1.3	–
D,L-Methionine	–	–	0.5
L-Arginine	–	–	1.0
Fish-oil (marine origin)	–	10.0	15.0
Vitamin E	–	0.2	In vitamin premix
Choline chloride	–	1.0	In vitamin premix
Mineral mix	4*	4.0†	8.0
Vitamin premix	9‡	2.0§	3.0
Oil premix	9‖	–	

* Mineral mix contains the following (g kg^{-1} premix): calcium bisphosphate, 135.7; calcium lactate, 326.9; ferric citrate, 29.7; magnesium sulphate, 132; potassium phosphate (dibasic), 239.8; sodium bisphosphate, 87.2; sodium chloride, 43.5; $AlCl_3.6H_2O$, 0.15; KI, 0.15; $CuCl_2$, 0.1; $MnSO_4.H_2O$, 0.8; $CoCl_2.H_2O$, 1.0; $ZnSO_4.H_2O$, 3.0.

† Mineral mix contains the following (g kg^{-1} premix): $CaCO_3$, 21; $CaHPO_4.2H_2O$, 735; K_2HPO_4, 81; K_2SO_4, 68; NaCl, 30.6; $Na_2HPO_4.6H_2O$, 21.4; MnO, 25.0; $FeC_6H_5O_7.3H_2O$, 5.58; $MnCO_3$, 4.18; $CuCO_2$, 0.34; $ZnCO_3$, 0.81; KI, 0.01; NaF, 0.02; $CoCl_2$, 0.2; citric acid, 6.88.

‡ Vitamin premix contains the following (g kg^{-1} premix): alpha-cellulose, 893; choline chloride, 56; inositol, 22; ascorbic acid, 11; niacin, 8.4; Ca-pantothenate, 5.6; riboflavin, 2.2; menadione, 0.45; pyridoxine.HCl, 0.56; thiamine-HCl, 0.56; folic acid, 0.17; biotin, 0.06; vitamin B_{12}, 10 mg.

§ Vitamin premix contains the following (g kg^{-1} premix): thiamine-HCl, 3.2; riboflavin, 7.2; niacinamide, 25.6; biotin, 0.08; Ca-pantothenate, 14.4; pyridoxine-HCl, 2.4; folic acid, 0.96; menadione, 0.8; vitamin B_{12}, 2.67; inositol, 125; ascorbic acid, 60.0; para-amino benzoic acid, 20; vitamin D_2, 0.4; BHA, 0.75; celite, 735.8.

‖ Oil premix contains the following (g kg^{-1} premix): maize oil, 664.7; cod-liver oil, 331.9; D,L-alpha-tocopheryl acetate, 4.4.

Nutrient Requirements

Rainbow trout require all of the nutrients found to be essential in the diets of pigs and poultry, plus ascorbic acid and several vitamins for which intestinal synthesis is sufficient to supply the needs of most birds and mammals, but not trout. The major difference between rainbow trout and other domestic (monogastric) animals in nutritional requirements is associated with essential minerals; trout can obtain a substantial portion of their mineral needs from their rearing water. Another major difference is the relative intolerance of rainbow trout to dietary carbohydrates; protein and lipids are preferred dietary energy sources for trout.

Protein, amino acids and energy

Protein and energy are essential dietary components for all animals, including rainbow trout, but there is no absolute dietary requirement for either *per se*. Rather, animals and fish require dietary sources of essential amino acids and calories. In addition, fish require a dietary supply of either non-essential amino acids or the precursors for their synthesis. In practical terms, sufficient dietary protein must be provided for fish to meet their essential amino acid requirement as well as to permit them to synthesize sufficient quantities of non-essential amino acids, and this must be done in a diet containing sufficient non-protein energy to minimize the use of ingested protein to meet metabolic energy demands. Given the fact that these demands vary with fish size, water temperature, activity level, reproductive status, dietary energy level and other variables, it is nearly impossible to make a precise estimate of dietary protein needs. Similarly, it is almost impossible to establish an absolute dietary energy requirement without taking into account many variables, including dietary protein content and protein quality (balance and apparent digestibility of amino acids). Nevertheless, practical rainbow-trout feeds generally are formulated to contain between 42% and 48% crude protein and 16–24% lipid, depending upon life-history stage (Table 14.3).

Rainbow trout require ten amino acids in their diet to grow and thrive; all other vertebrates require the same amino acids. However, estimates of dietary requirements have been made for only four of the ten amino acids: arginine, lysine, methionine and tryptophan. These are the four amino acids that are most likely to be limiting in nearly all conceivable diet formulations that can be developed from practical feed ingredients commonly used in trout diets. If diets are formulated to contain sufficient levels of these four amino acids, the other

Table 14.3. Recommended protein and fat levels in trout diets.

Production stage	Protein	Fat
Starter diet (fry)	45–50%	16–18%
Grower diet (fingerlings)	42–48%	20–24%
Brood-stock diet (maturing fish)	35–40%	14–16%

essential amino acids are very likely to be present in the diet at sufficient levels, based upon the known dietary requirements for these amino acids in closely related fish species, such as Pacific salmon. For the four amino acids for which estimates have been made, considerable differences exist in these estimates between various studies. For example, estimates of the dietary arginine requirement of rainbow trout range from 1.2% of the diet (Kaushik, 1979) to 2.8% of the diet (Ketola, 1983). Differences in dietary protein level and source in semipurified diets used in arginine requirement studies, in size of fish used in studies and in response criteria (and method of evaluating data) used to estimate the requirement are responsible for these differences (NRC, 1993). Similarly, for lysine, estimates of dietary requirements in rainbow trout vary from 1.3% to 2.9% of the diet. Despite these varying estimates, recommendations for all of the essential amino acids have been made (Table 14.4).

Lipids and fatty acids

Pioneering work conducted at Oregon State University demonstrated that rainbow trout require omega-3 fatty acids (n-3 fatty acids) in the diet to prevent deficiency signs and to support normal growth (Castell *et al.*, 1972a,b,c). Confirmation of these findings was offered by Watanabe *et al.*, (1974) and Takeuchi and Watanabe (1977) and was extended to maturing rainbow trout (Yu *et al.*, 1979). Trout require between 0.5% and 1% n-3 fatty acids in their diet, depending on dietary lipid level and life-history stage. Like most fish species, trout incorporate eicosapentaenoic (EPA) (C20:5) and docosahexaenoic acid (DHA) (C22:6) into phospholipids. Unlike marine species, however, trout can desaturate and elongate linolenic acid (C18:3) and thereby produce EPA and DHA from this shorter-chain precursor. Trout may also require small amounts of n-6 fatty acids, particularly arachidonic acid (C20:4), for phospholipid and prostaglandin synthesis (Cho and Cowey, 1991). Omega-3 fatty acids are normally supplied in the diet by adding marine fish-oil; 4–5% marine fish-oil supplies sufficient quantities of n-3 fatty acids to prevent deficiency. Providing

Table 14.4. Essential amino acid requirements of rainbow trout.

Amino acid	Per cent of diet	Per cent of protein
Arginine	2.0	5.0
Histidine	0.7	1.8
Isoleucine	0.8	2.0
Leucine	1.4	3.5
Lysine	1.8	4.5
Methionine + cystine	1.4	3.5
Phenylalanine + tyrosine	1.8	4.5
Threonine	0.8	2.0
Tryptophan	0.2	0.5
Valine	1.3	3.2

that the dietary n-3 fatty acid level is above 1%, trout grow normally when their feeds contain fats or oils of animal or plant origin.

Vitamins and minerals

Vitamins were the first nutrients shown to be essential in rainbow-trout diets (McLaren *et al.*, 1947a). Halver (1957) conducted feeding trials with chinook-salmon fingerlings fed a semipurified diet from which one vitamin at a time was deleted and demonstrated that thiamine, riboflavin, pyridoxine, pantothenic acid, niacin, inositol, folic acid, biotin and choline were essential dietary nutrients. Previously, thiamine had been shown to be essential for trout (Schneberger, 1941; Wolf, 1942; McLaren *et al.*, 1947b), as had niacin and biotin (Phillips and Brockway, 1947). Because of the similarity in life history and physiology between Pacific salmon (*Oncorhynchus* sp.) and rainbow tout, nutritionists assumed that the quantitative vitamin requirements of trout were similar to those of salmon fingerlings, and commercial trout diets were supplemented with vitamins accordingly. Determining the quantitative dietary requirements of rainbow trout for vitamins lagged behind that for Pacific salmon, but in the last two decades requirements for trout have been more or less quantified. Notably, response criteria used to establish quantitative vitamin requirements have become more sensitive, as criteria have switched from relatively insensitive ones, such as weight gain, absence of deficiency signs and maximum tissue storage, to sensitive ones, such as the activities of vitamin-dependent metabolic enzymes or concentrations of metabolically active forms in responsive tissues. In general, requirements determined using relatively insensitive response criteria are higher than those estimated from metabolic criteria.

Of the fat-soluble vitamins (A, D, E and K), only vitamin E has attracted much study in rainbow trout. The requirements for vitamins A and D are based upon single studies using weight gain and absence of deficiency signs as the response variables, and are listed as 2500 and 1600–2400 IU kg^{-1} diet, respectively (Kitamura *et al.*, 1967; Barnett *et al.*, 1979). No vitamin K requirement study has been conducted using rainbow trout, but lake trout (*Salvelinus namaycush*) are reported to require 0.5–1.0 mg kg^{-1} diet (Poston, 1976). The dietary requirement for vitamin E depends upon the concentration of polyunsaturated fatty acids in the diet and the degree of oxidation of dietary lipids. Estimates of the dietary requirement range from 25 to 100 mg kg^{-1} diet (Woodall *et al.*, 1964; Hung *et al.*, 1981; Watanabe *et al.*, 1981; Cowey *et al.*, 1983). Tissue levels can be greatly increased by feeding diets containing high amounts of vitamin E for relatively short periods to rainbow trout (Boggio *et al.*, 1985).

Estimates of the dietary requirements of rainbow trout for water-soluble vitamins are more common (NRC, 1993). Early estimates were based upon weight gain, absence of deficiency signs and maximum liver-storage levels, and were significantly higher than later estimates based upon enzyme activity. For example, McLaren *et al.* (1947a) estimated that the dietary requirement for thiamine was between 1 and 10 mg kg^{-1} diet, but a more recent estimate, based upon enzyme activity, placed the requirement at 1 mg kg^{-1} diet (Morito *et al.*, 1986). Similar revisions of earlier estimates have been made for pyridoxine, riboflavin, niacin, biotin and folic acid (Hughes, 1984; Poston and Wolfe, 1985; Woodward and Frigg, 1989; Cho and Woodward, 1990; Woodward, 1990). In contrast, more recent estimates of the dietary requirements of rainbow trout for pantothenic acid and choline have not resulted in lower estimates, but rather similar or higher ones compared with estimates from early studies (Cho and Woodward, 1990; Rumsey, 1991). The dietary requirements of rainbow trout for vitamin B_{12} and inositol have either not been established or are based on a single study over 50 years old. The current dietary vitamin requirements of rainbow trout are listed in Table 14.5.

Rainbow trout probably require the same minerals as terrestrial animals, but not necessarily in their diet (Table 14.6). Trout, like all fish, are adept at obtaining

Table 14.5. Vitamin requirements of rainbow trout (per kg dry diet).

Vitamin	Requirement
Vitamin A	2500 IU
Vitamin D	2400 IU
Vitamin E	50 IU*
Vitamin K	R
Thiamin	R
Riboflavin	4 mg
Pyridoxine	3 mg
Pantothenic acid	20 mg
Niacin	10 mg
Biotin	0.15 mg
Folic acid	1 m
Vitamin B	0.01 mg
Ascorbic acid	50 mg
Choline	1000 mg
Myo-inositol	300 mg

* May be higher in diets containing high proportion of PUFAs.
PUFAs, polyunsaturated fatty acids; R, required.

Table 14.6. Essential dietary minerals for rainbow trout.

Macrominerals	Microminerals (trace elements)
Calcium*	Iron
Phosphorus*	Manganese
Sodium*	Copper
Potassium*	Zinc
Chlorine*	Cobalt
Magnesium*	Selenium
Sulphur*	Iodine*

* Required in the diet, but generally not supplemented in practical feeds.

essential minerals from their rearing water, with the exception of phosphorus and iodine, which are not present in water in concentrations sufficient to supply the needs of the fish. Supplementation of minerals to trout diets has been limited to five trace minerals, i.e. copper, iodine, iron, manganese and zinc, because fish-meal-based diets supply other essential minerals in excess. As the proportion of fish-meal is reduced in trout diets, the necessity of supplying both macro- and trace minerals will increase. In addition, increasing use of grain- and oilseed-based protein sources increases the dietary content of phytate, the storage form of phosphorus in seeds. Phytate phosphorus is unavailable to monogastric animals, including fish, and further may reduce the availability of other minerals, especially zinc (Spinelli *et al.*, 1983; Richardson *et al.*, 1985). Several trace element premix formulations are used in trout feeds; their composition is shown in Table 14.7.

Phosphorus present in trout-farm discharge water is subject to regulatory restrictions in the state of Idaho. In contrast to restrictions in Europe, which are based upon upper limits of phosphorus in diets, regulations in Idaho are placed on each trout farm and based upon total maximum daily loads (TMDLs). Each farm is allocated a proportion of the total amount of phosphorus that the aquaculture industry is allowed to discharge each day, based upon each farm's water flow. It is up to the farm to decide how best to comply with their TMDL. Some farms choose to grow fewer fish, while others opt to capture and remove faeces and uneaten diet present in their effluent water. In all cases, farmers use low-phosphorus diets. Diet modification is the key to compliance with TMDLs, because diet is the source of phosphorus in discharge water, whether directly via uneaten diet, via the faeces (insoluble and unavailable phosphorus) or via urine (soluble phosphates). Ideally, all phosphorus in trout diets would be completely bioavailable, and the diets would be formulated to contain just enough phosphorus to meet the nutritional needs of the trout. The reality is that diets are made of complex feed ingredients containing phosphorus in a variety of chemical forms, not all of which is bioavailable. Fish bones, for example, are a constituent of fish-meal, and contain calcium phosphate. The bioavailability of phosphorus in fish bones varies with dietary level, ranging from *c.* 70% to < 20% as dietary bone content

Table 14.7. Mineral premix specifications for several open-formula salmonid diets.

Element (form*)	US Fish and Wildlife Service trace mineral premix no. 3 (g element kg^{-1} premix)	Ontario Ministry of Natural Resources mineral premix MIN-9504 (g element kg^{-1} premix)
Zn (as $ZnSO_4.7H_2O$)	75	12.0
Mn (as $MnSO_4$)	20	17.64
Cu (as $CuSO_4.5H_2O$)	1.54	1.5
I (as KIO_3) or I ($C_2H_8N_2.2HI$)	10	1.5
Fe (as $FeSO_4.7H_2O$)	–	2.6
NaCl	–	240

* Preferred compound.

increases from 0% to 10% (Sugiura *et al.*, 2000). Thus, restricting fish-bone content by using low-ash fish-meals is a viable formulation strategy for producing low-phosphorus diets. Other formulation approaches include the use of low-phytate grains (Sugiura *et al.*, 1999) and phytase (Cain and Garling, 1995; Rodehutscord and Pfeffer, 1995).

Carotenoid pigments

The dietary essentiality of carotenoid pigments in rainbow trout diets has not been established, but studies in Atlantic salmon suggest that carotenoid pigments, specifically astaxanthin, are essential components of brood-stock diets to ensure viable offspring (Christiansen *et al.*, 1995). Fry from female Atlantic salmon fed astaxanthin-free diets and fed starter diets devoid of carotenoid pigments exhibited high mortality, which was prevented by feeding diets containing astaxanthin to either brood-stock females or to fry. The dietary amount needed to prevent mortality was relatively low ($< 5\ \mu g\ g^{-1}$), but the results are likely to extend to other salmonids, including rainbow trout.

Feeds and Feeding Practices

Rainbow trout are typically fed small particles from first feeding until they reach the size at which they can ingest small pellets. Small particles are produced by crushing (crumbling) pellets and separating the resulting particles into size ranges by screening. Until recently, pellets smaller than 2.0 mm were not available, but < 1 mm pellets are now being produced, allowing fish to be switched from crumbles to pellets at 1.5–2.0 g. Fish are shifted to larger pellets as they grow, usually by mixing pellet sizes to ensure that the smallest fish within a tank or raceway do not get left behind. Generally, it is wise to feed trout the largest pellet that is recommended for a given size range of fish (Table 14.8). Pellet length should be slightly longer than pellet diameter for optimum results.

Table 14.8. Recommended diet particle sizes for rainbow trout. Pellet sizes are based upon extruded diets.

Granule or pellet size	Screen size	Fish weight (g per fish)
Starter granule	30–40	< 0.23
No. 1 granule	20–30	0.23–0.5
No. 2 granule	16–20	0.5–1.5
No. 3 granule or 1 mm	10–16	1.5–3.5
No. 4 granule or 2 mm	6–10	3.5–9
3 mm pellets		9–38
4 mm pellets		38–90
5 mm pellets		90–450
6 mm pellets		450–1500
8 mm pellets		> 1500

Rainbow-trout feeding is based upon the principle that one should overfeed fry and fingerlings to obtain the fastest growth possible, providing that over-feeding does not pollute rearing water, and feed post-juvenile fish for optimum growth and feed conversion ratios. The reasoning behind this is based upon economics; small fish consume < 5% of the total diet consumed during a production cycle, so any loss in feed efficiency during this stage of growth does not significantly affect the economics of trout production. However, any gain during this period resulting from the use of high-quality diets and feeding to maximum growth shortens the production cycle. As with all fish, feeding rates are based upon water temperature and fish size, decreasing as temperature decreases and fish size increases. Feeding charts are widely available for rainbow trout, but one must consider the energy density of the diet when following the recommendations of charts (Table 14.9). Feeding charts are not intended to be followed exactly; rather, they are guidelines. Farms should keep accurate records of fish growth and feed conversion ratios, and make adjustments to their feeding rates from actual farm records.

Pellets are delivered to trout by hand, by mechanical feeding systems and by demand feeders. Fry and fingerlings are generally fed frequently throughout the day, as much as four times per hour for first-feeding fry. Thus, mechanical feeders work best for fry and fingerlings. Once fish are placed in raceways, they are fed using demand feeders, at least in the USA. Demand feeders are simply diet hoppers with tapered, conical bottom outlets from which a metal rod is suspended into the rearing water. Within the tapered outlet, a platform slightly smaller in diameter than the outlet is attached to the suspended rod, and the feed rests upon this platform. When fish move the rod, the feed falls from the platform into the water. The delivery rate (amount of feed falling per rod movement) is adjusted by moving the platform up or down the rod within the tapered outlet. This increases or decreases the clearance between the edge of the platform and the sides of the conical outlet, thereby changing the quantity of feed that falls from the platform when it is moved. The benefits of demand feeders are several. First, labour costs are reduced. Secondly, fish feed when they choose and also consume the amount that they wish. Thirdly, feeding is spread throughout the day, minimizing frantic activity at feeding. This maintains dissolved oxygen levels in rearing water, and further allows less aggressive fish the opportunity to feed. The disadvantages of demand feeders revolve around diet waste. High wind can move the rod, causing feed to discharge and be wasted. Also, the platform must be adjusted properly to avoid wasting diet by delivering too much per rod movement.

Mechanical feeding and hand-feeding are employed in marine net pens where wave movements of the pens make demand feeders unsuitable. Mechanical feeding is done by a variety of machines, the most common being blowers that move diet through flexible tubes to pens. The frequency of feeding and amount of diet delivered per feeding are carefully controlled to obtain economic growth. Hand-feeding is used in areas where labour costs are relatively low. Care must be taken to train feeders properly and to ensure that they feed carefully.

Table 14.9. Example of a feeding chart for rainbow trout for extruded diet (from Nelson & Sons, Inc., with permission). Value in chart is multiplied by feed conversion ratio to obtain the daily amount to feed, expressed as percentage of biomass. For example, if expected feed conversion ratio is 1.2, then a 7.5 g fish grown at 12.8°C would receive 3.03 × 1.2 = 3.6% of its body weight, or 0.273 g diet day^{-1}. If a tank contained 5000 such fish, the tank would be fed 136.5 g feed day^{-1}.

	Water temperature (°C)									
Fish wt (g)	8.3	9.4	10.5	11.7	12.8	13.9	15.0	16.1	17.2	18.3
0.53	3.50	4.29	5.09	5.88	7.03	7.87	8.66	7.87	7.07	6.68
1.3	2.63	3.22	3.82	4.41	5.30	5.90	6.49	5.90	5.30	5.01
2.6	2.10	2.58	3.05	3.53	4.24	4.72	5.20	4.72	4.24	4.01
4.6	2.75	2.15	2.54	2.94	3.54	3.93	4.33	3.93	3.54	3.34
7.5	1.50	1.84	2.18	2.52	3.03	3.37	3.71	3.37	3.03	2.86
11.4	1.31	1.61	1.91	2.21	2.65	2.95	3.25	2.95	2.65	2.50
16.4	1.17	1.43	1.70	1.96	2.36	2.62	2.89	2.62	2.36	2.23
23	1.05	1.29	1.53	1.76	2.00	2.36	2.60	2.36	2.12	2.00
31	0.95	1.17	1.39	1.60	1.93	2.15	2.36	2.15	1.93	1.82
41	0.88	1.07	1.27	1.47	1.67	1.97	2.16	1.97	1.77	1.67
52	0.81	0.99	1.17	1.36	1.63	1.82	2.00	1.82	1.63	1.54
66	0.75	0.92	1.09	1.26	1.52	1.69	1.86	1.69	1.52	1.43
81	0.70	0.86	1.02	1.18	1.34	1.57	1.73	1.57	1.41	1.34
98	0.66	0.81	0.95	1.10	1.33	1.47	1.62	1.47	1.33	1.25
120	0.62	0.76	0.90	1.04	1.25	1.39	1.53	1.39	1.25	1.18
142	0.58	0.72	0.85	0.98	1.18	1.31	1.44	1.31	1.18	1.11
168	0.55	0.68	0.80	0.93	1.12	1.24	1.37	1.24	1.12	1.05
197	0.53	0.64	0.76	0.88	1.06	1.18	1.30	1.18	1.06	1.00
239	0.50	0.61	0.73	0.84	1.01	1.12	1.24	1.12	1.01	0.95
267	0.48	0.59	0.69	0.80	0.96	1.07	1.18	1.07	0.96	0.91
303	0.46	0.56	0.66	0.77	0.92	1.03	1.13	1.03	0.92	0.87
350	0.44	0.54	0.64	0.74	0.88	0.98	1.08	0.98	0.88	0.83
378	0.42	0.52	0.61	0.71	0.85	0.94	1.04	0.94	0.85	0.80
413	0.40	0.50	0.59	0.68	0.82	0.91	1.00	0.91	0.82	0.77
454	0.39	0.48	0.57	0.65	0.79	0.87	0.96	0.87	0.79	0.74
504	0.38	0.46	0.55	0.63	0.76	0.84	0.93	0.84	0.76	0.72
568	0.36	0.44	0.53	0.61	0.73	0.81	0.90	0.81	0.73	0.69
650	0.35	0.43	0.51	0.59	0.71	0.79	0.87	0.79	0.71	0.67
757	0.34	0.42	0.49	0.57	0.68	0.76	0.84	0.76	0.68	0.65
908	0.33	0.40	0.48	0.55	0.66	0.74	0.81	0.74	0.66	0.63

Environmental Considerations in Trout Farming

Most trout farms are not consumers of water, but water from the farms is enriched with nutrients that enrich lakes and rivers receiving farm water flows. These nutrients reduce water quality and increase the growth of algae and aquatic plants. Thus, rainbow-trout farms are subject to regulations limiting the levels of solids and nutrients in hatchery water effluents. As mentioned previously, phosphorus is a concern in hatchery effluents. Unassimilated nutrients originate in uneaten diet, pellet dust, faeces and metabolic excretions (urine and gill wastes). Many farms create quiescent zones at the ends of raceways where fish are excluded and particles settle without being disturbed. Settled material is regularly removed and applied to fields, sometimes after composting. Phosphorus in hatchery effluents is present in two forms: solid particles (bones and other insoluble forms) and soluble phosphorus excreted by fish in urine. Solid phosphorus can be collected and removed, but soluble phosphorus cannot be removed economically because it is present in very low concentrations in very large quantities of water. Thus, limiting the amount of digestible phosphorus in diets to the amount needed by the fish is the approach used in rainbow-trout diet formulation. Using this approach, the amount of soluble phosphorus excreted by fish has been reduced to very low levels. Reducing the amount of insoluble phosphorus in rainbow-trout diets requires two approaches: (i) using low-phosphorus diet ingredients; and (ii) increasing the bioavailability of phosphorus in diet ingredients. Several years ago there was little information on the bioavailability of phosphorus in common ingredients, but in recent years values for common ingredients have accumulated rapidly (Sugiura *et al.*, 1998; Bureau *et al.*, 1999; Sugiura and Hardy, 2000). A summary of these values is shown in Table 14.10.

Table 14.10. Availability of phosphorus (%) from practical diet ingredients for rainbow trout.

Ingredient	Total phosphorus	Available phosphorus	Availability
Herring fish-meal	2.05	0.91	44.4
Anchovy fish-meal	2.90	1.46	52.1
Menhaden fish-meal	3.43	1.25	36.5
Whitefish meal	3.50	0.60	17.2
Whitefish meal, deboned	1.69	0.79	46.8
Poultry by-product meal	2.36	0.90	38.3
Poultry by-product meal, low ash	1.68	1.06	63.0
Meat and bone-meal	2.68	0.58	21.8
Meat and bone-meal, low ash	2.49	0.87	35.0
Feather meal	1.26	1.00	79.4
Blood meal	0.72	0.74	103.5*
Wheat-gluten meal	0.18	0.13	74.7
Maize-gluten meal	0.54	0.05	8.5
Soybean meal	0.85	0.23	26.6
Wheat flour	0.32	0.15	47.0

* Obviously, the availability of a mineral cannot exceed 100%. This value is included to illustrate the difficulties associated with *in vivo* determination of availability of minerals in feed ingredients containing relatively low levels of the mineral of interest. In practical terms, it is unwise to rely on differences of less than 3–5% between ingredients.

Future Trends

Worldwide, farmed trout production is increasing, but at a lower rate than Atlantic salmon production (Tacon, 1998). Trout are an efficient fish with respect to converting diet ingredients not consumed by humans into human food. Currently, the conversion of trout diet into trout weight gain is about 1 : 1, meaning that for each tonne of diet, a tonne of trout is produced. Diets are composed of by-products of edible oil production (soybean meal), fish-meal, grain by-products and recovered protein from poultry and meat production. Trout yield more than 50% of edible product after processing, and this product is high in essential fatty acids and protein and low in saturated fats relative to animal proteins. Trout production is predicted to increase by 5% per year for the foreseeable future, and will probably maintain its place in the top 15 finfish and crustacean aquaculture species produced in the world and its place in the top ten species with respect to total value.

Over the past two decades, trout diets have changed in several ways. First, dietary protein and fat levels have increased, as has the quality of dietary protein sources, and this has resulted in an increase in the level of digestible protein in trout diets (Fig. 14.1). Secondly, trout diets are now produced mainly by cooking extrusion, rather than steam pelleting. Thirdly, emphasis in diet formulation is on low-polluting diets, which is leading to partial replacement of fish-meal with alternate protein sources, including low-ash rendered products and protein sources produced from grains and oil-seeds. Fish-meal, which typically constituted 50% of trout diet formulations a decade ago, now ranges between 25 and 40%, depending upon price. Alternative protein sources include blood meal, feather meal, low-ash poultry by-product meal, soybean meal and canola meal. Maize-gluten meal is another promising alternative ingredient, although in the

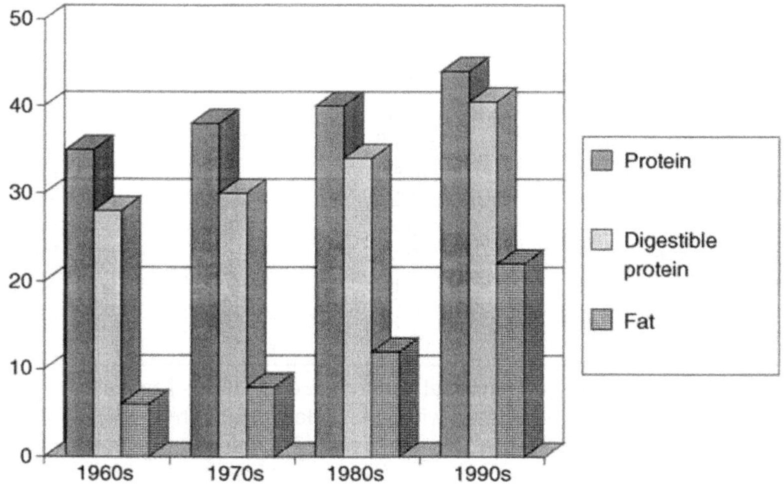

Fig. 14.1. Changes in protein and fat levels of trout feeds over the past 40 years.

USA maize gluten is limited to 5% of rainbow-trout diet formulations because of its content of xanthophyll pigments, which impart a yellow colour to trout muscle (Skonberg *et al.*, 1998). Development of alternative proteins for use in trout diets is likely to continue, especially in light of the finite nature of global fish-meal production and the likelihood of higher fish-meal prices over the next decade, as demand exceeds supply.

References

Barnett, B.J., Cho, C.Y. and Slinger, S.J. (1979) The essentiality of cholecalciferol in the diets of rainbow trout. *Comparative Biochemistry and Physiology* 63A, 291–297.

Behnke, R.J. (1992) *Native Trout of Western North America.* American Fisheries Society Monograph 6, American Fisheries Society, Bethesda, Maryland, 275 pp.

Boggio, S.M., Hardy, R.W., Babbitt, J.K. and Brannon, E.L. (1985) The influence of dietary lipid source and alpha-tocopheryl acetate level on product quality of rainbow trout (*Salmo gairdneri*). *Aquaculture* 51, 13–24.

Brannon, E.L. and Klontz, G.W. (1989) The Idaho aquaculture industry. *Northwest Environmental Journal* 5, 23–35.

Bureau, D.P., Harris, A.M. and Cho, C.Y. (1999) Apparent digestibility of rendered animal protein ingredients for rainbow trout (*Oncorhynchus mykiss*). *Aquaculture* 180, 345–358.

Cain, K.D. and Garling, D.L. (1995) Pretreatment of soybean meal with phytase for salmonid diets to reduce phosphorus concentrations in hatchery effluents. *Progressive Fish-Culturist* 57, 114–119.

Castell, J.D., Sinnhuber, R.E., Wales, J.H. and Lee, D.J. (1972a) Essential fatty acids in the diet of rainbow trout (*Salmo gairdneri*): growth, feed conversion and some gross deficiency symptoms. *Journal of Nutrition* 102, 77–85.

Castell, J.D., Sinnhuber, R.O., Lee, D.J. and Wales, J.H. (1972b) Essential fatty acids in the diet of rainbow trout (*Salmo gairdneri*): physiological symptoms of EFA deficiency. *Journal of Nutrition* 102, 87–92.

Castell, J.D., Lee, D.J. and Sinnhuber, R.E. (1972c) Essential fatty acids in the diet of rainbow trout (*Salmo gairdneri*): lipid metabolism and fatty acid composition. *Journal of Nutrition* 102, 93–99.

Cho, C.Y. and Cowey, C.B. (1991) Rainbow trout, *Oncorhynchus mykiss*. In: Wilson, R.P. (ed.) *Handbook of Nutrient Requirements of Finfish.* CRC Press, Boca Raton, pp. 131–143.

Cho, C.Y. and Woodward, B. (1990) Dietary pantothenic acid requirements of young rainbow trout (*Oncorhynchus mykiss*). *Journal of the Federation of American Societies for Experimental Biology* 4, 3747 [abstract].

Christiansen, R., Lie, O. and Torrissen, O.J. (1995) Growth and survival of Atlantic salmon, *Salmo salar* L., fed different dietary levels of astaxanthin: first-feeding fry. *Aquaculture Nutrition* 1, 189–198.

Cowey, C.B., Adron, J.W. and Youngson, A. (1983) The vitamin E requirement of rainbow trout (*Salmo gairdneri*) given diets containing polyunsaturated fatty acids derived from fish oil. *Aquaculture* 30, 85–93.

Halver, J.E. (1957a) Nutrition of salmonid fishes III. Water-soluble vitamin requirements of chinook salmon. *Journal of Nutrition* 62, 225–243.

Halver, J.E. (1957b) Nutrition of salmonid fishes IV. An amino acid test diet for chinook salmon. *Journal of Nutrition* 62, 245–254.

Halver, J.E. (1989) The vitamins. In: Halver, J.E. (ed.) *Fish Nutrition*. Academic Press, New York, pp. 31–109.

Halver, J.E. and Coates, J.A. (1957) A vitamin test diet for long-term feeding studies. *Progressive Fish-Culturist* 19, 112.

Hardy, R.W. (1991) Pacific salmon, *Oncorhynchus* spp. In: Wilson, R.P. (ed.) *Handbook of Nutrient Requirements of Finfish*. CRC Press, Boca Raton, pp. 105–121.

Hughes, S.G. (1984) Effect of excess dietary riboflavin on growth of rainbow trout. *Journal of Nutrition* 114, 1660–1663.

Hung, S.S.O., Cho, C.Y. and Slinger, S.J. (1981) Effect of oxidized fish oil, DL-α-tocopherol acetate and ethoxyquin supplementation on the vitamin E nutrition of rainbow trout (*Salmo gairdneri*) fed practical diets. *Journal of Nutrition* 111, 648–657.

Kaushik, S. (1979) Application of a biochemical method for the estimation of amino acid needs in fish: quantitative arginine requirements of rainbow trout in different salinities. In: Tiews, K. and Halver, J.E. (eds) *Finfish Nutrition and Fishfeed Technology*. Heenemann, Berlin, pp. 197–207.

Ketola, H.G. (1983) Requirement for dietary lysine and arginine by fry of rainbow trout. *Journal of Animal Science* 56, 101–107.

Kitamura, S., Suwa, T., Ohara, S. and Nakagawa, K. (1967) Studies on vitamin requirements of rainbow trout. III. Requirements for vitamin A and deficiency symptoms. *Bulletin of the Japanese Society for Scientific Fisheries* 33, 1126–1131.

Leitritz, E. and Lewis, R.C. (1980) *Trout and Salmon Culture*. California Fish Bulletin No. 164, Agricultural Sciences Publications, University of California, Berkeley, 197 pp.

McLaren, B.A., Keller, E., O'Donnell, D.J. and Elvehjem, C.A. (1947a) The nutrition of rainbow trout. I. Studies of vitamin requirements. *Archives of Biochemistry* 15, 169–178.

McLaren, B.A., Keller, E., O'Donnell, D.J. and Elvehjem, C.A. (1947b) The nutrition of rainbow trout: further studies with purified rations. *Archives of Biochemistry* 15, 179–185.

Morito, C.L.H., Conrad, D.H. and Hilton, J.W. (1986) The thiamin deficiency signs and requirement of rainbow trout (*Salmo gairdneri* Richardson). *Fish Physiology and Biochemistry* 1, 93–104.

NRC (National Research Council) (1993) *Nutrient Requirements of Fish*. National Academy Press, Washington, DC, 114 pp.

Phillips, A.M. and Brockway, D.R. (1947) The niacin and biotin requirements of trout. *Transactions of the American Fisheries Society* 77, 152–159.

Poston, H.A. (1976) Relative effect of two dietary water-soluble analogues of menaquinone on coagulation and packed cell volume of blood of lake trout. *Journal of the Fisheries Research Board of Canada* 33, 1791–1793.

Poston, H.A. and Wolfe, M.J. (1985) Niacin requirement for optimum growth, feed conversion and protection of rainbow trout, *Salmo gairdneri* Richardson, from ultraviolet-B irradiation. *Journal of Fish Diseases* 8, 451–460.

Richardson, N.L., Higgs, D.A., Beames, R.M. and McBride, J.R. (1985) Influence of dietary calcium, phosphorus, zinc and sodium phytate level on cataract incidence, growth and histopathology in juvenile chinook salmon (*Oncorhynchus tshawytscha*). *Journal of Nutrition* 115, 553–567.

Rodehutscord, M. and Pfeffer, E. (1995) Effects of supplemental microbial phytase on phosphorus digestibility and utilization in rainbow trout (*Oncorhynchus mykiss*). *Water Science and Technology* 31, 143–147.

Rumsey, G.L. (1991) Choline-betaine requirements of rainbow trout (*Oncorhynchus mykiss*). *Aquaculture* 95, 107–116.

Schneberger, E. (1941) Fishery research in Wisconsin. *Progressive Fish-Culturist* 56, 14–17.

Skonberg, D.I., Hardy, R.W., Barrows, F.T. and Dong, F.M. (1998) Color and flavor analyses of fillets from farm-raised rainbow trout (*Oncorhynchus mykiss*) fed low-phosphorus feeds containing corn or wheat gluten. *Aquaculture* 166, 269–277.

Spinelli, J., Houle, C.R. and Wekell, J.C. (1983) The effect of phytate on the growth of rainbow trout (*Salmo gairdneri*) fed purified diets containing varying quantities of calcium and magnesium. *Aquaculture* 30, 71–83.

Stickney, R.R. (1996) *Aquaculture in the United States, a Historical Survey.* John Wiley & Sons, New York, 372 pp.

Sugiura, S.H. and Hardy, R.W. (2000) Environmentally friendly feeds. In: Stickney, R.R. (ed.) *Encyclopedia of Aquaculture.* John Wiley & Sons, New York, pp. 299–310.

Sugiura, S.H., Dong, F.M., Rathbone, C.K. and Hardy, R.W. (1998) Apparent protein digestibility and mineral availabilities in various feed ingredients for salmonids. *Aquaculture* 159, 177–200.

Sugiura, S.H., Rayboy, V., Young, K.A., Dong, F.M. and Hardy, R.W. (1999) Availability of phosphorus and trace elements in low-phytate varieties of barley and corn for rainbow trout (*Oncorhynchus mykiss*). *Aquaculture* 170, 285–296.

Sugiura, S.H., Babbitt, J.K., Dong, F.M. and Hardy, R.W. (2000) Utilization of fish and animal by-product meals in low-pollution feeds for rainbow trout (*Oncorhynchus mykiss* Walbaum). *Aquaculture Research* 31, 585–593.

Tacon, A.G.J. (1998) *Trends in Global Aquaculture Production.* International Aqua Feed Directory and Buyer's Guide 1997/98, Turret Publishing, London.

Takeuchi, T. and Watanabe, T. (1977) Dietary levels of methyl laurate and essential fatty acid requirement of rainbow trout. *Bulletin of the Japanese Society for Scientific Fisheries* 43, 893–898.

Watanabe, T., Ogino, C., Koshiishi, Y. and Matsunaga, T. (1974) Requirement of rainbow trout for essential fatty acids. *Bulletin of the Japanese Society for Scientific Fisheries* 40, 493–499.

Watanabe, T., Takeuchi, T., Wada, M. and Vehara, R. (1981) The relationship between dietary lipid levels and α-tocopherol requirement of rainbow trout. *Bulletin of the Japanese Society for Scientific Fisheries* 47, 1463–1471.

Wolf, L.E. (1942) Fish diet disease of trout: a vitamin deficiency produced by diets containing raw fish. *New York State Conservation Department, Fisheries Research Bulletin* 4, 1–5.

Wolf, L.E. (1951) Diet experiments with trout. 1. Synthetic formula for dietary studies. *Progressive Fish-Culturist* 13, 17–24.

Woodall, A.N., Ashley, L.M., Halver, J.E., Olcott, H.S. and Van Der Veen, J. (1964) Nutrition of salmonid fishes XIII. The a-tocopherol requirement of chinook salmon. *Journal of Nutrition* 84, 125–135.

Woodward, B. (1990) Dietary B_6 requirements of young rainbow trout (*Oncorhynchus mykiss*). *Journal of the Federation of American Societies for Experimental Biology* 4, 3748 [abstract].

Woodward, B. and Frigg, M. (1989) Dietary biotin requirements of young rainbow trout (*Salmo gairdneri*) determined by weight gain, hepatic biotin concentration and maximal biotin-dependent enzyme activity in liver and white muscle. *Journal of Nutrition* 119, 54–60.

Yu, T.C., Sinnhuber, R.O. and Hendricks, J.D. (1979) Reproduction and survival of rainbow trout (*Salmo gairdneri*) fed linolenic acid as the only source of essential fatty acids. *Lipids* 14, 572–575.

Arctic Char, *Salvelinus alpinus* 15

Terry A. Dick[1] and X. Yang[2]
[1]Department of Zoology, University of Manitoba, Winnipeg, Manitoba R3T 2N2, Canada; [2]Department of Cancer Immunology and AIDS, Dana–Faber Cancer Institute and Harvard University Medical School, Boston, MA 02115, USA

Introduction

Artic char (*Salvelinus alpinus*) is a relative newcomer to the aquaculture industry and first impressions were that this would be a highly successful freshwater species because of its ready acceptance by the consumer, the anticipated high price predicted for it relative to its closest competitor, rainbow trout (*Oncorhynchus mykiss*), and the fact that it was easy to culture. A recent review of the potential for Arctic-char culture in Norway by Heggberget *et al.* (1994) reports that pre-smolt char grew better than Atlantic salmon (*Salmo salar*) and brown trout (*Salmo trutta*) and almost as fast as pink salmon (*Oncorhynchus gorbuscha*).

Most of the early culture efforts on char involved defining the growth characteristics of various char strains, such as the Canadian strains Nauyuk and Labrador and the Norwegian strains Sunndalsora, Hammerfest and Svalbard. Common features of the results from these studies were variation in size, early sexual maturation or precociousness of a component of all char populations and the reduced ability of char to absorb pigments. This meant that frequent grading was required and cultured char did not have the deep-red appearance of wild char. Nevertheless, there has been a steady effort in several parts of the northern hemisphere to understand nutritional requirements of char, using the rainbow-trout model.

Arctic char has the most northern distribution of all freshwater fish species and is widespread in the Arctic and subarctic regions of North America, Europe and Asia (Johnson, 1980). Slow-growing populations are restricted to land-locked systems, while the faster-growing and larger char feed in the marine environment during the open-water period. Char occur as monospecies in many Arctic lakes and also coexist with other freshwater species, such as lake trout (*Salvelinus namaycush*) and lake whitefish (*Coregonus clupeaformis*). In some regions, such as northern Labrador, Arctic char hybridize with lake trout and brook trout (Hammar *et al.*, 1991).

Nutrient Requirements

Protein and amino acids

Gurure *et al.* (1995a) found that the highest growth response of char occurred at 350 g kg^{-1} digestible crude protein (DCP), but diets with 390–430 g kg^{-1} did not produce any improvements in weight for protein gain. Differences between Tabachek (1986) and Jobling and Wandsvik (1983) were discussed by Jobling (1991), but reasons for the differences between char growth probably relate more to strains of char and diet formulations than to tank design, as other studies using the same tanks reported a specific growth rate (SGR) of 1.6% (Papst *et al.*, 1992). Ringo (1995) and Torissen and Shearer (1992) studied the relationship between growth, protein digestiblity and food conversion of char in fresh and sea water. Torissen and Shearer (1992) used the Hammerfest strain (fast-growing and anadromous) and the Skogseid strain (slow-growing and non-anadromous). Experimental conditions included salinity levels of 0 and 27 p.p.t., temperature of 6°C and 10°C and small (33–90 g) and large (200 g) char. Torissen and Shearer (1992) found protein digestibility for the slow-growing and fast-growing strains to be 82.9% and 80.6%, respectively, but SGR was significantly higher for the fast-growing strain at 6°C. Ringo (1995) studied the digestibility of two char diets, consisting of capelin roe and a commercial diet, using the chromic oxide technique. Char were grown in fresh water and sea water. The protein content of capelin and the commercial diets was 670 and 500 g kg^{-1}, respectively, and the apparent digestibility of capelin roe was higher at 92% and 88.2%, when char were grown in fresh water and salt water, respectively, compared with the commercial diet, which had values of 89% and 85.7% for char grown in fresh water and sea water, respectively (Ringo, 1995). Generally there was a reduction in the digestibility of protein from both diets for char grown in sea water.

A few studies have evaluated the amino acid requirements of char. Using the Labrador strain of char, Yang (1994) reported that concentrations of arginine of 34.9 and 21.9 g kg^{-1} did not markedly alter growth rates, haematocrit values, haemoglobin, moisture in the liver and muscle or the hepatic somatic index. The levels used by Yang (1994) are high relative to those listed for rainbow trout (1.5%) and Pacific salmon (*Oncorhynchus nerka*) (2.04%) by the National Research Council (NRC, 1993).

Simmons *et al.* (1997) reported that methionine at 18 g kg^{-1} of dietary protein was required for optimum growth, but 23 g kg^{-1} dietary protein was necessary to prevent nutritionally induced cataracts. In a further study, Simmons *et al.* (1999), using char with an initial weight of 20.5 g and grown over 16 weeks, found that the requirement of methionine for optimal growth was 17.6 g kg^{-1} of dietary protein (estimated using a quadratic regression), while the requirements based on carcass quality and energy gains were 18.8 and

17.8 g kg^{-1} of dietary protein, respectively. Simmons *et al.* (1999) found that 26.7 g kg^{-1} dietary protein was required to prevent cataracts in char.

Lysine levels were evaluated by Gurure *et al.* (1995b) in char weighing 3.8 g and 30 g and fed a diet containing 36% DCP, 15.1% digestible crude lipid and 15.5 MJ digestible energy kg^{-1}. Practical ingredients supplied 50% of the protein and the remainder was added in the form of synthetic glutamic acid and essential amino acids. Lysine was added to the feed to obtain dietary lysine levels of 10–34 g kg^{-1} diet. Gurure *et al.* (1995b) found that SGR for all treatments ranged from a high of 1.95 day^{-1} to a low of 0.65% day^{-1}, carcass moisture and ash were generally similar and final carcass protein went up with increasing lysine levels to a high of 19 g kg^{-1}. Feed conversion ratios, carcass lipid and net lipid accumulation decreased with increasing lysine up to 19 g kg^{-1}. Using broken-line analysis, Gurure *et al.* (1995b) estimated the requirements for lysine to be 21 g kg^{-1} diet (58 g kg^{-1} in protein).

Energy and digestibility

The energy requirement of food consumed has been estimated for natural food (Larsson and Berglund, 1998) and for pelleted diets (Tabachek, 1984; Larsson and Berglund, 1998). Tabachek reports values ranging from 3.15 to 3.72 kcal g^{-1} and Larsson and Berglund (1998) report values for *Neomysis integer* ranging from 20.0 to 21.0 MJ kg^{-1}. Gurure *et al.* (1995a) raised 2.56 g char at 9.5–13°C and fed practical diets containing 16.6 MJ digestible energy (DE) kg^{-1}. These char had specific growth rates of 2.86 day^{-1} and Gurure *et al.* (1995a) concluded that protein requirements of young growing char are similar to those of most salmonids. Digestible energy has been calculated for a number of experimental diets: 16.63–17.28 MJ kg^{-1} by Gurure *et al.* (1995a); 17.6 MJ kg^{-1} by Alanara and Kiessling (1996); 17.5 MJ kg^{-1} by Simmons *et al.* (1999); 18.36–19.07 kJ g^{-1} by Jobling and Wandsvik (1983). These values are similar to those of commercial rainbow-trout diets. Values ranging from 16.6 to 21 MJ kg^{-1} for formulated diets and a natural food source can be compared to the typical energy content of commercial rainbow trout diets of 3200 kcal DE kg^{-1} and Pacific salmon of 3600 kcal DE kg^{-1} (NRC, 1993).

Lipids and fatty acids

It is commonly known that the composition of lipids in fish is affected by the composition of dietary lipid (Sargent *et al.*, 1989). It is also well established for cold-water fish species that sparing of protein and lipids occurs because either can be used as a source of energy, and that polyunsaturated fatty acids (PUFAs) are important in cold-water fish diets (Dick and Yang, 1992, 1995). Marine fish feeding on various zooplankton and phytoplankton with high levels of n-3 PUFA, especially eicosapentaenoic acid (EPA) and docosahexaenoic acid (DHA), 20:5n-3 and 22:6n-3, respectively, have high levels of these lipids (Dick and

Yang, 1992). Freshwater fish generally feed on organisms with high levels of linoleic acid, 18:2n-6, and linolenic acid, 18:3n-3, and as a result have proportions of EPA and DHA lower than those for marine fish.

The fatty acid composition of muscle of wild char and other salmonids is an important starting-point for studying the dietary lipid requirements in fish (Castell, 1979; Table 15.1). The fatty acids 18:2n-6, EPA and DHA were higher in char than those reported for Atlantic salmon and rainbow trout, but 18:3n-3 was lower than that reported for Atlantic salmon and rainbow trout (Table 15.1). Values for char and Atlantic salmon are more similar than those for rainbow trout. Ringo and Nilsen (1987) found that the neutral lipid from the muscle of land-locked Arctic char had mostly 16:0, 16:1 and oleic acid (18:1n-9) fatty acids and Jobling (1991) reported that the PUFA content in the eggs was 35%, with EPA and DHA having the highest relative proportion, and in the first-feeding fry PUFA was about 40%. Ringo and Nilsen (1987) reported that the fatty acid composition of the neutral fraction of the wild char differed substantially from char fed on capelin and a commercial diet. Both 20:5n-3 and DHA increased in char muscle fed the capelin diet (Ringo and Nilsen, 1987).

The first studies on Arctic char nutrition reported that protein and lipid requirements ranged from 27.6% to 43.6% and from 11.4% to 17.3%, respectively (Jobling and Wandsvik, 1983). Tabachek (1984) reported that not all commercial trout or salmon diets produced optimal growth for char and suggested that char were more similar to coho salmon (*Oncorhynchus kisutch*) but different from trout in terms of the optimum ratio of dietary protein and lipids. The proportion of lipids in published char diets vary from 5 to 28% of the diet and a range of 2–16% has been reported for casein-based diets where information was sought on levels of essential fatty acids and possible interactions between fatty acids (Table 15.2).

Fish fed diets deficient in 18:2n-6 and 18:3n-3 usually exhibit poor growth, low feed efficiency, fatty livers (elevated total lipids and percentage of neutral lipids in the liver), increased water content in whole body or muscle, high hepatosomatic index and substantial accumulation of 20:3n-9 in tissue polar lipids (Henderson and Tocher, 1987). The general consensus is that freshwater fish are not capable of synthesizing 18:2n-6 and 18:3n-3 *de novo* but are able to desaturate and elongate these fatty acids to long-chain PUFAs, namely the C20 and C22 PUFAs. The process requires a common set of enzymes for desaturation and elongation which are influenced by levels of n-9, n-6 and n-3 in the diet (Henderson and Tocher, 1987).

Ringo (1989) reported that conventional commercial salmon diets supplemented with linoleic acid had negative effects on growth and protein and lipid digestibility in char. Olsen *et al.* (1991) reported that 18:3n-3 is the predominant PUFA in char. Olsen and Ringo (1992) found that char converted 18:2n-6 and 18:3n-3 into long-chain PUFA, but that 18:3n-3 was preferred to 18:2n-6 as a substrate. Yang and Dick (1993) reported that char fed increasing dietary 18:3n-3 from 0.1 to 2.0% had higher SGR and feed efficiency. Dietary 22:6n-3 was more effective than 18:3n-3 and 18:3n-3 was more effective than 18:2n-6 in promoting char growth. Yang and Dick (1993) concluded that dietary

Table 15.1. Composition of major fatty acids of muscle of wild Arctic char, Atlantic salmon and rainbow trout.

Fatty acid	Arctic char*	Arctic char†	Atlantic salmon‡	Rainbow trout§
14:0	0.97 ± 0.16	2.3	1.5 ± 0.01	3.8 ± 0.01
15:0	–	0.4	–	–
16:0	15.86 ± 0.6	12.9	14.2 ± 0.14	22.8 ± 1.8
16:1	–	25.7	–	–
16:1n-9	0.41 ± 0.03	–	5.5 ± 0.1	–
16:1n-7	3.66 ± 0.44	–	–	9.5 ± 0.8
17:0	–	0.2	–	–
18:0	5.48 ± 0.12	2.3	5.3 ± 0.04	4.3 ± 0.1
18:1	–	24.9	–	–
18:2	–	5.2	–	–
18:1n-9	11.27 ± 0.17	–	12.6 ± 0.19	21.6 ± 2.8
18:1n-7	3.93 ± 0.35	–	–	–
18:2n-6	6.76 ± 0.65	–	3.1 ± 0.06	4.9 ± 1.3
18:3n-6	0.02 ± 0.01	–	–	–
18:3n-3	1.81 ± 0.14	–	2.2 ± 0.05	6.9 ± 1.6
18:4n-3	0.43 ± 0.07	–	–	–
20:1	–	0.5	–	–
20:1n-9	0.53 ± 0.05	–	0.8 ± 0.05	0.8 ± 0.3
20:2	–	0.8	–	–
20:2n-6	0.54 ± 0.08	–	–	–
20:3n-6	0.71 ± 0.03	–	–	–
20:4n-6	8.26 ± 0.29	–	8 ± 0.06	1.7 ± 0.3
20:3n-3	0.16 ± 0.01	–	–	–
20:4n-3	0.46 ± 0.06	–	–	–
20:5n-3	8.03 ± 0.54	–	4.6 ± 0.01	5.3 ± 1.1
20:4w-3	–	0.8	–	–
20:5w-3	–	5	–	–
22:1	–	3	–	–
22:1n-9	0.09 ± 0.01	–	–	–
22:4	–	0.7	–	–
22:4n-6	0.68 ± 0.09	–	1.3 ± 0.05	–
22:5n-6	1.09 ± 0.13	–	2 ± 0.08	–
22:5n-3	3.3 ± 0.27	–	3.3 ± 0.15	–
22:5w-3	–	1.9	–	–
22:6n-3	18.26 ± 2.28	–	15.4 ± 0.27	11.7 ± 1.9
22:6w-3	–	4.6	–	–
24:1n-9	0.76 ± 0.12	–	–	–
Saturates	23.36 ± 0.95	–	24.4 ± 0.06	–
Monoenes	22.26 ± 0.79	–	26.5 ± 0.13	–
PUFA	50.20 ± 1.87	–	45.3	–
n-6	18.14 ± 0.31	–	15.8 ± 0.04	–
n-3	32.06 ± 2.1	–	27.2 ± 0.4	–

* Yang (1994).
† Ringo and Nilsen (1987).
‡ Ackman and Takeuchi (1986).
§ Suzuki *et al.* (1986).

Table 15.2. Ingredients for Arctic char casein-based diets.

Ingredients	Olsen *et al.* (1991) (% dry weight)	Olsen *et al.* (1999) (g kg^{-1})	Olsen *et al.* (2000) (g kg^{-1})	Yang and Dick (1993)* (% dry weight)
Casein	56	500	593–793	50
Dextrin	17	150	150	10
Gelatin	1.7	17	17	4
Fish protein	14.65	144.3	–	–
Mineral mix	3	–	–	–
Premix I, II, III	34.05	–	42	–
Premix I, II, III, IV	–	5.35	–	–
Coconut oil	–	–	–	–
Lipid supplement	5.3†	150‡	50–160§	5.0

* Also included starch, alpha-cellulose, carboxymethylcellulose, methionine, arginine, mineral premix and vitamin premix.
† 43% Coconut oil + 1% methylesters of 18:2n-6, 18:3n-3, 20:4n-6.
‡ Marine and/or coconut oil.
§ Linseed oil.

18:2n-6 had a minimal effect on growth if there were adequate amounts of 18:3n-3 (somewhere between 1.0 and 2.0%). Yang and Dick (1994b) showed that the conversion of 18:3n-3 to long-chain PUFA reaches a maximum of about 17 g dietary 18:3n-3 kg^{-1} diet in the liver of char. When equal levels of dietary 18:2n-6 and 18:3n-3 were fed to char, there were more products from 18:3n-3 conversion than from 18:2n-6, especially more DHA than arachidonic acid, 20:4n-6, in liver and polar lipids. This indicates that 18:3n-3 is preferentially elongated and desaturated. High levels of 18:3n-3 markedly inhibited the conversion of 18:2n-6, whereas inhibition of dietary 18:2n-6 on 18:3n-3 conversion was noted only when dietary 18:2n-6 to 18:3n-3 changed from 1.0 to 1.5%. Both Labrador and Nauyuk char require 1.0–1.7% 18:3n-3 in the presence of 0.5% 18:2n-6 (Yang and Dick, 1994a,b; Yang *et al.*, 1994). This suggests that juvenile-char lipid requirements are similar to those of Atlantic salmon, chum salmon (*Oncorhynchus keta*) and Chinook salmon (*Oncorhynchus tshawytscha*).

Olsen *et al.* (2000) concluded that feeding diets high in PUFA and low in saturated fats promotes the accumulation of lipid droplets and generates pathology, but, when 40 g kg^{-1} of 16:0 was added to 160 mg kg^{-1} of linseed oil, the enterocyte damage index was reduced significantly. Further support for the role of saturated fats is noted by Yang and Dick (1994b), where replacement of 12:0 with 18:1n-9 resulted in a decline in SGR and feed efficiency, and by Dick and Yang (1995), where char fed a diet containing 20% full-fat flax meal and 40% casein, plus 2.2% animal fat, had SGR of 3% day^{-1}. To summarize, the amount of n-6 and the amount and type of n-3 fatty acids and the ratio between n-6 and n-3 and saturated lipids are important for optimizing dietary lipid utilization by char.

Carbohydrates

Carbohydrates are sources of low-cost energy in fish diets but their utilization by cold-water species is not well understood. Generally, most diets developed for cold-water species have relatively low levels of carbohydrates. The carbohydrate content for several diets fed to char ranges from 17 to 25%, but for other studies where carbohydrate has not been determined directly both the total carbohydrate content and its source are more variable (Table 15.3) – for example, precooked grains (65% wheat and 35% oats) at 11% of the diet and maize meal at 11% of diet (Jobling and Wandsvik, 1983); wheat middlings at 31.9% (Tabachek, 1986); dextrin at 170 g kg^{-1} (Ringo and Olsen, 1991); maize starch (170 g kg^{-1}), wheat middlings (150 g kg^{-1}) and soy meal (Simmons *et al.*, 1999); maize gluten at 25–200 g kg^{-1}, wheat middlings at 111.5–362.1 g kg^{-1} and raw maize starch at 0–139 g kg^{-1} (Gurure *et al.*, 1995a); 15.2% starch and 10% dextrin in a semipurified diet (Yang and Dick, 1993); 20% full-fat ground flax, 10% wheat germ, 5% soybean meal, 8.5% glucose and 2% starch (Dick and Yang, 1995); 15% dextrin (Olsen *et al.*, 1999). Not all these studies were attempting to optimize growth nor was the carbohydrate component in the diet the main aspect of the study. Nevertheless, the data indicate that char seem to be able to handle a wide range of total carbohydrates, and in different forms.

Vitamins and minerals

Fish fed diets inadequate in vitamin C, zinc, copper or iron usually display retarded growth, poor feed utilization and anaemia (John *et al.*, 1979; Halver, 1989; Lall, 1989). Few studies have determined vitamin and mineral requirement for char. However, many of the studies on char nutrition report the amount of vitamins and mineral used. Table 15.4 outlines the levels of selected vitamins from a number of different laboratories and it is apparent that the amounts are variable. Only occasionally are the amounts an order of magnitude different (such as vitamin C). Generally the amount of vitamins and minerals used in diets for char is higher than that reported for rainbow trout and Pacific salmon (NRC, 1993).

Pigments

Red coloration of the flesh is a desirable consumer trait for salmonids and the factors considered to influence pigment deposition are size, growth rate and sexual maturation. Wild populations of char have a remarkable deep red colour of the flesh and skin. The predominant carotenoid in wild char is astaxanthin (Scalia *et al.*, 1989), but idoxanthin, a metabolite of astaxanthin, has been identified by Aas *et al.* (1997); and considered to be part of the char's normal development and is size-dependent, with the smallest fish having the highest concentration.

Table 15.3. Experimental diets for Arctic char.

	Hatlen *et al.* (1995)	Simmons *et al.* (1999)	Gurure *et al.* (1995a)	Ringo and Nilsen (1987)	Jobling and Wandsvik (1983)	Tabachek (1986)
Ingredients	(g 100 g^{-1})	Basal diet (g kg^{-1})	DCP 390‡ (g kg^{-1})	Capelin roe* (% weight)	MP diet (% weight)	Control diet (% weight)
Strain of char	Hammerfest		Labrador			Labrador
Agar	–	10	–	–	–	–
Amino acids	–	200	–	–	–	–
Maize gluten meal	–	–	90	–	–	–
Maize starch	–	170	6.4	–	–	–
Fish-meal	57	–	–	–	44	–
Fish silage	3.8	–	–	–	–	–
Herring meal	–	–	447.2	–	–	35
Gelatin	–	50	–	10	–	–
Maize meal	–	–	–	–	17	–
Torula yeast	–	–	–	–	7	–
Fish-oil	12.5	150	113.3	–	–	–
Cod-liver oil	–	–	–	–	12	–
Herring oil	–	110	–	–	–	10
Soybean meal	–	120	–	–	–	20
Wheat middlings	–	150	329.2	–	–	31.9
Micronutrients	–	–	–	–	–	–
Vitamins and minerals	–	40	–	–	1	–
Mineral	–	–	10	–	–	1
Vitamin	–		10	9.8	–	1.5
Choline chloride	–	–	–	–	–	0.4
Powdered chalk	–	–	–	–	1	–
Chromic oxide	–	–	–	–	1	–
D,L-Methionine	–	–	–	–	–	0.2
Wheat, crushed	26.1	–	–	–	–	–
Precooked grain	–	–	–	–	17	–
Micronutrient premix†	0.61	–	–	–	–	–
Ascorbic acid	–	–	1.3	–	–	–

* Capelin roe about 80% of the diet
† Medium protein (44%) diet
‡ Calculated digestible crude protein of 390 kg^{-1}
DCP, digestible crude protein; MP, medium protein.

Table 15.4. Composition of vitamin premixes used in Arctic char diets (mg kg^{-1} diet).

Ingredients	Yang and Dick (1993)	Tabachek (1986)	Olsen *et al.* (1999)	Simmons *et al.* (1999)	Olsen *et al.* (2000)
Biotin	1.2	1	0.28	0.75	1.4
Carnitine	300	–	–	–	–
Folic acid	6	5	2	15	13
Thiamine HCl	24	10	2	45	15.8
Pyridoxine HCl	18	10	10	45	22
Riboflavin	24	20	12	75	30
D-Capantothenate	–	40	40	225	150
Ascorbic acid	600	100	100	60	300
Myo-inositol	400	300	–	–	450
Niacin	48	150	20	300	200
BHA	–	–	–	–	1000
p-Amino benzoic acid	–	200	–	–	–
Vitamin K	48	10	2	45	26.3
Vitamin B_{12}	1	0.02	0.04	0.045	0.2
Vitamin A	1	3125	5000	7500	5000
Vitamin D_3	1	5000	4000	4500	2600
Vitamin E	–	–	100	150	250
Choline chloride	3000	3000	3000	–	2000

Jobling (1991) reported that char fed 40 mg cantaxanthin kg^{-1} diet resulted in 2 mg kg^{-1} in the flesh, but this was lower than the 3–4 mg kg^{-1} that is considered an acceptable level by the consumer. More recently, several studies have focused on the influence of the age and size of the char on pigmentaion. Hatlen *et al.* (1995) added astaxanthin at 63 and 114 mg kg^{-1} to the diets of 1-, 2- and 3-year-old char (Hammerfest strain). Significant differences were seen among year classes for flesh, skin and plasma carotenoid concentrations. At 13 weeks of feeding, the carotenoids in the flesh were lowest in the 1-year-old fish and highest in the 3-year-old fish. Even after 25 weeks, no differences were seen in pigmentation between 2- and 3-year-old fish but 1-year-old fish were still less pigmented. Metusalach *et al.* (1996) used canthaxanthin at concentrations of 65 mg kg^{-1} in diets fed to post-juvenile char (Labrador strain) weighing 150 g. The major pigments in the diet were canthaxathin, leutein and echinenone. After 24 weeks, flesh had the highest amount of carotenoids at 1.738–2.585 mg kg^{-1}, but concentrations of 9.38 mg kg^{-1} in the skin, 6.02 mg kg^{-1} in the flesh, 5.23 mg kg^{-1} in the liver and 2.16 mg kg^{-1} in the gonads were reported as well. Olsen and Mortensen (1997) evaluating the effects of temperature on the deposition of astaxanthin on flesh colour found a positive correlation between dietary astaxanthin up to 70 mg kg^{-1}, after which a plateau in pigmentation was reached. In addition, flesh coloration was positively correlated to specific growth and char maintained at 8°C had significantly higher pigmentation than char grown at higher temperatures.

Bjerkeng *et al.* (1999) found that sex steroids strongly influenced carotenoid distribution and that male fish contained more carotenoids in the fillets than females. Bjerkeng *et al.* (1999) recommended that the induction of sexual maturation should be avoided in a production operation and that perhaps growers should evaluate the possibility of using all-male or all-female char. Hatlen *et al.* (1995) concluded that char must reach a certain size before maximum pigmentation occurs. Hatlen *et al.* (1995) recommended that fish farmers planning to produce a 300 g char should include astaxanthin for the entire production period and for larger fish astaxanthin should be added to the diet when char are 200–300 g. Pigments are being used in production facilities at concentrations of 40 p.p.m. and at 60 p.p.m. with good success (Delabbio, 1995). To summarize, the process of metabolism and pigmentation of char tissues seems to be similar to that of salmon and rainbow trout and astaxanthin and cantaxanthin can be added to the diet. However, to optimize pigmentation, producers should consider the strain, size, sex and stage of sexual maturation of char and the grow-out temperature.

Practical Diets

There are a number of reports where char have been fed diets produced by commercial-diet companies (Table 15.5) and practical diets (Table 15.3). Among the first practical diets tested for Arctic char were those of Jobling and Wandsvik (1983) and Tabachek (1984). The diet by Jobling and Wandsvik (1983) consisted of fish-meal (28–60%), precooked grain (11–23%), maize meal (11–23%), torula yeast (7%), cod-liver oil (8–16%), powdered chalk (1%) and a vitamin/mineral mix (1%). Digestibility of these diets, estimated using chromic oxide, varied from 75.97% to 82.95%, but digestibility values for char were 5–10% lower than those for rainbow trout. Tabachek (1984, 1986) reported on the result of feeding trials using several commercial and one experimental diet. Char strains (Sunndalsora from Norway and Labrador strain from Canada) were evaluated for weight gain, feed conversion and mortality. More recently, practical diets have been reported for char by Dick and Yang (1995), Gurure *et al.* (1995a) and Simmons *et al.* (1999).

Particle size is known to have a significant effect on SGR of char. The highest growth rate for char was when particle size was generally 0.015–0.018 times the fork length for 73–109 mm (3–12 g) and 0.023–0.024 times fork length for 121–133 mm (16–21 g) char. This corresponds to 21% of mouth size for 3–7 g char, 23–25% of mouth size for 9–12 g char and 31–33% of mouth size for the 16–21 g char (Tabachek, 1988). Yang (1994) found particle sizes from semipurified diets for 3–5 g char as follows: 0.60–0.85 mm (17.8%), 0.85–1.0 mm (36.7%), 1.0–1.18 mm (45.8%) optimized food intake. Consumption time is shortest and the number of missed pellets is lowest when pellet length is approximately 2–3% of fork length (Linner and Brannas, 1994).

Table 15.5. Commercial diets fed to Arctic char.

Study	Char size at start (g)	Char strain	Temperature (°C)	SGR	Type diet	Source
Eales and Shostak (1985)	2 years	Labrador	12	0.65	Trout grower	EWOS 4-P pellets, Rundle Feed Mills, Ontario
Ringo and Nilsen (1987)	8	–	8	8 to 40 g in 75 days	NG	Skretting (Tess Elite)
Jorgensen and Jobling (1989)	–	–	–	–	NG	Ewos ST 40
Jobling *et al.* (1989)	49.1–69	–	6.1	0.388–1.333	NG	Ewos ST 40, 4 mm
	49.9–79.3	–	–	0.600–1.301	NG	
Miglavs and Jobling (1989)	5–10	–	6.5	1.5–2	NG	Tess Elite Plus, Skretting A/S
Christiansen *et al.* (1992)	10	Hammerfest	6.3	0.99–1.37	NG	EWOS ST 40/4 granulate
Torrissen and Shearer (1992)	33	Hammerfest	6.4	0.26 ± 0.03	NG	Tess Elite Plus
	88	Skogseid	6.4	0.12 ± 0.04	NG	T. Skretting A/S, Stavanger,
Jorgensen *et al.* (1993)	52–55	Hammerfest	6.2	0.8–1.1	NG	Norway FK start 3 mm pellet
	–	–	–	–	NG	–
Yang and Dick (1993)	–	Labrador	10	2.02	Rainbow trout	Martin Feed Mills, Elmira, Ontario
Hatlen *et al.* (1995)	202–980	Hammerfest	8.2	0.1	Arctic char	Royefor 6 mm pellets, Felleskjopet A/S, Stavanger, Norway containing 50 mg astaxanthin
Metusalach *et al.* (1996)	100	Labrador	4.9–6.3	–	NG	Moore-Clark Co. Inc., St Andrews, New Brunswick contained 65 μg conthaxanthin kg^{-1}
Alanara and Kiessling (1996)	~210	Hornavan	2.1–14.5	210 to 340 g in 60 days	NG	EWOS, Sodertalje, Sweden, Vextra Super
Olsen and Mortensen (1997)	130	–	8.1–12.5	0.74 (12.5)	Arctic char	Skretting Elite 3.5 mm
	–	–		0.62 (8)	NG	
Damsgaard *et al.* (1999)	1–2	Hammerfest	0.3–12.5	Up to 1.7	NG	Felleskjopet, Sandnes, Norway
	–	Svaldard		1.8	NG	FK – Vekst

NG, not given.

Char will take a larger pellet by expanding the length of the pellet compared with round pellets and there was a preference for large over small pellets, given a choice.

Feeding Practices

Feeding practices for char have been quite variable due to the nature of the research question and range from feeding to excess, to apparent satiation or the use of demand feeders. Generally, char are fed two to four times daily, with the amount of food calculated using commercial feeding tables, such as those of Hilton and Slinger (1981). For fish > 1 g, the ration is adjusted periodically based on changes in total wet weight of fish in the tank. Particle size for a specific size class of fish generally follows that for rainbow trout and is based on feeding guides produced by feed companies. However, for char at swim-up and small fry, it is better to hand-feed to apparent satiation rather than to follow rigidly a set of feeding tables. Due to the considerable variation in size of char shortly after swim-up, particle size should be variable to ensure that all individuals can feed. It is important to establish an adequate ration, as Papst (1994) reported that char (1 and 20 g initial weight) fed 50% of the ration produced the lowest variation in growth and char fed 100 and 150% of the ration produced the highest variation in growth and the highest SGR. There is some debate on whether char feed in the water column and/or on the bottom of the tank, but some of the confusion may be due to strain differences, as substantial numbers of the Nauyuk strain prefer the bottom of the tank. Most of the initial studies on char growth used sinking feeds but larger char grow equally well when fed either sinking or floating diets.

Alanara and Kiessling (1996) reported that char (mean weight 215 g) held under culture conditions were unable to adjust their demand-feeding activity based on the energy content of the food, but were able to adjust their demand-feeding activity to either low (0.33 g), medium (0.87 g) or high (1.53 g) rewards and thereby regulate their food supply. Alanara and Kiessling (1996) suggested an optimum reward level of 0.1 g kg^{-1} for char during spring and summer, but this value should be reduced by ten times during the autumn. Char stocked at low densities and subjected to current grew better than those in standing water, but char at high densities did not have increased growth rates.

Conclusions

Although many approaches have been used to the study char nutrition, char has not met expectations as a cultivable species due partly to its direct competitors in the market-place, rainbow trout and Atlantic salmon. This is surprising given that char can be grown at much higher densities, grow better at lower temperatures and exhibit what appear to be broad nutritional requirements. While it was known from initial culture efforts with char that rainbow trout and some Atlantic salmon diets were adequate for char growth, it is evident that char have some special dietary requirements, especially relating to lipid type and amounts and perhaps to type and amount of carbohydrates. The versatility of char in the utilization of key nutrients is evident by the varying proportions (30–75% protein, 5–15% lipid and 0–52% carbohydrates) reported in diets. This suggests that char may be able to utilize a wider range of ingredient types in their diets than other salmonids.

Acknowledgements

T.A. Dick thanks technicians M. Lu and D. Block for help with the manuscript and the Natural Sciences and Engineering Research Council of Canada for financial support through an operating grant.

References

Aas, G.H., Bjerkeng, B., Hatlen, B. and Storebakken, T. (1997) Idoxanthin, a major carotenoid in the flesh of Arctic charr (*Salvelinus alpinus*) fed diets containing astaxanthin. *Aquaculture* 150, 135–142.

Ackman, R.G. and Takeuchi, T. (1986) Comparison of fatty acids and lipids of smolting hatchery-fed and wild Atlantic salmon *Salmo salar*. *Lipids* 21, 117–120.

Alanara, A. and Kiessling, A. (1996) Changes in demand feeding behaviour in Arctic charr, *Salvelinus alpinus* L., caused by differences in dietary energy content and reward level. *Aquaculture Research* 27, 479–486.

Bjerkeng, B., Johnsen, K., Mayer, I., Storebakken, T. and Nilssen, K.J. (1999) Influence of 11-ketotestosterone, 17B-estradiol, and 3,5,3′-triiodo-L-thyronine on distribution and metabolism of carotenoids in arctic charr, *Salvelinus alpinus* L. *Fish Physiology and Biochemistry* 21, 353–364.

Castell, J.D. (1979) Review of lipid requirements of finfish. In: Halver, J.E. and Trews, K. (eds) *Finfish Nutrition and Finfish Technology*. Heenemon Verlagsgesellschaft, Berlin, Germany, pp. 59–84.

Christiansen, J.S., Svendsen, Y.S. and Jobling, M. (1992) The combined effects of stocking and density and sustained exercise on the behaviour, food intake, and growth of juvenile Arctic charr (*Salvelinus alpinus* L.). *Canadian Journal of Zoology* 70, 115–122.

Damsgard, B., Arnesen, A.M. and Jobling, M. (1999) Seasonal patterns of feed intake and growth of Hammerfest and Svalbard Arctic charr maturing at different ages. *Aquaculture* 171, 149–160.

Delabbio, J. (1995) Artic charr culture in Atlantic Canada. In: Boghen, A.D. (ed.) *Cold-water Aquaculture in Atlantic Canada*. Tribune Press, Sakville, New Brunswick, Canada, pp. 85–106.

Dick, T.A. and Yang, X. (1992) A-linolenic acid metabloism in fish. *Nutrition* 8, 218–220.

Dick, T.A. and Yang, X. (1995) Flaxseed in Arctic char and rainbow trout nutrition. In: Cunnane, S.C. and Thompson, L.U. (eds) *Flaxseed in Human Nutrition*. AOCS Press, Champaign, Illinois, pp. 295–314.

Eales, J.G. and Shostak, S. (1985) Correlations between food ration, somatic growth and thyroid function in Arctic charr, *Salvelinus alpinus* L. *Comparative Biochemistry and Physiology* 80A, 553–558.

Gurure, R.M., Moccia, R.D. and Atkinson, J.L. (1995a) Optimal protein requirements of young Arctic charr (*Salvelinus alpinus*) fed practical diets. *Aquaculture Nutrition* 1, 227–234.

Gurure, R., Moccia, R. and Atkinson, J. (1995b) Dietary lysine requirements for young growing arctic charr (*Salvelinus alpinus*) fed nine semi-practical diets. In: *87th Annual Meeting of the American Society for Animal Science*, Abstract 3. Orlando, Florida, p. 180.

Halver, J.E. (1989) The vitamins. In: Halver, J.E. (ed.) *Fish Nutrition*. Academic Press, San Diego, California, pp. 32–109.

Hammar, J., Dempson, J.B. and Verspool, E. (1991) Natural hybridization between arctic charr (*Salvelinus alpinus*) and brook trout (*S. fontinalis*) evidence from northern Labrador. *Canadian Journal of Fisheries and Aquatic Sciences* 48, 1437–1445.

Hatlen, B., Aas, G.H., Joergensen, E.H., Storebakken, T. and Goswami, U.C. (1995) Pigmentation of 1, 2 and 3 year old Arctic charr (*Salvelinus alpinus*) fed two different dietary astaxanthin concentrations. *Aquaculture* 138, 303–312.

Heggberget, T.G., Grotnes, P., Klemetsen, A., Richardsen, R. and Ugedal, O. (1994) Culture of Arctic char – possibilities in Norway. *Transactions of the American Fisheries Society* 123, 435–444.

Henderson, R.J. and Tocher, D.R. (1987) The lipid composition and biochemistry of freshwater fish. *Progressive Lipid Research* 26, 281–347.

Hilton, J.W. and Slinger, S.J. (1981) Nutrition and feeding of rainbow trout. *Canadian Special Publication of Fisheries and Aquatic Sciences* 55, 15.

Jobling, M. (1991) Arctic char, *Salvelinus alpinus* L. In: Wilson R.P. (ed.) *Handbook of Nutrient Requirements of Finfish*. CRC Press, Boca Raton, Florida, pp. 1–4.

Jobling, M. and Wandsvik, A. (1983) An investigation of factors controlling food intake in Arctic charr, *Salvelinus alpinus* L. *Journal of Fish Biology* 23, 397–404.

Jobling, M., Baardvik, B.M. and Jorgenson, E.H. (1989) Investigation of food-growth relationships of Arctic charr, *Salvelinus alpinus* L. using radiography. *Aquaculture* 81, 367–372.

John, T.M., George, J.C., Hilton, J.W. and Slinger, S.J. (1979) Influence of dietary ascorbic acid on plasma lipid levels in the rainbow trout. *International Journal Vitaminology and Nutrition Research* 49, 400–405.

Johnson, L. (1980) The Arctic char, *Salvelinus alpinus*. In: Balon, E.K. (ed.) *Charrs: Salmoide Fishes of the Genus* Salvelinus. W. Junk Publishers, The Hague, Belgium, pp. 15–98.

Jorgenson, E.H. and Jobling, M. (1989) Patterns of food intake in Arctic charr, *Salvelinus alpinus*, monitored by radiography. *Aquaculture* 81, 155–160.

Jorgensen, E.H., Christiansen, J.S. and Jobling, M. (1993) Effects of stocking density on food intake, growth performance and oxygen consumption in Arctic charr (*Salvelinus alpinus*). *Aquaculture* 110, 191–204.

Lall, S.P. (1989) The minerals. In: Halver, J.E. (ed.) *Fish Nutrition*. Academic Press, San Diego, California, pp. 219–257.

Larsson, S. and Berglund, I. (1998) Growth and food consumption of 0+ Arctic charr fed pelleted or natural food at six different temperatures. *Journal of Fish Biology* 52, 230–242.

Linner, J. and Brannas, E. (1994) Behavioral response to commercial food of different sizes and self-initiated food size selection by Arctic char. *Transactions of the American Fisheries Society* 123, 416–422.

Metusalach, Synowiecki, J., Brown, J. and Shahidi, F. (1996) Deposition and metabolism of dietary canthaxanthin in different organs of Arctic charr (*Salvelinus alpinus* L). *Aquaculture* 142, 99–106.

Miglavs, I. and Jobling, M. (1989) Effects of feeding regime on food consumption growth rates and tissue nucleic acids in juvenile Arctic charr, *Salvelinus alpinus*, with particular respect to compensatory growth. *Journal of Fish Biology* 34, 947–958.

NRC (National Research Council) (1993) *Nutrient Requirements of Fishes*. National Academy of Sciences, Washington, DC, 114 pp.

Olsen, R.E. and Mortensen, A. (1997) The influence of dietary astaxanthin and temperature on flesh colour in Arctic charr *Salvelinus alpinus* L. *Aquaculture Research* 28, 51–58.

Olsen, R.E. and Ringo, E. (1992) Lipids of Arctic charr, *Salvelinus alpinus* (L.) 2. Influence of dietary fatty acids on the elongation and desaturation of linoleic and linolenic acid. *Fish Physiology and Biochemistry* 9, 393–399.

Olsen, R.E., Henderson, R.J. and Ringo, E. (1991) Lipids of Arctic charr, *Salvelinus alpinus* (L.). 1. Dietary induced changes in lipid class and fatty acid composition. *Fish Physiology and Biochemistry* 9, 151–164.

Olsen, R.E., Loevaas, E. and Lie, O. (1999) The influence of temperature, dietary polyunsaturated fatty acids, alpha-tocopherol and spermine on fatty acid composition and indices of oxidative stress in juvenile Arctic char, *Salvelinus alpinus* (L.). *Fish Physiology and Biochemistry* 20, 12–29.

Olsen, R.E., Myklebust, R., Ringo, E. and Mayhew, T.M. (2000) The influences of dietary linseed oil and saturated fatty acids on caecal enterocytes in Arctic char (*Salvelinus alpinus* L.): a quantitative ultrastructural study. *Fish Physiology and Biochemistry* 22, 207–216.

Papst, M. (1994) Variation in growth of hatchery reared Arctic charr, *Salvelinus alpinus* (L.). PhD dissertation, University of Manitoba, Canada.

Papst, M.H., Dick, T.A., Aranason, A.N. and Engels, C.E. (1992) Effect of rearing density on the early growth and variation in growth of juvenile Arctic charr, *Salvelinus alpinus* (L.). *Journal of Aquaculture and Fisheries Management* 43, 21–27.

Ringo, E. (1989) The effect of linoleic acid (18:2 n-6) on lipid and protein digestibility and growth in Arctic char, *Salvelinus alpinus* (L.) *Physiology and Ecology Japan (Special Volume)* 1, 473–482.

Ringo, E. (1995) Hatchery-reared landlocked Arctic char, *Salvelinus alpinus* (L.), from Lake Takvatn reared in fresh and sea water: effect of salinity on digestibility of protein and individual amino acids in a capelin roe diet and commercial feed. *Aquaculture Research* 26, 221–223.

Ringo, E. and Nilsen, B. (1987) Hatchery-reared landlocked Arctic charr, *Salvelinus alpinus* (L.), from Lake Takvatn, reared in fresh and sea water. 1. Biochemical composition of food, and lipid composition of fish reared in fresh water. *Aquaculture* 67, 343–351.

Ringo, E. and Olsen, R.E. (1991) The effect of diet on aerobic bacterial flora associated with intestine of Arctic charr (*Salvelinus alpinus* L.). *Journal of Applied Microbiology* 86, 22–28.

Sargent, J., Henderson, R.J. and Tocher, D.R. (1989) The lipids. In: Halver, J.E. (ed.) *Fish Nutrition*. Academic Press, San Diego, California, pp. 154–209.

Scalia, S., Isaksen, M. and Francis, G.W. (1989) Carotenoids of the Arctic charr, *Salvelinus alpinus* (L.). *Journal of Fish Biology* 34, 969–970.

Simmons, L., Moccia, R.D., Bureau, D.P., Sivak, J.G. and Herbert, K. (1997) Dietary methionine requirement of juvenile Arctic char (*Salvelinus alpinus*). *Bulletin of the Aquaculture Association of Canada* 97(2), 36–38.

Simmons, L., Moccia, R.D., Bureau, D.P., Sivak, J.G. and Herbert, K. (1999) Dietary methionine requirement of juvenile Arctic charr *Salvelinus alpinus* (L.). *Aquaculture Nutrition* 5, 93–100.

Suzuki, H., Okazaki, K., Hayakawa, S., Wada, S. and Tamura, S. (1986) Influence of commercial dietary fatty acids on polyunsaturated fatty acids of cultured freshwater fish and comparison with those of wild fish of the same species. *Journal of Agricultural and Food Chemistry* 34, 58–60.

Tabachek, J.L. (1984) Evaluation of grower diets for intensive culture of two strains of Arctic charr (*Salvelinus alpinus* L.). *Canadian Technical Report of Fisheries and Aquatic Sciences* 1281, 25.

Tabachek, J.L. (1986) Influence of dietary protein and lipid levels on growth, body composition and utilization efficiencies of Arctic charr, *Salvelinus alpinus* L. *Journal of Fish Biology* 29, 139–151.

Tabachek, J.L. (1988) The effect of feed particle size on the growth and feed efficiency of Arctic charr (*Salvelinus alpinus* L.). *Aquaculture* 71, 319–330.

Torrissen, K.R. and Shearer, K.D. (1992) Protein digestion, growth and food conversion in Atlantic salmon and Arctic charr with different trypsin-like isozyme patterns. *Journal of Fish Biology* 41, 409–415.

Yang, X. (1994) Effects of dietary polyunsaturated fatty acids on growth and lipid metabolism of arctic charr, *Salvelinus alpinus* (L). PhD dissertation, University of Manitoba, Canada.

Yang, X. and Dick, T.A. (1993) Effects of dietary fatty acids on growth, feed efficiency and liver RNA and DNA content of Arctic charr, *Salvelinus alpinus* (L.). *Aquaculture* 116, 57–70.

Yang, X. and Dick, T.A. (1994a) Arctic char (*Salvelinus alpinus*) and rainbow trout (*Oncorhynchus mykiss*) differ in their growth and lipid metabolism in response to dietary polyunsaturated fatty acids. *Canadian Journal of Fisheries and Aquatic Sciences* 51, 1391–1400.

Yang, X. and Dick, T.A. (1994b) Dietary alpha-linolenic and linoleic acids competitively affect metabolism of polyunsaturated fatty acids in Arctic charr (*Salvelinus alpinus*). *Journal of Nutrition* 124, 1133–1145.

Yang, X., Tabachek, J.L. and Dick, T.A. (1994) Effects of dietary n-3 polyunsaturated fatty acids on lipid and fatty acid composition and haematology of juvenile Arctic charr *Salvelinus alpinus* (L.). *Fish Physiology and Biochemistry* 12, 409–420.

Percids

16

Paul B. Brown[1] and Frederic T. Barrows[2]

[1]Department of Forestry and Natural Resources, Purdue University, 1159 Forestry Building, West Lafayette, IN 47907-1159, USA; [2]US Fish and Wildlife Service, Bozeman Technology Center, Bozeman, MT 59715, USA

Introduction

Fishes of the family Percidae are found in temperate climates throughout the northern hemisphere. There are over 160 species in North America, but only three are of commercial or aquacultural interest. The yellow perch (*Perca flavescens*), sauger (*Stizostedion canadense*) and walleye (*Stizostedion vitreum vitreum*) are important foodfish in much of North America (Lindsay, 1980; Engle *et al.*, 1990; Cox and Karahadian, 1998). The blue pike (*Stizostedion vitreum glaucum*) once inhabited Lake Erie and contributed to the commercial harvest. However, this subspecies is thought to be extinct. There are two ecological equivalents in Europe, the Eurasian perch (*Perca fluviatilis*) and the pike-perch or zander (*Stizostedion leucioperca*), which are also commercially important and considered highly prized foodfish. The two perches, walleye and pike-perch are receiving significant interest as aquaculture species, as well as various hybrids (Lynch *et al.*, 1982; Held and Malison, 1996).

Current aquacultural production levels are not known, but are generally thought to be less than 5 million pounds (2.3 million kg) annually for each of the respective species. This production does not include propagation and rearing in public hatcheries where fish are destined for stocking in public bodies of water.

This chapter will focus on those species for which nutritional research has been conducted; yellow perch, walleye and Eurasian perch. Nutritional requirements for the pike-perch are unknown.

Nutrient Requirements

There are few published nutritional requirements for any of the percids. The quantitative requirements and recommended macronutrient concentrations that exist are presented in Table 16.1. For yellow perch, requirements for crude

Table 16.1. Quantitative nutritional requirements, initial weight of fish used, type of diet fed and references for percids. Values expressed as a function of dry matter in the diet.

Species	Nutrient	Requirement	Initial weight (g)	Type of diet	Reference
Yellow perch	Crude protein	21–27%	18–27	Semipurified	Ramseyer and Garling (1998)
	Protein : energy	20–22*	18–27	Semipurified	Ramseyer and Garling (1998)
	Lipid	6–8%	25	Purified	Cartwright *et al.* (1998)
	Lysine	1.1–1.3%	17	Purified	Twibell *et al.* (1998)
	Methionine†	1.0%	4	Purified	Twibell *et al.* (2000)
	Methionine‡	0.85%	8	Purified	Twibell *et al.* (2000)
	Arginine	1.4%	11	Purified	Twibell and Brown (1997)
	Choline	600 mg kg^{-1}	16	Purified	Twibell and Brown (2000)
Walleye	Crude protein	51%	8	Practical	Barrows *et al.* (1988)
	Protein : energy	144, 51/3530§	8	Practical	Barrows *et al.* (1988)
	Crude protein	42%	50	Practical	Barrows *et al.* (1988)
	Protein : energy	116, 42/3600§	50	Practical	Barrows *et al.* (1988)
	Crude protein	53%	127	Practical	Tidwell *et al.* (1999)
Eurasian perch	Crude protein	40–49%	3	Semipurified	Fiogbe *et al.* (1996)
	Lipid	12%	22	Practical	P. Kestemont, unpublished
	Lipid	15%	34	Practical	Xu *et al.* (2000)

* Expressed as g crude protein MJ^{-1} metabolizable energy.
† In the presence of 0.03% cyst(e)ine.
‡ Methionine requirement determined in diets containing 51 : 49 cyst(e)ine : methionine.
§ Expressed as mg crude protein $kcal^{-1}$ gross energy.

protein, protein : energy ratio, lipid, lysine, methionine, arginine and choline are available. Optimal dietary crude protein was reported to be in the range of 21–27% of the diet, although the highest weight gain was observed in fish fed 34% crude protein (Ramseyer and Garling, 1998). The optimum crude protein-to-metabolizable energy ratio was reported to be 20–22 g crude protein MJ^{-1} (Ramseyer and Garling, 1998). Metabolizable energy values from rainbow trout were used for calculation of the energy component as there are no values for perch. Ramseyer and Garling (1994) formulated diets to mimic the essential amino acid pattern of whole, juvenile yellow perch, and this is considered one of the better essential amino acid profiles for fish (Wilson and Poe, 1985). Determination of the minimum dietary crude-protein concentration for fish using the whole-body pattern of essential amino acids is a new approach and offers promise of more precise formulations and a rapid method of developing diets for new aquaculture species (Brown, 1995). Earlier, Calbert and Huh (1976) fed juvenile yellow perch dietary crude-protein levels of 27, 40 or 50% and reported no significant differences in weight gain. Some of the essential amino acid requirements of yellow perch were recently determined using diets containing 33–34% dietary crude protein.

The lysine requirement of yellow perch is one of the lower requirements reported for fish, whereas the requirements for methionine and arginine are more similar to those for other species (NRC, 1993). Twibell *et al.* (1998) reported that the dietary lysine requirement of juvenile yellow perch was 1.1–1.3% of the diet. A series of studies were conducted to determine the methionine requirement of yellow perch (Twibell *et al.*, 2000). In the first study, diets contained 0.03% cyst(e)ine, and the methionine requirement was 1.0–1.1% of the diet. In the second study, supplemental cyst(e)ine was added to the diet and a sparing effect on the methionine requirement was observed; 51% of the methionine requirement was spared by the addition of cyst(e)ine. A ratio of 51 : 49 cysteine (Cys) : methionine (Met) was used in the third study and the total sulphur amino acid requirement was 0.85% of the diet. The dietary arginine requirement for juvenile yellow perch was reported as 1.4% of the diet (Twibell and Brown, 1997).

Twibell and Brown (2000) conducted a study to determine the choline requirement of yellow perch using a ratio of 51 : 49 Cys : Met and total sulphur amino acid concentration of 0.85%. The conversion of sulphur amino acids to choline was thus limited (Kasper *et al.*, 2000). In the same study, phosphatidylcholine (PC) was evaluated as a choline supply. The choline requirement was approximately 600 mg kg^{-1} diet and dietary PC could meet the choline requirement. This study demonstrated that perch are capable of using PC as a source of choline, but, when the choline requirement was met by choline-chloride (Cl), there was no beneficial effect of dietary PC at the dietary concentrations used.

Cartwright *et al.* (1998) fed four lipid sources (menhaden, cold-pressed soybean, coconut and tallow) alone or in combination and reported that juvenile yellow perch gained significantly more weight when fed either menhaden fish-oil or cold-pressed soybean oil at 6% of the diet. Within each lipid source, weight gain significantly decreased as dietary lipid concentration increased to 12 and 18% of the diet. Additionally, there were no significant differences in weight gain of fish fed a 1 : 1 mixture of menhaden fish-oil and cold-pressed soybean oil at any of the dietary concentrations tested (6, 12 or 18%). Feeding rates recommended from these studies with yellow perch were 3–4% of body weight per day for juvenile fish.

While the essential fatty acid requirement of yellow perch has not been quantified, Dabrowski *et al.* (1993) identified docosahexaenoic acid (DHA) (22:6n-3) as the predominant fatty acid in phospholipids of yellow perch and reported a decrease as eggs hatched and fry grew. Further, both DHA and eicosapentaenoic acid (EPA) (20:5n-3) were higher in juvenile yellow perch than in their major food item, which was *Daphnia*. Thus, the long-chain n-3 fatty acids may be important in larval diets for yellow perch. Thongard *et al.* (1995) reported that juvenile yellow perch fed diets deficient in linolenic acid exhibited decreases in DHA in polar lipids. This finding supports the earlier findings that DHA in eggs and fry is important and that linolenic acid would be a logical area of study as the essential fatty acid in perch.

Low dietary ascorbic acid (vitamin C) intake of yellow perch resulted in decreased hepatic concentrations of this vitamin in 6 weeks, but overt signs of deficiency did not develop until 8 months (Dabrowski and Ciereszko, 1996). Thus, ascorbic acid is required in diets of yellow perch, but no quantitative requirement was identified. Spinal deformities have been relatively common in yellow perch reared in tank-culture situations from larvae and may indicate insufficient vitamin C in the selected prey items or starter diets.

Optimal dietary crude protein concentrations for juvenile walleye appear to decrease as the fish grow from 8 to 50 g. The optimal crude-protein level for fish with an initial weight of 8 g was 51%, while that for fish with an initial weight of 50 g was 42% (Barrows *et al.*, 1988). Similarly, the optimal dietary crude protein-to-energy ratio decreased from 144 mg $kcal^{-1}$ to 116 mg $kcal^{-1}$ as initial fish size increased from 8 to 50 g. Energy concentrations considered optimal in these studies were 14.7 and 15.1 MJ metabolizable energy, respectively. Metabolizable energy values were from feedstuffs fed to rainbow trout, as none have been determined for walleye.

Tidwell *et al.* (1999) compared the growth of walleye fingerlings reared in ponds and fed diets differing in crude-protein concentration. Fish with an average initial weight of 127 g were fed practical fish-meal-based diets with either 44% or 53% crude protein. Walleye fed the higher crude-protein concentration gained more weight (177 g per fish) than fish fed the lower concentration (129 g per fish). However, there was no statistical difference in total harvest weight (1910 and 1691 kg ha^{-1}, respectively). This study indicated an advantage of feeding higher-protein feeds for the grow-out phase of walleye production in ponds relative to the protein level found in commercial salmonid feeds.

The effect of the dietary ascorbic acid level on the performance of walleye fingerlings was evaluated by MacConnell and Barrows (1993). A dietary level of 100 or 200 mg kg^{-1} supported near-maximum growth of walleye with an initial weight of 3 g. This level of dietary ascorbic acid is similar to that targeted in practical diets for rainbow trout. Fish fed an unsupplemented purified diet gained significantly less weight and exhibited deficiency symptoms not previously reported for trout or catfish. In production situations, walleye were commonly observed with a broken isthmus, which resulted in death of the fish. This typically occurred during times of stress, such as elevated temperature, crowding or feeding outdated feeds. This study documented that low vitamin C intake resulted in a breaking of the cartilage that attaches the isthmus to the lower jaw. Other deficiency symptoms, such as twisting of the cartilage in the gill filaments, are also observed in trout and catfish. Spinal deformities and abnormalities are more common in trout and catfish, and breakage of the isthmus is the first macrosymptom observed in walleye.

The optimal dietary crude-protein requirement for juvenile Eurasian perch has been quantified. Fiogbe *et al.* (1996) estimated the requirement as 36–56% of the diet, depending on the mathematical model chosen, and recommended 40–49% in practical diets. Kestemont *et al.* (2001) (Universitaires Notre Dame de la Paix, Namur, Belgium) observed significantly better weight gain in juvenile Eurasian perch fed 12 or 18% dietary lipid as menhaden oil with ethoxyquin, compared with fish fed 6% menhaden oil or fish fed 12 or 18% lipid without ethoxyquin. In a subsequent study, juvenile perch gained significantly more weight when fed practical diets containing 19% dietary lipid compared with fish fed 12% (Xu *et al.*, 2000). Weight gain of fish fed 15% dietary lipid was not significantly different from fish fed either 12 or 19%. There are no other dietary requirement estimates for the Eurasian perch.

Diet Development

Our ability to continue quantifying nutritional requirements for percids has been enhanced in recent years by the identification of experimental diets that are accepted by percids and support rapid growth. In a series of studies, Brown *et al.* (1996) offered four different experimental formulations to juvenile yellow perch and compared the weight gain and feed efficiency of fish fed various commercial diets (Table 16.2). The weight gain of fish fed a purified diet containing approximately 10% intact protein in the form of vitamin-free casein and gelatin (9 : 1), with the remainder of the amino acids supplied in crystalline form, was not significantly different from that of fish fed the best practical diet. The weight gain of fish fed a diet containing vitamin-free casein with supplemental L-arginine-hydrochloric acid (HCl), dextrin and fish oil was 79% of the weight gain observed in fish fed the best commercial diet. In the same study, fish were offered two other experimental diets: one containing only vitamin-free casein as

Table 16.2. Mean weight gain and feed conversion ratio (FCR) of juvenile yellow perch (initial weight 5 g) fed a variety of experimental and commercial diets for 10 weeks. Values are means of three replications. Means within the same column and with the same letter designation were not significantly different.

	Weight gain*	FCR†
Experimental diets: predominant source of crude protein		
Casein/arginine	167.7b	3.1b
Crystalline amino acids	215.0a	1.8a
Commercial diets‡		
Catfish (36/6)	127.8c	2.9c
Trout (36/8)	174.7a,b	2.3a
Trout (45/15)	212.1a	2.3a

* Expressed as a percentage of initial body weight.
† Calculated as g dry weight of feed fed g^{-1} wet weight gain of fish.
‡ Values in parentheses are the guaranteed concentrations (%) of crude protein/fat supplied by the feed manufacturers, respectively.

the source of amino acids and one containing vitamin-free casein and gelatin. Fish fed both diets exhibited poor feed acceptance and these dietary treatments were terminated prior to completion of the study. This study was repeated with larger yellow perch with an average initial weight of 51 g.

The larger perch accepted all four experimental diets. The weight gain of perch fed the vitamin-free casein and L-arginine-HCl was significantly greater than that of fish fed the other diets, but not significantly different from that of fish fed the best practical diet (Table 16.3). The weight gain of fish fed the diet containing crystalline amino acids as the primary source of amino acids was not significantly different from that of fish fed vitamin-free casein or the vitamin-free casein and gelatin combination. The weight gain of perch fed the crystalline amino acid diet was 84% of the weight gain observed in fish fed the best practical diet, while weight gains of perch fed the other two experimental diets were 74–76% of that observed in fish fed the best practical diet. These results indicated that several experimental formulations are accepted by yellow perch and can be used in future nutritional research. Additionally, there appear to be ontogenic differences in acceptance of experimental diets and weight gains of fish.

These same experimental diets were also fed to juvenile walleye with an initial weight of 13 g (Bharadwaj *et al.*, 2000). The weight gain of fish fed a salmon grower diet was numerically higher than that of fish fed a trout grower diet, but the values were not significantly different. Similarly, the weight gain of

Table 16.3. Mean weight gain and feed conversion ratio (FCR) of yellow perch (average initial weight 51 g) fed a variety of experimental and commercial diets for 10 weeks. Values are the means of three replications. Means with the same letter designation were not significantly different.

	Weight gain*	FCR†
Experimental diets: predominant source of protein		
Casein	47.5b,c	5.0b,c
Casein/gelatin	48.1b,c	5.1c
Casein/arginine	64.6a	4.0a
Crystalline amino acids	53.1b	4.7b
Commercial diets‡		
Trout (33/8)	43.3c	5.9c
Trout (38/12)	38.7c,d	6.5c,d
Trout (40/10)	63.4a	4.4a
Trout (50/17.5)	46.3b,c	5.7b,c
Catfish (36/4)	46.0b,c	5.7b,c
Catfish (32/3.5)	10.2e	25.8e

* Expressed as a percentage of initial body weight.

† Calculated as g dry weight of feed fed g^{-1} wet weight gain of fish.

‡ Values in parentheses are the guaranteed concentrations (%) of crude protein/fat supplied by the feed manufacturers, respectively.

fish fed the diet containing a predominance of crystalline amino acids was not significantly different from that of fish fed the best commercial diets. However, the weight gain of walleye fed the other three experimental diets was significantly lower than that of fish fed the best diets: 13–20% of the weight gain observed in fish fed the best commercial diets.

Using the data developed with yellow perch, experimental diets for the Eurasian perch were formulated using casein and L-arginine-HCl as the basal supply of essential amino acids. These diets were not accepted well by the Eurasian perch (Kestemont *et al.*, 1997). Since that time, basal diets have included cod-muscle meal as a primary source of essential amino acids and as a flavour additive.

While several experimental diets have been identified for further nutritional research with percids, it seems clear that there are differences in the acceptance and utilization of the various formulations by members of the same family of fishes.

Larval Diets

All percids are comparatively small when they first hatch (~4–7 mm total length) and feeding small larval fish remains difficult. While several new dry larval diets have been developed in recent years, acceptance of these diets, resulting weight gain and survival remain highly variable and often comparatively low (NCRAC, 1994, 1996). Recently, in both laboratory and production-scale studies, survival rates of larval walleye fed only formulated diets have been comparable to rearing in fertilized ponds and switching to formulated diets (Barrows *et al.*, 1993; Moore *et al.*, 1994; Moore and Olson, 1998, 1999). Additionally, developments in larval-diet manufacturing methods offer promise for percids and a variety of small-egged species (Barrows and Hardy, 2000). A summary of the known studies on feeding larval percids was recently presented by Kestemont and Melard (2000) and will not be repeated here.

At present, the most common method of mass propagating larval percids in commercial aquaculture is in fertilized outdoor ponds. In these situations, larval percids feed on the available zooplankton. Rotifers and copepods are important nutritional sources at this stage, although mouth-gape measurements indicate that perch larvae can consume prey or food particles as large as 190 μm (Schael *et al.*, 1991). Thus, several taxonomic groups of zooplankton are potential prey items, including some strains of *Artemia* exhibiting smaller size of nauplii. After approximately 30–45 days, percids are of sufficient size and developmental stage to be trained to accept formulated diets. This transition can be accomplished in ponds or tanks by simply offering feed on a regular basis, often with an automatic feeder. Malison *et al.* (1994) took advantage of the photopositive nature of larval yellow perch and developed an automatic feeder that floats in ponds with attached lights. At night, the perch are attracted to the light and then learn to accept the diet offered at the site.

Production Diets

Practical diets fed to percids are usually formulated to meet the requirements for salmonids (Brown *et al.*, 1996; Melard *et al.*, 1996) or sea bass (Fontaine *et al.*, 1997), since nutrient requirement information for percids is limited and percids are, to varying degrees, piscivorous. Initial diets for percids in North America have been the soft-moist diets containing approximately 20% moisture and mostly fish products as ingredients (Brown *et al.*, 1996). Once perch and walleye accept formulated diets, the dry salmonid diets can be offered and variations of those formulations are most often used as grow-out diets.

Brown *et al.* (1996) conducted two studies evaluating various practical diets formulated for other species but fed to North American percids. Yellow perch with an initial weight of 5 g and fed a practical diet formulated for rainbow trout gained more weight than fish fed a similar diet formulated for channel catfish (Table 16.2). Both diets contained 36% crude protein. However, the trout diet contained 8% lipid and the catfish diet contained 6% lipid and a higher level of plant proteins. Feeding higher crude-protein (45%) and lipid (15%) concentrations in a feed formulated for trout did not result in higher weight gains. However, higher dietary crude-protein and lipid concentrations were effective with larger perch.

Perch with an initial weight of 51 g gained more weight when fed a commercial trout diet containing 40% crude protein and 10% lipid (40/10) than fish fed diets with 33/8, 38/12 or 50/17.5 (Table 16.3). The weight gain of the larger perch was generally better when fed diets formulated around the nutritional requirements for trout than those for catfish.

In a recent study conducted by Bharadwaj *et al.* (2000), walleye fed diets formulated for trout and salmon had greater weight gains than fish fed a practical trout grower diet that lacked fish-meal. Juvenile walleye in that study also gained more weight when fed the salmonid diets compared with fish fed a modification of a diet for walleye, WG-9206 (Barrows and Lellis, 1996). However, the WG-9206 diet was modified and contained a nutritionally complete mineral premix instead of the supplemental mineral premix used in the original formulation.

There are apparently no evaluations of practical diets fed to Eurasian perch or pike-perch. However, it seems clear that diets formulated around the nutritional requirements of trout or salmon will be better choices for practical diet selection in the short term than diets formulated for catfish. The nutrient-dense diets developed for salmonids in Europe have not been evaluated with percids. While salmonid or sea-bass diets seem adequate at this point, diets for percids may be quite different once the nutritional requirements for the target species have been quantified for the various life-history stages.

Feeding Practices

Current culture systems for percids include earthen ponds, raceways, cages and tank systems. However, all systems are relatively small in size when compared with earthen ponds used for catfish production in the southern USA or raceway systems for trout in the western USA. This is probably a reflection of the fact that percid aquaculture is in the early stages of development. With smaller culture systems, hand-feeding remains the most common method of dispensing diets to most percids. Most walleye diets are dispensed by way of automatic feeders, as walleye tend to be skittish or nervous. Additionally, the percids are generally slow feeders, particularly the yellow perch. Thus, feeding by hand tends to be more efficient. As mentioned earlier, special feeding practices are employed when feeding the early life-history stages.

Automatic feeders are virtually mandatory when feeding larval and early juvenile stages of percids. Not only are percids small at hatching, but also their digestive tract is not completely developed, necessitating continual dietary inputs. Recommendations for feeding larval walleye include the use of automatic feeders that deliver feed almost continuously throughout the entire day (24 h) (Barrows *et al.*, 1993; Moore *et al.*, 1994). As the fish grow, the feeding frequency decreases into the range of every hour throughout the entire day to eight to ten times during normal working times. Circular tanks with a strong water flow are considered necessary to prevent water fouling and to facilitate cleaning when high feeding rates and continuous feeding are employed.

References

Barrows, F.T. and Hardy, R.W. (2000) In: Wedemeyer, G. (ed.) *Nutrition and Feeding in Fish Hatchery Management*, 2nd edn. US Fish and Wildlife Service, Washington, DC.

Barrows, F.T. and Lellis, W.A. (1996) Diet and nutrition. In: Summerfelt, R.C. (ed.) *Walleye Culture Manual*. NCRAC Culture Series 101, North Central Regional Aquaculture Center Publications Office, Iowa State University, Ames, Iowa, pp. 315–321.

Barrows, F.T., Sell, J.L. and Nickum, J.G. (1988) The effect of dietary protein and energy level on weight gain, body composition and RNA/DNA ratios of fingerling walleye. *Progressive Fish-Culturist* 50, 211–218.

Barrows, F.T., Kindschi, G.A. and Zitzow, R.E. (1993) The effect of a surface water spray, diet, and phase feeding on survival and gas bladder inflation of intensively-reared larval walleye (*Stizostedion vitreum*). *Progressive Fish-Culturist* 55, 224–228.

Bharadwaj, A.S., Gould, N.L., Brown, P.B., White, M.R. and Moore, A. (2000) Evaluation of experimental and practical diets for walleye *Stizostedion vitreum*. *Journal of the World Aquaculture Society* (in press).

Brown, P.B. (1995) Using whole-body amino acid patterns and quantitative requirements to rapidly develop diets for new species such as striped bass (*Morone saxatilis*). *Journal of Applied Ichthyology* 11, 342–346.

Brown, P.B., Dabrowski, K. and Garling, D.L., Jr (1996) Nutrition and feeding of yellow perch (*Perca flavescens*). *Journal of Applied Ichthyology* 12, 171–174.

Calbert, H.E. and Huh, H.T. (1976) Culturing yellow perch *Perca flavescens* under controlled environmental conditions for the Upper Midwest market. *Proceedings of the World Mariculture Society* 7, 137–144.

Cartwright, D.D., Watkins, B., Burgess, J., Warner, K., White, R., Jonker, Y. and Brown, P.B. (1998) Dietary lipid requirements of juvenile yellow perch. In: *World Aquaculture '98*. Las Vegas, Nevada, p. 95.

Cox, D.H. and Karahadian, C. (1998) Evaluation of microbial counts, nucleotide degradation, and sensory attributes of cultured and wild yellow perch (*Perca flavescens*) during refrigerated storage. *Journal of Aquatic Food Product Technology* 7, 5–26.

Dabrowski, K. and Ciereszko, A. (1996) The dynamics of gonad growth and ascorbate status in yellow perch, *Perca flavescens*. *Aquaculture Research* 27, 539–542.

Dabrowski, K., Culver, D.A., Brooks, C.L. and Voss, A.C. (1993) Biochemical aspects of the early life history of yellow perch (*Perca flavescens*). In: Kaushik, S.J. and Luquet, P. (eds) *Fish Nutrition in Practice*. Les Colloques No. 61, Institut National de la Recherche Agronomique, Paris, France, pp. 531–540.

Engle, C., Capps, O., Dellenbarger, L., Dillard, J., Hatch, U., Kinnucan, H. and Pomeroy, R. (1990) *The US Market for Farm-raised Catfish*. Arkansas Agricultural Experiment Station, Fayetteville, Arkansas, 39 pp.

Fiogbe, E.D., Kestemont, P., Melard, C. and Micha, J.C. (1996) The effects of dietary crude protein on growth of the Eurasian perch *Perca fluviatilis*. *Aquaculture* 144, 239–249.

Fontaine, P., Gardeur, J.N., Kestemont, P. and Georges, A. (1997) Influence of feeding level on growth, intraspecific weight variability and sexual dimorphism of Eurasian perch *Perca fluviatilis* L. reared in a recirculating system. *Aquaculture* 157, 1–9.

Held, J.A. and Malison, J.A. (1996) Pond culture of hybrid walleye fingerlings. In: Summerfelt, R.C. (ed.) *Walleye Culture Manual*. NCRAC Culture Series 101, North Central Regional Aquaculture Center Publications Office, Iowa State University, Ames, Iowa, pp. 311–313.

Kasper, C.S., White, M.R. and Brown, P.B. (2000) Choline is required by tilapia when methionine is not in excess. *Journal of Nutrition* 130, 238–242.

Kestemont, P. and Melard, C. (2000) Aquaculture. In: Craig, J.F. (ed.) *Percid Fishes*. Blackwell Scientific Publishers, London, UK, pp. 191–224.

Kestemont, P., Vandeloise, E., Brown, P.B., Melard, C. and Fontaine, P. (1997) Effects of casein-based diets containing different levels of lipids on survival, growth and nutritional status of Eurasian perch *Perca fluviatilis*. In: *Martinique '97, Island and Tropical Aquaculture*. Les Trois Ilets, Martinique, French West Indies, p. 59.

Kestemont, P., Vandeloise, E., Mélard, C., Fontaine, P. and Brown, P.B. (2001) Growth and nutritional status of Eurasian perch *Perca fluviatilis* fed graded levels of dietary lipids with or without added ethoxyquin. *Aquaculture* (in press).

Lindsay, R.C. (1980) Comparative sensory analysis of aquacultured and wild yellow perch (*Perca flavescens*) fillets. *Journal of Food Quality* 3, 283–289.

Lynch, W.E., Jr, Johnson, D.L. and Schell, S.A. (1982) Survival, growth and food habits of walleye × sauger hybrids (saugeye) in ponds. *North American Journal of Fisheries Management* 2, 381–387.

MacConnell, E. and Barrows, F.T. (1993) Pathological changes associated with vitamin C deficiency in walleyes. *Journal of Aquatic Animal Health* 5, 287–293.

Malison, J.A., Northey, D.L., Held, J.A. and Kuczynski, T.E. (1994) Habituation of yellow perch (*Perca flavescens*) fingerlings to formulated feed in ponds using lights and vibrating feeders. In: *25th Annual Meeting of the World Aquaculture Society*. New Orleans, Louisiana, p. 218.

Melard, C., Kestemont, P. and Grignard, J.C. (1996) Intensive culture of juvenile and adult Eurasian perch (*P. fluviatilis*): effect of major biotic and abiotic factors on growth. *Journal of Applied Ichthyology* 12, 175–180.

Moore, A. and Olson, J. (1998) *Production Scale Testing of Formulated Diets for the Intensive Culture of Walleye and Muskellunge.* Federal Aid to Fish Restoration Annual Report, Project Number F-106-R, US Fish and Wildlife Service, Minneapolis, Minnesota, 46 pp.

Moore, A. and Olson, J. (1999) *Production Scale Testing of Formulated Diets for the Intensive Culture of Walleye and Muskellunge.* Federal Aid to Fish Restoration Annual Report, Project Number F-106-R, US Fish and Wildlife Service, Minneapolis, Minnesota, 37 pp.

Moore, A., Prange, M.A., Summerfelt, R.C. and Bushman, R.P. (1994) Evaluation of tank shape and a surface spray for intensive culture of larval walleyes fed formulated feed. *Progressive Fish-Culturist* 56, 100–110.

NCRAC (North Central Regional Aquaculture Center) (1994) *Annual Progress Report.* North Central Regional Aquaculture Center, East Lansing, Michigan, 94 pp.

NCRAC (North Central Regional Aquaculture Center) (1996) *Annual Progress Report.* North Central Regional Aquaculture Center, East Lansing, Michigan, 104 pp.

NRC (National Research Council) (1993) *Nutrient Requirements of Fish.* National Academy of Sciences, National Academy Press, Washington, DC, 114 pp.

Ramseyer, L.J. and Garling, D.L. (1994) Amino acid composition of the ovaries, muscle and whole body of yellow perch *Perca flavescens. Progressive Fish-Culturist* 56, 175–179.

Ramseyer, L.J. and Garling, D.L. (1998) Effects of dietary protein to metabolizable energy ratios and total protein concentrations on the performance of yellow perch, *Perca flavescens. Aquaculture Nutrition* 4, 217–223.

Schael, D.M., Rudstam, L.G. and Post, J.R. (1991) Gape limitation and prey selection in larval yellow perch (*Perca flavescens*), freshwater drum (*Aplodinotus grunniens*), and black crappie (*Pomoxis nigromaculatus*). *Canadian Journal of Fisheries and Aquatic Sciences* 48, 1919–1925.

Thongard, S., Dabrowski, K. and Ebeling, J. (1995) Lipid and fatty acid composition in gonad and liver of wild and captive broodstock of yellow perch (*Perca flavescens*) and consequently in their eggs and larvae. In: *World Aquaculture Society Annual Meeting.* San Diego, California, p. 93.

Tidwell, J.H., Webster, C.D., Coyle, S.D. and Barrows, F.T. (1999) Growth of juvenile walleye, *Stizostedion vitreum*, fed two dietary protein levels in ponds. *Journal of Applied Aquaculture* 9, 47–56.

Twibell, R.G. and Brown, P.B. (1997) Dietary arginine requirement of juvenile yellow perch. *Journal of Nutrition* 127, 1838–1841.

Twibell, R.G. and Brown, P.B. (2000) Dietary choline requirement of juvenile yellow perch (*Perca flavescens*). *Journal of Nutrition* 130, 95–99.

Twibell, R.G., Cartwright, D.D., Wilson, K.A. and Brown, P.B. (1998) Dietary lysine requirement of juvenile yellow perch. In: *World Aquaculture '98.* Las Vegas, Nevada, p. 553.

Twibell, R.G., Wilson, K.A. and Brown, P.B. (2000) Dietary sulfur amino acid requirement of juvenile yellow perch fed the maximum cystine replacement value for methionine. *Journal of Nutrition* 130, 612–616.

Wilson, R.P. and Poe, W.E. (1985) Relationship of whole body and egg essential amino acid patterns to amino acid requirement patterns in channel catfish, *Ictalurus punctatus. Comparative Biochemistry and Physiology* 80B, 385–388.

Xu, X.L., Fontaine, P., Melard, C. and Kestemont, P. (2000) Effects of different dietary fat levels on growth, feed efficiency, lipid composition and histological changes of Eurasian perch *Perca fluviatilis*. In: *Aqua 2000.* Nice, France, p. 758.

Coregonids 17

Konrad Dabrowski,[1] Sergiusz Czesny[1,*] and Malgorzata Matusiewicz[1,†]
[1]*School of Natural Resources, Ohio State University, Columbus, OH 43210, USA*

Introduction

Whitefish (*Coregonus* sp.) have a northern circumpolar distribution, occurring in North America, Europe and Asia (Bodaly *et al.*, 1991). From the viewpoint of aquaculture, it is a country with no native coregonids (Japan) and countries in the most southern distribution of this genus (Italy) that have shown the greatest interest in the potential of coregonids as a farm-raised fish. Coregonids in Japan originated from pond-cultured *Coregonus lavaretus maraena* and *Coregonus peled* from Czechoslovakia and were domesticated in Saku Station, Nagano Prefecture.

Interest in cage-farming methods increased during the 1970s and 1980s, primarily for the development of propagation programmes to mitigate the negative effects of lake eutrophication and spawning-habitat deterioration for coregonids. Results from these studies suggested that supplementary feeding of juveniles with commercial diets in cages is extremely beneficial to their growth and survival during the episodes of low zooplankton abundance in the epilimnetic layer of lakes (Mamcarz, 1990).

The nutritional requirements of coregonids are little known. Mass rearing of larval coregonids was resolved in 1984 (Dabrowski *et al.*, 1984) and subsequently several authors have reported success in growing coregonids fed formulated practical diets in Japan (Dabrowski *et al.*, 1986), Europe (Rosch and Dabrowski, 1986; Champigneulle, 1988; Dabrowski and Poczyczynski, 1988), the USA and Canada (Drouin *et al.*, 1986; Zitzow and Millard, 1988; Harris and Hulsman, 1991). Furthermore, Shiose *et al.* (1984) and Gillet (1991) reported production of offspring to maturity from semidomesticated whitefish reared on a dry diet in intensive systems in Japan and France, respectively.

Present addresses: *Illinois Natural History Survey, Lake Michigan Biological Station, Zion, IL 60099, USA; †Department of Biochemistry, Medical Academy, Wroclaw, Poland.

Nutrient Requirements

Protein

Koskela (1995) performed growth studies with European whitefish of 50 g initial body weight fed one of six experimental, fish-meal-based diets containing from 34% to 58.7% crude protein. However, lipid levels decreased from 30.7% to 16.6% as protein level increased, complicating the interpretation of the results. Fish increased body weight threefold during a 75-day-long study at water temperatures of 13.7–18.7°C; however, there were no significant differences in weight gain. The authors concluded that a protein level of 34% is sufficient to sustain growth of whitefish at a high rate, but this may be an underestimation as diet palatability most probably changed dramatically with decreasing fish-oil supplement from 21 to 5.9%.

Fukatsu (1985) has also used diets based on fish-meal with European whitefish domesticated in Japan's Nagano Prefecture for several generations. In these formulations, protein level increased from 27.2 to 46.5%, but the level of supplemented cod-liver oil was only 3.4–5%. Whitefish fingerlings of 4 g body weight were used and, based on the protein efficiency ratio, the optimum protein level was estimated at 34%.

In a study with Lake Erie *Coregonus clupeaformis* raised for 2 years in captivity on formulated, commercial diets, fish were allocated to one of five experimental, fish-meal-based diets with protein levels between 25.5 and 49.9% (Dabrowski, 1995). Diets were formulated (Table 17.1) with increasing proportions of 'basal' protein carrier and balanced for lipids and minerals in such a manner that an increase in 'basal' ingredient accompanied a decrease in fish-oil supplement (from 14 to 6%) and mineral and vitamin mixtures. Consequently, amounts of lipids and minerals in all diets were much more similar than would be the case with simple protein-ingredient replacement. Significant increases in weight gains were found when the dietary protein level was increased from 25.5 to 35% (Fig. 17.1). No significant differences were found in haematological parameters of whitefish fed different protein levels. The optimum protein level for growth of whitefish during the grow-out phase was determined as 36.5%.

In accompanying studies, the concentration of total free amino acids in blood plasma corresponded to fish growth; however, it indicated lower absorption of dietary amino acids in fish fed a diet with 49.5% protein. Several free amino acids in whitefish liver have shown a breakpoint at a dietary protein level maximizing growth (Fig. 17.2), but then the trends markedly differed among individual amino acids. The first panel in Fig. 17.2 may be an illustration of an inappropriate leucine/isoleucine ratio, whereas the panel at the bottom demonstrates a deficiency in methionine.

These results may indicate that methionine is the first limiting amino acid, followed by lysine and arginine (Fig. 17.2). Arzel *et al.* (1995) hypothesized that, at variable levels of protein in the diet, concentrations of some essential amino acids in fish tissues may increase at a protein value inferior to the requirements. This is further interrelated to a variable protein/lipid ratio. Fynn-Aikins *et al.*

Table 17.1. Examples of diet composition used in studies of protein, amino acid requirement and attractant supplements with lake whitefish (*C. clupeaformis*).

Ingredients	Protein requirement study (% dry wt)	Ingredients	Amino acid requirement study (% dry wt) (lysine)	Ingredients	Attractant study (% dry wt)
Basal*	40–80	Casein	20	Casein	7 or 10
Maize starch	12.35–43.75	Gelatin	7.15	Gelatin	3 or 4
Fish-oil	6–14	Dextrin	23.43	Dextrin	23.43
Mineral mix	1–1.4	Cellulose	18.97–21.12	Attractant†	4 or 8
Vitamin mix	0.5–0.7	Amino acid mixture‡	14.05	Amino acid mixture‡	14.05
Choline chloride	0.1	Cod-liver oil	4	Maize starch	10
Vitamin C	0.05	Maize oil	4	Cod-liver oil	4
Vitamin B_6	0.001	Vitamin mix	1	Maize oil	4
		Mineral mix	2	Vitamin mix	1
		Carboxymethyl cellulose	2	Mineral mix	2
				CMC	2
				Choline chloride	0.2
				Vitamin C	0.1
				Cellulose	23.22 or 27.22

*Basal (%): menhaden fish-meal 61.5, maize-gluten meal 16.5, blood meal 11, whey 11.
† Attractant: CPSC, concentrate des proteines soluble des poissons, Soporopeche S.A., Boulogne, France; whitefish muscle freeze-dried; krill meal; menhaden fish-meal
‡ Non-essential amino acids (%): alanine 15.74, aspartic acid 28.77, glycine 5.41, glutamic acid 27.56, proline 1.25, serine 21.27.

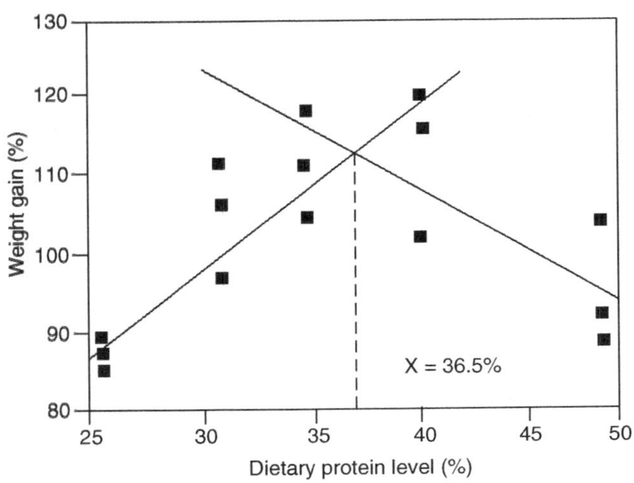

Fig. 17.1. Determination of protein requirement in lake whitefish (*C. clupeaformis*). The optimum level was estimated based on regression analysis at 36.5% (Dabrowski, 1995).

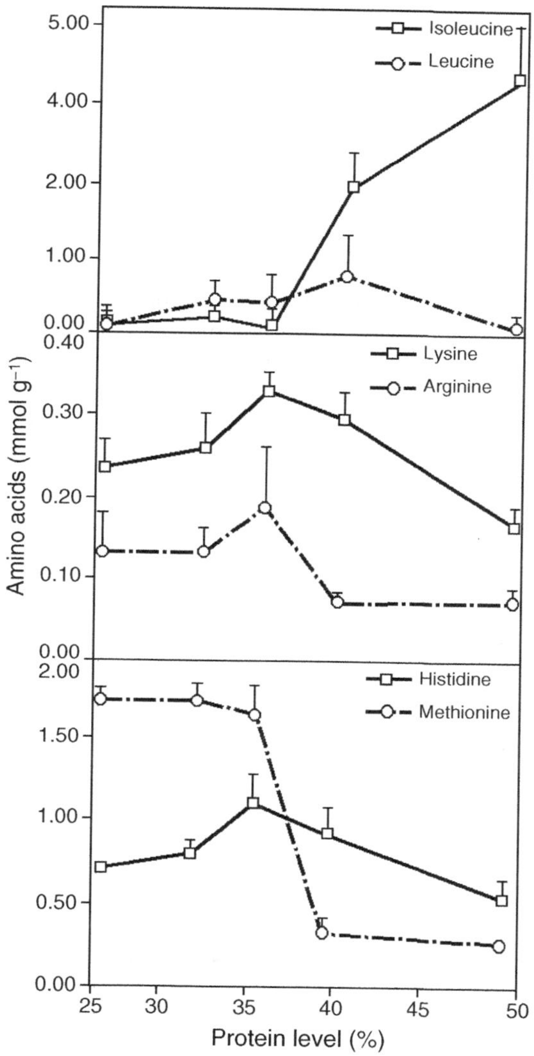

Fig. 17.2. The concentrations of free amino acids in lake whitefish liver following a feeding experiment with different protein levels. See Table 17.1 for diet formulations.

(1995) concluded that amino acid-catabolizing enzymes in salmonid fishes (closely related to coregonids) were not affected by variation in protein level, although different energy sources had a significant impact on protein retention. We examined the aspartate aminotransferase (AspAT) activity in whitefish liver (Ciereszko *et al.*, 1998). We found a significant decrease in AspAT activity only in fish fed a diet with the highest protein level (Dabrowski, 1995).

Preliminary studies were carried out for lysine and arginine requirements in lake whitefish juveniles (Dabrowski, 1995). Basal diets were formulated to contain 35% protein, using purified ingredients (Table 17.1). A mixture of crystalline L-amino acids (except lysine and arginine) was used to provide the balance so that the amino acid profile was similar to that of whole-egg protein (Santiago and Lovell, 1988). At the termination of the 24-week-long experiment, muscle samples from the dorsal part were taken and analysed for free amino acids (Fig. 17.3).

Weight increase analysis did not show significant differences among treatments, although these data subjected to a broken-line method indicated the dietary arginine and lysine requirements at 1.2 ± 0.1 and 1.5 ± 0.1%, respectively. Free amino acids have shown significantly higher concentrations at the dietary lysine level of 1.75% (Fig. 17.3). The latter is very close to an established maximum utilization level of lysine in rainbow trout, *Oncorhynchus mykiss*, 1.85% (Pfeffer *et al.*, 1992), and the requirement suggested for striped bass, *Morone saxatilis*, 1.5% (Griffin *et al.*, 1992).

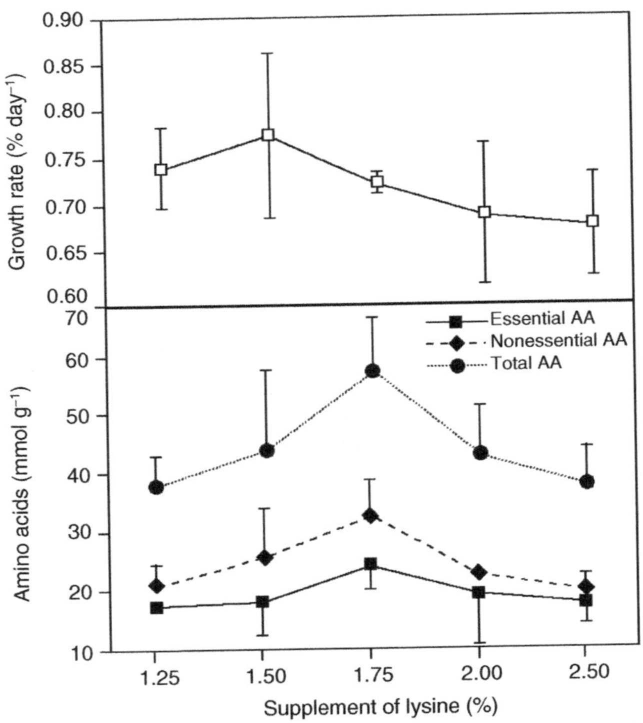

Fig. 17.3. Daily body increments and free amino acids (AA) in muscle of whitefish juveniles fed semipurified diets supplemented with L-lysine. Fish were 5–8 g body weight and were assigned to 30 conical tanks, in triplicate (20 fish) per treatment. Feed was provided twice a day at 2% body weight day^{-1}.

Energy

Different levels of protein offered in the experimental diets (Table 17.1) had no significant effect on the total lipid content in whitefish muscle. However, the polar/neutral lipid ratio in liver differed significantly after 4 and 7 months of feeding as well as between diets with different protein–energy ratios (Fig. 17.4). The high proportion of neutral lipids in the liver corresponds to accumulation of oleic acid (Fig. 17.4B) and saturated fatty acids (Fig. 17.4D). This indicates that more research is required in the area of protein/energy optimization as well as in respect of the detrimental effect of excessive dietary lipids in cases of low dietary protein.

Environmental manipulation can also influence the chemical composition of the body in coregonids. Delayed spawning, enforced by increased day length in the autumn, resulted in an extended period of feeding on high-energy-density diets and had a positive impact on the n-3/n-6 fatty acid ratio in gonad lipids of *C. peled* (Dabrowski *et al.*, 1992). In the deep-water Lake Baikal whitefish (*Coregonus autumnalis migratorius*), a dramatic decrease in the phospholipid–triglyceride ratio was observed with increasing age of the fish. As fish increased in age from 3 to 13 years, the proportion of docosahexaenoic acid (DHA) decreased from 33.2% to 19.7% (Ju *et al.*, 1997). This may indicate that protein–energy utilization changes considerably in coregonid ontogeny.

Lipid and fatty acids

Coregonid muscles contain high levels of lipids (Dabrowski, 1982a,b) and particularly high levels of the polyunsaturated fatty acids (PUFA) DHA, 22:6n-3 (8.4–10.1%) and eicosapentaenoic acid (EPA), 20:5n-3 (6.8–8.2%) (Kaitaranta and Linko, 1979), which are usually most abundant in lipids of marine fish. In addition, coregonid lipids had high proportions of linoleic acid, 18:2n-6 (4–5.2%) and linolenic acid, 18:3n-3 (4.3–5.1%).

Seasonally, the level of PUFA in the phospholipid fraction of coregonid flesh can increase to 39.7% for DHA and 10.3% for EPA (Linko *et al.*, 1992). The same authors also indicated that the plankton fatty acids are transferred into fish lipids without major modifications. The high level of PUFA is consequently reflected in the composition of fatty acids deposited in coregonid eggs and larvae (Soivio *et al.*, 1989). This composition is indicative of the structural importance of PUFA in phospholipid-containing membranes during embryonic development of coregonids in temperatures frequently close to freezing, 0.1–0.5°C (Luczynski *et al.*, 1984).

It has been observed that a difference exists between freshwater and marine fish in the fatty acid desaturation/elongation pathway both *in vivo* and in cultured cell lines (Tocher and Ghioni, 1999). The evidence obtained in studies with domesticated *C. lavaretus* by Watanabe *et al.* (1989) suggests that both n-6 and n-3 are essential for growth of coregonids, although the growth improvement after addition of only n-3 fatty acids to a diet containing exclusively oleic acid was

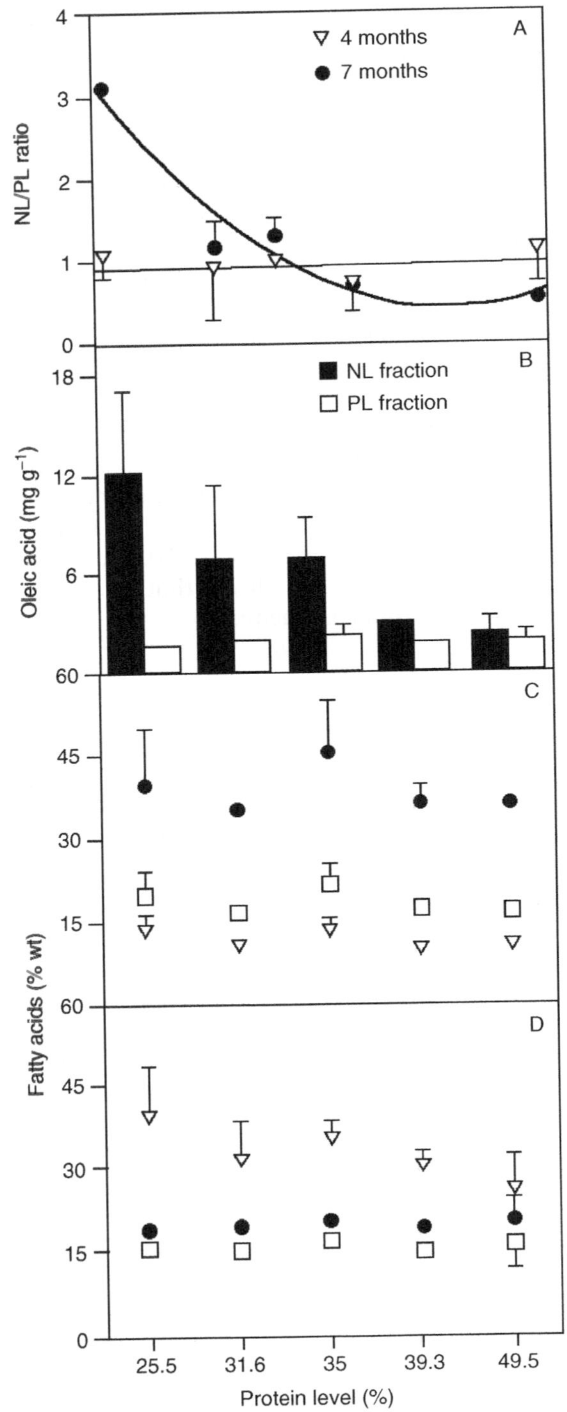
4
3
2
1
0
NL/PL ratio
4 months
7 months
A
18
12
6
Oleic acid (mg g^{-1})
NL fraction
PL fraction
B
60
45
30
15
C
60
45
30
15
0
Fatty acids (% wt)
D
25.5
31.6
35
39.3
49.5
Protein level (%)

highly significant. The conclusion drawn here differs from the original made by the authors, who claimed that linoleate 'was ineffective as an essential fatty acid'. Whitefish fed with a mixture of PUFA (34% EPA and 38% DHA) had higher weight gains than fish fed equivalent amounts of dietary linoleic acid, 18:2n-6, and linolenic acid, 18:3n-3. Furthermore, the levels of DHA in the polar lipid fraction of liver in fish fed a PUFA-supplemented diet was 26.9% in comparison with 7.2% in fish fed linoleic and linolenic acids. Therefore, Thongrod *et al.* (1989) addressed specifically the requirement for n-3 in the whitefish diet. Linolenic acid at a 1% level in a casein-based diet resulted in growth equivalent to 0.25% PUFA. This may suggest that the essential value of PUFA for whitefish is four times higher than that of linolenic acid. The efficiency of PUFA in rainbow trout, based on the growth rate and the liver lipid content, was twice as high as that of linolenate (Takeuchi and Watanabe, 1976). This may indicate that whitefish is closer to marine fish in its requirement for PUFA than to salmonids. Neither linoleic acid nor arachidonic acid, 20:4n-6, were included, however, for comparison with n-3 diets.

Takeuchi *et al.* (1983) reported that the saturated fatty acids, such as lauric acid, 12:0, depressed the appetite and reduced growth in Nile tilapia (*Tilapia nilotica*), whereas no such effect was observed in whitefish fed a diet with 4.5% lauric acid in comparison with oleic acid, 18:1n-9. The authors stated that there was no accumulation of lauric acid in liver lipids.

Vitamins and minerals

In all experiments with semipurified diets offered to coregonids, vitamin mixtures contained all required vitamins in accordance with the guidelines for salmonids (Watanabe *et al.*, 1989). However, in experiments with practical diets used for lake whitefish rearing by Zitzow and Millard (1988), lordosis, scoliosis and erratic swimming were observed. Addition of vitamin C to the processed diet eliminated the overt deficiency symptoms. Dabrowski (1990) demonstrated that European whitefish (*C. lavaretus*) fed with diets apparently deficient in vitamin C (9 and 118 mg kg^{-1}) exhibited a gradual decrease in body concentration of ascorbic acid, whereas fish offered feeds with vitamin C levels above 800 mg kg^{-1} maintained body ascorbate concentration at a relatively high level (76–88 μg g^{-1} wet weight).

When juveniles of lake whitefish of approximately 1 g body weight from Lake Erie were fed fish-meal-based diets supplemented with a graded level of ascorbyl monophosphate (Fig. 17.5), there were no significant differences in body-weight

Fig. 17.4. (Opposite) Changes in non-polar/polar (NL/PL) lipids ratio in liver of lake whitefish as a function of dietary protein levels (A). The amount of oleic acid found in both lipid fractions after 3 and 7 months of feeding (B). Polar lipids (C). Neutral lipids (D). □, Total saturated fatty acids; ▽, total monounsaturated fatty acids; ●, total polyunsaturated fatty acids.

gains, but the mortality of fish fed a vitamin C-free diet was considerably increased at week 4. At the end of the study (9 weeks), lordosis and extensive haemorrhages on ventral parts of the body, particularly around gill arches, were observed (K. Dabrowski and M. Matusiewicz, unpublished). We concluded that ascorbyl monophosphate is a bioavailable ester of ascorbic acid in coregonids and the required level in larval–juvenile diets seemed to be above 220 mg kg^{-1}. Coregonids manifested vitamin C deficiency much earlier than did salmonids – at a three- to fivefold body-weight increase in comparison with a 10–20-fold increase in salmonids (Matusiewicz *et al.*, 1994).

Dabrowski *et al.* (1987) demonstrated that, when concentrations of carotenoids varied in vendace (*Coregonus albula*) eggs from 2.57 to 9.15 μg g^{-1} dry weight, the concentration correlated inversely with survival of embryos exposed to visible light. These results in coregonids corresponded to Torrissen's (1984) conclusion that salmon-egg survival was highly dependent on astaxanthin level.

Dabrowski and Schwartz (1985) examined mineral concentration in larval diets for whitefish (*Coreganus shinzi palea*) and indicated that coregonids tolerated a wide variation of macro- and trace elements in their diets. *Artemia* nauplii had a low level of calcium (1.86 g kg^{-1} dry matter) in comparison with freshwater cladoceran zooplankton (30.7 g kg^{-1}), whereas the opposite applied to potassium (K) (14.56 and 2.2 g kg^{-1}, respectively). Trace elements, such as iron and copper, in whitefish juvenile bodies were at much lower levels in fish fed live food in comparison with those fed formulated diets. Excess minerals and interaction with other nutrients may be a concern in diet formulation for coregonids.

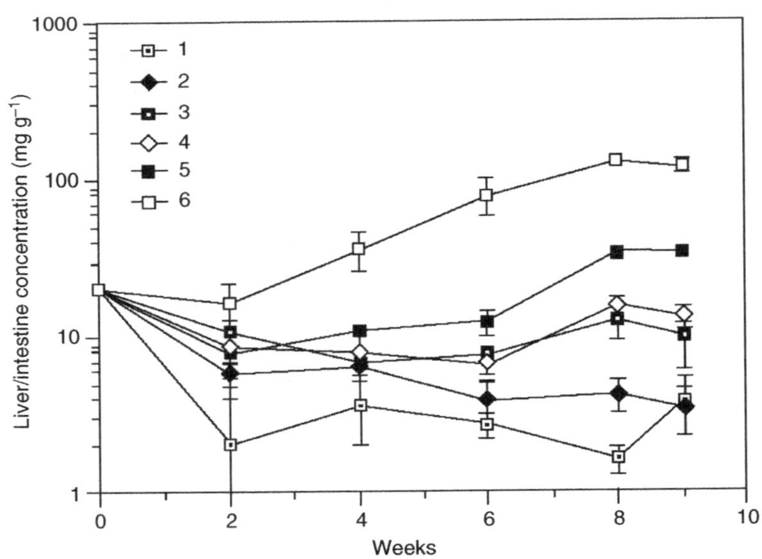

Fig. 17.5. Concentrations of ascorbic acid in tissues (liver plus intestine) of whitefish juveniles fed six experimental diets containing (1) 0, (2) 30, (3) 110, (4) 220, (5) 440 and (6) 870 mg kg^{-1} of ascorbate equivalents in the form of ascorbyl monophosphate.

Practical Diets

The first successful formulation of larval starter diets for coregonids was based on yeast (57%) and freeze-dried pork liver (28%) (Dabrowski *et al.*, 1984). In a subsequent study, Bergot *et al.* (1986) refined a diet formulation for larval whitefish and obtained juveniles of 0.275 g and 92% survival until 54 days of age. The optimized formulation contained 50% yeast, 35% dried beef liver, 5% cod-liver oil and 5% of both mineral and vitamin mixtures.

Watanabe *et al.* (1989) suggested supplementing a 2.47% amino acid mixture containing arginine (1.63), lysine (0.06), tryptophan (0.11), methionine (0.11), cystine (0.28) and histidine (0.28) as a feeding attractant for whitefish diets. However, no controlled experiments were performed to prove or optimize this attractant formulation. Studies performed with juveniles of lake whitefish (*C. clupeaformis*) provided evidence for significant growth enhancement from several animal proteins (Table 17.1; Fig. 17.6). Soluble fish-protein concentrate (CPSP) at 8% of the diet more than doubled the body-weight increase in juvenile coregonids during the 9-week-long experiment. This result was similar to the effect of CPSP in sturgeon diets (Moreau and Dabrowski, 1996).

Segner and Rosch (1990) indicated that large hepatocyte nuclei and well-developed endoplasmic reticulum, in concert with low glycogen stores, in

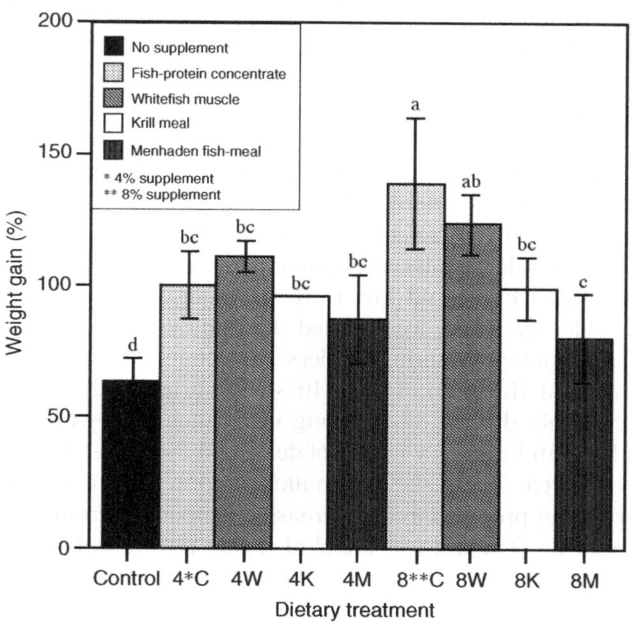

Fig. 17.6. The effect of attractant supplementation at the 4 and 8% level to semipurified diet for juvenile whitefish (Table 17.1). C, CPSP, soluble fish-protein concentrate, Sopropeche, Boulogne sur Mer 62204, France; W, freeze-dried whitefish muscle; K, krill meal; M, menhaden fish-meal. Bars having different letters indicate significant differences are at $P = 0.05$.

whitefish juveniles fed live food in comparison with smaller nuclei in hepatocytes of fish fed dry diets are relevant features for characterizing the nutritional value of the diets.

Feeding Practices

The food intake of coregonid larvae can be considerably improved by a high frequency of feeding (Dabrowski *et al.*, 1984). Bergot *et al.* (1986) enhanced the frequency of feeding by using automatic rotating dispensers, which delivered feed three times every 10 min during 12 h of illumination. At optimum water temperatures of 14°C, survival of larval Lake Leman coregonids was frequently over 95% during the first 54 days of rearing. Other investigators employed continuous feeding of *C. lavaretus* larvae using commercial diets and at lower water temperatures losses were considerably higher (50%), mostly due to mortality of non-feeding. Larvae (Segner *et al.*, 1988). Food dispensed during larval rearing of coregonids must be in excess of 10–15% biomass, as the specific growth rate of fish between 90 and 170 mg (individual weight) can be as high as 7.8% per day (Bergot *et al.*, 1986). Although no direct comparison of the effect of the duration of feeding (illumination) was carried out, an indirect comparison of growth rate in different experiments did not provide conclusive results on the advantage of 24 h illumination with continuous feeding (Dabrowski and Kaushik, 1985).

Growth of coregonid larvae from Late Hallwil (Switzerland) fed a Biokyowa diet (Biokyowa, Inc., Cape Girardeau, MO 63702) for 40 days was significantly better at 13°C than at 5°C; however, cumulative mortality for this experiment was 26% (Enz *et al.*, 2000). Mass rearing of European coregonids exclusively on dry diets throughout the first year is feasible, and larval diets can be gradually replaced with 'starter' salmonid diets (Champigneulle, 1988; Dabrowski and Poczyczynski, 1988). It is accepted that, at an individual weight of 50 mg, a transition can be made from larval diet to commercial, salmonid diets.

North American coregonid larvae (*C. clupeaformis*) were reared using a combination of decapsulated, freeze-dried *Artemia* cysts with commercial diets provided by automatic, vibrating feeders (Sweeney Enterprises, Boerne, TX 78006) (Harris and Hulsman, 1991). In some variants, the Biokyowa-B diet was used exclusively during 57-day-long experiments; however, growth was depressed and a high frequency (86%) of deformed fish, most missing opercular covers, was observed. These skeletal malformations were associated with the salmonid starter diet provided in an increasing proportion from the third to the fifth week of rearing. The authors concluded that a uniformly high survival of coregonids (97–98%) and individual weights obtained in 8 weeks (0.8–1 g) show that large-scale culture of this species are very practical options for farming coregonids.

Larger coregonids (initial weight 60–90 g) consumed over 70% of their daily intake during the first 3 h of feeding (Koskela *et al.*, 1997). Whitefish (*C. lavaretus*) maintained on gradually increasing feeding regimes from 6 to 24 h initially responded by decreased intake and depressed growth rate with the restricted time of feeding. Time-restricted feeding (6 h) apparently leads to an increase in gastric capacity and hypertrophy of digestive-tract tissue. In contrast to salmonids, restricted feeding in coregonids did not lead to the establishment of a feeding hierarchy, i.e. the presence of dominant and subordinate individuals (Jobling *et al.*, 1999). The authors observed that the catch-up growth of individual fish was directly related to the weight-gain decrease during the period of severe feed restriction. The full ration for whitefish of 38–44 g at 14°C amounted to 7–11 g feed kg^{-1} body weight day^{-1}.

Large whitefish (260–520 g) fed either a high-fat (27.5% lipid on dry basis) or a medium-fat (12.6%) diet did not show significant differences in feed intake (3.5–7.6 g kg^{-1} body weight day^{-1}) or growth (Koskela *et al.*, 1998). A high-fat diet increased lipids in the carcass of whitefish following 11 weeks of feeding to 12.7% in comparison with control (10.5%).

Acknowledgements

Part of this study was supported by a grant from the Saltonstall–Kennedy Program, National Oceanic and Atmospheric Administration, Silver Springs, Maryland. S. Czesny was supported by the Agency for International Development, Washington, DC, and M. Matusiewicz by F. Hoffman La Roche, Basle, Switzerland.

References

Arzel, J., Metailler, R., Kerteguer, C., Dellion, H. and Guillaume, J. (1995) The protein requirement of brown trout (*Salmo trutta*) fry. *Aquaculture* 130, 67–78.

Bergot, P., Charlon, N. and Durante, H. (1986) The effect of compound diets feeding on growth and survival of coregonid larvae. *Archiv für Hydrobiologie und Beiheft Ergebnisse der Limnologie* 22, 265–272.

Bodaly, R.A., Vuorinen, J., Wards, R.D., Luczynski, M. and Reist, J.D. (1991) Genetic comparisons of New and Old World coregonid fishes. *Journal of Fish Biology* 38, 37–51.

Champigneulle, A. (1988) A first experiment in mass-rearing of coregonid larvae in tanks with a dry food. *Aquaculture* 74, 249–261.

Ciereszko, A., Li, L. and Dabrowski, K. (1998) Optimal conditions for determination of aspartate aminotransferase activity in rainbow trout and whitefish. *Journal of Applied Ichthyology* 14, 57–63.

Dabrowski, K. (1982a) Seasonal changes in chemical composition of fish body and nutritional value of the muscle of the pollan (*Coregonus pollan* Thompson) from Lough Neagh, Northern Ireland. *Hydrobiologia (Hagua)* 87, 121–141.

Dabrowski, K. (1982b) Reproductive cycle of vendace (*Coregonus albula* L.) in relation to some chemical and biochemical changes in the body. *Hydrobiologia (Hagua)* 94, 3–15.

Dabrowski, K. (1990) Ascorbic acid status in the early life of whitefish (*Coregonus lavaretus* L.). *Aquaculture* 84, 61–70.

Dabrowski, K. (1995) *Domestication of Lake Whitefish* Coregonus clupeaformis*: Protein and Amino Acid Requirement to Optimize Feed Utilization and Growth.* National Oceanic and

Atmospheric Administration, Saltonstall–Kennedy Program, Final Report, Grant NA 36 FD 0099-01, Silver Springs, Maryland.

Dabrowski, K. and Kaushik, S.J. (1985) Rearing of coregonid (*Coregonus schinzi palea* Cuv. et Val.) larvae using dry and live food III. Growth of fish and developmental characteristics related to nutrition. *Aquaculture* 48, 123–135.

Dabrowski, K. and Poczyczynski, P. (1988) Laboratory experiment and mass rearing of coregonid fish fed exclusively on dry diets. *Aquaculture* 69, 307–316.

Dabrowski, K. and Schwartz, F.J. (1985) Rearing of coregonid (*Coregonus schinzi palea* Cuv. et Val.) larvae using dry and live food IV. Proximate and mineral composition of diets and fish. *Aquaculture* 48, 303–311.

Dabrowski, K., Charlon, N., Bergot, P. and Kaushik, S. (1984) Rearing of coregonid (*Coregonus schinzi palea* Cuv. et Val.) larvae using dry and live food. I. Preliminary data. *Aquaculture* 41, 11–20.

Dabrowski, K., Takashima, F., Strussmann, C. and Yamazaki, T. (1986) Rearing of coregonid larvae with live and dry food. *Bulletin of the Japanese Society for Scientific Fisheries* 51, 23–30.

Dabrowski, K., Luczynski, M., Czeczuga, B. and Falkowski, S. (1987) Relationships among coregonid fish reproductive effort, carotenoid content in eggs and survival of embryos. *Archiv für Hydrobiologie, Supplement* 79, 29–48.

Dabrowski, K., Yamazaki, T., Sato, S. and Takashima, F. (1992) Influence of delayed spawning time of *Coregonus peled* on the fatty acid composition of its tissues. *Polskie Archivum Hydrobiologii* 39, 553–561.

Drouin, M.A., Kidd, R.B. and Hynes, J.D. (1986) Intensive culture of lake whitefish (*Coregonus clupeaformis* Mitchell) using *Artemia* and artificial feed. *Aquaculture* 59, 107–118.

Enz, C.A., Schaffer, E. and Muller, R. (2000) Growth and survival of Lake Hallwil whitefish (*Coregonus* sp.) larvae reared on dry and live food. *Archiv für Hydrobiologie und Beiheft Ergebnisse der Limnologie* 148, 499–516.

Fukatsu, S. (1985) Optimum ratio of protein in diets of *Coregonus lararetus maraena*. *Bulletin of the Nagano Prefecture Fisheries Experiment Station*, pp. 12–13 (abstract from Dr T. Yamazaki).

Fynn-Aikins, K., Hughes, S.G. and Vanderberg, G.W. (1995) Protein retention and liver aminotransferase activities in Atlantic salmon fed diets containing different energy sources. *Comparative Biochemistry and Physiology* 111A, 163–170.

Gillet, C. (1991) Egg production in a whitefish (*Coregonus schinzi palea*) broodstock, effect of photoperiod on the timing of spawning and the quality of eggs. *Aquatic Living Resources* 4, 33–39.

Griffin, M.E., Brown, P.B. and Grant, A.L. (1992) The dietary lysine requirement of juvenile hybrid striped bass. *Journal of Nutrition* 122, 1332–1337.

Harris, K.C. and Hulsman, P.F. (1991) Intensive culture of lake whitefish (*Coregonus clupeaformis*) from larvae to yearling size using dry feeds. *Aquaculture* 96, 255–268.

Jobling, M., Koskela, J. and Winberg, S. (1999) Feeding and growth of whitefish fed restricted and abundant rations, influences on growth heterogeneity and brain serotonergic activity. *Journal of Fish Biology* 54, 437–449.

Ju, S.J., Kucklich, J.R., Kozlova, T. and Harvey, H.R. (1997) Lipid accumulation and fatty acid composition during maturation of three pelagic fish species in Lake Baikal. *Journal of Great Lakes Research* 23, 241–253.

Kaitaranta, J.K. and Linko, R.R. (1979) Fatty acids of a whitefish (*Coregonus albula*) flesh lipids. *Journal of Science, Food and Agriculture* 30, 921–926.

Koskela, J. (1995) Influence of dietary protein levels on growth and body composition of whitefish (*Coregonus lavaretus*). *Archiv für Hydrobiologie und Beiheft Ergebnisse der Limnologie* 46, 331–338.

Koskela, J., Jobling, M. and Pirhonen, J. (1997) Influence of the length of the daily feeding period on feed intake and growth of whitefish, *Coregonus lavaretus*. *Aquaculture* 156, 35–44.

Koskela, J., Jobling, M. and Savolainen, R. (1998) Influence of dietary fat level on feed intake growth and fat deposition in the whitefish *Coregonus lavaretus*. *Aquaculture International* 6, 95–102.

Linko, R.R., Rajasilta, M. and Hiltunen, R. (1992) Comparison of lipid and fatty acid composition in vendace (*Coregonus albula* L.) and available plankton feed. *Comparative Biochemistry and Physiology* 103A, 205–212.

Luczynski, M., Dlugosz, M., Szutkiewicz, B. and Kirklewska, A. (1984) The influence of the incubation temperature on the body length and the yolk sac volume of *Coregonus albula* (L.) eleutheroembryos. *Acta Hydrochimica et Hydrobiologica* 126, 615–628.

Mamcarz, A. (1990) Conditions for growth of *Coregonus peled* larvae in cage rearing. *Acta Academiae Agricultural Technitions Olstenensis* 17, 3–57.

Matusiewicz, M., Dabrowski, K., Volker, L. and Matusiewicz, K. (1994) Regulation of saturation and depletion of ascorbic acid in rainbow trout. *Journal of Nutritional Biochemistry* 5, 204–212.

Moreau, R. and Dabrowski, K. (1996) Feeding stimulants in semi-purified diets for juvenile lake sturgeon (*Acipenser fulvescens* Rafinesque). *Aquaculture Research* 27, 953–957.

Pfeffer, E., Al-Sabtg, H. and Haverkamp, R. (1992) Studies on lysine requirements of rainbow trout (*Oncorhynchus mykiss*) fed wheat gluten as only source of dietary protein. *Journal of Animal Physiology and Animal Nutrition* 67, 74–82.

Rosch, R. and Dabrowski, K. (1986) Tests of artificial food for larvae of *Coregonus lavaretus* from Lake Constanze. *Archiv für Hydrobiologie und Beiheft Ergebnisse der Limnologie* 22, 273–282.

Santiago, C.B. and Lovell, R.T. (1988) Amino acid requirements for growth of Nile tilapia. *Journal of Nutrition* 116, 1540–1546.

Segner, H. and Rosch, R. (1990) Development of dry food for larvae of *Coregonus lavaretus* L. II. Liver histology. *Aquaculture* 91, 117–130.

Segner, H., Rosch, R., Schmidt, H. and von Poeppinghausen, K.J. (1988) Studies on the suitability of commercial dry diets for rearing of larval *Coregonus lavaretus* from Lake Constance. *Aquatic Living Resource* 1, 231–238.

Shiose, J., Yamazaki, T. and Tominaga, M. (1984) Studies on culture techniques of coregonids. I. Import of the eggs and artificial rearing. *Bulletin of the Nagano Prefecture Fisheries Experiment Station* 1, 21–30.

Soivio, A., Niemisto, M. and Backstrom, M. (1989) Fatty acid composition of *Coregonus muksun* Pallas, changes during incubation, hatching, feeding and starvation. *Aquaculture* 79, 163–168.

Takeuchi, T. and Watanabe, T. (1976) Nutritive value of n3 highly unsaturated fatty acids in pollock liver oil for rainbow trout. *Nippon Suisan Gakkaishi* 42, 907–919.

Takeuchi, T., Satoh, S. and Watanabe, T. (1983) Requirement of *Tilapia nilotica* for essential fatty acids. *Bulletin of the Japanese Society for Scientific Fisheries* 49, 1127–1134.

Thongrod, S., Takeuchi, T., Satoh, S. and Watanabe, T. (1989) Requirement of fingerling White fish *Coregonus lavaretus maraena* for dietary n-3 fatty acids. *Nippon Suisan Gakkaishi* 55, 1983–1987.

Tocher, D.R. and Ghioni, C. (1999) Fatty acid metabolism in marine fish, low activity of fatty acyl D5 desaturation in Gilthead sea bream (*Sparus aurata*) cells. *Lipids* 34, 433–440.

Torrissen, O.J. (1984) Pigmentation of salmonids – effect of carotenoids in eggs and start-feeding diet on survival and growth rate. *Aquaculture* 43, 185–194.

Watanabe, T., Thongrod, S., Takeuchi, T., Satoh, S., Kubota, S.S., Fujimaki, Y. and Cho, C.Y. (1989) Effect of dietary n-6 and n-3 fatty acids on growth, fatty acid composition and histological changes of white fish *Coregonus lavaretus maraena*. *Nippon Suisan Gakkaishi* 55, 1977–1982.

Zitzow, R.E. and Millard, J.L. (1988) Survival and growth of lake whitefish (*Coregonus clupeaformis*) larvae fed only formulated dry diets. *Aquaculture* 69, 105–113.

Common Carp, *Cyprinus carpio* 18

Toshio Takeuchi, Shuichi Satoh and V. Kiron
Department of Aquatic Biosciences, Tokyo University of Fisheries, Tokyo 108-8477, Japan

Introduction

The common carp (*Cyprinus carpio*) belongs to the family Cyprinidae. In nature, carp live in the middle or lower reaches of a river with slow currents, or in marshes. Their habitats are usually weedy areas with a muddy bottom. Carp fry feed on zooplankton such as rotifers and copepods, but as they grow up they become benthic feeders, feeding on animals and other organic material.

The cyprinids have been farmed since ancient times and today they are undoubtedly the most important teleost family cultivated on a global scale, the current production figure reaching over 13 million tons year^{-1}. The common carp is an important culture species among the cyprinids, next only to silver carp and grass carp (Fig. 18.1), and its production has doubled over the last decade, reaching about 2.5 million tons in 1998, valued at US$2.8 billion (Fig. 18.2). A large percentage of this is from the Asian region, particularly China. While production of common carp is widely practised, only about 3% of the cyprinids are cultivated in intensive systems (Tacon, 1993).

Traditionally, carp are cultured in ponds or rice paddies, while advanced culture systems include irrigation ponds, running-water systems and net cages in lakes. The more organized culture techniques involve maintenance of breeders, fry production, yearling production and marketable fish production. The operations and feeding practices are outlined in Table 18.1. Relatively low-cost prepared diets are in vogue and little effort has been made to supply adequate amounts of nutrients from carefully selected ingredients, despite the existence of a great deal of scientific information.

Nutrient Requirements

The dietary requirements of common carp for protein, amino acids, lipids, fatty acids, carbohydrates (starch), vitamins, minerals, energy and protein/energy ratios have been investigated by many researchers and reviewed (Satoh, 1991;

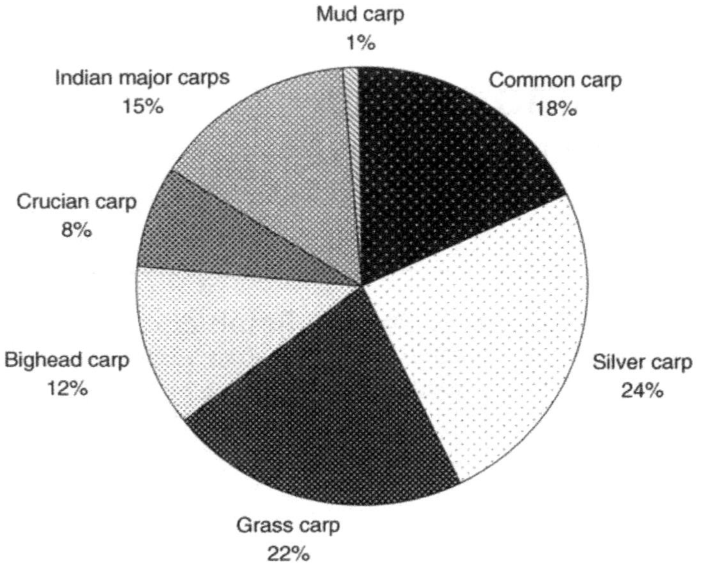

Fig. 18.1. The share of common carp among various farmed cyprinids.

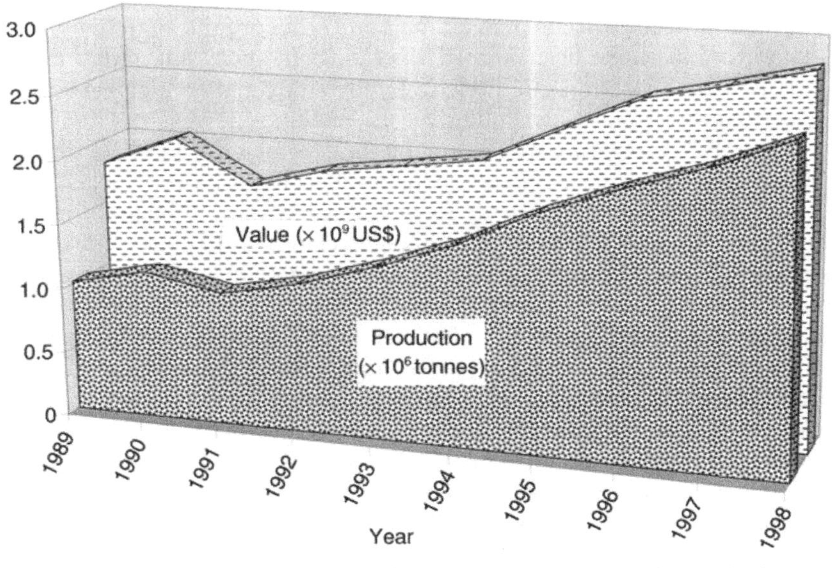

Fig. 18.2. The production and value of cultured common carp during the last decade.

Omae, 1992; Song, 1994; De Silva and Anderson, 1995; Kaushik, 1995; Takeuchi, 1999). This chapter incorporates some of the more recent information on nutrition of the carp. A snapshot of the macronutrient requirements of the common carp is provided in Table 18.2.

Table 18.1. Culture systems and feeding practices for different stages of common carp, *Cyprinus carpio* (modified from Kafuku, 1992).

Fish stage	Production system	Feeds/feeding
Parent fish	Spawning pond	Diets with 70% vegetable matter and 30% animal matter enriched with vitamins and minerals
Fry	Stagnant-water pond	Daphnia and other zooplankton
Yearlings	Stagnant-water pond	Artificial diets
	Half-running-water pond	Pellets fed over five times a day
	Farm pond	
Market-size fish	Farm pond, running-water pond	Artificial diets, silkworm pupae
	Spring-water pond	Vegetables, boiled wheat
	Circulating-water pond	
	Floating net cage	

Table 18.2. Macronutrient requirements of common carp, *Cyprinus carpio.*

Nutrient	Requirement
Protein	30–35 g 100 g^{-1}
Lipid	5–15 g 100 g^{-1} (related to energy)
Essential fatty acid	
Linoleate	1 g 100 g^{-1}
Linolenate	1 g 100 g^{-1}
Digestible energy	13–15 MJ kg^{-1} (310–360 kcal)
Carbohydrate (as starch)	30–40 g 100 g^{-1}

Protein and amino acids

Ogino and Chen (1973) and Ogino (1980b) reported that the daily requirement of common carp for protein is about 1 g kg^{-1} body weight for maintenance and 12 g kg^{-1} body weight for maximum protein retention. The efficiency of nitrogen utilization for growth, however, is highest with a protein intake of 7 to 8 g kg^{-1} body weight day^{-1}.

Investigations on the optimal requirement of common carp have demonstrated that crude protein levels ranging from 30 to 38% appear to satisfy the fish (Jauncey, 1982; Watanabe, 1988). Generally, this level has been determined by using semipurified diets containing a single high-quality protein source, such as casein, whole-egg protein or fish-meal. If sufficient digestible energy is contained in the diet, the optimal protein level can be effectively kept at 30–35% (Watanabe, 1982).

The whole-body amino acid profile of carp is not affected by variations in diet or by the age of the fish (Schwarz and Kirchgessner, 1988). The same ten essential amino acids (EAA) described for most fish are indispensable for carp growth too. The quantitative requirement for amino acids was established through different studies and is shown in Table 18.3. It should also be noted that there may be minor changes in the requirement of individual amino acids, depending on the growth stages (Baloguma, 1995). The lysine requirement at the fingerling stage is 2.25% of the diet (6% of protein) and decreases to 1.75% (5.4%) at the juvenile stage, whereas it does not change in the case of methionine. As has been recognized in other fish, cystine and tyrosine can spare or replace certain portions of dietary methionine and phenylalanine, respectively. In a recent observation on sulphur amino acid synthesis, Yokoyama *et al.* (2001) demonstrated that the hepatic cysteinesulphinate decarboxylase (CSD) (EC4.1.1.29) activity of carp is quite low compared with that of rainbow trout (*Oncorhynchus mykiss*) and tilapia (*Oreochromis niloticus*), where the values are 50 times higher. This is despite the fact that the carp muscle contains a high amount of taurine. CSD is an enzyme involved in the biosynthesis of taurine from cysteine and, if its activity is low, it would mean that carp requires dietary taurine supplements.

Ogino (1980a) reported that amino acid requirements could be estimated from data on the amino acid profile of the whole-body and daily body protein deposition. If a diet containing 35% protein with 80% of protein digestibility is fed daily at a level of 3% of the body weight, it can be assumed that the fish deposits 0.58 g of protein 100 g^{-1} of body weight daily. Data based on these assumptions are shown in Table 18.3, and the requirements determined by these two methods agree fairly well. However, the deposition rate does not account for metabolic pathways of amino acids that do not lead to protein synthesis. Further the

Table 18.3. Amino acid requirement of common carp, *Cyprinus carpio*.

Amino acid	Nose (1979) % in dietary protein	Nose (1979) % in diet	Ogino (1980a) (% in dietary protein)	Dabrowski (1983b) (mg kg^{-1} day^{-1})
Arginine	4.3	1.6	4.4	506
Histidine	2.1	0.8	1.5	145
Isoleucine	2.5	0.9	2.6	255
Leucine	3.3	1.3	4.8	429
Lysine	5.7	2.2	6.0	458
Methionine	2.1	0.8	1.8	105
Cystine	5.2	2.0	0.9	n/a
Phenylalanine	3.4	1.4	3.4	254
Tyrosine	2.6	1.0	2.3	190
Threonine	3.9	1.5	3.8	213
Valine	3.6	1.4	3.4	305
Tryptophan	0.8	0.3	0.8	n/a

n/a, Values not available.

absorption of individual amino acids differs greatly depending on protein source and time after feeding (Dabrowski, 1983a, 1986). In this context, the absorption rate describing metabolic amino acid requirements might prove useful, although it needs further study. Recently Akiyama *et al.* (1997) compared the A/E ratios (relative proportion of each EAA to the sum of the EAA) among fish species, and placed common carp adjacent to catla (*Catla catla*), both of which belong to the same family, Cyprinidae.

Energy

There is little information on the energy requirements of carp, compared with the volume of data on other aspects of nutrition. As described in other teleosts, both fasting metabolic rates and maintenance energy requirements are affected by water temperature. The resting metabolic rates at temperatures below 17°C are quite low (Kaushik, 1995). A linear relationship between nitrogen (N) intake and heat increment in feeding has also been proposed, the value being around 40 kJ g^{-1} N intake (Chakraborti *et al.*, 1992; Kaushik, 1995).

Protein and lipid requirements are related to digestible energy. The optimum range of the digestible energy/protein ratio for the maximum growth was 97–116 when based on the measured digestible energy (Takeuchi *et al.*, 1979b). A dietary energy budget was provided by Ohta and Watanabe (1996) for carp fed a practical diet comprising 25% fish-meal, 4% meat meal, 10% soybean meal and 8% maize-gluten meal as the main protein sources. The partitioning based on gross energy intake (100%) at the level required for maximum growth was: 29.9% lost as faecal energy, 1.5% as non-faecal energy, 31.9% as heat increment and 36.7% as net energy (including 12.6% for maintenance and activity and 24.1% as productive energy). The authors also reported that the digestible energy requirements for maximum growth were 285, 548 and 721 kJ kg^{-1} body weight day^{-1} (at feeding rates of 1.83, 3.60 and 5.17% of body weight day^{-1}, respectively), being influenced by both diet and fish size.

Lipids and fatty acids

Carp is an omnivorous fish and can utilize effectively both lipids and carbohydrates as dietary energy sources, and therefore the digestible energy content is more important than lipid content in the diet. It has been shown that the enrichment of the digestible energy content from 13 to 15 MJ kg^{-1} diet by addition of lipid at levels of 5–15% to diets did not result in improvement in growth performance nor net protein utilization (Takeuchi *et al.*, 1979a). Further, the negative aspect of increasing dietary lipids seems to be the increase in its body deposition, particularly as visceral fat (Zeitler *et al.*, 1983; Murai *et al.*, 1985).

As far as the essential fatty acids (EFA) are concerned, common carp and grass carp (*Ctenopharyngodon idella*) require both n-6 and n-3 fatty acids. It has been estimated that a supply of 1% of each of these fatty acids leads to best growth and feed efficiency in juvenile common carp (Takeuchi and Watanabe, 1977). However, there are other reports that the quantitative requirements may be lower (Kaushik, 1995). Though the deficiency symptoms related to EFA do not easily show up in common carp, poor growth, high mortality and skin depigmentation have been reported. On the other hand, in grass carp, a related species, the deficiency manifested as lordosis, resembling 'Sekoke disease', but is not characterized by apparent muscular dystrophy (Takeuchi *et al.*, 1992).

The role of medium-chain triglycerides (MCT) in the nutrition of carp larvae has been recently studied (Fontagné *et al.*, 1999, 2000). While tricaproin (C6:0), tricaprin (C10:0), trilaurin (C12:0) and triolein (C18:1) produced efficient growth and survival, tricaprylin (C8:0) proved to be a very poor source. The peculiarity of tricaprylin among the other MCT, which seemed to be well utilized up to 30 g 100 g^{-1} of total dietary fatty acid, was related to an unexplained disorganization of the fatty acid enzyme systems.

Phospholipids (PL) have numerous roles in larval feeding including an influence on texture, resistance to oxidation and water stability of food particles (Coutteau *et al.*, 1997). Attention should be paid to an adequate dietary PL supply when carp larvae are offered artificial diets instead of PL-rich live food (Geurden *et al.*, 1995). Dietary PL deficiency resulted in an accumulation of fat droplets in the enterocytes of the anterior intestine, an increase in the height of mucosal epithelium and a reduction in mean hepatocyte volume. However, supplementation of phosphatidylcholine (PC) from hen egg-yolk or from soybean prevented intestinal steatosis and increased the hepatocyte volume (Fontagné *et al.*, 1998). The foregoing study also suggested that PL are needed for absorption of neutral lipids although the benefits were independent of their emulsifying properties. On comparing several PL sources prepared from soybean lecithin, Geurden *et al.* (1998) found that while PC was important for obtaining high initial growth, phosphatidylinositol was responsible for normal development of carp larvae. They suggested that dietary PL supplements should preferably be a blend similar in composition to that of the larval body. In a later study Geurden *et al.* (1999) pointed out the conservative distribution of the PL classes in fish of a given size, irrespective of the diet, explaining that only limited remodelling is needed.

Carbohydrates

Several studies have been conducted on carbohydrate utilization in common carp. The amylase activity in the digestive tract and the digestibility of starch in fish are generally lower than those of terrestrial animals. Among fish, the intestinal activity of amylase is higher in omnivorous fish, including common carp, than in carnivorous fish. It has been found that the ratio of intestine length to body length in carp is 1.8–2.0, the values being four times greater than those of rainbow trout and eel (*Anguilla japonicus*), and this accounts for the better utilization of carbohydrates by carp. Murai *et al.* (1983) investigated the effects of various dietary carbohydrates and the frequency of feeding on patterns of feed utilization by carp. While the starch diet produced the highest weight gain and feed efficiency at two daily feedings, glucose and maltose were as efficiently utilized as starch when fed at least four times daily. This indicates that there is a drop in the absorption efficiency of glucose and maltose when large amounts are fed at a time.

Ogino *et al.* (1976) found that common carp used carbohydrate effectively as an energy source. Later Takeuchi *et al.* (1979a) also confirmed the dietary value of carbohydrates as energy sources. Based on the results of many studies, the optimum range of dietary carbohydrate may be considered to be 30–40% for common carp.

Vitamins and minerals

The qualitative and quantitative vitamin requirements of carp have been well investigated; a summary, including deficiency signs, has been provided in Table 18.4. Dietary requirements for folic acid and vitamins B_{12}, D and K have not

Table 18.4. Vitamin requirements of common carp and deprivation-induced symptoms (Satoh, 1991; NRC, 1993).

Vitamin	Requirement (mg kg^{-1})	Deficiency signs
Thiamine	0.5	Poor growth, nervousness, skin depigmentation, subcutaneous haemorrhage
Riboflavin	7	Anorexia, poor growth, haemorrhages in hepatopancreas, skin and fin, emaciation, photophobia, nervousness, anterior kidney necrosis
Pyridoxine	6	Anorexia, ascites, ataxia, exophthalmia, convulsions, nervous disorders, anaemia, low hepatopancreatic transferase
Pantothenate	30	Anorexia, poor growth, irritability, haemorrhages in skin, lethargy, exophthalmia
Niacin	28	Anorexia, poor growth, poor survival, haemorrhages in skin, high mortality
Biotin	1	Poor growth, erythrocyte fragility and fragmentation, lethargy, increased number of dermal mucous cells
Choline	500	Poor growth, fatty hepatopancreas, vacuolization of hepatic cells
Inositol	440	Anorexia, poor growth, dermatitis, loss of skin mucosa
Vitamin A	4000 IU	Anorexia, poor growth, exophthalmia, skin depigmentation, twisted opercula, haemorrhagic fin and skin
Vitamin E	100	Muscular dystrophy, exophthalmia, lordosis, kidney degeneration, pancreatic degeneration
Vitamin C	required	Caudal fin erosion and deformed gill arches in larval stage, poor growth

been observed yet, but some of these vitamins can be synthesized by intestinal microflora in carp and other freshwater fish (Lovell and Limsuwan, 1982).

Vitamin requirements of common carp may be affected by various factors, such as fish size, water temperature and diet composition. For example, juvenile or adult common carp do not require vitamin C because they can synthesize ascorbic acid from D-glucose. However, fry of common carp do show vitamin C deficiency signs, such as caudal fin erosion and deformed gill arches (Dabrowski *et al.*, 1988). Studies with first-feeding larvae have indicated that the level required for maximum tissue storage (270 mg ascorbic acid equivalent kg^{-1}) is higher than that needed for survival and growth (45 mg ascorbic acid equivalent kg^{-1}) (Gouillou-Coustans *et al.*, 1998). In regard to vitamin E, the requirement may increase corresponding to the level of polyunsaturated fatty acids in the diet.

Recently, extrusion techniques have been used to make floating fish diets and certain vitamins may be destroyed during diet manufacture and storage. To provide a safety margin, supplemental levels of vitamins in fish diets are always higher by two to five times the requirement levels.

Mineral requirements and their deficiency signs are summarized in Table 18.5. (It has been reported that common carp require cobalt, copper, iron magnesium, manganese, phosphorus and zinc.) Carp lack an acid-secreting stomach essential for digesting and solubilizing various compounds containing both calcium and phosphorus; thus the availability of phosphorus depends on the water solubility of the salt and ingredients (Satoh *et al.*, 1992, 1997). Phosphorus from tricalcium phosphate or fish-meal (FM) (white and/or brown FM) is less available than that from the more soluble mono- and dicalcium phosphates.

Table 18.5. Mineral requirements of common carp and deprivation-induced symptoms (Satoh, 1991; NRC, 1993; Kim *et al.*, 1998).

Mineral	Requirement	Deficiency signs
Phosphorus	6–8 g kg^{-1}	Poor growth, skeletal abnormality, low feed efficiency, low ash in whole body and vertebrae, increased visceral fat
Magnesium	0.4–0.5 g kg^{-1}	Poor growth, anorexia, high mortality, sluggishness and convulsions, cataracts, high mortality, high calcium content in bone, reduced magnesium in bone
Iron	150 mg kg^{-1}	Low specific gravity, haemoglobin content and haematocrit values, abnormal mean corpuscular diameter
Zinc	15–30 mg kg^{-1}	Poor growth, high mortality, erosion of fins and skin, low zinc content in bone
Manganese	13 mg kg^{-1}	Poor growth, dwarfism, skeletal abnormalities, high mortality, low calcium, magnesium, phosphorus, zinc and manganese in bone
Copper	3 mg kg^{-1}	Poor growth
Cobalt	0.1 mg kg^{-1}	Poor growth

Supplementation of monobasic phosphate to FM-based diets resulted in an increase in growth response of common carp. It has also been found that an exogenous supply of copper, manganese, magnesium and zinc is necessary for carp diets. However, it should be noted that an excess amount of tricalcium phosphate may inhibit the availability of trace elements, such as zinc and manganese, though to a much lesser extent than in rainbow trout (Satoh *et al.*, 1989). In a study examining the interaction between zinc deficiency and lipid intake, malabsorption of nutrients was observed and it was linked to the lipid deposition in the intestine (Taneja and Arya, 1994).

Practical Diets

Changing with the times, traditional ways of feeding carp with crude feeds prepared on site from local ingredients have given way to feeding commercially prepared diets. But in China, where carp culture predominates, there are still farms where natural food, fresh food and simple processed dry food are being used (Song, 1994). Several plant- and animal-based diets are available for carp. The former include oilcakes, beans, grains and brans, grasses and tree leaves, while the latter include FM, meat and bone-meal, silkworm pupae meal, blood meal, feather meal and chicken-farm waste. In his review Song (1994) has dealt extensively with the ingredients mentioned here, in the context of carp farming.

Prepared diets for carp are currently FM-based. Efforts are under way to produce alternative diets, aimed not only at reducing prices but also at making the diets more suitable to the environment. The expanded pellet is rapidly becoming popular for feeding many of the cultivated species, including carp. This pellet has characteristics between those of an extruded pellet and a steam pellet, along with being cost-effective. This section summarizes some of the recent success in practical diet development for common carp.

Digestibility of ingredients

It is well known that the apparent digestibility of protein (APD) and energy in several diet ingredients appears to be species-specific and temperature-dependent. This is particularly true for plant ingredients. Table 18.6 shows the apparent digestibility of protein and energy in various feedstuffs at different water temperatures for carp (Takeuchi, 1991). The digestibility values were not influenced by fish size, which ranged from 3 to 295 g.

In formulating diets for the carp, efforts are going on to replace FM with plant protein sources, taking advantage of the fact that carp utilize the latter efficiently. Takeuchi *et al.* (1990) reported that the availability of carbohydrate ingredients – potato starch, maize starch, maize, rye, wheat flour – was improved by the extrusion treatment, which elevated the gelatinization level as well as the energy value. Jeong *et al.* (1992) investigated the effect of the dietary gelatinized carbohydrate ratio and digestible energy content on carp fingerlings fed a diet

Table 18.6. The apparent digestibilities of protein and energy from various diet ingredients in carp at different temperatures (from Takeuchi, 1991).

	Proximate composition			Apparent digestibility (%)					
				Protein			Energy		
Ingredients	Moisture (%)	Protein (%)	Gross energy (MJ kg^{-1})	15°C	20°C	25°C	15°C	20°C	25°C
White fish-meal	8.8	65.8	0.17	94	90	92	86	89	88
Local fish-meal	10.8	63.5	0.18	90	90	93	87	87	86
Meat meal	11.8	69.0	0.22	91	91	95	76	79	82
Silkworm pupa meal	7.8	52.6	0.24	87	86	88	80	79	82
Maize-gluten meal	8.8	64.3	0.22	82	88	93	73	80	85
Defatted soybean meal	10.7	45.3	0.17	94	96	95	76	79	80
Defatted wheat-germ meal	10.6	31.3	0.17	92	93	94	70	74	77
Defatted rice bran	11.3	18.8	0.16	60	85	88	48	71	76
Wheat flour	13.1	15.8	0.16	78	79	80	49	73	79

containing raw and extruded potato starch. Growth performances were best when the gelatinization level was 40% in the low-energy diet (19.7 MJ kg^{-1}) and 20% in the high-energy diet (20.9 MJ kg^{-1}).

Soybean-protein concentrate (SPC), a promising diet ingredient for common carp, had superior apparent digestibility coefficients (ADC) for protein and lipid at 18°C compared with herring meal (HM), indicating the suitability of the plant ingredient for common carp (Kim *et al.*, 1998). However, the ADC of energy was not found to be different between SPC and HM. In another study evaluating the suitability of oilcakes, such as those made from sesame, mustard, linseed, groundnut and copra, incorporated at levels ranging from 25% to 75% of total protein, it was found that the APD were broadly similar, ranging from 78% to 90% (Hasan *et al.*, 1997).

Malt-protein flour (MPF), a brewery by-product, has been tested as yet another alternative protein source. MPF could substitute 40% or 60% of white FM (WFM) in fingerling carp diets, the latter level being possible when crystalline EAA are present, without any adverse effects on weight gain, feed efficiency or protein retention (Yamamoto *et al.*, 1996). Further, it was also demonstrated that a combination of MPF and soybean meal (SBM), at a replacement level of 60%, produced a significantly higher weight gain than did the WFM diet. Therefore MPF is a good choice as an alternative ingredient.

In formulating fish diets using multiple protein sources, information on protein digestibility as well as individual amino acid availability is essential. Though extrusion processing of a diet or ingredient improves the energy availability by gelatinization of the carbohydrate, it also generates high temperatures and pressures, which can cause losses of amino acids by oxidation and ultimately decrease the availability of some of them, such as lysine, through the Maillard reaction. Therefore, the influence of extrusion processing on the amino acid availability (AAA) from protein sources requires careful examination, as illustrated by the following study on common carp by Yamamoto *et al.* (1998). They compared the APD of several protein sources: WFM, SBM, extruded SBM (ExSBM), MPF, ExMPF and corn (maize)-gluten meal (CGM), and found that it was about 1.9% lower than true protein digestibility (TPD). Extrusion processing generally increased the apparent AAA of the ingredients. The differences between true and apparent AAA were as small as those observed between TPD and APD; however, relatively large differences were noted for several amino acids, such as lysine in CGM and threonine in ExMPF. These results demonstrated that WFM, SBM, ExSBM and MPF are good protein ingredients for carp diets, the exception being CGM. In the search for further alternative sources, it should be borne in mind that, although the individual AAA of a protein source approximates the protein digestibility value, the differences between true and apparent AAA are large for several amino acids. Thus, it is deemed essential to determine individual true AAA, which will allow a more precise assessment of the nutritional quality of diet ingredients and the formulation of practical and economic carp diets.

Tuna oil was evaluated as a lipid source in carp diets (Appleford and Anderson, 1997). There was a significant reduction in the ADC with the increase in inclusion level (from 83% to 59% for oil levels of 10% and 15%, respectively); however, an improvement in digestibility was recorded upon prolonged adaptation at the higher level. Comparing beef tallow and various hydrogenated fish-oils with differing melting-points, Takeuchi *et al.* (1979c) pointed out that the digestibility of fish-oil with a high melting-point (53°C) was very low, especially in carp weighing less than 10 g. When the melting-point of the hydrogenated fish oil was lower (38°C) it was effectively utilized, the digestibility being more than 70%, regardless of fish size and water temperature. Therefore, lipid ingredients of inferior quality might serve as a dietary energy source without any adverse effects, but only if additional lipids provide the necessary level of EFA for the fish.

Alternatives to fish-meal

Among all the tested alternatives, SBM has been the most widely used. A long-term study combining both laboratory- and field-based situations proved the efficacy of a hypoproteic soybean-based diet (Noble *et al.*, 1998).

A complete replacement of brown FM (BFM) with a combination of SBM, CGM and meat meal (MM) enriched with synthetic amino acids, such as lysine, methionine and threonine, to match the BFM diet could not produce comparable growth and diet utilization, although palatability was as good as the FM diet (Pongmaneerat *et al.*, 1993). Pongmaneerat *et al.* (1993) recommended a 56% replacement of BFM in the diet for carp by a combination of 25% SBM and 10% CGM at a 38% dietary protein level. This contrasts with a replacement level of about 90% at 44% dietary protein level achieved for rainbow trout (Fig. 18.3; Pongmaneerat, 1993). Although it is generally accepted that carp can utilize plant proteins more efficiently than can rainbow trout, the inferior levels attained in carp could be ascribed to the comparatively lower utilization of supplemental crystalline amino acids. It also remains to be tested if taurine was a limiting amino acid, as suggested earlier.

Experiments on replacing dietary FM also proved that environmental phosphorus loading could be lowered proportional to the FM levels. Based on total phosphorus and nitrogen loading, Jahan *et al.* (2000) suggested that a suitable level of FM in carp diet was 20–25% in combination with blood meal and defatted SBM. In another study, Kim *et al.* (1997) reported that dietary inclusion of fish-protein concentrate was also effective in decreasing phosphorus excretion.

Studies with carp fry indicated that 25% inclusion of linseed or groundnut oilcakes produced growth comparable to diets prepared with FM as the sole

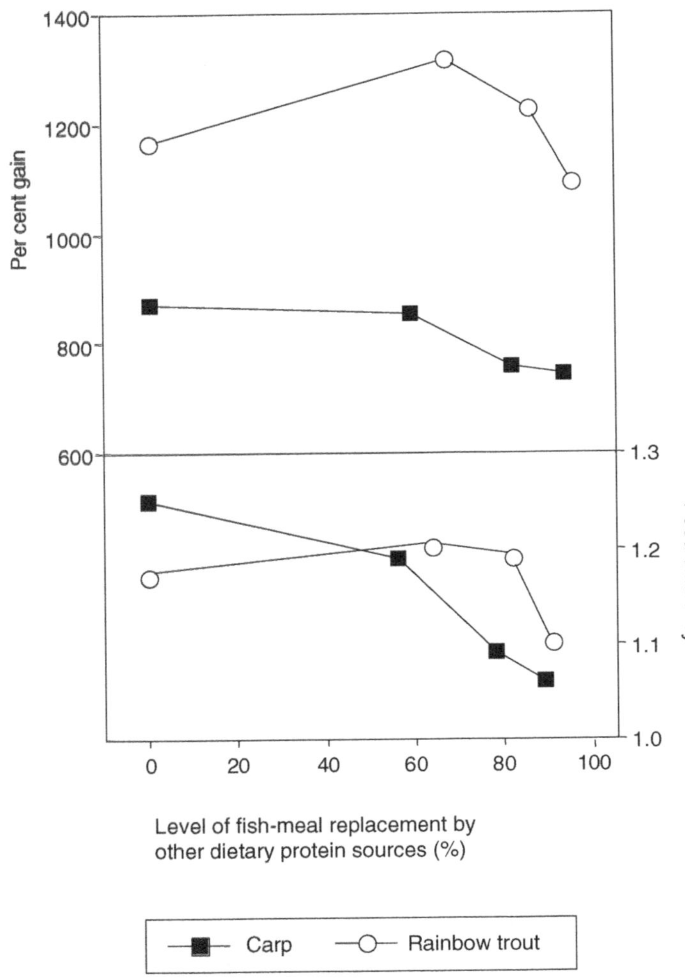

Fig. 18.3. Comparison of fish-meal replacement in dietary protein sources in relation to per cent gain and feed efficiency between carp and rainbow trout.

protein source, indicating the suitability of the two plant ingredients (Hasan *et al.*, 1997). Incorporation of SPC up to 40% did not adversely affect survival and growth of larvae; however, dietary levels of 60 or 70% resulted in growth retardation, which was not overcome by supplementation of sulphur amino acids (Escaffre *et al.*, 1997).

Feeding Practices

Intensification of carp culture should consider the impact of the use of proper feeding rates, not only in achieving optimum growth, but also in controlling aquatic wastes. Under semi-intensive culture, determination of variations in nutrient input and turnover from natural productivity using standard procedures is required. For intensive operations, appropriate formulations and feeding practices should be adopted, considering the potential digestive- and metabolic-waste discharge. In an elaborate study in Kasumigaura Lake, Japan, feeding rates have been established for commercial and regulated diets (Tables 18.7 and 18.8). The Kasumigaura diet is a low-protein, high-energy diet and as a result the feeding rates were higher for the smaller size ranges. Further, the rates were strongly dependent on the water temperature and fish size, highlighting the importance of determining the ration size for each and every culture operation. The preferred optimal temperatures for growth lie between 30 and 32°C, but retardation of growth occurs with a drop in temperature until around 10°C, when aphagia ensues. However, growth hormone treatment was found to stimulate appetite and growth during winter temperatures, making it possible to

Table 18.7. The temperature- and fish-size-based changes in daily feeding rate of carp fed on official standard-based commercial diets (from Miyatake, 1997).

Water temperature	Body weight (g)											
(°C)	2–5	5–10	10–20	20–30	30–40	40–50	50–100	100–200	200–400	400–600	600–800	800–1000
15	4.5	3.7	3.2	2.8	2.5	2.3	2.0	1.7	1.5	1.3	1.1	0.9
16	4.8	4.0	3.4	3.0	2.7	2.4	2.1	1.8	1.6	1.4	1.1	0.9
17	5.2	4.3	3.7	3.2	2.9	2.6	2.2	1.9	1.7	1.5	1.2	1.0
18	5.6	4.7	4.0	3.4	3.1	2.8	2.3	2.0	1.8	1.6	1.3	1.0
19	6.0	5.1	4.3	3.7	3.4	3.0	2.5	2.2	1.9	1.7	1.3	1.1
20	6.5	5.5	4.6	4.0	3.7	3.2	2.7	2.4	2.2	1.8	1.4	1.2
21	7.0	6.0	4.9	4.3	4.0	3.4	2.9	2.6	2.1	1.9	1.5	1.3
22	7.6	6.5	5.3	4.6	4.3	3.7	3.1	2.8	2.3	2.0	1.6	1.4
23	8.2	7.0	5.7	4.9	4.6	4.0	3.4	3.0	2.5	2.1	1.7	1.5
24	8.8	7.5	6.1	5.3	4.9	4.3	3.7	3.2	2.7	2.3	1.8	1.6
25	9.5	8.0	6.5	5.7	5.2	4.6	4.0	3.4	2.9	2.5	2.0	1.7
26	10.2	8.5	6.9	6.1	5.5	4.9	4.3	3.7	3.1	2.7	2.2	1.8
27	10.9	9.0	7.4	6.5	5.8	5.2	4.6	4.0	3.3	2.9	2.4	1.9
28	11.6	9.5	7.9	6.9	6.2	5.6	4.9	4.3	3.6	3.1	2.6	2.0
29	12.3	10.1	8.4	7.4	6.6	6.0	5.2	4.6	3.9	3.3	2.8	2.1
30	13.1	10.7	8.9	7.9	7.0	6.4	5.7	4.9	4.2	3.5	3.0	2.2

Table 18.8. The temperature and fish size based changes in daily feeding rate of carp fed on a low protein, high energy Kasumigaura regulated carp diet (from Miyatake, 1997).

Water temperature (°C)	Body weight (g)								
	1–5	5–10	10–50	50–100	100–200	200–400	400–600	600–800	800–1000
15	4.65	3.42	2.16	1.59	1.26	0.95	0.84	0.76	0.69
16	5.15	3.80	2.39	1.76	1.40	1.05	0.94	0.84	0.77
17	5.17	4.21	2.65	1.95	1.55	1.17	1.04	0.93	0.85
18	6.33	4.66	2.94	2.17	1.72	1.30	1.15	1.03	0.95
19	7.02	5.17	3.26	2.40	1.90	1.44	1.28	1.14	1.05
20	7.78	5.73	3.61	2.66	2.11	1.59	1.41	1.26	1.16
21	8.62	6.35	4.00	2.95	2.34	1.76	1.57	1.40	1.29
22	9.65	7.04	4.44	3.27	2.59	1.96	1.74	1.55	1.43
23	10.60	7.81	4.92	3.62	2.88	2.17	1.93	1.72	1.58
24	11.75	8.65	5.45	4.02	3.19	2.40	2.13	1.91	1.75
25 and over	13.02	9.60	6.04	4.45	3.53	2.66	2.37	2.11	1.94

culture the fish even under unfavourable climatic conditions (Teskeredžic *et al.*, 1995).

As carp culture continues to expand, more and more of the culture operations will become intensive, necessitating the development of advanced diet formulations, which could also address related issues of environmental concern.

References

Akiyama, T., Oohara, I. and Yamamoto, T. (1997) Comparison of essential amino acid requirements with A/E ratio among fish species. *Fisheries Science* 63, 963–970.

Appleford, P. and Anderson, T.A. (1997) Apparent digestibility of tuna oil for common carp, *Cyprinus carpio* – effect of inclusion level and adaptation time. *Aquaculture* 148, 143–151.

Baloguma, M. (1995) The prospects of indigenous alternative feed resources for carp culture in Nigeria. *Aquaculture* 129, 391 [abstract].

Chakraborti, S.C., Ross, L.G. and Ross, B. (1992) Specific dynamic action and feeding metabolism in common carp, *Cyprinus carpio* L. *Comparative Biochemistry and Physiology Part A* 103, 809–815.

Coutteau, P., Geurden, I., Camara, M.R., Bergot, P. and Sorgeloos, P. (1997) Review on the dietary effects of phospholipids in fish and crustacean larviculture. *Aquaculture* 155, 149–164.

Dabrowski, K. (1983a) Digestion of protein and amino acid absorption in stomachless fish, common carp (*Cyprinus carpio* L). *Comparative Biochemistry and Physiology Part A* 74, 409–415.

Dabrowski, K. (1983b) Comparative aspects of protein digestion and amino acid absorption in fish and other animals. *Comparative Biochemistry and Physiology Part A* 74, 416–425.

Dabrowski, K. (1986) Protein digestion and amino acid absorption along the intestine of the common carp (*Cyprinus carpio* L.), a stomachless fish: an *in vivo* study. *Reproduction Nutrition and Development* 26, 755–766.

Dabrowski, K., Hinterleitner, S., Sturmbauer, C., El-Fiky, N. and Wieser, W. (1988) Do carp larvae require vitamin C? *Aquaculture* 72, 295–306.

De Silva, S.S. and Anderson, T.A. (1995) *Fish Nutrition in Aquaculture*. Chapman & Hall, London, 319 pp.

Escaffre, A.-M., Zambonino Infante, J.L., Cahu, C.L., Mambrini, M., Bergot, P. and Kaushik, S.J. (1997) Nutrition value of soy protein concentrate for larvae of common carp (*Cyprinus carpio*) based on growth performance and digestive enzyme activities. *Aquaculture* 153, 63–80.

Fontagné, S., Geurden, I., Escaffre, A.-M. and Bergot, P. (1998) Histological changes induced by dietary phospholipids in intestine and liver of common carp (*Cyprinus carpio* L.) larvae. *Aquaculture* 161, 213–223.

Fontagné, S., Pruszynski, T., Corraze, G. and Bergot, P. (1999) Effect of coconut oil and tricaprylin vs. triolein on survival, growth and fatty acid composition of common carp (*Cyprinus carpio* L.) larvae. *Aquaculture* 179, 241–251.

Fontagné, S., Corraze, G. and Bergot, P. (2000) Tricaproin, tricaprin and trilaurin are utilized more efficiently than tricaprylin by carp (*Cyprinus carpio* L.). *Journal of Nutrition* 130, 2009–2015.

Geurden, I., Radünz-Neto, J. and Bergot, P. (1995) Essentiality of dietary phospholipids for carp (*Cyprinus carpio* L.) larvae. *Aquaculture* 131, 303–314.

Geurden, I., Marion, D., Charlon, N., Coutteau, P. and Bergot, P. (1998) Comparison of different soybean phospholipidic fractions as dietary supplements for common carp, *Cyprinus carpio*, larvae. *Aquaculture* 161, 225–235.

Geurden, I., Bergot, P., Van Ryckeghem, K. and Sorgeloos, P. (1999) Phospholipid composition of common carp (*Cyprinus carpio*) larvae starved or fed different phospholipid classes. *Aquaculture* 171, 93–107.

Gouillou-Coustans, M.-F., Bergot, P. and Kaushik, S.J. (1998) Dietary ascorbic acid needs of common carp (*Cyprinus carpio*) larvae. *Aquaculture* 161, 453–461.

Hasan, M.R., Macintosh, D.J. and Jauncey, K. (1997) Evaluation of some plant ingredients as dietary protein sources for common carp (*Cyprinus carpio* L.) fry. *Aquaculture* 151, 55–70.

Jahan, P., Watanabe, T., Satoh, S. and Kiron, V. (2000) Effect of dietary fish meal levels on environmental phosphorus loading from carp culture. *Fisheries Science* 66, 204–210.

Jauncey, K. (1982) Carp (*Cyprinus carpio* L.) nutrition – a review. In: Muir, J.F. and Roberts, R.J. (eds) *Recent Advances in Aquaculture*. Croom Helm, London, pp. 216–263.

Jeong, K-.S., Takeuchi, T., Okamoto, N. and Watanabe, T. (1992) The effect of dietary gelatinized ratios at different dietary energy levels on growth and characteristics of blood in carp fingerlings. *Nippon Suisan Gakkaishi* 58, 945–951.

Kafuku, T. (1992) Common carp (*Cyprinus carpio*). In: Ikenoue, H. and Kafuku, T. (eds) *Modern Methods of Aquaculture in Japan*. Kodansha Elsevier, Tokyo, pp. 65–78.

Kaushik, S.J. (1995) Nutrient requirements, supply and utilization in the context of carp culture. *Aquaculture* 129, 225–241.

Kim, J.D., Breque, J. and Kaushik, S.J. (1998) Apparent digestibilities of feed components from fish meal or plant protein based diets in common carp as affected by water temperature. *Aquatic Living Resources* 11, 269–272.

Kim, K.S., Kim, J.D., Jeong, K.S., Song, J.S., Lee, S.B., Woo, Y.B. and Lee, J.Y. (1997) Effects of partial substitution of dietary fish meal with fish protein concentrate or blood meal on growth and pollution loads of Israeli strain of common carp (*Cyprinus carpio*). *Korean Journal of Animal Nutrition and Feedstuffs* 21, 237–244 (in Korean with English abstract).

Lovell, R.T. and Limsuwan, T. (1982) Intestinal synthesis and dietary non-essentiality of vitamin B_{12} for *Tilapia nilotica*. *Transactions of the American Fisheries Society* 111, 485–490.

Miyatake, H. (1997) Carp. *Yoshoku* 34(5), 108–111 (in Japanese).

Murai, T., Akiyama, T. and Nose, T. (1983) Effects of glucose chain length of various carbohydrates and frequency of feeding on their utilization by fingerling carp. *Nippon Suisan Gakkaishi* 49, 1607–1611.

Murai, T., Akiyama, T., Takeuchi, T., Watanabe, T. and Nose, T. (1985) Effects of dietary protein and lipid levels on performance and carcass composition of fingerling carp. *Bulletin of the Japanese Society of Scientific Fisheries* 51, 605–608.

Noble, E., Demaël, A., Garin, D., Moulin, C. and Barré, H. (1998) Effects of an hypoproteic soybean based diet on the energy stores and growth of carp (*Cyprinus carpio* L.). *Comparative Biochemistry and Physiology Part A* 120, 157–161.

Nose, T. (1979) Summary report on the requirements of essential amino acids for carp. In: Tiews, K. and Halver, J.E. (eds) *Finfish Nutrition and Fishfeed Technology*. Heenemann, Berlin, pp. 145–156.

NRC (1993) *Nutrient Requirements of Fish*. National Academy Press, Washington, DC, 114 pp.

Ogino, C. (1980a) Requirements of carp and rainbow trout for essential amino acids. *Nippon Suisan Gakkaishi* 46, 171–175.

Ogino, C. (1980b) Protein requirements of carp and rainbow trout. *Nippon Suisan Gakkaishi* 46, 385–388.

Ogino, C. and Chen, M.-S. (1973) Protein nutrition in fish – V. Relation between biological value of dietary proteins and their utilization in carp. *Nippon Suisan Gakkaishi* 39, 955–959.

Ogino, C., Chiou, J.-Y. and Takeuchi, T. (1976) Protein utilization in fish – VI. Effects of dietary energy sources on the utilization of proteins by rainbow trout and carp. *Nippon Suisan Gakkaishi* 42, 213–218.

Ohta, M. and Watanabe, T. (1996) Dietary energy budgets in carp. *Fisheries Science* 62, 745–753.

Omae, H. (1992) Carp. *Yoshoku* 29(1s), 78–84 (in Japanese).

Pongmaneerat, J. (1993) Studies on the utilization of alternative protein sources in diets for rainbow trout and carp. Doctoral thesis, Tokyo University of Fisheries, Tokyo, Japan.

Pongmaneerat, J., Watanabe, T., Takeuchi, T. and Satoh, S. (1993) Use of different protein meals as partial or total substitution for fish meal in carp diets. *Nippon Suisan Gakkaishi* 59, 1249–1257.

Satoh, S. (1991) Common carp, *Cyprinus carpio*. In: Wilson, R.P. (ed.) *Handbook of Nutrient Requirements of Finfish*. CRC Press, Boca Raton, pp. 55–67.

Satoh, S., Izume, K., Takeuchi, T. and Watanabe, T. (1989) Availability to carp of manganese contained in various types of fish meal. *Nippon Suisan Gakkaishi* 55, 313–319.

Satoh, S., Viyakarn, V., Yamazaki, Y., Takeuchi, T. and Watanabe, T. (1992) A simple method for determination of available phosphorus content in fish diet. *Nippon Suisan Gakkaishi* 58, 2095–2100.

Satoh, S., Viyakarn, V., Takeuchi, T. and Watanabe, T. (1997) Availability of phosphorus in various phosphate to carp and rainbow trout determined by a simple fractionation method. *Fisheries Science* 63, 297–300.

Schwarz, F.J. and Kirchgessner, M. (1988) Amino acid composition of carp (*Cyprinus carpio* L.) with varying protein and energy supplies. *Aquaculture* 72, 307–317.

Song, T. (1994) Feeding and nutrition. In: Li, S. and Mathias, J. (eds) *Fresh Water Fish Culture in China: Principles and Practice*. Elsevier, Asterdam, pp. 79–127.

Tacon, A. (1993) *Feed Ingredients for Warm Water Fish: Fishmeal and Other Processed Feedstuffs*. FAO Fisheries Circular No. 856, FAO, Rome, 64 pp.

Takeuchi, T. (1991) Digestion and nutrition. In: Itazawa, Y. and Hanyu, I. (eds) *Fish Physiology*. Kouseisha-Kouseikaku, Tokyo, pp. 67–101 (in Japanese).

Takeuchi, T. (1999) Feeding the common carp in Japan. *International Aquafeed* 4, 27–32.

Takeuchi, T. and Watanabe, T. (1977) Requirement of carp for essential fatty acids. *Nippon Suisan Gakkaishi* 43, 541–551.

Takeuchi, T., Watanabe, T. and Ogino, C. (1979a) Availability of carbohydrate and lipid as dietary energy sources for carp. *Nippon Suisan Gakkaishi* 45, 977–982.

Takeuchi, T., Watanabe, T. and Ogino, C. (1979b) Optimum ratio of dietary energy to protein for carp. *Nippon Suisan Gakkaishi* 45, 983–987.

Takeuchi, T., Watanabe, T. and Ogino, C. (1979c) Digestibility of hydrogenated fish oils in carp and rainbow trout. *Nippon Suisan Gakkaishi* 45, 1521–1525.

Takeuchi, T., Jeong, K.-S. and Watanabe, T. (1990) Availability of extruded carbohydrate ingredients to rainbow trout *Oncorhynchus mykiss* and carp *Cyprinus carpio*. *Nippon Suisan Gakkaishi* 56, 1839–1845.

Takeuchi, T., Watanabe, K., Satoh, S. and Watanabe, T. (1992) Requirement of grass carp fingerlings for tocopherol. *Nippon Suisan Gakkaishi* 58, 1743–1749.

Taneja, S.K. and Arya, P. (1994) Lipid deposition in intestine as a possible cause of malabsorption of nutrients in zinc-deficient common carp (*Cyprinus carpio*). *British Journal of Nutrition* 72, 753–762.

Teskeredžic, E., Tomec, M., Hacmanjek, M., McLean, E., Teskeredžic, Z. and Donaldson, E. (1995) Growth and body composition of common carp (*Cyprinus carpio* L.) maintained under aphagic thermal conditions: response to porcine somatotropin therapy. *Aquaculture* 129, 343 [abstract].

Watanabe, T. (1982) Lipid nutrition in fish. *Comparative Biochemistry and Physiology Part A* 73, 3–15.

Watanabe, T. (1988) Nutrition and growth. In: Shepherd, C.J. and Bromage, N.R. (eds) *Intensive Fish Farming*. BSP Professional Books, London, pp. 154–197.

Yamamoto, T., Unuma, T., Akiyama, T. and Kishi, S. (1996) Utilization of malt protein flour in fingerling carp diets. *Fisheries Science* 62, 783–789.

Yamamoto, T., Akimoto, A., Kishi, S., Unuma, T. and Akiyama, T. (1998) Apparent and true availabilities of amino acids from several protein sources for fingerling rainbow trout, common carp, and red sea bream. *Fisheries Science* 64, 448–458.

Yokoyama, M., Takeuchi, T., Park, G. and Nakazoe, J. (2002) Hepatic cysteinesulfinate decarboxylase activity in fish. *Aquaculture Research* (in press).

Zeitler, M.H., Kirchgessner, M. and Schwarz, F.J. (1983) Effects of different protein and energy supplies on carcass composition of carp (*Cyprinus carpio* L.). *Aquaculture* 36, 37–48.

Single author books (also included in the reference section)

De Silva, S.S. and Anderson, T.A. (1995) *Fish Nutrition in Aquaculture*. Chapman & Hall, London, 319 pp.

NRC (1993) *Nutrient Requirements of Fish*. National Academy Press, Washington, DC, 114 pp.

Indian Major Carps 19

H. Shivananda Murthy
Department of Aquaculture, College of Fisheries, University of Agricultural Sciences, Mangalore 575002, India

Introduction

Indian major carps are commercially cultured in India and the Indian subcontinent; catla (*Catla catla*), rohu (*Labeo rohita*) and mrigal (*Cirrhinus mrigala*) are considered the three major carps of India. A few medium and minor carps that grow to an appreciable size are also being farmed in fresh water. Indian major carps are usually cultured together in 'polyculture' rather than monoculture. The principle behind this polyculture system is the utilization of all available natural food resources at different trophic levels to achieve maximum production per unit area of water body (Jhingran, 1991).

Major carps are cultured not only in India but also in the entire Indian subcontinent, particularly Pakistan, Bangladesh, Nepal, Sri Lanka, Myanmar and other adjoining countries. Carp culture in India has a long history and can be traced back several centuries. Most of the farmed carps are marketed within the country for domestic consumption. In recent years, fillets, flakes and other value-added products of carps have been exported to Middle East countries. Surprisingly, carp production was not reported in the Food and Agriculture Organization (FAO) fish-production statistics until recently. In the past few years, Indian carp production has registered substantial increases. India is second in world carp production, next only to China. Of the 2 million t of aquaculture production of the country, a majority is from the farming of carps.

Although Indian major carps are traditionally cultured in freshwater ponds, recently success has been achieved in culturing these species in brackish-water ponds (where penaeid shrimp were cultured) after gradual acclimatizaion to saline water (up to 15 parts per thousand salinity). Concerted research efforts were also made in nutrition, feed and feeding of Indian major carps, in addition to culture practices.

Nutrient Requirements

Protein and amino acids

The optimal dietary protein requirement is affected by the nutritional value of the dietary protein and the level of non-protein energy in the diet. When sufficient energy sources, such as lipid and carbohydrates, are available in the diet, most of the ingested protein is utilized for protein synthesis. Adults of Indian major carps require 30% dietary protein for proper growth and survival. Fingerlings and fry of these carps require 35% and 40% dietary protein, respectively, for good growth (Sen *et al.*, 1978; Renukaradhya and Varghese, 1986). Mondal *et al.* (2000) reported that catla fry require 38.5% dietary protein for optimal growth and survival.

Indian major carps, like other animals, do not have an absolute requirement for protein but require a balanced mixture of indispensable and dispensable amino acids (Murthy and Varghese, 1998). Qualitative and quantitative amino acid requirements of Indian major carps were determined by conducting growth studies by feeding graded levels of test amino acids and at the same time keeping all other essential amino acids at the required level. Qualitative amino acid requirements of carps could also be estimated by employing carbon-14 (^{14}C)-labelled isotopes. Carcass (body) composition of amino acids gives only a rough estimate of requirements, which could be used in the absence of information on dietary requirements. Studies on the qualitative dietary requirement of carps have indicated that they require all the ten essential amino acids (EAA) that are known as indispensable for other commercial finfish studied so far. Quantitative dietary amino acid requirements of catla, rohu and mrigal for all ten amino acids are presented in Tables 19.1, 19.2 and 19.3, respectively. The cystine replacement value was estimated to be 50% in Indian major carps.

Lipids and fatty acids

Lipids, or fats, are required as a source of energy and essential fatty acids. Further, lipids serve as a carrier for fat-soluble vitamins. Fatty acids and phospholipids help to maintain the structural integrity of cell membrane. The gross lipid requirement of Indian major carps is 7–8% of the diet. Young fish require relatively more fat and more protein than adults.

All three Indian major carps were found to grow well when the diet contained 1% n-3 and 1% n-6 fatty acids. Body composition of Indian major carps contains a high proportion of n-3 fatty acids as well as n-6 fatty acids. Limited studies have been carried out on the dietary fatty acid requirement of Indian major carps.

The essentiality of highly unsaturated fatty acids (HUFA), such as eicosapentaenoic acid (EPA) (20:5n-3) and docosahexaenoic acid (DHA) (22:6n-3), has not been reported for Indian major carps. These carps are basically freshwater species and found to grow well even in the absence of HUFA.

Table 19.1. Dietary amino acid requirements of *Catla catla* (% dietary protein).

Amino acid	Fry	Juvenile/adult
Arginine	4.80*	5.63†
Histidine	2.45*	2.38†
Isoleucine	2.35*	2.75†
Leucine	3.70*	4.38†
Lysine	6.23*	6.86‡
Methionine	3.55*	3.00†
Phenylalanine	3.70*	4.50†
Threonine	4.95*	4.50†
Tryptophan	0.95*	1.03§
Valine	3.55*	3.60†

* Ravi and Devaraj (1991).
† Based on the unpublished results of the author.
‡ Satheesha and Murthy (2000).
§ Satheesha and Murthy (1999).

Table 19.2. Dietary amino acid requirements of *Labeo rohita*.

Amino acid	Per cent of diet	Per cent of dietary protein	Reference
Arginine	2.30	5.75	Murthy and Varghese (1995)
Histidine	0.90	2.25	Murthy and Varghese (1995)
Isoleucine	1.20	3.00	Murthy and Varghese (1996a)
Leucine	1.85	4.63	Murthy and Varghese (1997a)
Lysine	2.27	5.58	Murthy and Varghese (1997b)
Methionine*	1.15	2.88	Murthy and Varghese (1998)
Phenylalanine†	1.60	4.00	Murthy and Varghese (1996b)
Threonine	1.71	4.28	Murthy and Varghese (1996c)
Tryptophan	0.45	1.13	Murthy and Varghese (1997c)
Valine	1.50	3.75	Murthy and Varghese (1997d)

* Total sulphur amino acid (methionine + cysteine) is 1.42% (3.55% dietary protein).
† Total aromatic amino acid (phenylalanine + tyrosine) is 2.31% (5.77% dietary protein).

Carbohydrates

Carbohydrate is the cheapest nutrient and also a less expensive energy source for carps. Being herbivorous/omnivorous feeders, Indian major carps easily digest appreciable quantities of carbohydrates in the diet. A dietary level of 22–30% of carbohydrate has been found to be optimum for the growth of Indian major carps. Growth retardation and reduced feed efficiency is observed when the carbohydrate level exceeds 35% of the diet. Starch and dextrin are readily

Table 19.3. Dietary amino acid requirement of *Cirrhinus mrigala*.*

Amino acid	Per cent of diet	Per cent of dietary protein
Arginine	2.10	5.25
Histidine	0.85	2.13
Isoleucine	1.10	2.75
Leucine	1.70	4.25
Lysine	2.35	5.88
Methionine	1.27	3.18
Phenylalanine	1.60	4.00
Threonine	1.65	4.13
Tryptophan	0.43	1.08
Valine	1.40	3.50

* Unpublished data from studies conducted by the author employing casein and gelatin as intact protein sources (40% protein), together with crystalline amino acids.

accepted by Indian major carps. However, in practical diets, wheat flour, tapioca flour and rice flour are used as cheap sources of carbohydrate in the diet formulation. These ingredients also serve as natural binders in the diet. Carbohydrates spare some protein when protein is not available in sufficient quantities in diets. Absence of adequate dietary carbohydrate in carp diets may result in the utilization of protein as an energy source.

Vitamins and minerals

Thiamine has been found to be an essential vitamin for the growth and survival of several freshwater fish species. No studies have been carried out on the dietary requirement of Indian major carps for thiamine. However, it is estimated that 8–12 mg kg^{-1} diet fulfils the dietary requirement, depending on the life stage. In commercial diets, higher levels are included to allow for leaching and anticipated losses during diet processing and storage. Riboflavin is essential for Indian major carps. A deficiency of riboflavin results in fin erosion, anorexia, loss of body colour and cataracts in fish. The riboflavin requirement of Indian major carps has been estimated to be about 6–8 mg kg^{-1} diet. However, in commercial diets, as high an amount as 10 mg kg^{-1} diet or more is incorporated, since it is water-soluble and part of it may be lost due to leaching. It is estimated that the three Indian major carps require 10–12 mg niacin kg^{-1} diet for normal growth and survival. The dietary requirement of Indian major carps for pantothenic acid has been estimated to be 9–11 mg kg^{-1} diet.

It is recommended to use stable forms of vitamin C, such as ascorbate 2-monophosphate or ascorbate 2-sulphate, in carp diets due its unstable nature and leaching problems. Few studies on the dietary requirement of ascorbic acid in Indian major carps have been carried out; however, a dietary level of 300 mg ascorbic acid kg^{-1} diet appears sufficient for the normal growth and survival of larvae and fry of rohu and mrigal. Tissue levels of vitamin C increased corresponding to dietary intake up to 1000 mg kg^{-1} diet in both rohu and mrigal. Disease resistance in respect of *Aeromonas hydrophila* was enhanced by vitamin C and lowest mortality was reported in rohu fed 600 mg vitamin C kg^{-1} diet and in mrigal fed 1000 mg kg^{-1} diet (Sobana, 1997). There are reports that ascorbic acid-free diets did not result in deficiency signs, mortality or significant growth variation in rohu when compared with diets with added ascorbic acid (Hasan *et al.*, 1993). In general, a dietary level of 100–150 mg ascorbic acid kg^{-1} diet has been found satisfactory for the normal growth of adults of Indian major carps. Deficiency of ascorbic acid leads to anorexia, poor growth, high mortality, fin necrosis, abnormal pigments, spinal curvature in mrigal (Agarwal and Mahajan, 1980; Mahajan and Agarwal, 1980a,b), haemorrhagic skin, microcytic anaemia and hypochronic anaemia in Indian major carps (Sobana, 1997).

Pyridoxine deficiency causes loss of appetite, oedema, nervous disorder, anorexia, hyperirritability and rapid gasping in fish. A dietary level of 6–8 mg pyridoxine kg^{-1} diet is recommended for Indian major carps. Dietary requirements for vitamin B_{12} have not been determined for Indian major carps, but a suggested dietary requirement is 0.01–0.02 mg kg^{-1} diet. The quantitative dietary requirements of Indian major carps for inositol range from 300 to 350 mg kg^{-1} diet. A dietary requirement of 5–8 mg biotin kg^{-1} diet was found satisfactory for Indian major carps. The quantitative dietary requirement of Indian major carps for folic acid appears to be 0.5–1 mg kg^{-1} diet. Dietary requirements of Indian major carps could be met at 500–600 mg choline kg^{-1} diet.

Vitamin A (retinoic acid) deficiency causes displacement of the eye lens and corneal thinning, poor growth, exophthalmia and depigmentation in carps. The dietary vitamin A requirement of Indian major carps appears to be 1500 IU. Vitamin D deficiency leads to depressed growth and muscle tetany in finfish, including carps. It is also important in the regulation of minerals, calcium and phosphorus. No quantitative requirement studies have been carried out on Indian major carps for vitamin D; however, a dietary level of 400–500 IU is suggested. The quantitative dietary requirement of vitamin E in Indian major carps appears to be 40–50 mg kg^{-1} diet. The dietary requirement of Indian major carps for vitamin K falls in the range of 5–10 mg kg^{-1} diet.

The dietary calcium requirements of Indian major carps has not been studied, but 4000–5000 mg calcium kg^{-1} diet is suggested. Calcium deficiency has not been detected in common carp (Ogino and Takeda, 1976; Lall *et al.*, 1985). Phosphorus deficiency results in poor food conversion, anorexia, poor bone mineralization, skeletal deformity, cranial deformity and other disorders in carps. Dietary levels ranging from 5000 to 6000 mg phosphorus kg^{-1} diet have been found to satisfy the requirement of Indian major carps. A dietary magnesium level of 500 mg kg^{-1} diet is optimum for Indian major carps, while 3–4 mg copper kg^{-1} diet has been found to satisfy their requirement. When supplemented at a level of 0.1 mg kg^{-1} diet in the diets of Indian major carp fry in nursery ponds, cobalt, in the form of cobalt chloride, enhanced growth and survival (Alikunhi, 1987). Quantitative dietary requirements of Indian major carps for other minerals have not been reported.

Practical Diets and Feeding

Supplementary feeding is essential to increase the production of carp fry in ponds. The average survival of Indian major carps during the early stages is rather low (about 30% from spawn to fry) and about 50% from fry to fingerling. This high mortality is due to lack of adequate and nutritionally balanced diets, and poor management practices.

The traditional or conventional diet used by small and rural fish farmers for the culture of Indian major carps is a mixture of rice bran and oilcake in equal proportions by weight. The diet consists of 1 : 1 rice bran and oilcake and usually contains 25–28% crude protein. However, this diet is not a nutritionally-balanced one. Most often, the oilcake is soaked in water for several hours and mixed with rice bran to make a dough and fed, in an uncooked form, twice daily at a rate of 10–20% of biomass in nursery ponds.

Nutritionally balanced diets were developed for carp juveniles and adults incorporating more ingredients than used prior to 1970 (Table 19.4). Varghese *et al.* (1976) formulated a pelleted diet for carps that contained 30% protein, incorporating fish-meal, and achieved 50% more production than with the traditional mixture of rice bran and oilcake. Good growth of common carp was recorded when feeding a pelleted diet containing silkworm pupae, prawn waste or fish-meal (Jeychandran and Paulraj, 1977). Jayaram and Shetty (1980) obtained higher growth of catla and common carp when they were fed with a diet containing silkworm pupae. A powder of dried duckweeds and cabbage leaves was incorporated in diets, but was found to be inferior as a complete diet for carp (Devaraj *et al.*, 1981). *Colocasia* leaf powder and fish silage were used, partially replacing fish-meal, in experimental diets for Indian major carps (Venugopal and Keshavanath, 1984). Silkworm faecal matter, slaughterhouse waste, soybean meal and squilla meal were used, replacing fish-meal, and a 25% higher growth than on fish-meal-based diets was recorded (Bhat *et al.*, 1986; Nandeesha *et al.*, 1989). Incorporation of earthworm meal, margarine, sardine oil and *Cassia tora* leaf powder in formulated diets enhanced the growth and survival of Indian major carps (Manissery *et al.*, 1988). Murthy and Devaraj (1990, 1991a,b) evaluated leaf powders of three floating aquatic weeds – namely, *Eichhornia, Pistia* and *Salvinia* – as low-cost ingredients partially replacing fish-meal in diets for carps and reported that the *Pistia*-based diet was found to be superior to the other two diets.

Fry of catla, rohu and mrigal 5–10 mm in size mainly feed on unicellular algae, while 10–20 mm fry feed on protozoans and other smaller zooplankton. Hatchlings of Indian major carps accept an artificial diet 2 days after their initiation to external food and grow well on oilcake, rice powder and black gram. Low-cost diets for carps were developed using locally available ingredients such as silkworm pupae, soybean (Chakraborthy *et al.*, 1973), fish and prawn powder (Mahajan and Yadav, 1974), petroleum protein and rice polish (Chakraborthy and Kar, 1975), powdered algae with fish-meal (Singh and Bhanot, 1988) and dried powder of *Nymphoides* and *Spirodella* weeds (Patnaik and Das, 1979).

Table 19.4. Important diet ingredients that are available for carp diets in India.

Ingredient	Moisture (%)	Crude protein (%)	Crude fat (%)	Crude fibre (%)	Ash (%)	Nitrogen-free extract (%)
Rice polish	12.6	14.5	17.3	7.5	n/a	n/a
Rice (broken)	10.1	12.6	11.3	19.3	10.2	36.5
Defatted rice bran	7.2	12.1	1.3	15.2	23.8	40.4
Wheat bran	12.3	15.8	4.3	8.7	n/a	n/a
Wheat (broken)	9.0	11.5	1.9	4.0	0.2	73.4
Wheat flour	12.6	14.5	3.7	2.7	2.3	64.2
Groundnut cake	10.0	42.0	7.3	13.0	2.5	25.2
Sunflower extract	8.0	31.0	2.1	18.4	1.5	39.0
Soybean meal	11.8	46.3	1.3	5.0	n/a	n/a
Rapeseed cake	11.0	35.9	0.9	13.2	6.9	32.1
Sesame cake	8.3	41.9	9.2	6.2	14.8	19.6
Mustard cake	8.5	30.8	9.3	6.2	10.3	34.9
Cottonseed cake	7.0	37.0	6.7	13.0	1.0	35.3
Gingely cake	9.0	34.0	7.8	7.9	3.1	38.2
Niger extract	7.0	35.0	2.0	19.0	3.5	33.5
Copra cake	12.0	22.0	6.5	12.2	5.2	42.1
Maize meal	13.5	9.5	4.0	4.0	1.5	67.5
Maize	10.4	4.6	7.8	3.5	1.0	72.7
Sorghum	10.0	9.0	2.8	3.0	0.1	75.1
Spirulina	8.7	50.5	1.0	2.1	11.0	26.7
Tapioca flour	11.5	3.1	2.3	2.0	2.3	78.8
Coffee pulp	2.3	14.0	1.2	20.8	8.2	43.5

n/a, not applicable.

A suspension diet was developed using rice bran, groundnut, squilla or small shrimp. The ingredients were mixed, cooked and passed through a 300–400 μm sieve and the filtrate was fed to carp spawn (hatchlings) several times a day (Alikunhi, 1987). In recent studies, leaf powders of mulberry and *Leucena leucocephala* and dried silkworm pupae were evaluated as effective ingredients for carp fry (Vijayakumaraswamy and Devaraj, 1994). Diets with graded protein levels for carp fry were developed using groundnut cake, rice bran and fish-meal as protein source and evaluated in the laboratory (Swamy *et al.*, 1988; Mohanty *et al.*, 1990). Mondal *et al.* (2000) have formulated diets with varied protein levels incorporating soy flour, fish-meal and rice bran as protein sources and evaluated them on catla fry. They obtained higher survival and food conversion when fish were fed a diet containing a 39% protein level.

A typical practical brood-stock diet formulation is given in Table 19.5. This diet contains 25–30% crude protein from locally available, less expensive ingredients and is fed at 2% of body weight day^{-1}. This diet helps in advancing the maturation and spawning of Indian major carps by 1–2 months. Further, it has resulted in increased fecundity and better seed quality.

Table 19.5. Proportion of different diet ingredients used in a typical brood-stock diet.

Ingredient	Percentage
Rice bran	25
Groundnut (peanut) cake	25
Fish-meal	10
Maize	10
Broken rice	10
Horse gram	10
Black gram	10
Vitamin and mineral mix	< 1

Several hormones and other products are used as diet additives or growth promoters in carp diets. Hormones are usually employed in the manipulation of sex in carps, while the same hormones incorporated at very low levels in diets are found to enhance the growth of carps. Hormones, such as 17α-methyltestosterone, when incorporated at 2.5–5 p.p.m. level in diets for common carp (Basavaraja *et al.*, 1989) and 1 p.p.m. for catla and rohu (Deb and Varghese, 1988), enhanced growth and survival. Other hormones, such as 19-norethisterone at a dietary level of 0.75 p.p.m. in rohu (Gangadhara *et al.*, 1998), diethylstilboestrol in common carp (Nanjundappa and Varghese, 1989) and human chorionic gonadotropic hormone at 5 p.p.m. in common carp (Keshavanath and Matty, 1994) and at 15 $\mu g\ g^{-1}$ diet in *Labeo fimbriatus* (Jayaprakas and Sherly, 1996), enhanced growth compared with that in control groups.

The most common method of feeding employed in carp culture in India is hand-feeding, particularly by small and artisanal carp farmers. Diet is fed once daily at a rate varying from 2% to 4% of the biomass. Semimoist diets of various sizes and shapes are broadcast manually over the pond surface. Another popular method of feeding carp is bag feeding. Diet is added to an empty plastic bag, several holes (1–2 cm diameter) being made at the bottom of the bag. The bags containing diet are suspended from bamboo poles, which have been driven into the pond bottom, so that the lower end of the bag is under water. The entire diet is consumed within 2–3 h. Empty bags are removed, washed and dried before reuse the next day.

References

Agarwal, N.K. and Mahajan, C.L. (1980) Nutritional difficiency disease in an Indian major carp, *Cirrhinus mrigala* (Hamilton) due to avitaminosis during early growth. *Journal of Fish Diseases* 3, 231–248.

Alikunhi, K.H. (1987) Technology of intensive rearing of carp spawn on processed diets. In: *USAID Training Programme on Freshwater Aquaculture, 14 July–13 August 1987.* Central Institute of Freshwater Aquaculture, Kausalyaganga, Bhubaneswar, India, pp. 351–359.

Basavaraja, N., Srikanta, C.K., Nandeesha, M.C. and Varghese, T.J. (1989) Impact of testosterone incorporated diets on growth, body composition and organoleptic quality of the common carp, *Cyprinus carpio* (Linn). *Journal of Aquaculture in the Tropics* 4, 21–28.

Bhat, G.K., Shetty, H.P.C. and Nandeesha, M.C. (1986) Formulation and evaluation of soybean meal and squilla meal based pelleted fish feeds. In: Maclean, J.L., Dzon, L.B. and Hosillos, L.V. (eds) *The First Asian Fisheries Forum, 25–31 May.* Asian Fisheries Society, Manila, the Philippines, pp. 539–541.

Chakraborthy, R.D. and Kar, S.L. (1975) Preliminary observations on the utility of petroleum protein as feed for carp spawn and fry. *Journal of Inland Fishery Society of India* 7, 116–119.

Chakraborthy, R.D., Sen, P.R. and Kowtal, G.V. (1973) Observation on the relative usefulness of different feeds for carp spawn and fry. *Journal of Inland Fishery Society of India* 5, 182–188.

Deb, D. and Varghese, T.J. (1988) Effect of 17 α-methyl testosterone on the growth and food conversion of major carps. In: Mohan Joseph, M. (ed.) *Proceedings of the First Indian Fisheries Forum.* Asian Fisheries Society, Indian Branch, Mangalore, pp. 69–71.

Devaraj, K.V., Krishna Rao, D.S. and Keshavappa, G.Y. (1981) Utilization of duckweed and waste cabbage leaves in the formulation of fish feed. *Mysore Journal of Agricultural Sciences* 15, 131–135.

Gangadhara, B., Nandeesha, M.C., Varghese, T.J. and Keshavanath, P. (1998) Effect of feeding 19-norethisterone on growth and body composition of rohu, *Labeo rohita. Asian Fisheries Science* 1, 51–58.

Hasan, M.R., Sahu, D.K., Das, P.M. and Hossain, M.A. (1993) Quantitative dietary ascorbic acid requirement of Indian major carp, *Labeo rohita* fingerlings. In: De Silva, S.S. (ed.) *Fish Nutrition Research in Asia. Proceedings of the 5th Asian Fish Nutrition Workshop.* Special Publication 9, Asian Fisheries Society, Manila, the Philippines, pp. 15–21.

Jayaprakas, V. and Sherly, D. (1996) Impact of HCG on growth responses of fringe-lipped carp, *Labeo fimbriatus* (Day). *Journal of Aquaculture in the Tropics* 3, 215–226.

Jayaram, M.G. and Shetty, H.P.C. (1980) Influence of different diets on the proximate body composition of *Catla catla, Labeo rohita,* and *Cyprinus carpio. Mysore Journal of Agricultural Sciences* 14, 381–384.

Jeychandran, P. and Paulraj, S. (1977) Formulation of pelleted feeds and feeding trials with common carp. *Journal of Inland Fishery Society in India* 9, 46–52.

Jhingran, V.G. (1991) *Fish and Fisheries of India,* revised and enlarged edition. Hindustan Publishing, New Delhi, 727 pp.

Keshavanath, P. and Matty, A.J. (1994) Growth promoting potential of human chorionic gonadotropin (HCG) in common carp. *Aquaculture and Fisheries Management* 26, 881–889.

Lall, S.P., Patterson, W.D., Hines, J.A. and Adams, N.J. (1985) Disease control through nutrition. *Journal of Fish Diseases* 8, 113–124.

Mahajan, C.L. and Agarwal, N.K. (1980a) Nutritional requirement of ascorbic acid by the Indian major carp, *Cirrhinus mrigala,* during early growth. *Aquaculture* 19, 37–48.

Mahajan, C.L. and Agarwal, N.K. (1980b) The role of vitamin C on calcium uptake by fish. *Aquaculture* 19, 287–294.

Mahajan, C.L. and Yadav, J.S. (1974) Experiment on synthetic diets for carps – some preliminary observations and their relevance to aquaculture. *Journal of the Inland Fishery Society in India* 6, 178–184.

Manissery, Joseph, K., Narayana Gowda, B.S., Shankar, K.M. and Varghese, T.J. (1988) Growth of common carp, *Cyprinus carpio* fed on food incorporated with *Cassia tora* leaf. *Indian Journal of Animal Science* 58(6), 737–740.

Mohanty, S.N., Swamy, D.N. and Tripathi, S.D. (1990) Growth, nutritional indices and carcass composition of the Indian major carps fry, *Catla catla* and *Cirrhinus mrigala*, fed four dietary protein levels. *Aquaculture Hungarica* 6, 211–217.

Mondal, A.M., Murthy, H.S. and Satheesha, S.K. (2000) Effect of soya flour based diets and varied protein levels on the growth and survival of Indian major carp, *Catla catla* fry in rearing ponds. *Mysore Journal of Agricultural Sciences* 34, 257–262.

Murthy, H.S. and Devaraj, K.V. (1990) Effect of *Echhornia* based diet on the growth of carps. In: *Second Indian Fisheries Forum Proceedings*. Asian Fisheries Society, Indian Branch, Mangalore, India, pp. 9–11.

Murthy, H.S. and Devaraj, K.V. (1991a) Utility of pistia (*Pistia stratiotes*) in the diet of carps. *Journal of Aquaculture in the Tropics* 6, 9–14.

Murthy, H.S. and Devaraj, K.V. (1991b) Comparison of growth of carps fed *Salvinia* based feed and conventional feed. *Fishery Technology* 28, 106–110.

Murthy, H.S. and Varghese, T.J. (1995) Arginine and histidine requirements for the Indian major carp, *Labeo rohita* (Hamilton). *Aquaculture Nutrition* 1, 235–239.

Murthy, H.S. and Varghese, T.J. (1996a) Quantitative dietary isoleucine requirement for growth and survival of Indian major carp, *Labeo rohita* (Hamilton) fry. *Indian Journal of Experimental Biology* 34, 1141–1143.

Murthy, H.S. and Varghese, T.J. (1996b) Dietary requirements of the Indian major carp, *Labeo rohita* for total aromatic amino acids. *Israeli Journal of Aquaculture – Bamidgeh* 48(2), 78–83.

Murthy, H.S. and Varghese, T.J. (1996c) Quantitative dietary requirement of threonine for growth in the Indian major carp, *Labeo rohita* (Hamilton). *Journal of Aquaculture in the Tropics* 11, 1–7.

Murthy, H.S. and Varghese, T.J. (1997a) Dietary requirement of the Indian major carp, *Labeo rohita* for the essential amino acid leucine. *Mysore Journal of Agricultural Sciences* 31, 348–353.

Murthy, H.S. and Varghese, T.J. (1997b) Dietary requirements of juveniles of the Indian major carp, *Labeo rohita* for essential amino acid lysine. *Israeli Journal of Aquaculture – Bamedgeh* 49(1), 19–24.

Murthy, H.S. and Varghese, T.J. (1997c) Tryptophan requirement of juveniles of the Indian major carp, *Labeo rohita*. *Journal of Applied Aquaculture* 7, 71–79.

Murthy, H.S. and Varghese, T.J. (1997d) Quantitative dietary requirements of the Indian major carp, *Labeo rohita* for the essential amino acid valine. *Indian Journal of Animal Sciences* 67(11), 1028–1030.

Murthy, H.S. and Varghese, T.J. (1998) Total sulphur amino acid requirement of the Indian major carp, *Labeo rohita* (Hamilton). *Aquaculture Nutrition* 4, 61–65.

Nandeesha, M.C., Basavaraja, N., Keshavanath, P., Varghese, T.J. and Shetty, H.P.C. (1989) Influence of soybean meal and squilla meal based diets enriched with sardine oil on the growth and organoleptic quality of common carp, *Cyprinus carpio* (Linn.). *Biological Wastes* 30, 61–69.

Nanjundappa, T. and Varghese, T.J. (1989) Effect of diethylstilbestrol on growth and food conversion of common carp, *Cyprinus carpio* (L.). *Proceedings of Indian Academy of Sciences (Animal Sciences)* 93(2), 85–88.

Ogino, C. and Takeda, H. (1976) Mineral requirements of fish. III. Calcium and phosphorus requirements in carp. *Bulletin of Japanese Society Scientific Fisheries* 42, 793–799.

Patnaik, S. and Das, K.M. (1979) Utilization of some aquatic weeds as feed for rearing carp spawn and fry. In: *Symposium on Inland Aquaculture (Abstracts)*. CIFRI, Barrackpore, India, p. 116.

Ravi, J. and Devaraj, K.V. (1991) Quantitative essential amino acid requirements for growth of catla, *Catla catla* (Hamilton). *Aquaculture* 96, 281–291.

Renukaradhya, K.M. and Varghese, T.J. (1986) Protein requirement of the carps, *Catla catla* (Hamilton) and *Labeo rohita* (Hamilton). *Proceedings of Indian Academy of Sciences (Animal Sciences)* 95(1), 105–107.

Satheesha, S.K. and Murthy, H.S. (1999) Dietary requirement of the Indian major carp, *Catla catla* for the essential amino acid lysine. *Indian Journal of Animal Science* 69(4), 277–279.

Satheesha, S.K. and Murthy, H.S. (2000) Dietary requirement of the Indian major carp, *Catla catla* juveniles for the essential amino acid tryptophan. *Mysore Journal of Agricultural Sciences* 34, 263–267.

Sen, P.R., Rao, N.G.S., Ghosh, S.R. and Rout, M. (1978) Observation on protein and carbohydrate requirements of carps. *Aquaculture* 13, 245–255.

Singh, B. and Bhanot, K.K. (1988) Observation on the protein requirement of *Catla catla* (Ham.) fry. In: Mohan Joseph, M. (ed.) *The First Indian Fisheries Forum Proceedings*. Asian Fisheries Society, Indian Branch, Mangalore, pp. 77–78.

Sobana, K.S. (1997) Modulatory effects of vitamin C on disease resistance mechanism of Indian major carps. PhD thesis, University of Agricultural Sciences, Bangalore, India.

Swamy, D.N. and Mohanty, S.N. (1988) Response of rohu (*Labeo rohita* Ham.) fingerlings to the diets of different protein contents. In: *Asian Third Fish Nutrition Workshop, Held at Asian Institute of Technology, Bangkok*. Asian Fisheries Society (Indian Branch), Mangalore, India, pp. 81–83.

Varghese, T.J., Devaraj, K.V., Santharam, B. and Shetty, H.P.C. (1976) Growth response of the common carp (*Cyprinus carpio* var. *communis*) to protein rich pelleted feed. In: *Symposium on Development and Utilization of Inland Fishery Resources*, Colombo, Sri Lanka. FAO Regional Office for Asia and Far East, Bangkok, Thailand, pp. 408–416.

Venugopal, M.N. and Keshavanath, P. (1984) Influence of supplementary feeds on the biochemical composition of flesh of freshwater carps, *Catla catla, Cirrhinus mrigala* and *Cyprinus carpio*. *Indian Journal of Animal Science* 54, 555–559.

Vijayakumaraswamy, H.V. and Devaraj, K.V. (1994) Growth response of fry of *Catla catla* and the three formulated feeds. *Environment and Ecology* 12(3), 519–523.

Tilapia, *Oreochromis* spp.

20

Shi-Yen Shiau
Department of Food Science, National Taiwan Ocean University, Keelung, Taiwan 202, Republic of China

Introduction

Tilapias are mainly lacustrine fish and are well adapted to enclosed water. They are fast-growing, resistant to disease and handling, easy to reproduce in captivity and able to tolerate a wide range of environmental conditions. They are widely cultured in tropical and subtropical regions of the world and constitute the third largest group of farmed finfish, with an annual production growth rate of about 11.5% (El-Sayed, 1999). Global production of farmed tilapia has increased more than threefold since 1984, from 186,544 Mt to 659,000 Mt, representing 4.48% of total farmed finfish in 1995 (Tacon, 1997).

Most cultured tilapias are grouped into two genera (Trewavas, 1982): *Tilapia*, which are macrophagous and substrate spawners, and *Oreochromis*, which are microphagous and mouth-breeders. The main cultured species are *Oreochromis niloticus*, *Oreochromis aureus*, hybrid *O. niloticus* × *O. aureus*, *Oreochromis mossambicus* and *Tilapia zillii*. For the mouth-breeders of the *Oreochromis* genus, there exists a marked delay in the growth of females and monosex culture of males is preferred. Thus, it is important to indicate the sex of the cultured fish, and whether sex control was obtained through manual sorting, sex-reversal by hormone treatment or hybridization.

As the industry expands and technology development continues, traditional extensive culture of tilapia is being replaced by semi-intensive and intensive production systems. Natural pond organisms are the only source of nutrients for the growth of fish in extensive culture. In semi-intensive farming systems, supplemental diets, which consist of locally available, low-cost single foodstuffs, such as rice bran, maize meal, copra meal, coffee pulp, brewery by-products and/or their combination, are generally used as supplements to natural food (Lim, 1989). As stocking rates increase, the contribution of natural food decreases and more nutritionally complete diets are needed. In semi-intensive and intensive culture systems, diet is the most expensive cost item, often ranging from 30% to 60% of the total variable expenses, depending on the intensity of the culture operation. Thus, the use of least-cost, nutritionally balanced diets and good feeding management are two most important requisites for successful fish production.

Nutrient Requirements

Protein and amino acids

Proteins are indispensable nutrients for the structure and function of all living organisms, including tilapia. Since proteins are continually being used for maintenance, growth and reproduction, a continuous supply of proteins or their component amino acids is needed. Fish do not have a true protein requirement, but need a well-balanced mixture of indispensable and dispensable amino acids. Inadequate intake of protein will result in retardation or cessation of growth due to the withdrawal of protein from less vital tissues to maintain the function of more vital ones. If too much protein is supplied, however, only part will be used to synthesize new tissues and the remainder will be converted to energy (NRC, 1983).

Many investigators have used semipurified and purified diets to estimate the optimum dietary protein level for tilapia. Most of these values have been derived from dose–response curves yielding the minimum of dietary protein necessary for maximal growth. The optimum dietary protein level for tilapia appears to be influenced by the size or age of the fish and ranges from 28% to 50% (Table 20.1). For fry (first feeding to approximately 0.5 g), dietary protein levels ranging from 36 to 50% have been shown to produce maximum growth (Davis and Stickney,

Table 20.1. Protein requirements of tilapia cultured in fresh water.

Species	Fish size (g)	Requirement (%)	Reference
O. mossambicus	1.0–2.5	29–38	Cruz and Laudencia (1977)
	Fry	50	Jauncey and Ross (1982)
	0.5–1.0	40	
	6–30	30–35	
	1.8	40	Jauncey (1982)
O. niloticus	1.5–7.5	36	Kubaryk (1980)
	3.2, 3.7	30	Wang *et al.* (1985)
	0.838	40	Siddiqui *et al.* (1988)
	40	30	
	24	27.5–35	Wee and Tuan (1988)
	0.012	45	El-Sayed and Teshima (1992)
O. aureus	0.3–0.5	36	Davis and Stickney (1978)
	0.16	40	Santiago and Laron (1991)
Tilapia zillii	1.7	35–40	Teshima *et al.* (1978)
	1.65	35	Mazid *et al.* (1979)
O. niloticus × *O. aureus*	0.6–1.1	32	Shiau and Peng (1993)
	21	28	Twibell and Brown (1998)

1978; Jauncey and Ross, 1982; Santiago and Laron, 1991; El-Sayed and Teshima, 1992); for juveniles (approximately 0.5–5 g), 29–40% has been determined to produce optimal growth (Cruz and Laudencia, 1977; Teshima *et al.*, 1978; Mazid *et al.*, 1979; Kubaryk, 1980; Jauncey, 1982; Jauncey and Ross, 1982; Wang *et al.*, 1985; Siddiqui *et al.*, 1988; Shiau and Peng, 1993); and, for young adult fish (up to 40 g), 27.5–35% appears to be optimal (Jauncey and Ross, 1982; Siddiqui *et al.*, 1988; Wee and Tuan, 1988; Twibell and Brown, 1998).

Practical diets for grow-out of tilapias usually contain 25–35% crude protein. In ponds, however, fish may have access to natural food that is rich in protein; thus dietary protein levels as low as 20–25% have been estimated to be adequate (Newman *et al.*, 1979; Lovell, 1980; Wannigama *et al.*, 1985). Water salinity appears to affect the protein requirements, being lower at higher salinity (Table 20.2).

Tilapias require the same ten essential amino acids (arginine, histidine, isoleucine, leucine, lysine, methionine, phenylalanine, threonine, tryptophan and valine) as other fish and terrestrial animals. The quantitative requirements for these essential amino acids for growth of young Nile tilapia (*O. niloticus*) determined by Santiago and Lovell (1988) are presented in Table 20.3.

Non-essential amino acids can be adequately synthesized by fish but their presence in the diet has nutritional significance because it reduces the need for fish to synthesize them. Two specific examples of sparing action are the conversion of methionine to cystine and phenylalanine to tyrosine. These non-essential amino acids can only be synthesized from the essential amino acid precursor (NRC, 1983). Tilapias actually have a requirement for sulphur-containing amino acids, which can be met by either methionine alone or the proper mixture of methionine and cystine. Dietary cystine can replace up to 50% of the total sulphur-containing amino acid requirement for *O. mossambicus* (Jauncey and Ross, 1982). A similar relationship exists between aromatic amino acids. The presence of tyrosine in the diet will reduce some of the requirement for phenylalanine (NRC, 1993).

In general, protein sources having an essential amino acid profile that closely matches the essential amino acid requirements are likely to have a high nutritional value. Fish-meal is used as the main conventional protein source in aquaculture diets. Attempts have been made to partially or totally replace

Table 20.2. Protein requirements of tilapia at different salinities.

Species	Fish size (g)	Salinity (p.p.t.)	Requirement (%)	Reference
O. niloticus	0.024	0	30.4	De Silva and Perera (1985)
		5	30.4	
		10	28.0	
		15	28.0	
O. niloticus × *O. aureus*	2.88	32–34	24.0	Shiau and Huang (1989)
Florida red tilapia	10.6	37	20.0	Clark, A.E. *et al.* (1990)

Table 20.3. Essential amino acid requirements of tilapia.

Amino acids	Requirement (% of dietary protein)*
Arginine	4.20
Histidine	1.72
Isoleucine	3.11
Leucine	3.39
Lysine	5.12
Methionine	2.68†
Phenylalanine	3.75‡
Threonine	3.75
Tryptophan	1.00
Valine	2.80

* Data resulting from growth experiments (Santiago and Lovell, 1988).
† Total sulphur amino acid (methionine plus cystine) requirement is 3.21% of the protein.
‡ Total aromatic amino acid (phenylalanine plus tyrosine) requirement is 5.54% of the protein.

fish-meal with less expensive, locally available protein sources for tilapia. Soybean meal has been studied extensively, with various degree of success. Factors such as antinutritional factors (Tacon, 1993), limiting amino acids (Jackson *et al.*, 1982; Tacon *et al.*, 1983; Viola and Arieli, 1983; Teshima and Kanazawa, 1988; Shiau *et al.*, 1989a), minerals (Viola *et al.*, 1986, 1988) and dietary protein levels (Shiau *et al.*, 1987, 1989a; Viola *et al.*, 1994a,b) have been discussed. A comprehensive review of alternative dietary protein sources, including animal by-products, oil-seed plants, aquatic plants, single-cell proteins, grain legumes and plant protein concentrates, for tilapia is available (El-Sayed, 1999).

Energy

Energy is not a nutrient but is a property of nutrients that are released during the metabolic oxidation of proteins, carbohydrates and lipids. Generally, protein is given the first priority in formulating fish diets because it is the most expensive component of the prepared diets. However, energy should be the first nutritional consideration in diet formulation since tilapias, like other fish, eat to satisfy their energy needs. If insufficient non-protein energy is available, part of the protein is used as an energy source. However, excess energy may produce fatty fish, reduced diet consumption (reducing total protein intake) and inhibit proper utilization of other feedstuffs. Kubaryk (1980) observed that, as the digestible energy (DE) content in the diet increased, consumption by *O. niloticus* decreased, but the amount of protein in the diet did not affect consumption.

Available data on energy requirements of tilapias have been reported in terms of gross energy (GE), DE or metabolizable energy (ME) in relation to the

Table 20.4. Optimum protein-to-energy ratio of tilapia.

Species	Fish size (g)	Optimum P/E ratio	Reference
O. mossambicus	1.80	116.6 mg $kcal^{-1}$ ME	Jauncey (1982)
	5.19	99.48 mg $kcal^{-1}$ DE	El-Dahhar and Lovell (1995)
O. niloticus	0.012	110 mg $kcal^{-1}$ GE	El-Sayed and Teshima (1992)
	1.7	120 mg $kcal^{-1}$ DE	Kubaryk (1980)
O. aureus	2.5	123 mg $kcal^{-1}$ DE	Winfree and Stickney (1981)
	7.5	108 mg $kcal^{-1}$ DE	Winfree and Stickney (1981)
Tilapia zillii	1.65	95.3 mg $kcal^{-1}$ DE	Mazid *et al.* (1979)
	50	103 mg $kcal^{-1}$ DE	El-Sayed (1987)
O. niloticus × *O. aureus*	0.16	111 mg $kcal^{-1}$ DE	Santiago and Laron (1991)

dietary level of protein (P) (Table 20.4). Generally, the dietary protein-to-energy ratio required for maximum growth decreases with increasing size of fish. For fry (to about 0.5 g) and juveniles (from 0.5 to 5 g), a protein/energy (P/E) ratio ranging from 95.3 to 123 mg protein $kcal^{-1}$ DE (Mazid *et al.*, 1979; Kubaryk, 1980; Winfree and Stickney, 1981; Jauncey, 1982; Santiago and Laron, 1991; El-Sayed and Teshima, 1992; Table 20.4) and, for adult fish (5–50 g), 99.48–108 mg protein $kcal^{-1}$ DE appears to be optimal (Winfree and Stickney, 1981; El-Sayed, 1987; El-Dahhar and Lovell, 1995). Shiau and Huang (1990) reported maximum growth of *O. niloticus* × *O. aureus* (1.60 g) reared in sea water (32–34 p.p.t. salinity) when fed a 21% and 24% protein diet with P/ME ratio of 67.74 and 104.35 mg protein $kcal^{-1}$.

Lipids and fatty acids

Protein is responsible for a large part of the cost of most prepared diets. The protein fraction should be utilized for growth rather than serve as an energy source for fish. Knowledge of the protein-sparing effects of non-protein nutrients such as lipids is necessary and should be used to reduce diet costs and maximize nitrogen retention. Moreover, dietary lipids are the only source of essential fatty acids needed by fish for normal growth and development. They are also important carriers and assist in the absorption of fat-soluble vitamins. Lipids, especially phospholipids, are important for cellular structure and maintenance of membrane flexibility and permeability. Lipids also serve as precursors of steroid hormones and prostaglandins, improve the flavour of diets and affect the diet texture and fatty acid composition of fish.

Studies on *T. zillii* (Teshima *et al.*, 1978; El-Sayed and Garling, 1988) indicated that increasing dietary lipid up to 15% resulted in a significant improvement in protein efficiency ratio (PER) and protein production value (PPV). Similar results were reported by Teshima *et al.* (1985) with *O. niloticus*. Hanley (1991) indicated that tilapia were able to store significant quantities of lipid in their carcass and viscera, but could not utilize this energy source to improve growth. However, this study was conducted outdoors, where natural productivity in the culture system may have influenced the results. Dietary lipids have been shown to have a sparing effect on the utilization of dietary protein. The level of protein in the diet of *O. niloticus* can be reduced from 33.2% to 25.7% by increasing dietary lipid from 5.7% to 9.4% and carbohydrate from 31.9% to 36.9% (Li *et al.*, 1991). A dietary lipid level in excess of 12% depressed growth of juvenile *O. aureus* × *O. niloticus* hybrids (Jauncey and Ross, 1982). The growth of *O. aureus* can be substantially improved when menhaden oil or fish-oil is provided at 7.5–10% of the diet as compared with lower levels. However, best performance was obtained with menhaden oil at 10% of the diet (Stickney and Wurts, 1986). The optimal dietary lipid level for tilapia was quantified by Chou and Shiau (1996). Five isoenergetic and isonitrogenous purified diets containing 0–20% lipid (maize oil, cod-liver oil and pork lard at a 1 : 1 : 1 ratio) in 5% increments were fed to the juvenile tilapia hybrid, *O. niloticus* × *O. aureus*. Results indicated that 5% dietary lipid appeared to be sufficient to meet the minimal requirement of juvenile tilapia, but a level of 12% was needed for maximal growth.

Cold-water fish have a higher requirement for n-3 polyunsaturated fatty acid (PUFA), whereas warm-water fish tend to require greater quantities of n-6 fatty acids. Several studies have reported the fatty acid requirements of various tilapia. Kanazawa *et al.* (1980a) indicated that *T. zillii* is unable to convert much 18:2n-6 to 20:4n-6. However, Kanazawa *et al.* (1980b) suggested that this same species probably does convert dietary 18:2n-6 to 20:4n-6. Stickney and McGeachin (1985) indicated that growth of blue tilapia, *O. aureus*, was not affected by dietary linoleic acid levels as high as 2%. When blue tilapia were fed diets containing soybean oil, growth improved as the percentage of linoleic acid increased (Stickney *et al.*, 1985). Takeuchi *et al.* (1983) found that the growth of the tilapia *O. niloticus* was significantly reduced with a fish-oil-containing diet (pollock liver oil) as compared with maize oil or soybean oil diets. Ramachandran Nair and Gopakumar (1981) reported unusually high levels of 22:6n-3 in eggs of *O. mossambica*, suggesting some important role of docosahexaenoic acid in reproduction and embryo development. However, Santiago and Reyes (1993) indicated that, although a fish-oil (cod-liver oil) high in 22:6n-3 promoted the highest weight gain in Nile tilapia, this same lipid resulted in the poorest reproductive performance of this species. Blue tilapia have been reported to grow well on practical diets containing either 10% soybean oil, which is high in 18:2 n-6, or 10% menhaden oil, which is high in 20:5n-3 and 22:6n-3 (Stickney and McGeachin, 1983). Stickney and Hardy (1989), however, reported that *O. aureus* have a requirement for a relatively high level of n-6 fatty acids, though the requirement can be reduced when n-3 fatty acids are present. Stickney and Wurts (1986) compared the growth of blue tilapia fed diets containing graded levels of catfish and menhaden oil, and the best growth was obtained with the 10% menhaden-oil diet. These studies gave contradictory results as to the requirement of tilapia for dietary n-3 and n-6 PUFA. Recently, Chou and Shiau (1999) demonstrated that both n-3 and n-6 highly unsaturated fatty acids are essential for maximum growth of hybrid tilapia (*O. niloticus* × *O. aureus*). Further quantification of essential fatty acid requirements for tilapia species and hybrids is needed.

Carbohydrates

Carbohydrates are the least expensive form of dietary energy for human and domestic animals, but their utilization by fish varies and remains somewhat obscure. Fish in general utilize dietary carbohydrate poorly. No dietary requirement of carbohydrate has been demonstrated in fish. It has been reported that the utilization of starch was significantly higher than that of glucose by tilapia (Anderson *et al.*, 1984; Tung and Shiau, 1991; Shiau and Chen, 1993; Shiau and Lin, 1993). It has also been reported that tilapia utilizes disaccharides better than glucose but more poorly than starch. Of the disaccharides, maltose was better utilized, followed by sucrose and lactose (Shiau and Chuang, 1995). Lin and Shiau (1995) reported that malic enzyme, glucose-6-phosphate dehydrogenase (G-6-PD) and phosphogluconate dehydrogenase (PGD) activities were higher in tilapia fed the starch diet than in fish fed the glucose diet. Changing the diet from starch to glucose decreased malic enzyme, G-6-PD and PGD activities, whereas changing the diet from glucose to starch increased these enzyme activities in fish liver. The authors suggest that lipogenic enzyme activity in tilapia liver can adapt to dietary carbohydrates.

Tung and Shiau (1991) have studied the effect of daily feeding frequency on the utilization of carbohydrate by tilapia. Diets containing 44% of glucose, dextrin and starch were fed either twice or six times a day. The authors found that fish fed all carbohydrate sources six times a day had significantly higher weight gain, diet efficiency, protein deposition and energy retention than fish fed these diets twice a day. This study also demonstrated that, as the feeding frequency increased from twice a day to six times a day, fish fed the starch and dextrin diets gained more weight than fish fed the glucose diet, but the differences were not statistically significant. It was also noted that the growth performance of tilapia fed the glucose diet six times a day was higher than that of fish fed the starch or dextrin diets twice a day. The effect of feeding strategy on carbohydrate utilization by tilapia has recently been confirmed by Shiau and Lei (1999), when they re-evaluated the effects of continuous feeding and two meals a day on carbohydrate utilization by tilapia. Results indicated that continuous feeding improves carbohydrate utilization.

Shiau *et al.* (1988) studied the influence of carboxymethylcellulose (CMC) on growth, digestion and gastric emptying time of tilapia. In their studies, CMC was incorporated at 2, 6, 10 and 14% of the diet. Dextrin was used as the carbohydrate source. Both weight gain and diet conversion ratio decreased as the dietary CMC level increased. Shiau *et al.* (1989b) studied the effects of five different dietary fibres (guar gum, agar, carrageenan, CMC and cellulose) on the utilization of dextrin and glucose in tilapia. Each fibre was added at 10% of the diet. Results indicated that the weight gain was significantly lower for tilapia fed fibre-containing diets than those fed the dextrin or glucose diet. The intestinal absorption of carbohydrate and the blood-sugar content were low in fish fed diets containing fibre, regardless of source. Recently, Shiau and Yu (1999) demonstrated that dietary supplementation of chitin or chitosan depresses growth in tilapia, regardless of the supplementation level (2, 5 and 10%).

Niacin is a precursor of certain coenzymes needed for carbohydrate metabolism. Shiau and Suen (1992) demonstrated that the amount of niacin needed in tilapia varies with the sources of dietary carbohydrate. They found that the adequate dietary niacin level for maximal growth of tilapia fingerlings was 26 mg kg^{-1} diet in fish fed the glucose diet and 121 mg kg^{-1} diet in fish fed the dextrin diet, indicating that the dietary niacin requirement for tilapia is affected by the type of dietary carbohydrates.

Shiau and Peng (1993) conducted a study to evaluate the possible protein-sparing effects of carbohydrates in tilapia. Three dietary protein levels (32, 28, 24%) were achieved by substitution with three levels (33, 37, 41%) and three sources (glucose, dextrin, starch) of dietary carbohydrates. Results indicated that the protein-sparing effect of carbohydrates (dextrin or starch) only occurred when the dietary protein level was suboptimal. Shiau and Chuang (1996) also reported that carbohydrate utilization by tilapia is affected by dietary protein source.

Other nutrients, such as chromium, have also been reported to affect carbohydrate utilization by fish. Shiau and Lin (1993) first demonstrated that dietary chromium supplementation in the form of chromium chloride significantly improved glucose but not starch utilization by tilapia. Shiau and Chen (1993) later reported that the improvement of glucose utilization by tilapia fed diets supplemented with chromic oxide was markedly higher than with other forms of chromium, including chromium chloride. Shiau and Liang (1995) reported that the level of chromic oxide in the diet alters glucose utilization by tilapia. Fish fed a glucose diet with 5000 mg chromic oxide kg^{-1} diet had a higher weight gain than fish fed a glucose diet with 20,000 mg chromic oxide kg^{-1} diet. They speculated that the poorer growth performance of tilapia fed the higher level of chromic oxide may be due to a toxic effect of dietary chromium. More recently, Shiau and Shy (1998) reported that maximum growth and glucose utilization in hybrid tilapia were achieved with a glucose diet containing 204.4 mg chromic oxide kg^{-1} diet.

Lin *et al.* (1995) indicated that plasma insulin concentration in tilapia ingested with glucose was higher than that ingested with starch and that the plasma insulin pattern of tilapia was similar to that of humans. However, the exact mechanism of insulin in the utilization of carbohydrate by tilapia is still not clear.

Ammonia excretion and oxygen consumption of tilapia are affected by the type of carbohydrate ingested. Shiau and Cheng (1999) fed tilapia diets containing three levels (33, 37 and 41%) of carbohydrates (starch and glucose) for 8 weeks. Fish fed a starch diet had lower total ammonia excretion than fish fed a glucose diet at all three carbohydrate levels. Ammonia excretion in the glucose-fed fish decreased as the carbohydrate inclusion level increased. In the starch-fed fish, ammonia excretion was significantly lower at the 41% inclusion level than at the 33 and 37% inclusion levels. Total oxygen consumption was significantly higher in the starch-fed fish than in the glucose-fed fish. In both the glucose- and starch-fed fish, higher oxygen consumption was observed at the 37 and 41% inclusion level than at the 33% inclusion level.

Biotic factors, such as physiological stages, may also influence the utilization of carbohydrates by tilapia. Tung and Shiau (1993) compared the utilization of two carbohydrate sources (glucose, starch) by two sizes (4.55, 0.46 g) of tilapia. They found that the larger fish gained significantly more body weight and had a better diet conversion ratio, protein and energy deposition than the smaller fish when fed a glucose diet. However, these parameters were similar in the larger and smaller fish when the starch diet was fed.

Vitamins and minerals

Vitamin supplements are often not included in practical diets for tilapias stocked at moderate densities in fertilized ponds. In intensive systems where limited or no natural foods are available, supplemental vitamins must be added.

Table 20.5 presents the vitamin requirements of various tilapia. Water-soluble vitamin requirements of tilapias that have been studied are:

Table 20.5. Vitamin requirements of various tilapia.

	Requirement (mg kg^{-1} diet)				
Vitamins	*O. niloticus* × *O. aureus*	*O. aureus*	*O. niloticus*	*O. mossambicus* × *O. niloticus*	*O. spilurus*
Thiamine	–	–	–	2.5[a]	–
Riboflavin	–	6[b]	–	5[c]	–
Pyridoxine	1.7–9.5 (28% C.P.)[d] 15–16.5 (36% C.P.)[d]	–	–	3[e]	–
Vitamin B_{12}	Dispensable[f]	–	Dispensable[g,h]	–	–
Niacin	121[i]	–	–	–	–
Biotin	0.06[j]	–	–	–	–
Folic acid	–	–	–	–	–
Inositol	–	–	–	–	–
Choline	1000[k]	Dispensable[l,m]	–	–	–
Pantothenic acid	–	6–10[n,o]	–	–	–
Ascorbic acid	79 (C1)[p] 41–48 (C2S)[t] 37–42 (C2MP-Mg)[t] 63.4 (C2MP-Na)[u]	50 (Cl)[q]	420 (C1)[r]	–	100–200 (C2S)[s]
A	–	–	–	–	–
D	0.00937 (374.8 IU kg^{-1})[v]	Dispensable[w]	–	–	–
E	42–44 (5% lipid)[x] 60–66 (12% lipid)[x]	10 (3% lipid)[y] 25 (6% lipid)[y]	50–100 (5% lipid)[z] 500 (10~15% lipid)[z]	–	–
K	–	–	–	–	–

[a]Lim and LeaMaster (1991); [b]Soliman and Wilson (1992b); [c]Lim *et al.* (1993); [d]Shiau and Hsieh (1997); [e]Lim *et al.* (1995); [f]Shiau and Lung (1993); [g]Lovell and Limsuwan (1982); [h]Sugita *et al.* (1990); [i]Shiau and Suen (1992); [j]Shiau and Chin (1999); [k]Shiau and Lo (2000); [l]Roem *et al.* (1990a); [m]Roem *et al.* (1990b); [n]Roem *et al.* (1991); [o]Soliman and Wilson (1992a); [p]Shiau and Jan (1992); [q]Stickney *et al.* (1984); [r]Soliman *et al.* (1994); [s]Al-Amoudi *et al.* (1992); [t]Shiau and Hsu (1995); [u]Shiau and Hsu (1999); [v]Shiau and Hwang (1993); [w]O'Connell and Gatlin (1994); [x]Shiau and Shiau (2001); [y]Roem *et al.* (1990c); [z]Satoh *et al.* (1987).

thiamine, riboflavin, pyridoxine, vitamin B_{12}, niacin, biotin, choline, pantothenic acid and ascorbic acid. Thiamine deficiency signs observed in red hybrid (*O. mossambicus* × *O. niloticus*) fingerlings cultured in sea water (32 p.p.t. salinity) were reduced growth and diet efficiency and a low haematocrit. A dietary thiamine level of 2.5 mg kg^{-1} diet was sufficient for maximum growth and prevention of deficiency signs (Lim and LeaMaster, 1991). Typical deficiency signs reported for tilapias fed a riboflavin-free diet were anorexia, poor growth, high mortality, fin erosion, loss of normal body colour, short body dwarfism and lens cataracts. The dietary riboflavin requirements were 6 mg kg^{-1} diet for juvenile *O. aureus* grown in fresh water (Soliman and Wilson, 1992b) and 5 mg kg^{-1} diet for red tilapia (*O. mossambicus* × *O. niloticus*) grown in 32 p.p.t. sea water (Lim *et al.*, 1993).

Fish fed a diet without pyridoxine supplementation developed abnormal neurological signs, anorexia, convulsions, caudal fin erosion, mouth lesions, poor growth and high mortality. Weight gain and hepatic alanine aminotransferase activity analysed by broken-line regression indicated that the optimum dietary pyridoxine requirements in juvenile *O. niloticus* × *O. aureus* reared in fresh water were 1.7–9.5 mg kg^{-1} diet and 15.0–16.5 mg kg^{-1} diet containing 28% and 36% protein, respectively (Shiau and Hsieh, 1997). A dietary pyridoxine level of 3 mg kg^{-1} diet was reported to be adequate for *O. mossambicus* × *O. niloticus* fed 38% protein diets and reared in sea water. Tilapias produced vitamin B_{12} in their gastrointestinal tract through bacterial synthesis and did not have a dietary requirement for this vitamin (Lovell and Limsuwan, 1982; Sugita *et al.*, 1990; Shiau and Lung, 1993).

Niacin is a dietary essential for the *O. niloticus* × *O. aureus* hybrid but the level required varies depending on the source of dietary carbohydrate. Optimum dietary levels for maximum growth have been reported to be 26 mg kg^{-1} diet for fish fed a glucose diet and 121 mg kg^{-1} diet for fish fed a dextrin diet. Fish deprived of dietary niacin developed haemorrhages, a deformed snout, gill oedema and skin, fin and mouth lesions (Shiau and Suen, 1992). The dietary biotin requirement for maximal growth of *O. niloticus* × *O. aureus* has been estimated to be 0.06 mg kg^{-1} diet (Shiau and Chin, 1999). While it was suggested that a high level of dietary methionine may have partially satisfied choline requirement in *O. aureus* (Roem *et al.*, 1990a,b), Shiau and Lo (2000) recently established that the optimal dietary choline requirement for *O. niloticus* × *O. aureus* was 1000 mg kg^{-1} diet.

Pantothenic acid deficiency causes poor growth, haemorrhage, sluggishness, high mortality, anaemia and severe hyperplasia of the epithelial cells of gill lamellae in *O. aureus*. A dietary level of 10 mg of calcium-pantothenate kg^{-1} diet was sufficient to prevent these deficiency signs (Soliman and Wilson, 1992a). Roem *et al.* (1990b) reported that *O. aureus* could satisfy their requirement for pantothenic acid by feeding on bacteria in a recirculating system.

Tilapias showed classical vitamin C deficiency signs when fed a vitamin-deficient diet in the absence of natural foods. The requirement for normal growth of *O. aureus* has been reported to be 50 mg ascorbic acid kg^{-1} diet (Stickney *et al.*, 1984). In juvenile *O. niloticus* × *O. aureus*, a dietary level of 79 mg ascorbic acid

kg^{-1} diet is needed for maximum growth (Shiau and Jan, 1992). A recommended dietary vitamin C inclusion level for *O. niloticus* is 420 mg kg^{-1} diet (Soliman *et al.*, 1994). In all these studies, L-ascorbic acid was used as the source of vitamin C. L-Ascorbic acid (C1) is unstable and most of its activity in practical diets is lost during processing and storage. Shiau and Hsu (1993) found that about 75% of the initial amount of supplemented C1 in aquatic diets can be lost during processing and storage at ambient temperature. Because of the unstable nature of C1, the use of more stable forms of ascorbic acid is necessary. Several derivatives of C1 have been shown to have antiscorbutic activity for tilapia. L-Ascorbyl-2-sulphate (C2S) has been shown to have equal antiscorbutic activity to L-ascorbyl-2-monophosphate-magnesium (C2MP-Mg) for *O. niloticus* × *O. aureus*. The optimum dietary level is 41–48 mg of C2S kg^{-1} diet and 37–42 mg of C2MP-Mg kg^{-1} diet (Shiau and Hsu, 1995). The optimum dietary L-ascorbyl-2-monophosphate-sodium (C2MP-Na) for *O. niloticus* × *O. aureus* is 63.4 mg kg^{-1} diet. It was suggested that C2MP-Mg is about 85% as effective as C2MP-Na in meeting the vitamin C requirement for tilapia (Shiau and Hsu, 1999). The requirement for optimum growth of *Oreochromis spilurus* in sea water has been determined to be between 100 to 200 mg C2S kg^{-1} diet (Al-Amoudi *et al.*, 1992).

Available information on fat-soluble vitamins for tilapias concerns only vitamin D and E. The optimum vitamin D_3 (cholecalciferol) requirement for maximum growth of *O. niloticus* × *O. aureus* is 374.8 IU kg^{-1} diet (Shiau and Hwang, 1993). O'Connell and Gatlin (1994) reported that vitamin D_3 is not a dietary essential for *O. aureus*. Vitamin E-deficient *O. aureus* exhibited anorexia, reduced weight gain and diet efficiency, skin haemorrhages, impaired erythropoiesis, muscle degeneration, ceroid in liver and spleen and abnormal skin coloration. The dietary vitamin E requirement increased with increasing levels of dietary lipid. The requirement of *O. aureus* was estimated at 10 mg and 25 mg D,L-α-tocopheryl acetate kg^{-1} diet in diets containing 3% and 6% dietary lipid, respectively, or 3–4 mg α-tocopheryl acetate for each per cent of dietary lipid (Roem *et al.*, 1990c). The dietary vitamin E requirement of *O. niloticus* was reported to be 50–100 mg kg^{-1} diet for a diet containing 5% lipid, increasing to 500 mg kg^{-1} diet for a diet containing 10–15% lipid (Satoh *et al.*, 1987). Recently, tilapia has been shown to require both n-3 highly unsaturated fatty acid and n-6 fatty acid for maximal growth (Chou and Shiau, 1999). Consequently, the vitamin E requirement of tilapia has been re-evaluated to be 42–44 mg kg^{-1} and 60–66 mg kg^{-1} in 5% and 12% lipid diets, respectively (Shiau and Shiau, 2001). No studies have been conducted on the requirements of vitamin A and K for tilapias.

There is little information on mineral requirements of tilapias. Five minerals – namely, calcium, phosphorus, magnesium, zinc and potassium – have been quantified for their requirements in tilapias (Table 20.6). In low-calcium water, supplementation of 7.0 g (0.7%) calcium and 5.0 g (0.5%) phosphorus kg^{-1} diet are required for normal growth and bone mineralization of *O. aureus* (Robinson *et al.*, 1984, 1987). The requirement of phosphorus for maximum growth and normal bone mineralization for *O. nilotica* was reported to be less than 0.9% in the diet (Watanabe *et al.*, 1980). For *O. aureus*, levels of 0.05% magnesium (Reigh *et al.*, 1991) and 0.002% zinc (McClain and Gatlin, 1988) are needed for optimum growth, whereas requirements of 0.059–0.077% magnesium (Dabrowska *et al.*, 1989) and 0.003% zinc (Eid and Ghonim, 1994) were reported for *O. niloticus*. The dietary potassium requirement for *O. niloticus* × *O. aureus* is about 0.2–0.3% (Shiau and Hsieh, 2001). Dietary levels of 0.0012% manganese and 0.0003–0.0004% copper were recommended for *O. niloticus* (Watanabe *et al.*, 1988). However, the quantitative data on the requirement of these two minerals are not available.

Practical Diets

Tilapia have been successfully cultured on diets prepared for other species, such as carp and catfish. The practical formulations commonly consist of only three or four feedstuffs, mainly of plant origin. Commercial diets formulated for tilapia usually contain 24–28% protein. Natural pond foods contribute a significant amount of protein, so this level is assumed to be high enough (Lim, 1989). However, few pond studies have been conducted to compare various diet formulations for semi-intensive and intensive culture of tilapias. A 25% protein pellet composed of 15% fish-meal, 20% soybean meal, 20% ground wheat and 45% ground sorghum has been used successfully in Israel in this type of production (Viola and Arieli, 1983). The importance of micronutrient supplementation in pond diets for tilapia is not well known. Due to the extreme variation in the culture practices used, formulation of practical diets to efficiently supplement the nutrient contribution of the natural food is practically impossible. In intensive culture, such as in raceways or cages, tilapia rely solely on the prepared diets as a source of nutrients. Thus, a nutritionally complete diet containing all essential nutrients is required.

Tilapia accept a variety of diets, in meal form and in moist, sinking and floating pellets. Crude feedstuffs, such as rice bran, are offered in meal form, whereas the compounded diets are most often processed into pellets. Crude feedstuffs may be uneconomical when pelleted for pond feeding of tilapia; however, high-quality

Table 20.6. Mineral requirements of tilapia.

	Requirement of diet (% of diet)		
Minerals	*O. aureus*	*O. niloticus*	*O. niloticus* × *O. aureus*
Ca	0.7[a]	–	–
P	0.5[b]	< 0.9[c]	–
Mg	0.05[d]	0.059–0.077[e]	–
Zn	0.002[f]	0.003[g]	–
K	–	–	0.21–0.33[h]

[a]Robinson *et al.* (1984); [b]Robinson *et al.* (1987); [c]Watanabe *et al.* (1980); [d]Reigh *et al.* (1991); [e]Dabrowska *et al.* (1989); [f]McClain and Gatlin (1988); [g]Eid and Ghonim (1994); [h]Shiau and Hsieh (2001).

diets should be processed into pellets to minimize ingredient separation, selective feeding on a single ingredient, loss of water-soluble nutrients and waste.

Tilapia seem to prefer smaller pellets than channel catfish and salmonids of comparable size. Recommended pellet sizes for tilapias are available (Jauncey and Ross, 1982). For feeding tilapias to a marketable size of 500 g, the most common pellet size is approximately 3–4 mm in diameter and 6–10 mm in length. Diets in meal or crumble forms are used for fry and fingerlings. These are made by first pelleting or extruding the diet mixture and then reducing the particles to a desirable size by crumbling.

Feeding Practices

It is generally recognized that smaller fish consume more diet on a per cent weight basis than larger fish. Suggested feeding rates for different sizes of tilapias are given in Table 20.7. Water temperature influences the metabolic rate and energy expenditure, thus having profound effects on the feeding rate. Tilapia will consume less diet in cold weather than in warmer weather. Luquet (1991) suggested that feeding tilapia should be stopped at a water temperature below 16°C. In semi-intensive culture systems, natural food can make a significant contribution to the nutrient requirements of fish. Under these conditions, the amount of diet used should be less than that for fish grown under intensive systems.

Since there are many factors that influence diet consumption of fish, satiation feeding (feeding as much as the fish can eat) may be a better alternative to feeding a prescribed amount based on the percentage of fish biomass. Clark, J.H. *et al.* (1990) obtained maximum growth of Florida red tilapia *Oreochromis urolepis hornorum* × *O. mossambicus* reared in marine cages at feeding rates near satiation (90%). Diet cost, however, could be reduced without significant growth reduction by feeding at 70% of satiation.

Diets are offered to fish by hand, blower, automatic feeder or demand feeder. Hand-feeding is labour-intensive but has advantages over other methods of feeding in that it allows the feeder to observe feeding activity and behaviour and thus to regulate the amount of diet fed to prevent overfeeding. However, this method of feeding is not feasible in large commercial farms. The most common method of feeding with large ponds is blowing the diet on to the water surface

Table 20.7. Feeding rate and frequencies for various sizes of tilapia at 28°C (from Lim, 1997).

Size	Daily feeding rate (% of body weight)	Daily feeding frequency
2 days old to 1 g	30–10	8
1–5 g	10–6	6
5–20 g	6–4	4
20–100 g	4–3	3–4
> 100 g	3–2	2–3

using mechanical devices that are either mounted on or pulled behind vehicles. Automatic feeders, driven by time-clock or electrical devices, allow farmers to preset the amount of diet to be fed at various time intervals. Demand feeders consist of a hopper with a top opening for loading the diet and a bottom opening, which serves as a movable gate for diet delivery. Attached to the gate is a rod whose tip extends down into the water, where it can be activated by the fish. As long as the fish continue to hit the rod, diet will continue to flow out. Clark J.H. *et al.* (1990) indicated that the demand feeder may prove to be the best method for feeding tilapia in marine cages as it provides good growth and feed conversion with reduced labour.

Frequency of feeding varies with fish size. Feeding frequency decreases as the fish grows (Table 20.7). Feeding three to four times per day is common for smaller fingerlings but decreases to two to three times per day for grow-out. Because of the continuous feeding behaviour and smaller stomach capacity, tilapias respond better to more frequent feeding than do channel catfish and salmonids. *Oreochromis niloticus* fingerlings grew faster when fed four times a day than when fed twice a day but did not grow faster when fed eight times a day (Kubaryk, 1980).

Acknowledgements

I wish to thank my graduate students, Min-Hsien Lee, Ming-Hsun Wu, Yu-Hung Lin, Shih-Ying Huang, Yu-Ming Lin, Yi-Ting Wang and Li-Wen Su, for their valuable assistance during the preparation of this chapter.

References

Al-Amoudi, M.M., El-Nakkadi, A.M.N. and El-Nouman, B.M. (1992) Evaluation of optimum dietary requirement of vitamin C for the growth of *Oreochromis spilurus* fingerlings in water from the red sea. *Aquaculture* 105, 165–173.

Anderson, J., Jackson, A.J., Matty, A.J. and Capper, B.S. (1984) Effects of dietary carbohydrate and fibre on the tilapia *Oreochromis niloticus* (Linn.). *Aquaculture* 37, 303–314.

Chou, B.S. and Shiau, S.-Y. (1996) Optimal dietary lipid level for growth of juvenile hybrid tilapia, *Oreochromis niloticus* × *Oreochromis aureus*. *Aquaculture* 143, 185–195.

Chou, B.S. and Shiau, S.-Y. (1999) Both n-6 and n-3 fatty acids are required for maximal growth of juvenile hybrid tilapia. *North American Journal of Aquaculture* 61, 13–20.

Clark, A.E., Watanabe, W.O., Olla, B.L. and Wicklund, R.I. (1990) Growth, diet conversion and protein utilization of Florida red tilapia fed isocaloric diets with different protein levels in seawater pools. *Aquaculture* 88, 75–85.

Clark, J.H., Watanabe, W.O. and Ernest, D.H. (1990) Effect of feeding rate on growth and diet conversion of Florida red tilapia reared in floating marine cages. *Journal of the World Aquaculture Society* 21, 16–24.

Cruz, E.M. and Laudencia, I.R. (1977) Protein requirements of *Tilapia mossambica* fingerlings. *Philippine Journal of Biology* 6, 177–182.

Dabrowska, H., Meyer-Burgdorff, K. and Gunther, K.D. (1989) Interaction between dietary protein and magnesium level in tilapia (*Oreochromis niloticus*). *Aquaculture* 76, 277–291.

Davis, A.T. and Stickney, R.R. (1978) Growth responses of *Tilapia aurea* to dietary protein quality and quantity. *Transactions of the American Fisheries Society* 107, 479–483.

De Silva, S.S. and Perera, P.A.B. (1985) Effects of dietary protein level on growth, food conversion and protein use in young *Tilapia nilotica* at four salinities. *Transactions of the American Fisheries Society* 114, 685–689.

Eid, A.E. and Ghonim, S.I. (1994) Dietary zinc requirement of fingerling *Oreochromis niloticus*. *Aquaculture* 119, 259–264.

El-Dahhar, A.A. and Lovell, R.T. (1995) Effect of protein to energy ratio in purified diets on growth performance, diet utilization and body composition of Mozambique tilapia, *Oreochromis mossambicus* (Peters). *Aquaculture Research* 26, 451–457.

El-Sayed, A.-F.M. (1987) Protein and energy requirements of *Tilapia zillii*. Doctoral dissertation, Michigan State University, East Lansing, Michigan.

El-Sayed, A.-F.M. (1999) Alternative dietary protein sources for farmed tilapia, *Oreochromis* spp. *Aquaculture* 179, 149–168.

El-Sayed, A.-F.M. and Garling, D.L. Jr (1988) Carbohydrate-to-lipid ratio in diets for *Tilapia zillii* fingerlings. *Aquaculture* 73, 157–163.

El-Sayed, A.-F.M. and Teshima, S.-I. (1992) Protein and energy requirements of Nile tilapia, *Oreochromis niloticus*, fry. *Aquaculture* 103, 55–63.

Hanley, F. (1991) Effects of feeding supplementary diets containing varying levels of lipid on growth, food conversion, and body composition of Nile tilapia *Oreochromis niloticus* (L.). *Aquaculture* 93, 323–334.

Jackson, A.J., Capper, B.S. and Matty, A.J. (1982) Evaluation of some plant proteins in complete diets for the tilapia, *Sarotherodon mossambicus*. *Aquaculture* 27, 97–109.

Jauncey, K. (1982) The effects of varying dietary protein level on the growth, food conversion, protein utilization and body composition of juvenile tilapias (*Sarotherodon mossambicus*). *Aquaculture* 27, 43–54.

Jauncey, K. and Ross, B. (1982) *A Guide to Tilapia Feeds and Feeding*. Institute of Aquaculture, University of Stirling, Scotland, 111 pp.

Kanazawa, A., Teshima, S.I. and Imaij, K. (1980a) Biosynthesis of fatty acids in *Tilapia zillii* and the puffer fish, *Fuga rubripes*. *Memoir of the Faculty of Fisheries, Kagoshima University* 29, 313–318.

Kanazawa, A., Teshima, S.I., Sakamoto, M. and Awali, M.A. (1980b) Requirement of *Tilapia zillii* for essential fatty acids. *Bulletin of the Japanese Society of Scientific Fisheries* 46, 1353–1356.

Kubaryk, J.M. (1980) Effect of diet, feeding schedule and sex on food consumption, growth and retention of protein and energy by tilapia. Doctoral dissertation, Auburn University, Auburn, Alabama.

Li, Z., Lei, W., Ye, J. and He, X. (1991) The nutritional value of commercial diet ingredients for Nile tilapia (*Oreochromis* L.) in China. In: De Silva, S.S. (ed.) *Fish Nutrition Research in Asia*. Special Publication No. 5, Asian Fisheries Society, Manila, the Philippines, pp. 101–106.

Lim, C. (1989) Practical feeding – tilapias. In: Lovell, T. (ed.) *Nutrition and Feeding of Fish*. Van Nostrand Reinhold, New York, pp. 163–183.

Lim, C. (1997) Nutrition and feeding of tilapias. In: Alston, D.E., Green, B.W. and Clifford, H.C. (eds) *IV Symposium on Aquaculture in Central America: Focusing on Shrimp and Tilapia, 22–24 April 1997, Tegucigalpa, Honduras*. Associacion Nacional de

Acuicultores de Honduras and the Latin American Chapter of the World Aquaculture Society, pp. 94–107.

Lim, C. and LeaMaster, B. (1991) In: *Program and Abstracts, World Aquaculture Society 22nd Annual Conference and Exposition, 16–20 June 1991*. San Juan, Puerto Rico, p. 39.

Lim, C., LeaMaster, B. and Brock, J.A. (1993) Riboflavin requirement of fingerling red hybrid tilapia grown in seawater. *Journal of the World Aquaculture Society* 24, 451–458.

Lim, C., LeaMaster, B. and Brock, J.A. (1995) Pyridoxine requirement of fingerling red hybrid tilapia grown in seawater. *Journal of Applied Aquaculture* 5, 49–60.

Lin, J.H. and Shiau, S.-Y. (1995) Hepatic enzyme adaptation to different dietary carbohydrates in juvenile tilapia *Oreochromis niloticus* × *O. aureus*. *Fish Physiology and Biochemistry* 14, 165–170.

Lin, J.H., Ho, L.T. and Shiau, S.-Y. (1995) Plasma glucose and insulin concentration in tilapia after oral administration of glucose and starch. *Fisheries Science* 61, 986–989.

Lovell, R.T. (1980) Feeding tilapia. *Aquaculture Magazine* 7, 42–43.

Lovell, R.T. and Limsuwan, T. (1982) Intestinal synthesis and dietary nonessentiality of vitamin B_{12} for *Tilapia nilotica*. *Transactions of the American Fisheries Society* 11, 485–490.

Luquet, P. (1991) Tilapia, *Oreochromis* spp. In: Wilson, R.P. (ed.) *Handbook of Nutrient Requirement of Finfish*. CRC Press, Boca Raton, Florida, pp. 169–179.

McClain, W.R. and Gatlin, D.M. (1988) Dietary zinc requirement of *Oreochromis aureus* and effects of dietary calcium and phytate on zinc bioavailability. *Journal of the World Aquaculture Society* 19, 103–108.

Mazid, M.A., Tanaka, Y., Katayama, T., Rahman, M.A., Simpson, K.L. and Chichester, C.O. (1979) Growth response of *Tilapia zillii* fingerling fed isocaloric diets with variable protein levels. *Aquaculture* 18, 115–122.

Newman, M.W., Huezo, H.E. and Hughes, D.G. (1979) The response of all-male tilapia hybrids to four levels of protein in isocaloric diets. *Proceedings of World Mariculture Society* 10, 788–792.

NRC (National Research Council) (1983) *Nutrient Requirement of Warmwater Fishes and Shellfishes*. National Academy Press, Washington, DC.

NRC (National Research Council) (1993) *Nutrient Requirement of Fish*. National Academy Press, Washington, DC.

O'Connell, J.P. and Gatlin, D.M. (1994) Effects of dietary calcium and vitamin D_3 on weight gain and mineral composition of the blue tilapia (*Oreochromis aureus*) in low-calcium water. *Aquaculture* 125, 107–117.

Ramachandran Nair, K.G. and Gopakumar, K. (1981) Influence of sex, spawning, starvation and water temperature on fatty acid composition in *Tilapia mossambica*. *Fishery Technology* 18, 123–127.

Reigh, R.C., Robinson, E.H. and Brown, P.B. (1991) Effect of dietary magnesium on growth and tissue magnesium content of blue tilapia *Oreochromis aureus*. *Journal of the World Aquaculture Society* 22, 192–200.

Robinson, E.H., Rawles, S.D., Yette, H.E. and Greene, L.W. (1984) An estimate of the dietary calcium requirement of fingerling *Tilapia aureus* reared in calcium-free water. *Aquaculture* 41, 389–393.

Robinson, E.H., LaBomascus, D., Brown, P.B. and Linton, T.L. (1987) Dietary calcium and phosphorus requirements of *Oreochromis aureus* reared in calcium-free water. *Aquaculture* 64, 267–276.

Roem, A.J., Kohler, C.C. and Stickney, R.R. (1990a) Inability to detect a choline requirement for the blue tilapia *Oreochromis aureus*. *Journal of the World Aquaculture Society* 21, 238–240.

Roem, A.J., Kohler, C.C. and Stickney, R.R. (1990b) Vitamin requirements of blue tilapias in a recirculating water system. *Progressive Fish-Culturist* 52, 15–18.

Roem, A.J., Kohler, C.C. and Stickney, R.R. (1990c) Vitamin E requirements of the blue tilapia, *Oreochromis aureus* (Steindachner), in relation to dietary lipid level. *Aquaculture* 87, 155–164.

Roem, A.J., Stickney, R.R. and Kohler, C.C. (1991) Dietary pantothenic acid requirement of the blue tilapia. *Progressive Fish-Culturist* 53, 216–219.

Santiago, C.B. and Laron, M.A. (1991) Growth response and carcass composition of red tilapia fry fed diets with varying protein levels and protein to energy ratio. In: De Silva, S.S. (ed.) *Fish Nutrition Research in Asia*. Special Publication No. 5, Asian Fisheries Society, Manila, the Philippines, pp. 55–62.

Santiago, C.B. and Lovell, R.T. (1988) Amino acid requirements for growth of Nile tilapia. *Journal of Nutrition* 118, 1540–1546.

Santiago, C.B. and Reyes, O.S. (1993) Effects of dietary lipid source on reproductive performance and tissue lipid levels of Nile tilapia *Oreochromis niloticus* Linnaeus broodstock. *Journal of Applied Ichthyology* 9, 33–40.

Satoh, S.T., Takeuchi, T. and Watanabe, T. (1987) Requirement of tilapia for alpha tocopherol. *Nippon Suisan Gakkaishi* 53, 119–124.

Shiau, S.-Y. and Chen, M.J. (1993) Carbohydrate utilization by tilapia (*Oreochromis niloticus* × *O. aureus*) as influenced by different chromium sources. *Journal of Nutrition* 123, 1747–1753.

Shiau, S.-Y. and Cheng, D.J. (1999) Ammonia excretion and oxygen consumption of tilapia are affected by different carbohydrate ingestion. *Fisheries Science* 65, 321–322.

Shiau, S.-Y. and Chin, Y.H. (1999) Estimation of the dietary biotin requirement of juvenile hybrid tilapia, *Oreochromis niloticus* × *O. aureus*. *Aquaculture* 170, 71–78.

Shiau, S.-Y. and Chuang, J.C. (1995) Utilization of disaccharides by juvenile tilapia, *Oreochromis niloticus* × *O. aureus*. *Aquaculture* 133, 249–256.

Shiau, S.-Y. and Chuang, J.C. (1996) The influences of dietary protein source on carbohydrate utilization by tilapia, *Oreochromis niloticus* × *O. aureus*. *Journal of the Fisheries Society of Taiwan* 23, 49–57.

Shiau, S.-Y. and Hsieh, H.L. (1997) Vitamin B_6 requirements of tilapia *Oreochromis niloticus* × *O. aureus* fed two dietary protein concentrations. *Fisheries Science* 63, 1002–1007.

Shiau, S.-Y. and Hsieh, J.F. (2001) Quantifying the dietary potassium requirement of juvenile hybrid tilapia, *Oreochromis niloticus* × *O. aureus*. *British Journal of Nutrition* 85, 213–218.

Shiau, S.-Y. and Hsu, T.S. (1993) Stability of ascorbic acid in shrimp diet during analysis. *Nippon Suisan Gakkaishi* 59, 1535–1537.

Shiau, S.-Y. and Hsu, T.S. (1995) L-Ascorbyl-2-sulfate has equal antiscorbutic activity as L-ascorbyl-2-monophosphate for tilapia, *Oreochromis niloticus* × *O. aureus*. *Aquaculture* 133, 147–157.

Shiau, S.-Y. and Hsu, T.S. (1999) Quantification of vitamin C requirement for juvenile hybrid tilapia, *Oreochromis niloticus* × *O. aureus*, with L-ascorbyl-2-monophosphate-Na and L-ascorbyl-2-monophosphate-Mg. *Aquaculture* 175, 317–326.

Shiau, S.-Y. and Huang, S.L. (1989) Optimal dietary protein level for hybrid tilapia (*Oreochromis niloticus* × *O. aureus*) reared in seawater. *Aquaculture* 81, 119–127.

Shiau, S.-Y. and Huang, S.L. (1990) Influence of varying energy levels with two protein concentrations in diets for hybrid tilapia (*Oreochromis niloticus* × *O. aureus*) reared in seawater. *Aquaculture* 91, 143–152.

Shiau, S.-Y. and Hwang, J.Y. (1993) Vitamin D requirement of juvenile hybrid tilapia *Oreochromis niloticus* × *O. aureus*. *Nippon Suisan Gakkaishi* 59, 553–558.

Shiau, S.-Y. and Jan, F.L. (1992) Dietary ascorbic acid requirement of juvenile tilapia *Oreochromis niloticus* × *O. aureus*. *Nippon Suisan Gakkaishi* 58, 671–675.

Shiau, S.-Y. and Lei, M.S. (1999) Feeding strategy does affect carbohydrate utilization by hybrid tilapia *Oreochromis niloticus* × *O. aureus*. *Fisheries Science* 65, 553–557.

Shiau, S.-Y. and Liang, H.S. (1995) Carbohydrate utilization and digestibility by tilapia, *Oreochromis niloticus* × *O. aureus*, are affected by chromic oxide inclusion in the diet. *Journal of Nutrition* 125, 976–982.

Shiau, S.-Y. and Lin, S.F. (1993) Effect of supplemental dietary chromium and vanadium on the utilization of different carbohydrates in tilapia, *Oreochromis niloticus* × *O. aureus*. *Aquaculture* 110, 321–330.

Shiau, S.-Y. and Lo, P.S. (2000) Dietary choline requirements of juvenile hybrid tilapia, *Oreochromis niloticus* × *O. aureus*. *Journal of Nutrition* 130, 100–103.

Shiau, S.-Y. and Lung, C.Q. (1993) No dietary vitamin B_{12} required for juvenile tilapia *Oreochromis niloticus* × *O. aureus*. *Comparative Biochemistry and Physiology* 105A, 147–150.

Shiau, S.-Y. and Peng, C.Y. (1993) Protein-sparing effect by carbohydrates in diets for tilapia, *Oreochromis niloticus* × *O. aureus*. *Aquaculture* 117, 327–334.

Shiau, S.-Y. and Shiau, L.F. (2001) Reevaluation of the vitamin E requirement of juvenile tilapia, *Oreochromis niloticus* × *O. aureus*. *Animal Science* 72, 529–534.

Shiau, S.-Y. and Shy, S.M. (1998) Dietary chromic oxide inclusion level required to maximize glucose utilization in hybrid tilapia, *Oreochromis niloticus* × *O. aureus*. *Aquaculture* 161, 357–364.

Shiau, S.-Y. and Suen, G.S. (1992) Estimation of the niacin requirements for tilapia fed diets containing glucose or dextrin. *Journal of Nutrition* 122, 2030–2036.

Shiau, S.-Y. and Yu, Y.P. (1999) Dietary supplementation of chitin and chitosan depresses growth in tilapia, *Oreochromis niloticus* × *O. aureus*. *Aquaculture* 179, 439–446.

Shiau, S.-Y., Chuang, J.L. and Sun, C.L. (1987) Inclusion of soybean meal in tilapia (*Oreochromis niloticus* × *O. aureus*) diets at two protein levels. *Aquaculture* 65, 251–261.

Shiau, S.-Y., Yu, H.L., Hwa, S., Chen, S.Y. and Hsu, S.I. (1988) The influence of carboxymethylcellulose on growth, digestion, gastric emptying time and body composition of tilapia. *Aquaculture* 70, 345–354.

Shiau, S.-Y., Kwok, C.C., Huang, J.Y., Chen, C.M. and Lee, S.L. (1989a) Replacement of fish meal with soybean meal in male tilapia (*Oreochromis niloticus* × *O. aureus*) fingerling diets at a suboptimal protein level. *Journal of the World Aquaculture Society* 20, 230–235.

Shiau, S.-Y., Kwok, C.C., Chen, C.J., Hong, H.T. and Hsieh, H.B. (1989b) Effects of dietary fiber on the intestinal absorption of dextrin, blood sugar level and growth of tilapia, *Oreochromis niloticus* × *O. aureus*. *Journal of Fish Biology* 34, 929–935.

Siddiqui, A.Q., Howlander, M.S. and Adam, A.A. (1988) Effects of dietary protein levels on growth, diet conversion and protein utilization in fry and young Nile tilapia, *Oreochromis niloticus*. *Aquaculture* 70, 63–73.

Soliman, A.K. and Wilson, R.P. (1992a) Water-soluble vitamin requirements of tilapia. 1. Pantothenic acid requirement of blue tilapia, *Oreochromis aureus*. *Aquaculture* 104, 121–126.

Soliman, A.K. and Wilson, R.P. (1992b) Water-soluble vitamin requirements of tilapia. 2. Riboflavin requirement of blue tilapia, *Oreochromis aureus*. *Aquaculture* 104, 309–314.

Soliman, A.K., Jauncey, K. and Roberts, R.J. (1994) Water-soluble vitamin requirements of tilapia: ascorbic acid (vitamin C) requirement of Nile tilapia, *Oreochromis niloticus* (L.). *Aquaculture and Fisheries Management* 25, 269–278.

Stickney, R.R. and Hardy, R.W. (1989) Lipid requirements of some warmwater species. *Aquaculture* 79, 145–156.

Stickney, R.R. and McGeachin, R.B. (1983) Responses of *Tilapia aurea* to semipurified diets of differing fatty acid composition. In: Fishelson, L. and Yaron, Z. (eds) *Proceedings of the International Symposium on Tilapia in Aquaculture*. Tel Aviv University Press, Tel Aviv, Israel, pp. 346–355.

Stickney, R.R. and McGeachin, R.B. (1985) Effects of dietary lipid quality on growth and food conversion of tilapia. *Proceedings of the Annual Conference Southeastern Association of Fish and Wildlife Agencies* 37(1983), 352–357.

Stickney, R.R. and Wurts, W.A. (1986) Growth response of blue tilapias to selected levels of dietary menhaden and catfish oils. *Progressive Fish-Culturist* 48, 107–109.

Stickney, R.R., McGeachin, R.B., Lewis, D.H., Marks, J., Sis, R.F., Robinson, E.H. and Wurts, W. (1984) Responses of *Tilapia aurea* to dietary vitamin C. *Journal of the World Aquaculture Society* 15, 179–185.

Stickney, R.R., McGeachin, R.B., Robinson, E.H., Arnold, G.G. and Suter, L. (1985) Growth of *Tilapia aurea* as a function of degree of dietary lipid saturation. In: *Proceedings of the Annual Conference Southeastern Association of Fish and Wildlife Agencies* 36 (1982), 172–181.

Sugita, H., Miyajima, C. and Deguchi, Y. (1990) The vitamin B_{12}-producing ability of intestinal bacteria isolated from tilapia and channel catfish. *Nippon Suisan Gakkaishi* 56, 701.

Tacon, A.G.J. (1993) *Diet Ingredients for Warmwater Fish: Fish Meal and Other Processed Feedstuffs*. FAO Fisheries Circular No. 856, FAO, Rome, Italy, 64 pp.

Tacon, A.G.J. (1997) Global trends in aquaculture and aquafeed production 1984–1995. In: *International Aquafeed Directory and Buyer's Guide 1987/1998*. Turret RAI Plc. Publ, pp. 5–37.

Tacon, A.G.J., Jauncey, K., Falaye, A., Pantah, M., MacGowen, I. and Stafford, E. (1983) The use of meat and bone meal, hydrolyzed feather meal and soybean meal in practical fry and fingerling diets for *Oreochromis niloticus*. In: Fishelson, J. and Yaron, Z. (eds) *Proceedings of the International Symposium on Tilapia in Aquaculture*. Tel Aviv University Press, Tel Aviv, Israel, pp. 356–365.

Takeuchi, T., Satoh, S. and Watanabe, T. (1983) Requirement of *Tilapia nilotica* for essential fatty acids. *Bulletin of the Japanese Society of Scientific Fisheries* 49, 1127–1134.

Teshima, S. and Kanazawa, A. (1988) Nutritive value of methionine-enriched soy plastein for *Oreochromis niloticus* fry. In: Pullin, R.S.V., Bhukaswan, T., Tonguthai, K. and Maclean, J.L. (eds) *Second International Symposium on Tilapia in Aquaculture*. Conference Proceeding No. 15, ICLARM, Manila, the Philippines, pp. 393–399.

Teshima, S., Gonzalez, G. and Kanazawa, A. (1978) Nutrition requirements of tilapia: utilization of dietary protein by *Tilapia zillii*. *Memoirs of the Faculty of Fisheries, Kagoshima University* 27, 49–57.

Teshima, S., Kanazawa, A. and Uchiyama, Y. (1985) Effects of dietary protein, lipid and digestible carbohydrate levels on the weight gain, feed conversion efficiency and protein efficiency ratio of *Tilapia niloticus*. *Memoirs of the Kagoshima University, Research Center for the South Pacific* 6, 56–71.

Trewavas, E. (1982) Tilapia: taxonomy and speciation. In: Pullin, R.S.V. and Lowe-McConnel, R.H. (eds) *The Biology and Culture of Tilapias*. ICLARM, Manila, the Philippines, pp. 3–14.

Tung, P.H. and Shiau, S.-Y. (1991) Effect of meal frequency on growth performance of hybrid tilapia, *Oreochromis niloticus* × *O. aureus*, fed different carbohydrate diets. *Aquaculture* 92, 343–350.

Tung, P.H. and Shiau, S.-Y. (1993) Carbohydrate utilization versus body size in tilapia, *Oreochromis niloticus* × *O. aureus*. *Comparative Biochemistry and Physiology* 104A, 585–588.

Twibell, R.G. and Brown, P.B. (1998) Optimal dietary protein concentration for hybrid tilapia (*Oreochromis niloticus* × *Oreochromis aureus*) fed all-plant diets. *Journal of the World Aquaculture Society* 29, 9–16.

Viola, S. and Arieli, Y. (1983) Nutrition studies with tilapia (*Sarotherodon*): 1. Replacement of fishmeal by soybean meal in diets for intensive tilapia culture. *Bamidgeh* 35, 9–17.

Viola, S., Zohar, G. and Arieli, Y. (1986) Phosphorus requirements and its availability from different sources for intensive pond culture species in Israel: Part 1. Tilapia. *Bamidgeh* 38, 3–12.

Viola, S., Arieli, Y. and Zohar, G. (1988) Animal-protein-free diets for hybrid tilapia (*Oreochromis niloticus* × *O. aureus*) in intensive culture. *Aquaculture* 75, 115–125.

Viola, S., Angeoni, H. and Lahav, E. (1994a) Present limits of protein sparing by amino acid supplementation of practical carp and tilapia diets. *Israel Journal of Aquaculture – Bamidgeh* 46, 203–211.

Viola, S., Angeoni, H., Gur, N. and Lahav, E. (1994b) Growth performance, protein and energy balances of hybrid tilapia fed two levels of lysine at three levels of protein. *Israel Journal of Aquaculture – Bamidgeh* 46, 212–222.

Wang, K., Takeuchi, T. and Watanabe, T. (1985) Effect of dietary protein levels on growth of *Tilapia nilotica*. *Bulletin of the Japanese Society of Scientific Fisheries* 51, 133–140.

Wannigama, N.D., Weerakoon, D.E.M. and Muthukumarana, G. (1985) Cage culture of *S. niloticus* in Sri Lanka: effect of stocking density and dietary crude protein levels on growth. In: Cho, Y., Cowey, C.B. and Watanabe, T. (eds) *Finfish Nutrition in Asia. Methodological Approaches to Research and Development*. IDRC-233e, Ottawa, Ontario Canada, pp. 113–117.

Watanabe, T., Takeuchi, T., Murakami, A. and Ogino, C. (1980) The availability to *Tilapia nilotica* of phosphorus in white fish meal. *Bulletin of the Japanese Society of Scientific Fisheries* 46, 897–899.

Watanabe, T., Satoh, S. and Takeuchi, T. (1988) Availability of minerals in fish meal to fish. *Asian Fisheries Science* 1, 175–195.

Wee, K.L. and Tuan, N.A. (1988) Effects of dietary protein level on growth and reproduction of Nile tilapia (*Oreochromis niloticus*). In: Pullin, R.S.V., Bhukaswan, T., Tonguthai, K. and Maclean, J.L. (eds) *The Second International Symposium on Tilapia in Aquaculture*. ICLARM Conference Proceedings 15, Department of Fisheries, Bangkok, Thailand and International Center for Living Aquatic Resources Management, Manila, the Philippines, pp. 401–410.

Winfree, R.A. and Stickney, R.R. (1981) Effects of dietary protein and energy on growth, feed conversion efficiency and body composition of *Tilapia aureus*. *Journal of Nutrition* 111, 1001–1012.

21 Channel Catfish, *Ictalurus punctatus*

Edwin H. Robinson and Menghe H. Li
Thad Cochran National Warmwater Aquaculture Center, Mississippi State University, PO Box 197, Stoneville, MS 38776, USA

Introduction

Farm-raised channel catfish are a highly prized foodfish in the USA, as evidenced by the fact that per capita consumption has doubled since 1985, reaching an all-time high of 0.45 kg in 1997. It is now the fifth most popular fish in the USA behind tuna, pollock, salmon and cod. The popularity of farm-raised channel catfish is due to its consistent quality, delicate flavour, firm texture, versatility, year-round availability and high nutritional value. Production of farm-raised channel catfish exceeded 270,000 t in 1999 (USDA, 2000) and accounted for about three-quarters of the annual aquacultural production in the USA (Johnson, 1999).

The farm-raised channel-catfish industry is centred in the south-eastern USA, primarily on the Mississippi River flood-plain (called the Delta), where a unique combination of physical and socio-economic factors were favourable for development of the industry. Four states – Alabama, Arkansas, Louisiana and Mississippi – account for 95% of channel-catfish production, with Mississippi leading with 70% of total production. The industry employs over 13,000 people in production, processing, feed manufacturing and related support industries. Sales of farm-raised channel catfish total about US$600 million annually, but the total impact on the economies of the four major catfish-producing states exceeds US$4 billion annually.

Channel catfish are typically raised in earthen ponds, which are generally 4–6 ha in size and 0.9–1.5 m deep. Most of the ponds in the Delta are built on flat land, with dirt removed to build levees around the perimeter of the pond, and use groundwater supplied by wells as a water source. Some ponds outside the Delta are watershed ponds, which collect rainfall as a water source. There are about 70,000 ha of catfish ponds currently in production. The production cycle for channel catfish includes four distinct phases: spawning, hatching, fingerling grow-out and foodfish grow-out.

Male and female broodfish are placed into earthen ponds and, when the water temperature reaches about 20°C (generally around May in the southeastern USA), the fish spawn. Various types of containers are placed in the pond for the fish to spawn in. Female broodfish lay eggs in the container and the male fertilizes the eggs. The fertilized eggs are removed from the container and taken into the hatchery. The eggs hatch in a controlled environment and remain in the hatchery for 5–10 days and develop into a stage referred to as fry. Fry are taken from the hatchery and placed in nursery ponds for fingerling grow-out. Fry are fed a high-protein diet that is ground into a fine powder for the first few weeks after being introduced into the pond. Once the fry come to the water surface seeking feed, a small-diameter pelleted feed is used. Fry stocked in the summer are of suitable size (20–40 g) for stocking into foodfish grow-out ponds by late autumn or early winter. The fish may be stocked at this time or stocking may take place in the spring of the following year. Fingerlings are stocked at a rate of 10,000–30,000 ha^{-1} to grow out to foodfish. The fish are fed a high-protein diet (28–32% protein) daily for 150–180 days. Channel catfish feed at temperatures above 15°C, but as the temperature increases to about 30°C feed consumption increases. The prime feeding months are from May to October. Feeding rates generally range from 85 to 140 kg ha^{-1} day^{-1}. Once the fish reaches 0.45 kg in weight, it is suitable for processing. At this time, all fish can be removed from the pond (clean harvest) and the pond drained for restocking. This type of cropping system, referred to as ‘clean harvesting’, is not used extensively. More typically, the fish are selectively harvested using nets with the proper mesh size to remove fish of 0.45 kg and above and leaving the smaller fish in the pond. This harvest scenario is commonly referred to as ‘topping’. Under this scenario, once the ponds are ‘topped’, new fingerlings are placed in the pond to replace those that were removed and to replace any losses that may have occurred (‘understocking’). Topping and understocking are repeated for several years without draining the ponds. Thus, several different year classes of fish are in the same pond at any given time. Once the fish are harvested, they are placed into aerated tanks on trucks and taken to the processing plant alive. They remain alive until they are slaughtered, which usually occurs within 30 min after arriving at the processing plant.

The farm-raised channel-catfish industry has enjoyed phenomenal growth over the past two decades and, although the growth rate has slowed, the industry is still growing at a rate of 5–10% annually. This rapid growth of the industry can be attributed in large part to the efforts of the Catfish Institute (TCI), a marketing organization of the farm-raised catfish industry. Considering the potential for market expansion and the abundance of available natural resources, the potential for further expansion of the farm-raised channel-catfish industry is good.

Nutrient Requirements

Channel catfish require the same nutrients as other fish species for normal growth and metabolic functions. However, the specified amount of a particular nutrient needed by channel catfish may differ from that of other fish. Nutrient requirements of channel catfish have generally been based on studies with small fish conducted under laboratory conditions. Weight gain and feed efficiency are the main criteria used to determine nutrient requirements, although other measurements, such as specific enzyme activity and tissue nutrient reserve, are also used. Studies have been also conducted to define nutrient requirements of channel catfish under conditions that reflect commercial culture practices.

Protein and amino acids

Theoretically, animals do not have a protein requirement, but rather require indispensable (essential) amino acids contained in protein for normal growth. However, dietary protein requirements for channel catfish have been determined in various studies, which vary considerably, ranging from 24% to 55% (NRC, 1993). The wide variation in protein requirements for channel catfish is not surprising because of the different conditions under which the studies were conducted. Factors that may affect the dietary protein requirement include fish size, feeding rate, protein quality/amino acid composition, presence of natural foods, digestible energy (DE), water temperature and stocking density.

Dietary protein requirements for channel catfish appear to decrease as fish grow larger. Winfree and Stickney (1984) reported that channel-catfish fry raised from 0.02 to 0.2 g required about 55% protein in the diet for normal growth. Page and Andrews (1973) reported that channel catfish required 35% protein in the diet for maximum growth for fish raised from 14 to 100 g, while a diet containing 25% protein is adequate for fish grown from 114 to 500 g.

The protein requirement of channel catfish is affected by DE in the diet. Mangalik (1986) demonstrated that small channel catfish grew as well on a 27% protein diet as on a 38% protein diet when the energy level in the diet was low, but when the energy level increased feed consumption decreased and the low-protein diet did not support maximum growth.

Feeding rate may have a profound effect on the dietary protein requirement of channel catfish. Minton (1978) reported that weight gain of pond-raised channel catfish was not different when the fish were fed to satiation with 30% and 36% protein diets, but when the fish were fed at approximately 75% of satiation the fish fed the 36% protein resulted in a higher weight gain. Similarly, Li and Lovell (1992) reported that a dietary protein concentration of 26% was adequate for optimum weight gain when fish were fed as much as they would consume, while for fish fed at a predetermined maximum level of 60 kg ha^{-1} day^{-1} a minimum dietary protein level of 32% was necessary for optimum growth. The reason for the interaction between dietary protein level and feeding rate may be that the protein requirement for maintenance accounts for a higher proportion of the total protein requirement in fish fed a low-protein diet compared with a higher-protein diet when fish are underfed. In a recent study with pond-raised channel catfish, Robinson and Li (1999b) showed that a 28% protein diet resulted in a weight gain equivalent to that of a 32% protein diet when the fish were fed at a predetermined maximum feeding rate of 90 kg ha^{-1} day^{-1} and above.

The presence of natural food organisms may affect the dietary protein requirement of channel catfish, but this effect may be insignificant in large fish. Although natural food organisms are abundant in fish-ponds, their contribution to the growth of foodfish stocked intensively is relatively small. It has been estimated that only 2.5% of the protein requirement and 0.8% of the energy needed for channel catfish grown at a moderate density in ponds were obtained from natural foods (Wiang, 1977). There are indications that natural foods are significant sources of micronutrients, such as vitamins, minerals and essential fatty acids (Robinson and Li, 1996; Robinson *et al.*, 1998).

Based on pond studies, the minimum dietary protein concentration for maximum growth of channel catfish raised from advanced fingerlings to marketable size appears to be about 24% if the fish are fed to satiation. However, the minimum dietary protein concentration for optimum growth, processing yield and body composition appears to be about 28%. A 24–26% protein (P) diet apparently supports maximum growth of pond-raised channel catfish fed to satiation, but tends to increase body fat as compared with diets containing 28% or 32% protein, mainly due to the relatively lower P/DE ratio in the low-protein diet.

Channel catfish require the same ten indispensable amino acids as most simple-stomach animals (Table 21.1). Cystine and tyrosine are classified as dispensable amino acids; however, their presence in the diet can reduce dietary requirements for methionine and phenylalanine, respectively. Cystine can replace 60% of the methionine requirement and tyrosine about 50% of the phenylalanine requirement for channel catfish (NRC, 1993).

Amino acid requirements may be met by feeding an excess of protein, supplementing deficient proteins with crystalline amino acids or feeding a mixture of complementary proteins. Feeding a mixture of complementary

Table 21.1. Amino acid requirements of channel catfish.

Amino acid	Requirement* and recommendation (% of dietary protein)
Arginine	4.3
Histidine	1.5
Isoleucine	2.6
Leucine	3.5
Lysine	5.1
Methionine + cystine	2.3
Phenylalanine + tyrosine	5.0
Threonine	2.0
Tryptophan	0.5
Valine	3.0

* From NRC (1993).

proteins appears to be the most economical choice. To mix protein effectively to meet the amino acid requirements of fish, one needs to know the amino acid composition of the dietary ingredients and the biological availability of amino acids from each ingredient. Amino acid composition data for commonly used feedstuffs are generally available in nutrient composition tables. Amino acid availability data for several commonly used feedstuffs in channel-catfish diets are also known (Wilson *et al.*, 1981). Lysine is considered to be the first limiting amino acid for channel catfish. If diets are formulated to meet a minimum lysine requirement, all other amino acids are met or exceeded if traditional ingredients are used. Lysine is likely to be the only supplemental amino acid used in commercial channel-catfish diets.

The practice of using supplemental crystalline amino acids to improve the quality of inferior protein sources in fish diets has been a subject of debate by fish researchers. Early studies with common carp (*Cyprinus carpio*) have shown that supplemental crystalline amino acids improved the quality of inferior protein sources only if the fish were fed several times daily. Amino acids from the crystalline amino acid test diets appear to be absorbed more rapidly from the intestine of common carp than the amino acids from a casein diet. Zarate and Lovell (1997) demonstrated that bioavailability of crystalline lysine by channel catfish was 57–68% of that of the intact lysine in soybean meal. Crystalline lysine apparently passed out of the stomach of channel catfish earlier than protein-bound lysine (Zarate, 1997); thus, the crystalline lysine would be absorbed from the intestine before the protein-bound lysine was available for absorption. However, when a lysine-deficient diet was supplemented with crystalline lysine to bring the available lysine level (assuming the crystalline lysine was 100% available) to 5% of the protein, a similar growth rate in channel catfish was achieved to that on a lysine-adequate diet without supplemental lysine when fed once or twice daily (Robinson *et al.*, 1980; Robinson, 1991; Robinson and Li, 1994). This indicates that supplemental crystalline lysine is well utilized by channel catfish. Zarate (1997) found no differences in the utilization of crystalline lysine by channel catfish fed two and five times daily. Unlike common carp, feeding frequency apparently has little effect on lysine utilization in channel catfish.

In certain animals, supplemental amino acids improve weight gain and feed efficiency in low-protein, amino acid-replete diets. This does not seem to be the case for channel catfish. Li and Robinson (1998) were unable to demonstrate any benefit of lysine, methionine or both when supplemented in low-protein, amino acid-sufficient diets for channel catfish.

Protein digestibility coefficients for feedstuffs (Table 21.2) are useful in formulating diets, but a more precise diet formulation can be achieved if one uses amino acid availability values (Table 21.3) as the basis for formulating diets rather than protein digestibility. For example, the protein digestibility of cottonseed meal to channel catfish is about 82%, whereas the lysine availability is only about 66%. If diets are formulated based on protein digestibility using cottonseed meal, a lysine deficiency may result.

Table 21.2. Average percentage apparent protein, fat, carbohydrate and energy digestibility coefficients of feedstuffs determined for channel catfish.

Feedstuffs	International feed number	Protein	Fat	Carbohydrate	Energy
Blood meal (81%*)	5-00-380	74‡			
Cottonseed meal (41%)	5-01-621	81,† 83§	81†	17†	56,† 80§
Fish-meal (anchovy) (65%)	5-01-985	90†	97†		
Fish-meal (menhaden) (61%)	5-02-009	87,† 70–86,‡ 85§			85,† 92§
Fish-oil			97†		
Groundnut meal (49%)	5-03-650	86,‡ 74§			76§
Lucerne meal (17%)	1-00-023	13†			16†
Maize-gluten meal (43%)	5-04-900	92‡			
Maize grain (10%)	4-02-935	60,† 97§	76†	59–66†	26,† 57§
Maize grain (cooked) (10%)		66†	96†	62–78†	59†
Meat meal and bone-meal (50%)	5-00-388	75,† 82,‡ 61§	77†		81,† 76§
Poultry by-product meal (61%)	5-04-798	65‡			
Poultry feather meal (84%)	5-03-795	74†	83†		67†
Rice bran (13%)	4-03-928	73§			50§
Rice mill feed (9%)		63§			14§
Soybean meal (44%)	5-04-604	77†	81†		56†
Soybean meal (48%)	5-04-612	84,† 85,‡ 97§			72†
Wheat bran (16%)	4-05-190	82†			56†
Wheat grain (13%)	4-05-268	84,† 92§	96†	59†	60,† 63§
Wheat shorts (17%)	4-05-201	72†			

* Crude protein.
† From Cruz (1975).
‡ From Brown *et al.* (1985).
§ From Wilson and Poe (1985).

Energy

Energy requirements of channel catfish were largely neglected in the early stages of channel-catfish diet development, primarily because an imbalance in dietary energy does not appreciably affect the health of the fish. Also, feed prepared from feedstuffs typically used in channel-catfish diets, such as soybean meal, maize and fish-meal, are unlikely to be extreme in respect of energy balance. However, correct balance of dietary energy is an important consideration when formulating channel-catfish diets because too much energy can result in a reduction in diet

Table 21.3. Average apparent amino acid availabilities (%) for various feedstuffs determined for channel catfish (from Wilson *et al.*, 1981).

Amino acid	Groundnut meal (5-03-650)*	Soybean meal (5-04-612)	Meat and bone-meal (5-00-388)	Menhaden fish-meal (5-02-009)	Maize grain (4-02-935)	Cottonseed meal (5-01-621)	Rice bran (4-03-928)	Wheat middlings (4-05-204)
Ala	88.9	79.0	70.9	87.3	78.2	70.4	82.0	84.9
Arg	96.6	95.4	86.1	89.2	74.2	89.6	91.0	91.7
Asp	88.0	79.3	57.3	74.1	53.9	79.3	82.4	82.8
Glu	90.3	81.9	72.6	82.6	81.4	84.1	88.8	92.3
Gly	78.4	71.9	65.6	83.1	53.1	73.5	80.0	85.2
His	83.0	83.6	74.8	79.3	78.4	77.2	70.4	87.4
Ile	89.7	77.6	77.0	84.8	57.3	68.9	81.4	81.8
Leu	91.9	81.0	79.4	86.2	81.8	73.5	84.1	84.6
Lys	85.9	90.9	81.6	82.5	69.1	66.2	81.3	85.9
Met	84.8	80.4	76.4	80.8	61.7	72.5	81.9	76.7
Phe	93.2	81.3	82.2	84.1	73.1	81.4	82.9	87.2
Pro	88.0	77.1	76.1	80.0	78.4	73.4	79.5	88.3
Ser	87.3	85.0	63.7	80.7	63.9	77.4	82.0	83.0
Thr	86.6	77.5	69.9	83.3	53.9	71.8	77.3	78.8
Tyr	91.4	78.7	77.6	84.8	68.7	69.2	86.7	83.0
Val	89.6	75.5	77.5	84.0	64.9	73.2	83.2	84.5
Average	88.4	81.0	74.3	82.9	68.3	75.1	82.2	84.9

* International feed number.
Ala, alanine; Arg, arginine; Asp, aspartic acid; Glu; glutamic acid; Gly, glycine; His, histidine; Ile, isoleucine; Leu, leucine; Lys, lysine; Met, methionine; Phe, phenylalanine; Pro, proline; Ser, serine; Thr, threonine; Tyr, tyrosine; Val, valine.

intake and thus reduce nutrient intake. Also, an excess in dietary energy may result in an increased deposition of body fat. If the dietary energy level is too low, protein will be used for energy instead of tissue synthesis. Based on current information, it appears that a DE level of 2.8 kcal g^{-1} for 28% and 32% protein diets is adequate for diets used for foodfish grow-out. Reducing DE may reduce body fat but weight gain is also likely to decrease. In commercial catfish diet formulation there is no practical way to adjust the DE and maintain dietary fat and fibre levels in the desired ranges. Thus, for a diet containing certain percentage of protein, the DE/P ratio is usually fixed. The DE/P ratio for a 28% and a 32% protein diet is about 11.8 and 10.3 kcal g^{-1} of digestible protein (or a P/DE ratio of 85 and 97 mg digestible protein $kcal^{-1}$), respectively. DE/P ratios of this range appear to be optimal for maximal growth and result in an acceptable body-fat level.

Energy values of feedstuffs should be expressed in a manner that reflects available (utilizable) energy. Gross energy, which is a measure of the heat liberated on complete oxidation (combustion) of a compound, is not a practical indicator of usable energy because certain compounds are not as digestible as others. DE is often used to express the dietary energy of channel catfish diets. Metabolizable energy, DE minus energy losses from the urine in livestock or urine and gills in fish, is used to express energy content of feeds for livestock. Theoretically, using metabolizable energy to express dietary energy may be more desirable than using DE, since metabolizable energy is a more precise measure of available energy for metabolism. However, in a practical sense, there is little advantage in using metabolizable energy values rather than DE values in formulating fish diets because losses in digestion account for most of the variation in losses of gross energy. Also, energy losses through the gills and urine by fish are smaller than non-faecal losses in other animals, and these losses do not vary among feedstuffs as much as faecal losses.

Energy digestibility coefficients of commonly used feedstuffs are listed in Table 21.2. Lipids are a good source of energy for channel catfish. Starches have lower DE values than lipids and proteins. Generally, animal feedstuffs have higher DE values than plant feedstuffs.

Lipids and fatty acids

Channel catfish appear to have the ability to synthesize most of their fatty acids; thus, nutritionally there may be no 'best' level of dietary lipid except that needed to provide essential fatty acids (EFA). Channel catfish have been fed diets containing up to 16% lipid without conclusive evidence as to which level is the best for optimum growth. Even so, there is probably an optimum level of lipid to be used in the diet in respect of protein sparing, product quality and the constraints of diet manufacture. Lipid levels in commercial diets for fingerlings and foodfish rarely exceed 5–6%. About 3–4% of the lipid is inherent in the dietary ingredients, with the remaining 1–2% being sprayed on to the finished pellets to control diet dust. Too much dietary lipid may result in excessive fat deposition in the body cavity and tissues, which may adversely affect yield, quality and storage of processed products.

EFA requirements for channel catfish and most other warm-water fish have not been precisely defined, but channel catfish appear to require a small amount of n-3 fatty acids (Satoh *et al.*, 1989). The n-6 fatty acids appear to be non-essential to channel-catfish growth, but may be required for the proper immune functions of channel catfish (Fracalossi and Lovell, 1994). Because channel catfish apparently have the capability to elongate and desaturate linolenic acid (18:3n-3) to synthesize n-3 highly unsaturated fatty acids (HUFA), it appears that 1–2% dietary linolenic acid or 0.5–0.75% n-3 HUFA will satisfy the n-3 EFA requirement (Satoh *et al.*, 1989). The n-3 EFA requirement can be supplied by marine fish-oil, such as menhaden oil. Natural food organisms in the pond may also be a source of EFA.

Fish, animal or vegetable oils can be used in commercial channel-catfish diets. Marine fish-oils may impart 'fishy' flavours to the fish flesh if fed at high levels. In addition, there is evidence that levels of menhaden oil of 2–7% of the diet may be immunosuppressive to channel catfish, probably because of high levels of n-3 HUFA in menhaden oil (Fracalossi and Lovell, 1994; Li *et al.*, 1994). Channel-catfish diets manufactured in the Mississippi Delta have been generally sprayed with catfish oil, a local product extracted from catfish offal, or with a blend of catfish and menhaden oils. A mixture of animal and plant oils can also be used.

Fish-oil is 97% digestible to channel catfish (Table 21.2). Digestibility of other concentrated lipid sources has not been determined for channel catfish, but one would expect that lipids are highly digestible.

Carbohydrates

Fresh- and warm-water fish, including channel catfish, can use much higher levels of dietary digestible carbohydrate than cold-water or marine fish (NRC, 1993; Wilson, 1994). This is probably because warm-water fish have much higher intestinal amylase activity than marine fish and their cold-water counterparts (Wilson, 1994). Although no dietary requirement for carbohydrate has been demonstrated in fish, it is important to provide the appropriate amounts of digestible carbohydrate in fish diets because carbohydrates are the least expensive energy source. Enzymes for digestion and metabolism of carbohydrate have been detected in several fish species. However, hormonal and metabolic control of carbohydrate metabolism in fish remains unclear and may differ from that of mammals. *In vitro* studies have shown that insulin release in channel-catfish islets responds to glucose in a similar manner to that in mammals, but responds to only a few of the common stimuli that cause the release of insulin in mammals (Ronner and Scarpa, 1987).

Channel catfish are able to utilize the polysaccharides dextrin and starch (Wilson and Poe, 1987) as effectively as lipids as an energy source within certain carbohydrate-to-lipid ratios. However, utilization of mono- (glucose and fructose) and disaccharides (maltose or sucrose) by channel catfish is not as effective. Oral administration of glucose and maltose in channel catfish resulted in persistent hyperglycaemia, an indication of diabetes. However, circulating insulin levels in fish are similar or higher than those observed in mammals (Mommsen and Plisetskaya, 1991). The relative intolerance of fish to large doses of glucose, despite the high levels of circulating insulin, has been suggested to resemble non-insulin-dependent rather than insulin-dependent diabetic conditions (Hertz *et al.*, 1989). Studies using radioisotopic techniques revealed that channel catfish metabolize glucose in a manner similar to that of mammals, but at a much slower rate (Saad, 1989). Apparently, channel catfish do not have adequate enzyme or endocrine systems for the rapid metabolism of glucose.

A typical commercial channel-catfish diet contains 25% or more soluble (digestible) carbohydrate. An additional 3–6% carbohydrate is generally present as crude fibre. Fibre is considered to be indigestible by channel catfish; thus, it is not desirable in the diet because indigestible materials may 'pollute' the water. However, there is always some fibre inherent in practical ingredients. Channel catfish do not utilize starches as well as lipids, but digestibility of soluble carbohydrates by channel catfish (Table 21.2) is about 60% or higher, which is much higher than values reported for cold-water fish.

Vitamins

Vitamin requirements for channel catfish (Table 21.4) have generally been determined for juvenile fish fed chemically defined diets under controlled conditions in the laboratory. Requirements determined for young fish are considered to be adequate to meet the needs of larger fish. Commercial channel-catfish diets

Table 21.4. Vitamin deficiency signs, minimum dietary levels required to prevent signs of deficiency in channel catfish* and recommended levels in foodfish diet.

Vitamin	Deficiency signs	Requirement (unit kg^{-1} diet)	Recommendation (unit kg^{-1} diet)
A	Exophthalmia, oedema, haemorrhagic kidney, skin depigmentation	1000–2000 IU	2200 IU
D	Low body ash, calcium and phosphorus	250–1000 IU	1100 IU
E	Muscular dystrophy, exudative diathesis, skin depigmentation, erythrocyte haemolysis, splenic and pancreatic haemosiderosis, fatty liver, ceroid deposition	25–50 IU	30 IU
K	Haemorrhagic skin	R	4.4 mg
Thiamine	Loss of equilibrium, nervousness, dark skin colour	1 mg	2.5 mg
Riboflavin	Short-body dwarfism	6†–9 mg	6 mg
Pyridoxine	Greenish-blue coloration, tetany, nervous disorders, erratic swimming	3 mg	5 mg
Pantothenic acid	Clubbed gills, emaciation, anaemia, eroded epidermis	10–15 mg	15 mg
Niacin	Skin and fin lesions, exophthalmia, deformed jaws, anaemia	7.4‡–14 mg	None
Biotin	Hypersensitive, skin depigmentation, reduced liver pyruvate carboxylase activity	R	None
Folic acid	Anaemia	1.5 mg	2.2 mg
B_{12}	Anaemia	R	0.01 mg
Choline	Fatty liver, haemorrhagic kidney and intestine	400 mg	None
Inositol	Not demonstrated	NR	None
Ascorbic acid	Scoliosis, lordosis, internal and external haemorrhage, fin erosion, reduced bone collagen formation	11–60 mg	50 mg§

* Requirements and deficiency signs are based on Robinson (1989) and NRC (1993) unless indicated otherwise. Anorexia, reduced weight gain, and mortality are not listed as deficiency signs since they are common vitamin deficiency signs.

† Serrinni *et al.* (1996).

‡ Ng *et al.* (1997).

§ In finished diet – amount needed depends on type of vitamin C added.

R, required; NR, not required.

are generally supplemented with a vitamin premix that contains all essential vitamins in sufficient quantities to meet the requirement, including losses due to diet processing. The vitamins present in the dietary ingredients and natural food organisms in the pond water have usually not been considered due to the lack of information on the vitamin bioavailability of these sources. New data have indicated that vitamins inherent in dietary ingredients make a significant contribution to the vitamin nutrition of pond-raised channel catfish (Ng *et al.*, 1998; Zhang and Wilson, 1999). There is a sizeable body of evidence that supplementation of some vitamins may be reduced or eliminated in diets for pond-raised channel catfish (Robinson *et al.*, 1998).

Vitamin deficiency signs (Table 21.4) can be induced in channel catfish fed chemically defined diets deficient in a particular vitamin under laboratory conditions. The common vitamin deficiency signs are anorexia, reduced growth and increased mortality. However, vitamin deficiencies rarely occur in wild fish populations because fish growth in the wild is relatively slow and natural foods contain adequate amounts of all vitamins to meet the needs of the fish. Although cultured channel catfish grow rapidly, vitamin deficiencies are uncommon because commercial diets contain a vitamin supplement that is generally more than adequate to meet the fish's needs. There has been considerable interest among channel-catfish researchers and producers concerning the use of megadose levels of certain vitamins, particularly vitamin C, to enhance the disease resistance of the fish. Early studies indicated that high levels of vitamin C (ten times or more than the level needed for normal growth) reduced mortality due to certain bacterial diseases affecting channel catfish. More recent studies did not show an advantage to using high levels of dietary vitamin C for disease resistance in channel catfish (Li *et al.*, 1993, 1998). Data from these studies indicate that channel catfish respond to dietary vitamin C during disease challenge in an 'all-or-none' type of response – that is, if vitamin C is not present mortalities are increased during disease challenge, but if vitamin C is present in the diet mortalities are significantly reduced. Concentrations as low as 25 mg kg^{-1} vitamin C have been shown to enhance the survival of channel catfish during challenge with the bacterium *Edwardsiella ictaluri*. There is evidence that the vitamin C requirement of channel catfish is as low as 11–15 mg kg^{-1} of diet. Commercial channel-catfish diets manufactured in the Mississippi Delta generally contain about 50–100 mg kg^{-1} vitamin C in the final diet, which is sufficient for the optimum growth and health of the fish.

Minerals

Mineral nutrition studies with fish are complicated by dissolved minerals found in the water. For example, a dietary calcium requirement can only be demonstrated in channel-catfish fingerlings reared in calcium-free water. In water containing sufficient calcium, channel catfish can meet their calcium requirement by absorption of calcium from the water. Fourteen minerals are considered to be essential for channel catfish. Although mineral studies with fish are difficult to conduct, deficiency signs and quantitative requirements for these 14 minerals have been determined for channel catfish (Table 21.5).

Phosphorus is a particularly important mineral that is required in a relatively large quantity by fish. Feedstuffs, especially those of plant origin, are poor sources of biologically available phosphorus and fish do not obtain significant amounts of phosphorus from pond water. As a result, channel-catfish diets are usually supplemented with phosphorus. Dicalcium and defluorinated phosphates are

Table 21.5. Mineral deficiency signs, minimum dietary levels required to prevent deficiency signs in channel catfish and recommended levels in foodfish diet. Requirements and deficiency signs are based on Robinson (1989) and NRC (1993). Anorexia, reduced weight gain and mortality are not listed as deficiency signs since they are common mineral deficiency signs. Minerals listed as not determined (ND) are assumed to be required.

Mineral	Deficiency signs	Requirement ($unit\ kg^{-1}$ diet)	Recommendation ($unit\ kg^{-1}$ diet)
Ca*	Reduced bone ash	None	None
P†	Reduced bone ash, Ca and P	3–4 g	3.5 g
Mg	Sluggishness, muscle flaccidity, reduced body Mg	0.2‡–0.4 g	None
Na	ND	ND	None
K§	None	2.6 g	None
Cl	ND	ND	None
S	ND	ND	None
Co‖	ND	ND	0.05 mg
I‖	ND	ND	2.4 mg
Zn‖¶	Reduced serum zinc and serum alkaline phosphatase activity, reduced bone zinc and calcium concentrations	20 mg	200 mg
Se	Reduced liver and plasma selenium-dependent glutathione peroxidase activity	0.25 mg	0.1 mg
Mn‖	None	2.4 mg	25 mg
Fe‖	Reduced haemoglobin, haematocrit, erythrocyte count, reduced serum iron and transferrin saturation levels	20 mg	30 mg
Cu‖	Reduced heart cytochrome c oxidase, reduced hepatic Cu–Zn superoxide dismutase activities	4.8 mg	5 mg

* Deficiency cannot be demonstrated in channel catfish reared in water containing sufficient calcium.
† Requirement expressed on an available basis.
‡ From Lim and Klesius (1999).
§ From Wilson and El Naggar (1992). Requirement based on whole-body potassium balance.
‖ A supplement may not be needed in diets containing 4% animal protein and above.
¶ Requirement will increase in presence of phytic acid.

commonly used as phosphorus supplements in channel-catfish diets. Growth data from both laboratory and pond studies have indicated that defluorinated phosphates had essentially the same nutritional value as dicalcium phosphate for channel catfish. Approximately two-thirds of the phosphorus in feedstuffs of plant origin is in the form of phytate, a bound form of phosphorus that is poorly available to fish. Studies have demonstrated that phytase enzymes can be used in channel-catfish diets to release phytate phosphorus, making it available for use (Jackson *et al.*, 1996; Li and Robinson, 1997). Phosphorus availability of phosphorus sources and commonly used feedstuffs is listed in Table 21.6.

Channel-catfish diets may be supplemented with a trace-mineral premix that contains trace minerals in sufficient amounts to meet or exceed dietary requirements of channel catfish. However, there is evidence that some supplemental trace minerals are not needed, particularly in diets containing animal protein, such as fish-meal or meat, bone and blood meal.

Recently, chelated minerals have received some attention in animal nutrition research. It is assumed that chelation of minerals with protein or an amino acid protects the element from forming insoluble complexes with phytic acid or other minerals in the digestive tract. Zinc from zinc methionine has been

Table 21.6. Average percentage apparent phosphorus availability of feedstuffs determined for channel catfish.

Feedstuffs	International feed number	Availability (%)
Calcium phosphate		
Monobasic	6-01-082	94*
Dibasic	6-01-080	65,* 82†
Defluorinated	6-01-780	82†
Sodium phosphate, monobasic	6-04-288	90*
Casein	5-01-162	90‡
Egg albumin		71‡
Meat and bone/blood meal		84†
Fish-meal (anchovy)	5-01-985	40*
Fish-meal (menhaden)	5-02-009	39,* 75†
Maize grain	4-02-935	25*
Cottonseed meal (41%)	5-01-621	43†
Soybean meal (44%)	5-04-604	50*
Soybean meal (48%)	5-04-612	54,* 49,† 29‡
Wheat middlings	4-05-205	28,* 38†

* From Lovell (1978). Based on digestibility trial using chromium oxide as an indicator.

† From Li and Robinson (1996b). Based on weight gain of fish compared with that on a reference diet containing 0.4% available phosphorus from monobasic sodium phosphate. Weight gain appeared to be a more reliable indicator than did bone phosphorus concentrations.

‡ From Wilson *et al.* (1982). Based on digestibility trial using chromium oxide as an indicator.

reported to be several times more available than that from zinc sulphate to channel catfish fed either chemically defined or practical diets, as measured by weight gain and bone zinc concentration (Paripatananont and Lovell, 1995). In contrast, Li and Robinson (1996a) found that zinc methionine was equally but not more available than zinc sulphate to channel catfish fed practical diets. The basal diet containing about 60 mg kg^{-1} endogenous zinc appeared to be adequate for maximum growth. Lim *et al.* (1996) showed that iron methionine and iron sulphate were equally effective in preventing anaemia in channel catfish.

Practical Diets

Fry and fingerling diets

Diets of a small particle size (flours, meals or crumbles) are needed for feeding channel-catfish fry and small fingerlings. Flour- or meal-type diets are usually prepared by either reducing the particle size of a steam-pelleted or extruded diet by grinding and screening to the appropriate size or by finely grinding ingredients to a particle size of less than 0.5 mm and mixing the ground ingredients. Crumbles are usually prepared by crushing (crumbling) pelleted or extruded diets and screening for proper size. If flour- or meal-type diets are pelleted or extruded and then reground to the proper particle size instead of simply grinding and mixing ingredients, water-soluble nutrients are less likely to be lost to the water. Since water-soluble vitamins are easily lost to the water, fry diets should be overfortified with these vitamins. Spraying fat on the surface of meal or crumble diets improves water stability and reduces nutrient loss to the water.

A typical channel-catfish fry diet contains 45–50% crude protein, which is mainly composed of fish-meal (Table 21.7). Fines from 28–32% protein diets may be fed to fry in nursery ponds because diet mainly serves to fertilize the ponds. Fingerling channel catfish are typically fed crumbles or small floating pellets (3 mm diameter). Typical fingerling diets contain 35% crude protein (Table 21.7), a part of which is supplied by fish-meal or a mixture of fish-meal and other animal products. Some producers feed fingerlings the same diet they feed to foodfish; even though the pellets are large, the fingerlings nibble the diet after the pellets are softened and begin to break up in the water. Fingerlings appear to grow well using this feeding strategy, but nutrient losses are likely, due to leaching, because of the extended time the pellet is in contact with the water.

Foodfish diets

Channel catfish grown from advanced fingerlings to marketable size are typically fed a 28% or a 32% protein floating diet of approximately 5 mm in diameter (Table 21.7). Low-protein diets (24–26%) can be used if the fish are fed to satiation, but high-protein diets are required for maximum production if diet is

Table 21.7. Examples of typical channel-catfish diets.

	% of diet							
	Fry diet	Fingerling	Foodfish diet					
Ingredient	(50%)*	diet (35%)	(32%)	(32%)	(32%)	(28%)	(28%)	(28%)
Soybean meal (48%)*	–	38.8	35.0	34.6	48.4	24.4	29.7	37.8
Cottonseed meal (41%)	–	10.0	10.0	12.0	–	10.0	10.0	–
Menhaden meal (61%)	60.2	6.0	4.0	–	–	4.0	–	–
Meat/bone/blood (65%)	15.3	6.0	4.0	8.0	4.0	4.0	4.0	4.0
Maize grain	–	16.1	29.9	30.3	30.2	35.5	34.0	35.8
Wheat middlings	19.0	20.0	15.0	15.0	15.0	20.0	20.0	20.0
Dicalcium phosphate	–	1.0	0.5	0.5	0.75	0.5	0.75	0.75
Catfish vitamin mix†	Include	Include	Include	Include	Include	Include	Include	Include
Catfish mineral mix†	Include	Include	Include	Include	Include	Include	Include	Include
Fat/oil‡	5.0	2.0	1.5	1.5	1.5	1.5	1.5	1.5

* Percentage protein.

† Commercial mix that meets or exceeds all requirements for channel catfish.

‡ Sprayed on finished diet pellet to reduce diet dust ('fines').

severely restricted (Li and Lovell, 1992). However, when channel catfish were fed at a maximum rate of 90 kg ha^{-1} day^{-1}, a 28% protein diet provided satisfactory production, processing yield and body composition characteristics comparable to those of fish fed a 32% protein diet (Robinson and Li, 1999b). Fish fed diets containing 24–26% protein or below generally have higher body fat and lower carcass yield, primarily due to an imbalance of the DE/P ratio in the diet. Because management practices vary greatly throughout the channel-catfish industry, the choice of dietary protein percentage to use is up to the individual producer.

Broodfish diets

Brood channel catfish are usually fed the same type of diet as that used for foodfish. Some producers prefer feeding sinking or slow-sinking diets because large broodfish may be hesitant to feed at the surface. However, because broodfish generally feed slowly, sinking pellets may disintegrate before they can be consumed.

Winter diets

During the winter, if producers choose to feed, they may feed the same diet that they use during the summer to feed foodfish. However, some producers feed a 25% protein, slow-sinking feed (Table 21.8) in cool weather because the fish seem reluctant to feed on the water surface, and wind may blow floating feed ashore before it is eaten.

Medicated diets

Although the efficacy of using medicated diets to treat bacterial infections is debatable, medicated diets have been used to treat diseased fish for a number of years. Two antibiotics, Romet® (sulphadimethoxine-ormetoprim, Alpharma, Inc., Animal Health Division, Fort Lee, New Jersey, USA) and Terramycin® (oxytetracycline, Philbro Animal Health, Fort Lee, New Jersey, USA) are registered by the US Food and Drug Administration (FDA) to treat bacterial infections of channel catfish through their incorporation into diets. Romet® is registered for control of enteric septicaemia of catfish (ESC) and has also been shown to be effective in treating motile aeromonad septicaemia (MAS), caused by *Aeromonas hydrophila*, and systemic columnaris infections. The Romet®-medicated diet (Table 21.8) is fed at a rate (dependent on the formulation of Romet® used) sufficient to deliver 5.1 g of antibiotic 100 kg^{-1} fish day^{-1}. Romet® is heat-stable, so it can be used in a floating diet. Research indicates that the level of fish-meal should be increased to 16% to improve the palatability of diets containing

Table 21.8. Typical channel-catfish winter diet and Romet®-medicated diet.

	% of diet	
Ingredient	Winter diet (25% protein slow-sinking)	Medicated diet* (32% protein with Romet®)
Soybean meal (48%)†	18.3	26.8
Cottonseed meal (41%)	10.0	10.0
Menhaden meal (61%)	4.0	16.0
Meat/bone/blood (65%)	4.0	–
Maize grain	35.1	23.0
Wheat middlings	25.0	20.0
Dicalcium phosphate	1.0	1.0
Catfish vitamin mix‡	Include	Include
Catfish mineral mix‡	Include	Include
Fat/oil§	2.5	1.5
Romet®	–	1.65

* Protein levels not critical. Could be lowered, but fish-meal needs to remain at 16%.
† Percentage protein.
‡ Commercial mix that meets or exceeds all requirements for channel catfish.
§ Sprayed on after extrusion to reduce diet dust.

Romet®. Romet® is registered by the FDA to be fed at the prescribed rate for 5 consecutive days. If the majority of fish affected by the disease in the pond are fingerlings, feeding crumbles or 3-mm diameter pellets may be beneficial. If mortality does not decrease after treatment, additional sick fish should be diagnosed. An additional 5-day period of medicated diet may be prescribed. A 3-day mandatory withdrawal period is required before fish can be slaughtered.

Terramycin® is a broad-spectrum antibiotic that is registered by the FDA to treat MAS infections. Terramycin® has also been shown to be effective in treating other aeromonad infections, ESC and systemic columnaris infections. The Terramycin®-medicated diet is fed at a rate sufficient to deliver 5.5 g active oxytetracycline 100 kg^{-1} fish day^{-1}. Terramycin®-medicated diets have been manufactured as sinking pellets because the antibiotic is heat-labile and does not withstand the high temperatures required to make floating pellets. However, a new 'cold' extrusion process has been developed to make floating Terramycin® diets, which allow the feeder to observe the fish feed during a bacterial disease episode. The Terramycin®-medicated diet is primarily recommended to treat systemic columnaris infections or ESC infections caused by strains of *E. ictaluri* that are resistant to Romet®. Terramycin® is registered to be fed for 7–10 consecutive days. A 21-day withdrawal period is required before fish are slaughtered and processed.

All-plant diets

Feedstuffs of animal origin, especially marine fish-meals, are generally considered to be of higher quality than feedstuffs of plant origin. However, because animal protein is expensive and its availability is often limited, efforts are being made to reduce or eliminate animal protein in diets for foodfish grow-out. Animal protein appears to be required for normal growth and survival in the diets of channel-catfish fry and fingerlings. There is a sizeable amount of evidence that all-plant protein diets composed of soybean meal, cottonseed meal, maize, wheat middlings and vitamin and mineral premixes are satisfactory for normal growth of channel catfish raised from advanced fingerling to marketable size in ponds under single-batch (Robinson and Li, 1998, 1999a) or multi-batch cropping systems (Reigh, 1999). Currently, all-plant channel-catfish diets are not available commercially. However, diets are available that contain only 1% or 2% animal protein.

Feeding Practices

Although considerable research has been conducted on feeding channel catfish, feeding is far from an exact science. It is a highly subjective process that differs greatly among channel-catfish producers. Basically there does not appear to be one 'best' method for feeding channel catfish, particularly considering that numerous factors (most of which cannot be controlled) affect feeding.

Feed allowance/feeding rates

Channel catfish have been cultured for many years, but there is still considerable variation in feeding rates on commercial catfish farms. Computer programs, which generally determine feeding rate based on a percentage of fish body weight, are available and are used by some producers. Feeding a prescribed amount of diet based on fish biomass in a particular pond works best when the biomass in each pond is known and a fairly accurate estimate of feed conversion can be made. However, since most channel-catfish producers do not clean-harvest but rather remove only harvestable-size fish and replace the harvested fish with fingerlings, after several harvests and restockings it is difficult to accurately determine biomass. In fact, many channel-catfish producers judge their inventory by the amount of diet fed. As a result, channel catfish are generally fed once daily to what is commonly called satiation, i.e. feeding the fish all they will ingest in a reasonable period of time. However, feeding to satiation is highly subjective and is often difficult to achieve in ponds containing a large standing crop of fish without adversely affecting water quality. Many producers limit diet to a pre-set (cut-off) level because of concern about deteriorating water quality.

Because channel catfish are most often cultured using a multiple-batch production system, in which several sizes of fish are present in the pond, it is recommended that they be fed to satiation. Actively feeding fish should be satiated within 20–30 min. Offering as much diet as possible (without wasting diet) provides a better opportunity for the smaller, less aggressive fish to receive diet. Satiation feeding appears to be particularly important when channel catfish are fed less frequently than on a daily basis. Although it is recommended that channel catfish typically be fed as much diet as they will consume, at large standing crops of fish it may be impossible to satiate the fish and maintain water quality at an acceptable standard. Feeding rates should not exceed what can be assimilated by organisms in the pond and not require excessive use of aeration. This is difficult to judge, but generally the long-term average daily feed allowance should not exceed about 120 kg ha^{-1}. Feeding levels higher than this on occasion is not harmful. Overfeeding should be avoided because wasted diet increases production cost by reducing feed efficiency and it also contributes to deterioration of water quality.

Feeding methods

In the hatchery, channel-catfish fry are usually fed either by hand or by using an automatic dispenser. For fingerlings and foodfish on large farms, diet is typically blown on to the surface of the water using pneumatic dispensers mounted on or pulled by vehicles. Diets should be scattered over a large area to provide feeding opportunities for as many fish as possible. It is desirable to feed on all sides of the pond, but this is often not practical because prevailing winds dictate that diet must be distributed along the upwind side to prevent it from washing ashore.

Feeding frequency

Newly hatched fry are usually held in indoor troughs and tanks for about 5–10 days before release into outdoor nursery ponds. During this period, fry should be fed at about 25% body weight divided into eight to ten feedings daily. Automatic feeders can be used. Once released into nursery ponds, fry may be fed once or twice daily because at this stage natural foods play a major role in fry nutrition and diet serves primarily to fertilize the pond. It is difficult to effectively feed fry recently stocked in large nursery ponds. The tiny fish spread out over the pond and are relatively weak swimmers so they are not able to move rapidly to areas where manufactured diets are offered.

Typically, channel-catfish fingerlings and foodfish are fed once a day, 7 days a week, during the growing season. Feeding twice a day appears to improve growth and feed efficiency, especially for fingerlings. However, the logistics of multiple feedings on large farms may make this impractical. During certain bacterial disease episodes, it may be beneficial to feed every other day or every third day. Winter feeding frequency of channel catfish is mainly dictated by water temperatures. Feeding diseased fish and overwintering fish is discussed in detail in later sections.

Feeding schedule

The best time of day to feed is still debated, but the point is more or less academic. On large catfish farms, the time fish are fed is largely dictated by the logistics required to feed many hectares of ponds in a limited time period. As a result, many producers start feeding early in the morning as soon as dissolved oxygen levels begin to increase. Research has shown that there are no significant differences in weight gain, feed consumption, feed conversion and survival among channel catfish fed to satiation at 0800, 1600 or 2000 h (Robinson *et al.*, 1995). There are also no differences in emergency aeration time among treatments. However, feeding in the late afternoon or at night in large commercial ponds is not recommended because peak oxygen demand by the fish occurs 6–12 h after feeding, when dissolved oxygen levels in the pond are low. Thus, if aeration is insufficient, fish may be stressed or die. During warm weather, it appears most practical to begin feeding in the morning as the dissolved oxygen begins to increase. In cool weather, during autumn, spring and winter, water temperature is usually higher in the afternoon and fish will feed better.

Winter feeding

Although channel catfish feed inconsistently at water temperatures below 20°C, a winter feeding programme appears to be beneficial to prevent weight loss and maintain fish health. Research has shown that significant increases in weight gain can be obtained in fish that are fed during the winter months as compared with fish that are not fed. This appears to be particularly true with fingerlings. The amount of gain or loss realized during winter depends on the severity of the winter. Fish gain or lose more weight during a mild winter than during a cold winter. Research has shown that channel catfish raised under a single-batch cropping system that are not fed during the coldest winter months (December, January and February) can catch up with fish that are fed during the winter months when satiation feeding is resumed in the spring and summer (Kim and Lovell, 1995). However, under a multi-batch cropping system with various sizes of fish present, this may not be the case because the larger, more aggressive fish typically consume a disproportionate amount of diet. Thus, smaller fish may be unable to consume enough diet to overcome weight loss experienced during the winter months. If fish are to be marketed during winter, it would appear prudent to feed in the winter, particularly during a mild winter; otherwise weight loss and reduction of processing yield may be inevitable.

The benefit of winter feeding for health is less well defined than for weight gain. Logically, one would expect fish fed during the winter to be in better condition and perhaps more resistant to disease-causing organisms than fish that were not fed. However, research results on the effect of winter feeding on fish health are inconclusive. Some studies indicate that large channel catfish that are not fed during the winter are more resistant to *E. ictaluri* challenge, but small (fingerling) channel catfish not fed during the winter are less resistant to *E. ictaluri* (Kim and Lovell, 1995; Okwoche and Lovell, 1997). Studies conducted at the National Warmwater Aquaculture Center at Stoneville, Mississippi, USA, indicate that the effect on immune response of withholding diet from fish may be immediate – that is, withholding diet from fish immediately following bacterial exposure increases the chance of survival over fish that are continuously fed during the exposure (Wise *et al.*, 1999). Withholding diet from fish prior to the development of infection has no significant effect on the survival of fish following *E. ictaluri* exposure.

Many channel-catfish producers do not feed during the winter months. Often fish are not fed in the winter because inclement weather may prevent access to pond levees. It is recommended that channel catfish be fed during the winter, when temperatures are warm and pond levees are accessible, to prevent weight losses. Several schedules for winter feeding of fingerlings and foodfish have been suggested. Generally, all schedules are such that water temperature dictates feeding frequency. A typical winter feeding schedule is shown in Table 21.9. As a rule of thumb, one should basically feed on days that are warm enough for fish to be active.

While it is important that broodfish receive adequate nutrition throughout the year, it is especially important during the winter months. It is at this time of the year that eggs that were produced by females the previous summer after spawning are developing yolks and maturing. This process requires that broodfish receive adequate nutrition on a regular basis, which makes winter feeding of broodfish important. Feeding rates and frequencies used in winter feeding of broodfish (Table 21.9), as with fingerlings and foodfish, are based on water temperatures. Research has shown that broodfish that are not fed or are

Table 21.9. Winter feeding schedule for fingerling, food and brood channel catfish.

	Fingerlings		Foodfish		Broodfish	
Temperature (°C)	% Body weight	Frequency	% Body weight	Frequency	% Body weight	Frequency
<7		Do not feed		Do not feed		Do not feed
7–13	0.5–1.0	Once or twice a week	0.25–0.5	Once a week	0.25–0.5	Once a week
13–20	1.0–2.5	Daily or every other day	0.5–1.0	Every other day	0.5–1.0	Two or three times a week

fed a nutritionally deficient diet reproduce poorly (Lovell, 1989). Feeding forage fish (e.g. fathead minnows) has no advantages for broodfish over feeding a nutritionally complete diet. However, some producers stock forage fish in broodfish ponds to ensure that adequate food is available during the winter.

Feeding diseased fish

Feeding diseased fish is difficult because fish that are sick feed poorly, if at all. However, offering medication through the diet is generally the only method available to treat bacterial infections. There is considerable debate over the efficacy of medicated diets and the best method to treat diseased fish. Research has indicated that diet restriction reduces mortality of channel-catfish fingerlings exposed to *E. ictaluri* (Wise and Johnson, 1998). Not feeding or restricting diet during an ESC episode is a common practice of many channel-catfish producers, but it should be conducted on a pond-to-pond basis and not applied across an entire farm. Several important considerations should be taken into account when treating fish with medicated diets. An early and accurate diagnosis of the specific disease affecting the fish population must first be obtained if effective treatment is expected. Some strains of bacteria causing diseases in channel catfish are resistant to currently available antibiotics. If so, an alternative treatment strategy may be used. Resistance of bacteria to antibiotics may result from indiscriminately feeding medicated diets or from using non-prescribed feeding schedules. It is important to remember that, once a bacterial strain becomes resistant to an antibiotic, it may be impossible to treat future disease outbreaks due to the lack of effective legal antibiotics.

Feeding caged fish

Although the vast majority of channel catfish are cultured in open ponds, a small portion of channel catfish are reared in cages suspended in ponds that would be difficult to harvest by seine. In most cases, caged fish do not receive natural foods and therefore must be fed a nutritionally complete diet. Caged channel catfish are typically fed a floating diet containing 32–36% protein (Masser, 1997). Early studies appear to indicate that the dietary protein requirement of cage-raised channel catfish is higher than that of pond-cultured fish (Lovell, 1973; Webster *et al.*, 1992). A more recent study showed that caged channel catfish can be fed a 27% protein diet without adverse effects on fish growth and feed efficiency (Webster *et al.*, 1994). However, fish fed the 27% protein diet had a lower carcass yield and a higher body fat than fish fed higher-protein diets.

Caged channel catfish should be fed to approximate satiation (what they will consume in about 20–30 min) to achieve optimum production. Feeding fish based on percentage of biomass can also be used. Small fingerlings can be fed at 4–5% of their body weight day^{-1}. After they reach 15–20 cm in length, feeding rate will decrease to about 3% of body weight and at harvest size will reduce to about 2% or less. Fish biomass in cages can be estimated by total amount of diet fed and the feed conversion ratio, which is based on records from the previous year. A feed conversion ratio of 1.8 is common in cage culture of channel catfish. Feeding rate should be adjusted weekly based on fish feeding activity and growth. Some studies indicate that feeding twice a day increases fish growth and feed efficiency of cage-cultured channel catfish (Masser, 1997). In contrast, Webster *et al.* (1992) found that feeding once or twice daily did not affect growth and feed efficiency of channel catfish raised in cages.

Like pond-raised fish, cage-reared fish feed most aggressively at optimum temperature and oxygen levels. During the warm growing season, caged fish should be fed in the morning after sunrise when dissolved oxygen starts to increase. In early spring and autumn, fish feed better when the temperature rises in the late afternoon. Recent research on cage culture of channel catfish indicate that feeding at dusk increases fish growth and feed efficiency, probably because dusk is the natural feeding time for channel catfish (Masser, 1997). Dusk feeding is not recommended if aeration is not available.

In cage culture, diet is generally delivered by hand into the feeding ring located inside the cage. Demand feeders have been used successfully by some experienced producers. Demand feeders have some advantages, such as less labour required and optimized feeding rate, but disadvantages include training fish to use the feeder and frequent adjustments of the feeder. Feeding activity and water quality should be closely monitored because these factors are often neglected when using demand feeders. Demand feeders should not be used unless aeration is available.

Acknowledgements

This chapter is approved for publication as Book Chapter No. BC-9674 of the Mississippi Agricultural and Forestry Experiment Station (MAFES), Mississippi State University. The project is supported under MAFES Project Number MIS-0811.

References

Brown, P.B., Strange, R.J. and Robbins, K.R. (1985) Protein digestion coefficients for yearling channel catfish fed high protein feedstuffs. *Progressive Fish-Culturist* 47, 94–97.

Cruz, E.M. (1975) Determination of nutrient digestibility in various classes of natural and purified feed materials for channel catfish. PhD dissertation, Auburn University, Alabama.

Fracalossi, D.M. and Lovell, R.T. (1994) Dietary lipid sources influence responses of channel catfish (*Ictalurus punctatus*) to challenge with the pathogen *Edwardsiella ictaluri. Aquaculture* 119, 287–298.

Hertz, Y., Epstein, N., Abraham, M., Madar, Z., Hepher, B. and Gertler. A. (1989) Effects of metformin on plasma insulin, glucose metabolism and protein synthesis in the common carp (*Cyprinus carpio* L.). *Aquaculture* 80, 175–187.

Jackson, L.S., Li, M.H. and Robinson, E.H. (1996) Use of microbial phytase in channel catfish *Ictalurus punctatus* diets to improve utilization of phytate phosphorus. *Journal of the World Aquaculture Society* 27, 309–313.

Johnson, H.M. (1999) *Annual Report on the United States Seafood Industry*, 7th edn. H.J. Johnson and Associates, Bellevue, Washington.

Kim, M.K. and Lovell, R.T. (1995) Effect of overwinter feeding regimen on body weight, body composition, and resistance to *E. ictaluri* in channel catfish, *Ictalurus punctatus. Aquaculture* 134, 237–246.

Li, M. and Lovell, R.T. (1992) Comparison of satiate feeding and restricted feeding of channel catfish with various concentrations of dietary protein in production ponds. *Aquaculture* 103, 165–175.

Li, M.H. and Robinson, E.H. (1996a) Comparison of chelated zinc and zinc sulfate as zinc sources for growth and bone mineralization of channel catfish *Ictalurus punctatus* fed practical diets. *Aquaculture* 146, 237–243.

Li, M.H. and Robinson, E.H. (1996b) Phosphorus availability (digestibility) of common feedstuffs to channel catfish *Ictalurus punctatus* as measured by weight gain and bone mineralization. *Journal of the World Aquaculture Society* 27, 297–302.

Li, M.H. and Robinson, E.H. (1997) Microbial phytase can replace inorganic phosphorus supplements in channel catfish *Ictalurus punctatus* diets. *Journal of the World Aquaculture Society* 28, 402–406.

Li, M.H. and Robinson, E.H. (1998) Effects of supplemental lysine and methionine in low protein diets on performance of channel catfish *Ictalurus punctatus. Aquaculture* 163, 297–307.

Li, M.H., Johnson, M.R. and Robinson, E.H. (1993) Elevated dietary vitamin C concentrations did not increase resistance of channel catfish, *Ictalurus punctatus* against *Edwardsiella ictaluri* infection. *Aquaculture* 117, 303–312.

Li, M.H., Wise, D.J., Johnson, M.R. and Robinson, E.H. (1994) Dietary menhaden oil reduced resistance of channel catfish, *Ictalurus punctatus*, to *Edwardsiella ictaluri. Aquaculture* 128, 335–344.

Li, M.H., Wise, D.J. and Robinson, E.H. (1998) Effect of dietary vitamin C on weight gain, tissue ascorbate concentration, stress response, and disease resistance of channel catfish *Ictalurus punctatus. Journal of the World Aquaculture Society* 29, 1–8.

Lim, C. and Klesius, P.H. (1999) Influence of dietary levels of magnesium on growth, tissue mineral content and resistance of channel catfish challenged with *Edwardsiella ictaluri*. In: *27th Fish Feed and Nutrition Workshop*. Portland, Oregon [abstract], p. 15.

Lim, C., Sealey, W.M. and Klesius, P.H. (1996) Iron methionine and iron sulfate as sources of dietary iron for channel catfish *Ictalurus punctatus*. *Journal of the World Aquaculture Society* 27, 290–296.

Lovell, R.T. (1973) Protein requirements of cage-cultured channel catfish. *Proceedings of the Annual Conference Southeastern Association of Game and Fish Commissioners* 26, 357–361.

Lovell, R.T. (1978) Dietary phosphorus requirement of channel catfish (*Ictalurus punctatus*). *Transactions of the American Fisheries Society* 197, 617–621.

Lovell, R.T. (1989) *Nutrition and Feeding of Fish*. Van Nostrand Reinhold, New York, 260 pp.

Mangalik, A. (1986) Dietary energy requirement of channel catfish. PhD dissertation, Auburn University, Alabama.

Masser, M.P. (1997) *Cage Culture: Handling and Feeding Caged Fish*. Publication No. 164, Southern Regional Aquaculture Center, Stoneville, Mississippi.

Minton, R.V. (1978) Responses of channel catfish fed diets of two nutrient concentrations at three rates in ponds. MSc thesis, Auburn University, Alabama.

Mommsen, T.P. and Plisetskaya, E.M. (1991) Insulin in fishes and agnathans: history, structure, and metabolic regulation. *Reviews in Aquatic Sciences* 4, 225–259.

Ng, W.K., Serrini, G., Zhang, Z. and Wilson, R.P. (1997) Niacin requirement and inability of tryptophan to act as a precursor of NAD^+ in channel catfish *Ictalurus punctatus*. *Aquaculture* 152, 273–285.

Ng, W.K., Keembiyehetty, C.N. and Wilson, R.P. (1998) Bioavailability of niacin from feed ingredients commonly used in feeds for channel catfish, *Ictalurus punctatus*. *Aquaculture* 161, 393–404.

NRC (National Research Council) (1993) *Nutritional Requirements of Fish*. National Academy Press, Washington, DC, 114 pp.

Okwoche, V.O. and Lovell, R.T. (1997) Cool weather feeding influences responses of channel catfish to *Edwardsiella ictaluri* challenge. *Journal of Aquatic Animal Health* 9, 163–171.

Page, J.W. and Andrews, J.W. (1973) Interactions of dietary levels of protein and energy on channel catfish *Ictalurus punctatus*. *Journal of Nutrition* 103, 1339–1346.

Paripatananont, T. and Lovell, R.T. (1995) Chelated zinc reduces the dietary zinc requirement of channel catfish, *Ictalurus punctatus*. *Aquaculture* 133, 73–82.

Reigh, R.C. (1999) Production characteristics of pond-raised channel catfish *Ictalurus punctatus* fed diets with and without animal protein for three growing seasons. *Journal of the World Aquaculture Society* 30, 154–160.

Robinson, E.H. (1989) Channel catfish nutrition. *Reviews in Aquatic Sciences* 1, 365–391.

Robinson, E.H. (1991) Improvement of cottonseed meal protein with supplemental lysine in feeds for channel catfish. *Journal of Applied Aquaculture* 1(2), 1–14.

Robinson, E.H. and Li, M.H. (1994) Use of plant proteins in catfish feeds: replacement of soybean meal with cottonseed meal and replacement of fish meal with soybean meal and cottonseed meal. *Journal of the World Aquaculture Society* 25, 271–276.

Robinson, E.H. and Li, M.H. (1996) *A Practical Guide to Nutrition, Feeds, and Feeding of Catfish*. Bulletin 1041, Mississippi Agricultural and Forestry Experiment Station, Mississippi State, Mississippi.

Robinson, E.H. and Li, M.H. (1998) Comparison of practical diets with and without animal protein at various concentrations of dietary protein for performance of channel catfish *Ictalurus punctatus* raised in earthen ponds. *Journal of the World Aquaculture Society* 29, 273–280.

Robinson, E.H. and Li, M.H. (1999a) Evaluation of practical diets with various levels of dietary protein and animal protein for pond-raised channel catfish *Ictalurus punctatus. Journal of the World Aquaculture Society* 30, 147–153.

Robinson, E.H. and Li, M.H. (1999b) Effect of dietary protein concentration and feeding rate on weight gain, feed efficiency, and body composition of pond-raised channel catfish *Ictalurus punctatus. Journal of the World Aquaculture Society* 30, 311–318.

Robinson, E.H., Wilson, R.P. and Poe, W.E. (1980) Re-evaluation of the lysine requirement and lysine utilization by fingerling channel catfish. *Journal of Nutrition* 110, 2313–2316.

Robinson, E.H., Jackson, L.S., Li, M.H., Kingsbury, S.K. and Tucker, C.S. (1995) Effect of time of feeding on growth of channel catfish. *Journal of the World Aquaculture Society* 26, 320–322.

Robinson, E.H., Li, M.H. and Oberle, D. (1998) *Catfish Vitamin Nutrition.* Bulletin 1078, Mississippi Agricultural and Forestry Experiment Station, Mississippi State, Mississippi.

Ronner, P. and Scarpa, A. (1987) Secretagogues to pancreatic hormone release in the channel catfish (*Ictalurus punctatus*). *General Comparative Endocrinology* 65, 354–362.

Saad, C.R.B. (1989) Carbohydrate metabolism in channel catfish. PhD dissertation, Auburn University, Alabama.

Satoh, S., Poe, W.E. and Wilson, R.P. (1989) Effect of dietary n-3 fatty acids on weight gain and liver polar lipid fatty acid composition of fingerling channel catfish. *Journal of Nutrition* 119, 23–28.

Serrini, G., Zhang, Z. and Wilson, R.P. (1996) Dietary riboflavin requirement of fingerling channel catfish (*Ictalurus punctatus*). *Aquaculture* 139, 285–290.

USDA (United States Department of Agriculture) (2000) *Aquaculture Outlook, March 2000.* Economic Research Service, USDA, Washington, DC.

Webster, C.D., Tidwell, J.H. and Yancey, D.H. (1992) Effect of protein level and feeding frequency on growth and body composition of cage-reared channel catfish. *Progressive Fish-Culturist* 54, 92–96.

Webster, C.D., Goodgame-Tiu, L.S., Tidwell, J.H. and Reed, E.B. (1994) Effects of dietary protein level on growth and body composition of channel catfish reared in cages. *Journal of Applied Aquaculture* 4(2), 73–86.

Wiang, C. (1977) Nutritional contribution of natural pond organisms to channel catfish growth in intensively fed ponds. PhD dissertation, Auburn University, Alabama.

Wilson, R.P. (1994) Review: utilization of dietary carbohydrate by fish. *Aquaculture* 124, 67–80.

Wilson, R.P. and El Naggar, G. (1992) Potassium requirement of fingerling channel catfish, *Ictalurus punctatus. Aquaculture* 108, 169–175.

Wilson, R.P. and Poe, W.E. (1985) Apparent digestible protein and energy coefficients of common feed ingredients for channel catfish. *Progressive Fish-Culturist* 47, 154–158.

Wilson, R.P. and Poe, W.E. (1987) Apparent inability of channel catfish to utilize dietary mono- and disaccharides as energy sources. *Journal of Nutrition* 117, 280–285.

Wilson, R.P., Robinson, E.H. and Poe, W.E. (1981) Apparent and true availability of amino acids from common feed ingredients for channel catfish. *Journal of Nutrition* 111, 923–929.

Wilson, R.P., Robinson, E.H., Gatlin, D.M., III and Poe, W.E. (1982) Dietary phosphorus requirement of channel catfish. *Journal of Nutrition* 112, 1197–1202.

Winfree, R.A. and Stickney, R.R. (1984) Starter diets for channel catfish: effect of dietary protein on growth and carcass composition. *Progressive Fish-Culturist* 46, 79–86.

Wise, D.J. and Johnson, M.R. (1998) Effect of feeding frequency and Romet-mediated feed on survival, antibody response, and weight gain of fingerling channel catfish *Ictalurus punctatus* after natural exposure to *Edwardsiella ictaluri*. *Journal of the World Aquaculture Society* 29, 169–175.

Wise, D.J., Li, M.H. and Robinson, E.H. (1999) Effect of starvation on the development of enteric septicemia in channel catfish prior to and following exposure to *Edwardsiella ictaluri*. In: *Mississippi Agricultural and Forestry Experiment Station Information Bulletin 357*. Mississippi State University, Mississippi State, pp. 37–41.

Zarate, D.D. (1997) Utilization of free versus protein-bound lysine by channel catfish. PhD dissertation, Auburn University, Alabama.

Zarate, D.D. and Lovell, R.T. (1997) Free lysine (L-lysine HCl) is utilized for growth less efficiently than protein-bound lysine (soybean meal) in practical diets by channel catfish *Ictalurus punctatus*. *Aquaculture* 159, 87–100.

Zhang, Z. and Wilson, R.P. (1999) Reevaluation of the choline requirement of fingerling channel catfish (*Ictalurus punctatus*) and determination of the availability of choline in common feed ingredients. *Aquaculture* 180, 89–98.

Eel, *Anguilla* spp.

22

Shuichi Satoh
Department of Aquatic Biosciences, Tokyo University of Fisheries, Konan, Minato, Tokyo 108-8477, Japan

Introduction

Eel is one of the major aquaculture species, and its global production was estimated to have reached above 200,000 t in 1997. The four major cultivated species are *Anguilla japonica* in the Far East, *Anguilla anguilla* in Europe, *Anguilla rostrata* in North America and *Anguilla australis* in Australia and New Zealand. Among these, *A. japonica* is the dominant species and is cultivated mainly in China and Japan.

Arai (1991) reviewed the nutritional requirement of eels. Since then, some new information has arisen. This chapter attempts to collate the recent research and to cover fundamental aspects of eel nutrition.

Nutritional Requirements

Protein and amino acids

Eels are carnivorous and require a higher amount of protein than their herbivorous and omnivorous counterparts. Nose and Arai (1972) reported that young *A. japonica* require more than 45% dietary protein when a purified casein-based diet was used. Recently, Tibbets *et al.* (2000) estimated that the optimum level of dietary protein for juvenile American eel (*A. rostrata*) was 47% when the diet was herring meal-based. De la Higuera *et al.* (1989) reported that the optimal daily protein intake for maximum growth was about 1.4 g 100 g^{-1} fish for *A. anguilla*. This value is much higher than that for rainbow trout (*Oncorhynchus mykiss*) and common carp (*Cyprinus carpio*), which was reported to be 1.16 g 100 g^{-1} fish by Ogino (1980). The protein requirement of eel (45–47%) is relatively higher than that of other fish species.

Arai *et al.* (1972b) determined the qualitative amino acid requirements of *A. japonica* and *A. anguilla*, using a diet deficient in specific amino acids. When the eels were deprived of dietary arginine, histidine, isoleucine, leucine, lysine, methionine, phenylalanine, threonine, tryptophan and valine, respectively, they showed lower growth than fish on the control diet (bovine liver profile), which contained all these amino acids (Fig. 22.1). A recovery in growth was noted when the amino acid-deficient diets were replaced by the control diet. Thus, eels require the ten essential amino acids, as with other fish species.

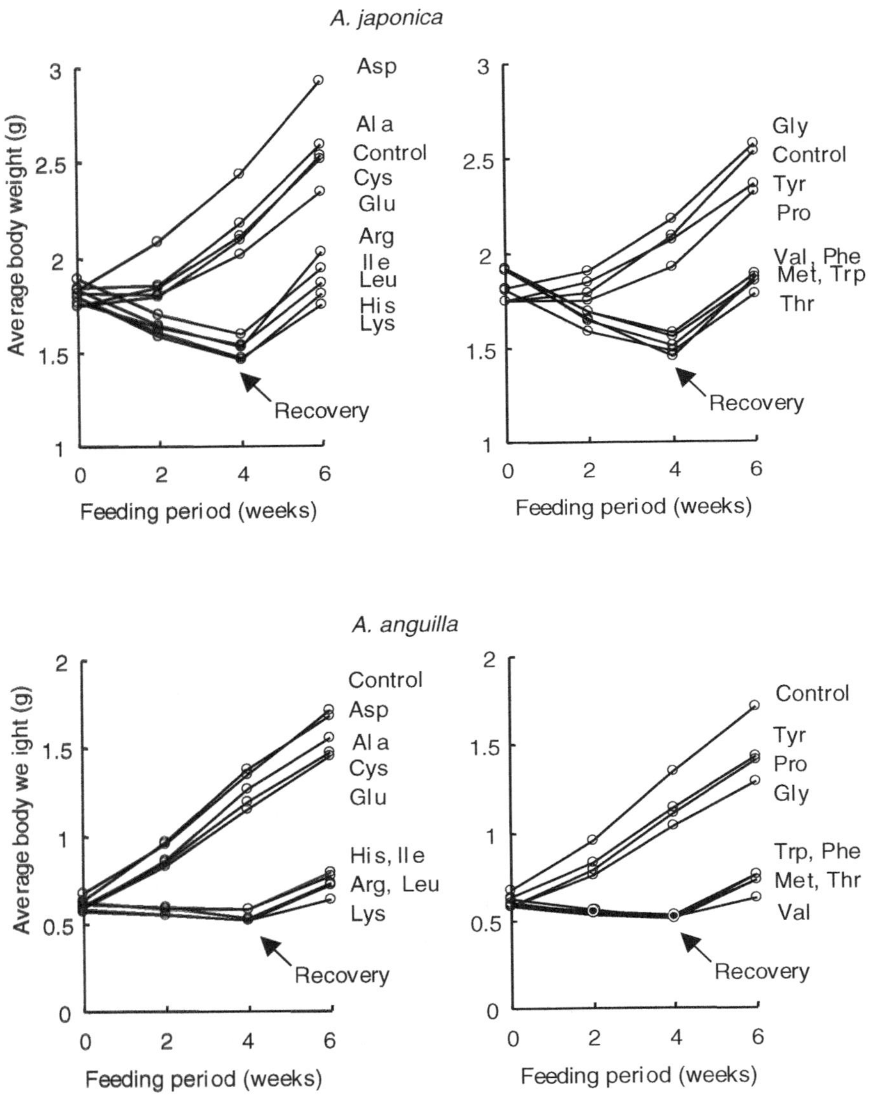

Fig. 22.1. Growth of eel (*A. japonica* and *A. anguilla*) fed amino acid-deficient (–○–) and complete (–●–) diets. Diets deficient in: Ala, alanine; Arg, arginine; Asp, aspartic acid; Cys, cystine; Glu, glutamine; Gly, glycine; His, histidine; Ile, isoleucine; Leu, leucine; Lys, lysine; Met, methionine; Phe, phenylalanine; Pro, proline; Thr, threonine; Trp, tryptophan; Tyr, tyrosine; Val, valine.

Nose (1979) carried out growth studies to determine the quantitative requirement of Japanese eel for each essential amino acid (Table 22.1). The arginine requirement of eel was lower than that of chinook salmon (*Oncorhynchus tshawytscha*). However, isoleucine and leucine requirements were higher than those of carp, channel catfish (*Ictalurus punctatus*) and Nile tilapia (*Oreochromis niloticus*). Further, the requirement of tryptophan was higher than that of the other species except tilapia. Recently, Akiyama *et al.* (1997) compared the essential amino acid requirements with the ratio of each essential amino acid content per total essential acid content (A/E ratio) among fish species, and inferred that Japanese eel was similar to milkfish (*Chanos chanos*) on the basis of its dietary requirements, while it ranked close to Nile tilapia in body amino acid profiles.

High-quality fish-meal has traditionally been a major ingredient in eel diets because of growth and palatability, averaging about 75% of the diet content (Gallagher and Degani, 1988). Among alternative protein sources, it was found that raw chicken meat was better than chicken blood in diets for European eel, and it was suggested that chicken meal and raw chicken wastes could be utilized in eel diets (Degani *et al.*, 1984). Recently, Lee and Bai (1997) reported that fish-meal can be replaced by haemoglobin powder up to 50% without supplementation of the three essential amino acids methionine, isoleucine and arginine for Japanese eel.

Schmitz *et al.* (1984) examined the digestibility of protein in the ingredients casein, gelatin, fish-meal, bacteria protein, soybean meal and soy-protein concentrate and reported values are 99, 94, 94, 89, 96 and 94%, respectively.

Lipids and fatty acids

Eel requires essential fatty acids, similarly to rainbow trout and common carp. Arai *et al.* (1971) examined the effects of supplemental lipids on the growth of

Table 22.1. Amino acid requirement of juvenile Japanese eel (from Nose, 1979).

Amino acid	Requirement as percentage of dietary protein*	Requirement as percentage of dry diet
Arginine	4.5	1.7
Histidine	2.1	0.8
Isoleucine	4.0	1.5
Leucine	5.3	2.0
Lysine	5.3	2.0
Methionine	3.2	1.2
Phenylalanine	5.8	2.2
Threonine	4.0	1.5
Tryptophan	1.1	0.4
Valine	4.0	1.5

* The experimental diet contained 38% crude protein.

young Japanese eel using purified casein-gelatin diets, and found that a mixture of maize oil and cod-liver oil at a 2 : 1 ratio was the most favourable for the growth of eel. Later Takeuchi *et al.* (1980) determined the essential fatty acid requirements of Japanese eel using diets containing methyl esters of fatty acids. They reported that the eel required both linoleic acid (18:2n-6) and linolenic acid (18:3n-3) in the same proportion as common carp but at a lower level in the diet – 0.5% rather than 1% of each.

Lipids play an important role as a source of energy in fish diets, especially for carnivorous species like eel, which utilize carbohydrates poorly as an energy source. Addition of lipids to a diet improved the utilization of dietary protein (protein-sparing effect). Burgos *et al.* (1989) studied the influence of partial replacement of protein content of diets by fat. They reported that decreasing dietary protein levels did not produce a decrease in body-weight gain, but instead body weight increased with increasing dietary fat level, indicating the possibility of limited protein replacement by fat. Degani *et al.* (1987) reported that large eels (40–120 g) grew well when fed diets containing a protein-to-energy ratio above 16.7 g MJ^{-1}. Gallego *et al.* (1993) also investigated the sparing effect of lipids on protein utilization. They obtained the best performance when fish were fed a diet containing 30% protein and 20% lipid, namely a protein/energy ratio of 16.1 g MJ^{-1} gross energy and a total energy content of 19.0 MJ kg^{-1} diet. Tibbets *et al.* (2000) estimated that the optimum protein/energy ratio for American eel was 22 g MJ^{-1}.

Fish-oil is the main lipid source in eel diet. Its superiority has been proved in a study by Gallagher and Degani (1988). They evaluated poultry meal and poultry oil as replacements for fish-meal and/or fish-oil in the diet of juvenile eels, and found that those fed the diet supplemented with 10% fish-oil had a significantly higher weight gain than those receiving either 5% fish-oil or 5 or 10% poultry oil.

Carbohydrates

As mentioned earlier, eels are carnivorous and therefore have limited ability to utilize dietary carbohydrates. Though supporting information is scanty, it has also been reported that the digestibility of gelatinized starch was quite high, ranging from 78 to 98%, when it was included in the diet at 20–60%. Gelatinized starch serves an important role as a binder for eel diet. It is an indispensable ingredient, irrespective of its doubtful utilization rate by eels.

Degani (1987) determined the influence of several dietary carbohydrate sources on enzyme activity and glucose content in the muscle and liver of eel. They reported that eels fed 30% bread flour or soluble maize starch had a higher glucose content in muscle and liver, and that aldolase activity was high in the fish on wheat meal, bread flour or soluble starch. Garcia-Gallego *et al.* (1991) evaluated the possibility of including high carbohydrate levels in commercial diets for eel, and suggested that the better utilization rates permit its inclusion at levels of 40% and above.

Vitamins

Arai *et al.* (1972a) reported the qualitative requirements of young *A. japonica* for water-soluble vitamins and listed their deficiency symptoms (Table 22.2). Japanese eel requires 11 water-soluble vitamins; however, deficiency was not evident in eels on a diet without *p*-aminobenzoic acid. Among the vitamins, deficiency of pyridoxine, pantothenic acid and choline produced effects earlier than the others, an indication of their relative importance. Most of the deficiency signs disappeared soon after feeding the related vitamin.

Yamakawa *et al.* (1975) described the signs of vitamin E deficiency and quantified the requirement of α-tocopherol using vitamin-free casein as the main protein source and fatty acid methyl esters as lipid sources. Deficiency signs observed were poor appetite and growth, haemorrhage and congestion in the fins and dermatitis on the skin. They estimated that the minimal requirement of α-tocopherol for growth of young eels was 200 mg kg^{-1} diet.

Minerals

Nose and Arai (1979) determined the mineral requirement of Japanese eel employing purified diets. Eels fed calcium or phosphorus-deficient diets gradually lost their appetite after a week of feeding, followed by reduced growth. However, when fed magnesium- or iron-deficient diets they showed poor appetite after 3–4 weeks. Eels fed the iron-deficient diet had hypochromic microcytic anaemia. The optimum requirements for calcium, magnesium, phosphorus and iron of young

Table 22.2. Vitamin deficiency signs in eels (from Arai *et al.*, 1972a).

Vitamin	Deficiency signs
Thiamine	Trunk-winding, ataxia, haemorrhage and congestion in fins, dark coloration and sluggish movement
Riboflavin	Haemorrhage and congestion in fins, dermatitis, photophobia and sluggish movement
Pyridoxine	Nervous disorders, epileptiform fits and convulsions
Pantothenic acid	Abnormal swimming, ataxia, mortality, haemorrhage in epidermis and dermatitis
Niacin	Abnormal swimming, ataxia, anaemia, haemorrhage in epidermis and dermatitis
Biotin	Abnormal swimming and dark coloration
Folic acid	Dark coloration and poor growth
Cyanocobalamin	Poor growth
Choline	White-grey intestine
Inositol	White-grey intestine
Ascorbic acid	Haemorrhage in fins, head and skin, and lower jaw erosion
p-Aminobenzoic acid	None
α-Tocopherol	Haemorrhage in fins and skin and dermatitis

Table 22.3. Proximate composition (%) of a typical commercial feed for eel (from Tsuda, 1997).

Type of feed	Eel size (g)		Moisture	Protein	Lipid
Paste type	Elver (just after catch)	0.2–0.5	82	14	0.8
Type A	Elver	0.4–1.0	7	65	5
	Elver	0.5–10	7	54	5
Type B	Elver	10–20	7	52	6
	Juvenile to growing	20–200	7	49	5
Extruded floating	Growing	10–200	7	49	16

eel were estimated to be about 2700, 400, 2500–3200 and 170 mg kg^{-1} diet, respectively.

Park and Shimizu (1989a,b) reported that optimal supplemental levels of aluminium and zinc to pollock meal-based diets were 15 and 50–100 mg kg^{-1} diet, respectively.

Practical Diets

Commercial diets for eel in Japan – dough-type feeds (A and B) – were developed almost 35 years ago. Recent innovations have been a paste-type diet developed to replace the *Tubifex* worm (*Tubifex hattai*) and a floating diet employed as a substitute for the dough-type diet. The paste-type diet is a weaning diet and includes fish flesh to improve its palatability for elvers; it is provided in a frozen condition. The paste diet is replaced by dough-type A diet, which contains wheat gluten as a binder. Dough-type B is the main diet for eel culture, and is used during the whole cultivation period after complete acclimatization to an artificial diet. Type B diet contains gelatinized starch as a binder. Table 22.3 provides the proximate composition of representative commercial diets for eel in Japan. The B-type commercial diet contains 5–6% lipid, which is derived from fish-meal. Since lipid is an important source of energy, besides its protein-sparing effect, it would be beneficial and effective to increase dietary lipid in the dough-type diet (B). It has been reported that the best growth and feed efficiency were obtained by increasing dietary oil to 15% (Tsuda, 1997).

Although a dry-type floating diet has been developed for eel, it has not gained wide market acceptance due to the slow growth of eel. The dough-type diet is still the dominant diet, although it is difficult and laborious to prepare. There is still a great need to develop a high-performance dry diet for eel.

Reference

Akiyama, T., Oohara, I. and Yamamoto, T. (1997) Comparison of essential amino acid requirements with A/E ratio among fish species (review paper). *Fisheries Science* 63, 963–970.

Arai, S. (1991) Eel, *Anguilla* spp. In: Wilson, R.P. (ed.) *Handbook of Nutrient Requirements of Finfish*. CRC Press, Boca Raton, Florida, pp. 69–75.

Arai, S., Nose, T. and Hashimoto, Y. (1971) A purified test diet for the eel, *Anguilla japonica*. *Bulletin of the Freshwater Fisheries Research Laboratory* 21, 161–178.

Arai, S., Nose, T. and Hashimoto, Y. (1972a) Qualitative requirements of young eels *Anguilla japonica* for water soluble vitamins and their deficiency symptoms. *Bulletin of the Freshwater Fisheries Research Laboratory* 22, 69–83.

Arai, S., Nose, T. and Hashimoto, Y. (1972b) Amino acids essential for the growth of eels, *Anguilla anguilla* and *A. japonica*. *Bulletin of the Japanese Society of Scientific Fisheries* 38, 753–759.

Burgos, C., Castillo, M., Zafra, M.F. and Garcia-Peregrin, E. (1989) *Influence of Protein Replacement by Fat in Eel Diets on Growth and Body Composition*. Special Publication No. 10, European Aquaculture Society, Bredene, Belgium, 281 pp.

Degani, G. (1987) Effect of dietary carbohydrate source on soluble protein glucose concentration and enzyme activity (aldolase) of European eels (*Anguilla anguilla* L.). *Comparative Biochemistry and Physiology A* 87A, 27–30.

Degani, G., Levanon, D. and Trieger, G. (1984) Preliminary study on the influence of difference of different feeds on mortality and growth of eels (*Anguilla anguilla* L.) in the initial period. *Bamidgeh* 36, 47–52.

Degani, G., Gallagher, M.L. and Levanon, D. (1987) Effect of dietary protein and energy levels on growth of European eels (*Anguilla anguilla*). *Boletim de Fisiologia Animal da Universidale de São Paulo* 11, 95–100.

de la Higuera, M., Garcia-Gallego, M., Sanz, A., Hidalgo, M.C. and Suarez, M.D. (1989) Utilization of dietary protein by the eel (*Anguilla anguilla*): optimum dietary protein levels. *Aquaculture* 79, 53–61.

Gallagher, M.L. and Degani, G. (1988) Poultry meal and poultry oil as sources of protein and lipid in the diet of European eels (*Anguilla anguilla*). *Aquaculture* 73, 177–187.

Gallego, M.G., Hidalgo, M.C., Suarez, M.D., Sanz, A. and de la Higuera, M. (1993) Feeding of the European eel *Anguilla anguilla*, 2. Influence of dietary lipid levels. *Comparative Biochemistry and Physiology A* 105A, 171–175.

Garcia-Gallego, M., Bazoco, J., Sanz, A. and Suarez, M.D. (1991) A comparative study of the nutritive utilization of dietary carbohydrates by eel and trout. In: Kaushik, S. and Luquet, P. (eds) *Fish Nutrition in Practice*. Colloques 61, INRA, Paris, pp. 939–943.

Lee, K.-J. and Bai, S.C. (1997) Haemoglobin powder as a dietary fish meal replacer in juvenile Japanese eel, *Anguilla japonica* (Temminck et Schlegel). *Aquaculture Research* 28, 509–516.

Nose, T. (1979) Summary report on the requirements of essential amino acids for carp. In: Tiews, K. and Halver, J.E. (eds) *Finfish Nutrition and Fishfeed Technology*. Heenemann, Berlin, pp. 145–156.

Nose, T. and Arai, S. (1972) Optimum level of protein in purified diet for eel, *Anguilla japonica*. *Bulletin of the Freshwater Fisheries Research Laboratory* 22, 145–155.

Nose, T. and Arai, S. (1979) Recent advances in studies on mineral nutrition of fish in Japan. In: Pillay, T.V.R. and Dill, W.A. (eds) *Advances in Aquaculture*. Fishing News, Farnham, England, pp. 584–590.

Ogino, C. (1980) Protein. In: Ogino, C. (ed.) *Fish Nutrition and Feed*. Koseisha-Koseikaku, Tokyo, Japan, pp. 111–135.

Park, C.W. and Shimizu, C. (1989a) Quantitative requirements of aluminum and iron in the formulated diets and its interrelation with other minerals in young eel. *Nippon Suisan Gakkaishi* 55, 111–116.

Park, C.W. and Shimizu, C. (1989b) Suitable level of zinc supplementation to the formulated diets in young eel. *Nippon Suisan Gakkaishi* 55, 2137–2141.

Schmitz, O., Greuel, E. and Pfeffer, E. (1984) Digestibility of crude protein and organic matter of potential sources of dietary protein for eels (*Anguilla anguilla*, L.). *Aquaculture* 41, 21–30.

Takeuchi, T., Arai, S., Watanabe, T. and Shimma, Y. (1980) Requirement of eel *Anguilla japonica* for essential fatty acids. *Bulletin of the Japanese Society of Scientific Fisheries* 49, 345–353.

Tibbets, S.M., Lall, S.P. and Anderson, D.M. (2000) Dietary protein requirement of juvenile American eel (*Anguilla rostrata*) fed practical diets. *Aquaculture* 186, 145–155.

Tsuda, J. (1997) Eel. In: *Best Guide for Aquatic Feed.* Yoshoku, Special Issue, Midori Shobo, Tokyo, Japan, pp. 93–98.

Yamakawa, T., Arai, S., Watanabe, T. and Shimma, Y. (1975) Vitamin E requirement for Japanese eel. *Vitamins* 49, 62.

23 Hybrid Striped Bass

Carl D. Webster
Aquaculture Research Center, Kentucky State University, Frankfort, KY 40601, USA

Introduction

Hybrid striped bass is an aquaculture species of commercial importance in the USA. In 1987, production of hybrid striped bass as a foodfish was 180,000 kg, with a majority of the fish being grown by one producer in California. However, by 1993, production was estimated at about 2.3 million kg with a projected farm value of $11 million; total production for 1996 was 3.6 million kg and in 1999 an estimated 4.6 million kg were produced. These values do not include propagation and growing of hybrid striped bass in state and federal hatcheries for stocking in public lakes and rivers. There are five major growers of hybrid striped bass in the USA and they produce approximately 70% of the product, while other smaller companies and growers comprise the remaining 30%.

Hybrid striped bass are produced by crossing the striped bass (*Morone saxatilis*) with white bass (*Morone chrysops*). Two crosses are most commonly utilized by aquaculturists; one is called the 'original cross' (or palmetto bass), which is produced by crossing a female striped bass with a male white bass. The other cross is called the 'reciprocal cross' (or sunshine bass), which is produced by crossing the female white bass with the male striped bass. The latter cross is used by many producers due to the ease of obtaining suitable white bass females for spawning and the difficulties in obtaining adequate numbers of brood-stock female striped bass.

Hybrid striped bass require warm water temperatures to achieve optimal growth (25–30°C). Most hybrid striped bass fingerlings for sale for grow-out are grown in ponds, while many fish for sale as foodfish are produced in tanks; however, they can be cultured using numerous means, including ponds, raceways and cages. Hybrid striped bass culture is divided into four phases of production: the hatchery phase; phase I culture, in which the fish are grown from fry (0.5 cm) to fingerlings (1.5–2.0 cm); phase II culture, where the fish are reared for 5–9 months and are grown to 10–14 cm in length; and phase III culture, in which the fish are grown out to market size (700–1200 g).

Hybrid striped bass can be grown using intensive systems, such as raceways, cages or indoor recirculating systems; however, extra care must be used when growing fish by these methods. Circular tanks allow for water circulation to all areas within the tank and water currents help to remove wastes. Rectangular tanks (raceways) are also used to culture hybrid striped bass; however, corners tend to accumulate waste and do not allow for a continuous swimming path for the fish.

Cages have been used to grow hybrid striped bass for many years and provide producers with a method for growing fish in ponds that may not be suitable for conventional pond-culture methods; the pond may be irregularly shaped or too deep to be seined. Thus, use of cages allows producers to grow fish in areas that are more difficult to access and/or harvest. Recirculating systems require biological and mechanical filtration so that solids and metabolites are removed from the culture water. Mechanical filtration can be accomplished through the use of sand filters, diatomaceous earth, coral, charcoal or various human-made 'fabrics' or plastic beads.

Nutrient Requirements

Nutrient requirements and practical diet formulations for hybrid striped bass have only recently been investigated and additional research needs to be conducted. It may be erroneous to assume that striped bass, palmetto bass and sunshine bass require similar nutrients and diets. For example, there appear to be many similarities in the requirements of the amino acids arginine, threonine and the total sulphur amino acids between striped bass and the two hybrid striped bass; however, lysine requirements appear to be different (Small and Soares, 1998).

Protein and amino acids

Dietary protein requirements can vary with the age of the fish, energy level of the diet, amino acid balance and culture conditions. Percentage dietary protein for striped bass and hybrid striped bass has ranged from 36% to 55%. Millikin (1983) reported that a diet with 47% protein and 12% lipid produced optimal growth in small (2.0 g) striped bass. Diets with higher amounts of protein (55%) did not improve growth, while feeding a diet with 37% protein and 7% lipid resulted in reduced growth. Optimal protein level has been reported to be 41% when sunshine bass were fed diets containing menhaden fish-meal as the sole protein source (Brown *et al.*, 1992).

Hybrid striped bass are presumed to require the same ten essential amino acids as most fish and terrestrial animals: arginine, histidine, isoleucine, leucine, lysine, methionine, phenylalanine, threonine, tryptophan and valine. However, unlike some cultured species, the requirements for several of the amino acids are not known for hybrid striped bass. Lysine and methionine are generally considered to be two of the first limiting amino acids in fish diets. The lysine requirement has been estimated to be 1.4% of the diet (4.0% of dietary protein) when sunshine bass have been fed either semipurified (Keembiyehetty and Gatlin, 1992) or purified (Griffin *et al.*, 1992) diets. The total sulphur amino acid (TSAA) requirement (methionine and cystine) of sunshine bass has been reported as 0.87% of the diet as methionine with 0.13% of the diet as cystine (Keembiyehetty and Gatlin, 1993), resulting in a TSAA requirement of 1.0% of the diet (2.9% of dietary protein). Keembiyehetty and Gatlin (1993) also found that cystine spared less than 40% of the methionine. However, when using a broken-line analysis of weight-gain data, Griffin *et al.* (1994a) reported a TSAA requirement of 0.73% of the diet (2.1% of dietary protein) when palmetto bass were fed a purified diet. This was verified when they analysed serum free methionine concentrations of fish and reported that fish fed 0.6% TSAA had lower serum free methionine concentrations than fish fed diets containing 0.8–1.8% TSAA. Griffin *et al.* (1994a) also reported that cystine could comprise 40% of the TSAA requirement in palmetto bass. The arginine requirement for sunshine bass using broken-line analyses of weight-gain data was calculated to be 1.55% of the diet (4.4% of dietary protein) when fed a purified diet containing 35% crude protein (Griffin *et al.*, 1994b). Keembiyehetty and Gatlin (1997a) reported that the dietary threonine requirement of juvenile sunshine bass was 0.97% of the diet (2.6% of the dietary protein) when fed a semipurified diet in aquariums. The values for hybrid striped bass are compared with those for other fish species in Table 23.1.

Apparent protein digestibility for practical feed ingredients for palmetto bass ranges from 71% to 93% (Table 23.2). Wheat flour, wheat middlings, menhaden fish-meal, maize grain, cottonseed meal and blood meal all had protein digestibilities greater than 83% (Sullivan and Reigh, 1995). Rice bran and meat-and-bone meal had the lowest protein digestibilities of 71% and

Table 23.1. Arginine, lysine, total sulphur amino acids (methionine/cystine) and threonine requirements of hybrid striped bass (HSTB) and several aquaculture species.

	Fish species							
Amino acid	HSTB	Channel catfish	Rainbow trout	Common carp	Nile tilapia	Red drum	Gilt-head sea bream	Atlantic salmon
Arginine	1.0(4.3)[a]	1.0(4.4)[b]	1.2(3.1)[c]	1.7(4.4)[d]	1.2(4.2)[e]	–	(2.6)[f]	1.6(4.1)[g]
Lysine	1.4(4.0)[h, i]	1.2(5.0)[j]	1.9(4.7)[k]	2.2(5.7)[d]	1.4(5.1)[e]	1.6–2.0 (4.6–5.7)[m]	(5.0)[f]	1.8(4.0)[l]
TSAA	0.73[n] 1.0(2.9)[s]	0.5(2.3)[o]	0.8(2.3)[p]	1.2(3.1)[d]	0.8(2.7)[e]	1.1(3.0)[q]	(4.0)[f]	1.1(2.4)[r]
Threonine	0.97(2.6)[t]	0.5(2.2)[u]	(3.3)[v]	1.5(3.9)[d]	1.05(3.75)[e]	(2.3)[w]	–	–

[a]Griffin *et al.* (1994b); [b]Robinson *et al.* (1981); [c]Kaushik *et al.* (1988); [d]Nose (1979); [e]Santiago and Lovell (1988); [f]Kissil (1991); [g]Lall *et al.* (1994); [h]Griffin *et al.* (1992); [i]Keembiyehetty and Gatlin (1992); [j]Wilson *et al.* (1977); [k]Kim *et al.* (1992); [l]Anderson *et al.* (1993); [m]Brown *et al.* (1988); [n]Griffin *et al.* (1994a); [o]Harding *et al.* (1977); [p]Rumsey *et al.* (1983); [q]Moon and Gatlin (1991); [r]Rollin *et al.* (1994); [s]Keembiyehetty and Gatlin (1993); [t]Keembiyehetty and Gatlin (1997a); [u]Wilson *et al.* (1978); [v]Ogino (1980); [w]Boren and Gatlin (1995).

Table 23.2. Average percentage apparent protein and energy digestibility coefficients for various feed ingredients for palmetto bass, channel catfish, and rainbow trout.

Ingredient	Palmetto bass*	Channel catfish†/‡	Rainbow trout
Protein			
Fish-meal, menhaden	88	86/76	n/a
Blood meal	86	74/n/a	69§
Meat-and-bone meal	73	82/64	82‖
Cottonseed meal	84	n/a/80	76§
Wheat middlings	92	n/a/88	76§
Wheat flour (grain)	93	n/a	n/a
Soybean meal	80	85/93	83§
Rice bran	71	n/a/78	n/a
Maize grain	87	92/96	95‖
Energy			
Fish-meal, menhaden	96	95	
Blood meal	n/a	n/a	
Meat-and-bone meal	80	70	
Cottonseed meal	73	65	
Wheat middlings	61	49	
Wheat flour (grain)	54	n/a	
Soybean meal	55	73	
Rice bran	47	53	
Maize grain	41	79	

* Sullivan and Reigh (1995).
† Brown *et al.* (1985).
‡ Wilson and Poe (1985).
§ Smith *et al.* (1980).
‖ NRC (1993).
n/a, not available.

73%, respectively. The high protein digestibility values of many plant-protein ingredients for hybrid striped bass indicate that diets for these fish can be formulated with high levels of plant-protein ingredients. This may reduce the cost of diets used to feed hybrid striped bass.

Energy

Since protein is the most expensive component in a diet, nutritionists need to supply levels of protein sufficient to meet essential amino acid requirements, but not in excess of requirements, because protein may be used for energy. Dietary lipids and carbohydrates can spare protein; however, excessive amounts of energy can lead to increased fat deposition and possibly reduced growth. Fish have lower dietary energy requirements than terrestrial animals because they do not need to expend as much energy to maintain their position in water, and they excrete nitrogenous wastes as ammonia instead of urea or uric acid, both of which require energy to form.

The level of energy in a diet affects the amount of diet consumed by fish, and the protein-to-energy (P : E) ratio of the diet will influence growth response and conversion efficiency. An excessively high ratio may cause fish to utilize protein as an energy source, which is expensive, whereas fish fed a diet with a low P : E ratio may increase fat deposition. Fattiness in cultured fish is generally undesirable because consumers eat fish as part of a low-fat diet. Fat in fish may not only reduce consumer acceptance, but may also reduce storage life and decrease processed yield. There are several factors that affect the amount of fat in fish, such as feeding rate, size of fish, culture conditions and the P : E ratio in the diet. The optimal dietary E : P ratio for sunshine bass has been reported as 6–9 kcal available energy (AE) g^{-1} protein for a diet that contained 35–45% crude protein (Nematipour *et al.*, 1992a).

Composition of muscle from sunshine bass was not affected by dietary E : P ratios; however, whole-body protein concentration decreased with increasing E : P ratio, while whole-body lipid and dry matter increased. Increased lipid deposition in sunshine bass fed a diet with an increasing E : P ratio was due to an increase in abdominal fat. This is in agreement with Woods *et al.* (1995), who reported that a diet with 7 kcal AE g^{-1} of protein was adequate for growth in juvenile striped bass. Webster *et al.* (1995) fed practical diets to sunshine bass grown in cages and reported that a diet with 40% protein and between 8 and 10 kcal AE g^{-1} of protein was found to be adequate for sunshine bass. If the AE : P ratio was higher than 10 kcal AE g^{-1} of protein, an increase in lipid deposition and a decrease in protein level in sunshine bass were reported. Keemibiyehetty and Wilson (1998) reported that juvenile sunshine bass fed semipurified diets containing 6, 7 and 9 kcal AE g^{-1} of protein had a significantly higher percentage weight gain compared with fish fed 8, 10 and 11 kcal AE g^{-1} of protein. Diets that supported the highest growth rates had protein levels of either 40% or 45% and lipid levels of either 5% or 15%.

It appears that carbohydrate and lipid in practical diets may be utilized equally well by sunshine bass as energy sources (Webster *et al.*, 1995). However, Gallagher (1999) reported that liver alanine aminotransferase activities of sunshine bass fed diets with increased lipid levels indicated that there was a substantial amount of deamination of amino acids for non-protein use. Therefore, it was concluded that lipids may not spare protein as well as carbohydrates. Use of higher levels of dietary carbohydrate, instead of lipid, may reduce storage and spoilage concerns in practical hybrid striped-bass diets. Hybrid striped bass do not utilize maize grain and rice bran well for energy needs (Table 23.2), but utilized menhaden fish-meal (96%) and meat-and-bone meal (80%) efficiently as energy sources (Sullivan and Reigh, 1995). Cottonseed meal, wheat middling, wheat flour and soybean meal all had energy digestibility values between 55% and 73% (Sullivan and Reigh, 1995).

Lipids and fatty acids

Essential fatty acids (EFA) are important for membrane fluidity and proper neurological and visual development and act as precursors for eicosanoids and prostaglandins. Highly unsaturated fatty acids (HUFA) of the n-3 group are essential for the proper growth and development of marine-fish larvae. This is because most marine-fish larvae have a limited ability, if any, to elongate (add carbon units) and desaturate (create molecular carbon-to-carbon double bonds) shorter (18 carbons in length) fatty acid chains. Thus, marine-fish larvae must rely on obtaining these EFA from their food. Like marine fish, striped-bass and palmetto-bass larvae appear to have a limited ability to biosynthesize n-3 HUFA and require a dietary source.

Two of these HUFA are eicosapentaenoic acid (EPA), 20:5n-3, and docosahexaenoic acid (DHA), 22:6n-3. It appears that EPA levels higher than 5% of the total fatty acids are required for good growth and survival of striped-bass larvae (Webster and Lovell, 1990). Palmetto-bass larvae appear to be more susceptible to HUFA deficiency than striped-bass larvae during metamorphosis (Tuncer and Harrell, 1992). Since HUFA are important components of phospholipids, which in turn are important constituents of biomembranes, during metamorphosis there is extensive tissue development and reorganization by the larvae. Phospholipids containing HUFA fatty acids appear to be vital for these tissue changes. High mortality rates occur if adequate levels of EPA or DHA are not present in the food of hybrid striped-bass larvae at metamorphosis.

As fish grow, their EFA requirement may change. Sunshine bass (13 g) fed diets with 1.1% n-3 HUFA had higher weight gains and protein efficiency ratios than fish fed diets with less than 1.1% or greater than 3.2% n-3 HUFA (Nematipour and Gatlin, 1993). However, palmetto bass (72 g) fed practical diets containing 40% crude protein, approximately 30% menhaden fish meal, and with either 0% or 4% menhaden oil did not show differences in growth when grown in ponds (Zhang *et al.*, 1994). This is in agreement with Robinette *et al.* (1997), who reported that large (> 60 g) juvenile palmetto bass grown in ponds had similar total weight, mean weight gain, percentage weight gain, fillet dress-out percentage, proximate composition of whole-body and fillet tissue, and abdominal fat when fed diets containing 30% menhaden fish-meal and either 2.5% catfish oil or 2.5% menhaden oil. They concluded that the addition of 30% menhaden fish-meal to a diet provided greater than the 1% n-3 EFA required by hybrid striped bass and that the use of marine fish-oils for hybrid striped bass grown in ponds may be unnecessary when 30% menhaden fish-meal is added to a diet. However, if marine fish-meal is not added, marine fish-oils may need to be added to ensure that the n-3 HUFA requirement is met.

Carbohydrates

Since protein is the most expensive energy source and lipid sources, especially marine fish-oils, may be expensive or difficult to obtain, utilization of carbohydrate as an inexpensive energy source in diets of aquaculture species would be advantageous. Nematipour *et al.* (1992b) fed four isocaloric (3.4 kcal) and isonitrogenous (35% protein) semipurified diets to juvenile sunshine bass with differing carbohydrate-to-lipid ratios (CHO : L), ranging from 25 : 10 to 42 : 2.5. Percentage weight gain, feed efficiency, protein efficiency ratio, hepatosomatic index and muscle ratio were not different ($P > 0.05$) in sunshine bass fed any of the diets. However, abdominal fat levels increased as dietary CHO : L decreased. These data indicate that sunshine bass can utilize both carbohydrate and lipid as energy sources. A similar conclusion was reached by Webster *et al.* (1995), who fed sunshine bass practical diets with differing levels of protein, lipid and carbohydrate.

In a later study, Hutchins *et al.* (1998) stated that weight gain, protein efficiency and feed efficiency were highest in sunshine bass fed a diet containing 20% carbohydrate compared with fish fed diets containing 40% or 0% carbohydrate. They reported that the type of carbohydrate (glucose, maltose or dextrin) had no significant ($P > 0.05$) effect on growth or body composition parameters, except that lipid levels in muscle tissue of fish fed diets with dextrin and maltose were higher than those in fish fed diets containing glucose. This is in agreement with Rawles and Gatlin (1998), who reported that carbohydrate type had little effect on growth of sunshine bass when fed diets containing 25% carbohydrate as either glucose, maltose or dextrin. However, the authors stated that sunshine bass had higher growth rates than striped bass fed the same diets. Thus, sunshine bass may have a lower metabolic rate than striped bass, which would result in more energy being available for growth.

Inclusion of 25% soluble carbohydrate in a diet for sunshine bass had some effect on nutrient digestibility (Rawles and Gatlin, 1998). Protein digestibility was significantly ($P < 0.05$) reduced when glucose was added to the diet; however, no statistical differences were reported when maltose or dextrin was added. Apparent digestibility coefficients (ADC) of both lipid and dietary energy were not different ($P > 0.05$) among the three carbohydrate sources. However, carbohydrate ADC was lower ($P < 0.05$) in sunshine bass fed a diet containing dextrin as the carbohydrate source compared with fish fed diets with glucose or maltose. Based upon these reports, it appears that sunshine bass can effectively utilize diets with 25% dietary carbohydrate.

Vitamins and minerals

Very little information is available on vitamin requirements of hybrid striped bass. One reason is that until recently, hybrid striped bass were fed commercial salmonid diets, which have high levels of vitamins. Research needs to be conducted into the quantitative vitamin requirements of the fish. Sealey and Gatlin (1999) fed semipurified diets with graded levels of a stable form of vitamin C (ascorbyl-2-polyphosphate, 25% ascorbic acid activity) to juvenile sunshine bass in aquariums. Diets were supplemented with either 0, 10, 20, 30, 45, 60, 75 or 150 mg vitamin C kg^{-1}. Non-linear least-squares regression analysis of weight-gain data indicated that the dietary vitamin C requirement was 22 (± 6) mg vitamin C kg^{-1} of diet.

Choline has been found to be essential for sunshine bass. Using purified diets, Griffin *et al.* (1994c) showed that the choline requirement for juvenile sunshine bass, based on weight gain and feed efficiency data, was 500 mg kg^{-1} of diet.

Lipid-soluble vitamins are also essential to fish. Feeding semipurified diets to juvenile sunshine bass with graded levels of vitamin E, Kocabos and Gatlin (1999) reported that the dietary vitamin E requirement for sunshine bass was 28 (± 3) mg kg^{-1} of diet based on weight-gain data and 26 (± 3) mg kg^{-1} of diet based on feed efficiency data. In view of the scarcity of information on vitamin requirements of hybrid striped bass, the vitamin allowances presented in Table 23.3 are recommended. The amounts of each vitamin listed, which allow for processing and storage losses, are based primarily on the requirements of salmonids and some warm-water species. These levels have been used successfully in commercial and experimental diets.

Hybrid striped bass probably require the same minerals as other fish for proper tissue and bone formation and metabolic activities and to maintain the osmotic balance between their body fluids and the surrounding water. Fish can

Table 23.3. Recommended vitamin and mineral levels for a practical hybrid striped bass diet (per kg of diet).

	Dietary level
Vitamins	
A	6000 IU
D	2200 IU
E	28 mg (100–150 IU)
K	10 mg
Niacin	200 mg
Pantothenic acid	60 mg
Thiamine	30 mg
Riboflavin	20 mg
Pyridoxine	20 mg
Folic acid	5 mg
B_{12}	0.1 mg
Biotin	2 mg
C	22 mg (22–100 mg)
Choline	500 mg
Minerals	
Phosphorus	540 mg
Manganese	180 mg
Copper	8 mg
Cobalt	1.5 mg
Iron	66 mg
Zinc	150 mg
Iodine	6 mg
Selenium	0.3 mg

derive a large proportion of certain minerals, such as calcium, from the water; however, many minerals must be supplied in the diet since water is not the major source. Since there is a lack of data on the mineral requirements for hybrid striped bass, the levels presented in Table 23.3 can be used. In a feeding study using semipurified diets, sunshine bass were reported to require 0.54% phosphorus in a diet for proper growth and bone and scale mineralization (Brown *et al.*, 1993a).

Practical Diet Formulation

When formulating diets for any fish species, careful consideration must be given to ingredient selection. Not only must the diets meet the protein (amino acid), essential fatty acid, vitamin and mineral requirements of the fish, but the diets must be palatable so that the fish will eat them. Further, ingredients must be relatively digestible so that the nutrients can be utilized. During the early 1990s, hybrid striped-bass producers would feed either modified catfish or salmonid diets because little research had been conducted on the nutrient requirements of the fish or on formulating economical, nutritious, practical diets. Hybrid striped bass consume either a floating or sinking pellet; however, use of a floating pellet allows the producer to observe the feeding activity of the fish. This is advantageous for determining how much diet to feed to the fish, as well as the feeding activity of the fish. The physical properties of the diet are important, especially water stability and pellet size. Since hybrid striped bass are generally aggressive in their eating habits, water stability of the diet is not as important as it would be for crustacean diets; however, the diet must remain intact for a period of at least several minutes to allow the fish time to eat.

Hybrid striped bass have an ability to digest carbohydrates in some plant ingredients, such as soybean meal and cottonseed meal (Zhang *et al.*, 1994; Sullivan and Reigh, 1995; Webster *et al.*, 1995). This may allow nutritionists to formulate diets with higher levels of these plant ingredients and lower levels of fish-meal for hybrid striped bass compared with those for salmonids. Sunshine bass (1.5 g) are able to utilize carbohydrate and lipid for energy and for sparing protein, which can be utilized to meet essential amino acid requirements. Feeding diets having CHO : L of between 25 : 10 and 42 : 2.5 resulted in similar growth; however, whole-body lipid (as abdominal fat) increased as lipid increased (Nematipour *et al.*, 1992b).

Likewise, it appears that hybrid striped bass can digest protein from plant and animal sources similarly to channel catfish (*Ictalurus punctatus*) but not as effectively as rainbow trout (*Oncorhynchus mykiss*). They can digest dietary starch more effectively than rainbow trout, but not as well as channel catfish. Thus, digestibility coefficients for hybrid striped bass appear to be somewhere between those for an omnivorous species, such as the channel catfish, and those for a carnivorous species, such as rainbow trout.

Fish-meal is an important ingredient in aquaculture diets because of its high protein quality and palatability. However, the long-term availability of fish-meal may be uncertain. Further, fish-meal is the most expensive ingredient in fish diets. Recently, nutritionists have begun to determine the minimum level of fish-meal in diets for hybrid striped bass. Small palmetto bass (5 g) appear to require between 17 and 36% fish-meal in a diet, while larger fish (> 150 g) may need a diet in which soybean meal can replace 75% of the fish-meal (17% fish-meal in the diet) when the diet contains 35% protein (Gallagher, 1994). Brown *et al.* (1997) stated that palmetto bass fed a diet with 23% fish-meal and 40% soybean meal had similar growth to that of fish fed a diet with 46% fish-meal and 0% soybean meal. Keembiyehetty and Gatlin (1997b) stated that a diet with 14% fish-meal, 56% soybean meal and either 0.3% D,L-methionine, 0.3% D,L-methionine or 0.3% acetylmethionine gave similar weight gain, feed efficiency and whole-body composition for sunshine bass compared with fish fed a diet containing 57% fish-meal. However, fish fed a diet without supplemental methionine had significantly ($P < 0.05$) reduced growth, as had been reported in a previous study by Keembiyehetty and Gatlin (1995).

Fish-oil is often added to hybrid striped-bass diets not only as a source of EFA but also to improve the palatability of aquaculture diets. When a diet is unpalatable, hybrid striped bass take the pellet into the mouth and then quickly eject it back into the water. This action might be repeated several times before the fish either rejects or accepts the diet. Hybrid striped bass do not find a purified diet (Table 23.4) palatable and feeding stimulants have to be added to the diet. However, menhaden fish-oil has not been found to be an effective attractant for sunshine bass (5–10 g), but fish-meal was found to be an attractant at a level of 10% (Brown *et al.*, 1993b). The use of lyophilized fish flesh (red-drum muscle) has been successful in formulating a semipurified diet that is consumed by hybrid striped bass (Table 23.4). While fish-oil may not be a strong attractant, it does provide essential n-3 fatty acids. However, the use of marine fish-oils, especially at high percentages, may affect the flavour of the fish.

While diets with low (15%) fish-meal and supplemental methionine appear to provide for good growth, the use of a proper combination of plant- and animal-protein sources may allow nutritionists to formulate diets that do not require supplemental crystalline amino acids. When small (20 g) palmetto bass were fed diets containing 40% protein, it was found that diets with 15, 30 and 45% fish-meal resulted in similar growth, but a diet with 0% fish-meal resulted in reduced growth, possibly due to reduced palatability (Webster *et al.*, 1997). However, in a subsequent study, Webster *et al.* (1999) reported that sunshine bass fed diets containing 32% soybean meal, 28% poultry by-product meal and 0% fish-meal had similar percentage weight gain, specific growth rate, feed conversion ratio and carcass proximate composition to those of fish fed a diet with 30% menhaden fish-meal. This is in agreement with Webster *et al.* (2000), who reported that sunshine bass fed a diet containing 35% soybean meal, 35% meat-and-bone meal and 0% fish-meal or one containing 27% soybean meal, 27% meat-and-bone meal, 20% hempseed meal and 0% fish-meal had similar growth and body composition to those of fish fed a diet with 30% menhaden fish-meal. Practical diet formulations that have totally replaced fish-meal with a combination of plant- and animal-protein ingredients,

Table 23.4. Formulations for a purified diet (adapted from Brown *et al.*, 1993b), a semipurified diet (adapted from Sealey and Gatlin, 1999) and practical diets (adapted from Webster *et al.*, 1999) that have been fed to hybrid striped bass.

Purified diet			
Casein	43.4		
Dextrin	20.0		
Menhaden oil	12.0		
Cellulose	14.1		
Other ingredients	10.5		
Semipurified diet			
Red-drum muscle	19.6		
Amino acid premix	23.0		
Dextrin	17.5		
Menhaden oil	7.9		
Cellulose	22.0		
Other ingredients	10.0		
Practical diets	Diet 1	Diet 2	Diet 3
Menhaden fish-meal	0.0	0.0	30.0
Soybean meal	32.0	32.0	25.5
Meat-and-bone meal	32.0	0.0	0.0
Poultry by-product meal	0.0	28.0	0.0
Wheat middlings	15.0	15.0	15.0
Maize meal	15.0	16.4	24.2
Menhaden oil	4.5	4.5	2.0
Other ingredients	1.5	4.1	3.3

which appear to provide adequate growth in sunshine bass are presented in Table 23.4.

Commercial diets may have an effect on the flavour of the fish. While grains and many plant-protein and animal by-product meals do not usually impart flavour to cultured fish, some ingredients, such as oils/fats and fish-meals, can affect the flavour, appearance, composition and storage quality of fish. Bett *et al.* (1998) stated that it was the marine fish-oil added to diets for sunshine bass that caused the fillets to taste 'fishy', not the fish-meal. A diet with 30% anchovy fish-meal and 10% menhaden oil had a stronger 'fishy' taste than fillets of fish that were fed a diet with 40% anchovy meal and 0% menhaden oil. This indicates that it is the oil that contributes to the fishy flavour, while the fish-meal may not impart that strong a flavour.

Replacement of fish-meal with oilseed meals can also affect the flavour of the fish. When soybean meal was added to diets for palmetto bass, the flavour and texture were affected compared with fillets of fish fed a diet with a high percentage (47%) of fish (herring) meal and 0% soybean meal (Postel *et al.*, 1996). While the taste differences between fish fed a diet with fish-meal and no soybean meal from those fed diets that contained soybean meal could be detected, all fillets were judged to be acceptable in terms of flavour quality.

Feeding Practices

Larval

Selection of a brine-shrimp strain with the proper fatty acid composition (high levels of n-3 HUFA) is necessary to ensure the proper growth, development and survival of larval palmetto bass (Clawson and Lovell, 1992). However, it has been found that enrichment of brine-shrimp nauplii with marine fish-oils is a method to add n-3 HUFA to brine-shrimp nauplii that may have originally been deficient in EPA and/or DHA (Clawson and Lovell, 1992; Tuncer and Harrell, 1992). Fatty acid enrichment is accomplished by allowing newly hatched brine-shrimp nauplii to grow to the third nauplii stage (24 h after hatching) and then place them into a container with the fatty acid emulsion. This emulsion can be made by mixing the n-3 HUFA lipid, salt (sea) water and gum xanthum. Baker's yeast can also be added to the mixture. Strong aeration is provided to keep the nauplii suspended in the emulsion for up to 24 h, allowing the nauplii to consume the yeast or the lipid droplets.

Larval hybrid striped bass do not utilize prepared diets as well as live foods. There have been numerous attempts to feed hybrid striped-bass larvae prepared diets but all have resulted in poor growth and survival. It has been thought that the inability of larval fish to utilize prepared diets was due to a lack of digestive enzymes. However, larval striped bass appear to have the necessary enzymes (trypsin, chymotrypsin, pepsin, carboxypeptidase A and α-amylase) to digest food (Baraji and Lovell, 1986). Tuncer *et al.* (1990) recommended that brine-shrimp nauplii be fed for 21 days and then gradually change to a prepared diet after that time period. For sunshine bass larvae, rotifer culture is required for feeding because of the smaller mouth of the sunshine bass, which prohibits the eating of brine-shrimp nauplii until the larva is larger.

Juvenile

Feeding small palmetto-bass juveniles all they will eat produces larger fish than feeding a restricted diet (2.5% and 5.0% of body weight) (Tuncer *et al.*, 1990). Diets used to feed phase II (juvenile) fish are generally high in protein (38–50% protein) and, while hatcheries tended to use a sinking diet in the past, floating pellets can be used equally well. As the fish grows, pellet size needs to be increased. Some producers feed smaller fish up to six times per day. However, more frequent feedings are labour-intensive. On large farms, it may be practical to feed only once per day. Recently, Thompson *et al.* (2000) fed sunshine bass once a day, twice a day, once every other day or twice every other day in aquariums and reported that sunshine bass fed to satiation twice a day had significantly ($P < 0.05$) higher final weight and percentage weight gain compared with fish fed once a day, once every other day and twice every other day, while percentages of abdominal fat and fillet were not different ($P > 0.05$) compared with fish fed once a day and twice every other day. Fillet proximate composition (wet-weight basis) was not different among treatments. Sunshine bass fed twice a day did have a higher feed conversion ratio (FCR) compared with fish fed once a day and once every other day. This is in agreement with the results of our laboratory, which show that sunshine bass fed twice a day had higher ($P < 0.05$) final weight, percentage weight gain, specific growth rate (SGR) and FCR compared with fish fed once a day, once every other day and twice every other day, when grown in cages.

No information is available on which time of day may be more advantageous to feed (e.g. dawn, midday or dusk). Optimal temperature for growth has been reported to be 27°C when palmetto bass was fed to satiation (Woiwode and Adelman, 1991), while maximum food consumption occurred when water temperatures were 28°C. Keembiyehetty and Wilson (1998) found that sunshine bass grown at 26°C had higher growth rates and appetite compared with fish grown at 32°C.

Fish can be fed by various methods, including feeding by hand, use of blowers, feeding by use of automatic feeders and use of demand feeders. Each method of feeding has advantages and disadvantages. The advantage of feeding by hand or by blower is that the person feeding can observe feeding activity and assess fish health. Use of a blower allows the diet to be spread over a larger area of the pond and can be used when feeding fish stocked in large ponds. Automatic feeders are connected to timers and can distribute diet at a predetermined time of day as frequently as needed. Demand feeders drop diet into the water when the fish activates the dispensing mechanism of the unit. Both automatic and demand feeders can be used effectively to feed fish growing in tanks, cages or raceways.

While a feeding chart can be used to feed hybrid striped bass, feeding all the diet the fish will consume is most often used. Use of a floating diet allows the feeder to observe feeding activity and determine the amount of diet to be fed. While use of feeding charts is an acceptable method of feeding fish, if fish do not consume all the diet fed or would consume more diet than fed, waste of diet or less than optimal growth of fish could result.

Broodfish

There has been little research conducted on the nutritional requirements of striped-bass and/or white-bass brood-stock. Most brood-stock used to produce hybrid striped bass are captured from the wild. If brood-stock are reared in captivity, diets that supply all essential fatty acids, amino acids and vitamins and minerals are generally fed. Some nutrient requirements of striped-bass brood-stock are being, or have been, investigated (L.C. Woods, University of Maryland, personal communication). Small *et al.* (1999) reported that mature striped bass required the same levels of the ten essential amino acids as juvenile fish (% of diet): arginine, 1.4; histidine, 0.6; isoleucine, 0.9; leucine, 1.9; lysine, 2.2; TSAA, 1.0; phenylalanine/tyrosine, 1.7; threonine, 1.1; tryptophan, 0.3; and valine, 1.0. Lopez and Woods (1999) suggested that, based upon the fatty acid composition of striped bass eggs, females require dietary sources of EPA, DHA and possibly arachidonic acid (20:4n-6).

References

Anderson, T.S., Call, S.P., Anderson, D.M. and McNiven, M. (1993) Quantitative lysine requirement for Atlantic salmon (*Salmo salar*) fingerlings. *Canadian Journal of Fisheries and Aquatic Sciences* 50, 316–322.

Baraji, V. and Lovell, R.T. (1986) Digestive enzyme activities in striped bass from first feeding through larvae development. *Transactions of the American Fisheries Society* 115, 478–484.

Bett, K.L., Johnsen, P.B., Webster, C.D., Tiu, L.G., Xiong, Y.L. and Decker, E.A. (1998) Sensory and chemical evaluation of sunshine bass (*Morone chrysops* × *M. saxatilis*) fillets during frozen storage. *Journal of Applied Aquaculture* 8(1), 53–67.

Boren, R.S. and Gatlin, D.M., III (1995) Dietary threonine requirement of juvenile red drum, *Sciaeneps ocellatus*. *Journal of the World Aquaculture Society* 26, 279–283.

Brown, P.B., Strange, R.J. and Robbins, K.R. (1985) Protein digestibility coefficients for yearling channel catfish fed high protein feedstuffs. *Progressive Fish-Culturist* 47, 94–97.

Brown, P.B., Davis, D.A. and Robinson, E.H. (1988) An estimate of the dietary lysine requirement of juvenile red drum, *Sciaenops ocellatus*. *Journal of the World Aquaculture Society* 19, 109–112.

Brown, M.L., Nematipour, G.R. and Getlin, D.M., III (1992) Dietary protein requirements of sunshine bass at different salinities. *Progressive Fish-Culturist* 54, 148–156.

Brown, M.L., Jaramillo, F. and Gatlin, D.M., III (1993a) Dietary phosphorus requirement of juvenile sunshine bass, *Morone chrysops* × *M. saxatilis*. *Aquaculture* 113, 355–363.

Brown, P.B., Griffin, M.E. and White, M.R. (1993b) Experimental and practical diet evaluations with juvenile hybrid striped bass. *Journal of the World Aquaculture Society* 24, 80–89.

Brown, P.B., Twidell, R., Jonker, Y. and Wilson, K.A. (1997) Evaluation of three soybean products in diets fed to juvenile hybrid striped bass *Morone saxatilis* × *M. chrysops*. *Journal of the World Aquaculture Society* 28, 215–223.

Clawson, J.A. and Lovell, R.T. (1992) Improvement of nutritional value of *Artemia* for hybrid striped bass/white bass (*Morone saxatilis* × *M. chrysops*) larvae by n-3 HUFA enrichment of nauplii with menhaden oil. *Aquaculture* 108, 125–134.

Gallagher, M.L. (1994) The use of soybean meal as a replacement for fish meal in diets for hybrid striped bass (*Morone saxatilis* × *M. chrysops*). *Aquaculture* 126, 119–127.

Gallagher, M.L. (1999) Growth responses, tissue composition, and liver enzyme changes in juvenile sunshine bass, *Morone chrysops* × *M. saxatilis*, associated with dietary protein and lipid level. *Journal of Applied Aquaculture* 9(4), 41–51.

Griffin, M.E., Brown, P.B. and Grant, A.C. (1992) The dietary lysine requirement of juvenile hybrid striped bass. *Journal of Nutrition* 122, 1332–1337.

Griffin, M.E., Brown, P.B. and White, M.R. (1994a) Total sulfur amino acid requirement and cysteine replacement value for juvenile hybrid striped bass. *Comparative Biochemistry and Physiology* 108A, 423–429.

Griffin, M.E., Wilson, K.A. and Brown, P.B. (1994b) Dietary arginine requirement of juvenile hybrid striped bass. *Journal of Nutrition* 124, 888–893.

Griffin, M.E., Wilson, K.A., White, M.R. and Brown, P.B. (1994c) Dietary choline requirement of juvenile hybrid striped bass. *Journal of Nutrition* 124, 1685–1689.

Harding, D.E., Allen, O.W. and Wilson, R.D. (1977) Sulfur amino acid requirement of channel catfish in L-methionine and L-cystine. *Journal of Nutrition* 107, 2031–2035.

Hutchins, C.G., Rawles, S.D. and Gatlin, D.M. III (1998) Effects of dietary carbohydrate kind and level or growth, body composition and glycemic response of juvenile sunshine bass (*Morone chrysops* × *M. saxatilis*). *Aquaculture* 161, 187–199.

Kaushik, S.J., Fauconneau, B., Terrier, L. and Greg, J. (1988) Arginine requirement and status assessed by different biochemical indices in rainbow trout (*Salmo gairdneri*). *Aquaculture* 70, 75–95.

Keembiyehetty, C.N. and Gatlin, D.M., III (1992) Dietary lysine requirement of juvenile hybrid striped bass (*Morone chrysops* × *M. saxatilis*). *Aquaculture* 104, 271–277.

Keembiyehetty, C.N. and Gatlin, D.M., III (1993) Total sulfur amino acid requirement of juvenile hybrid striped bass (*Morone chrysops* × *M. saxatilis*). *Aquaculture* 110, 331–339.

Keembiyehetty, C.N. and Gatlin, D.M., III (1995) Evaluation of different sulfur compounds in the diet for juvenile sunshine bass (*Morone chrysops* × *M. saxatilis*). *Comparative Biochemistry and Physiology* 121A, 155–159.

Keembiyehetty, C.N. and Gatlin, D.M., III (1997a) A dietary threonine requirement of juvenile hybird striped bass (*Morone chrysops* × *M. saxatilis*). *Aquaculture Nutrition* 3, 217–221.

Keembiyehetty, C.N. and Gatlin, D.M., III (1997b) Performance of sunshine bass fed soybean-meal bases diets supplemented with different methionine compounds. *Progressive Fish-Culturist* 59, 28–30.

Keembiyehetty, C.N. and Wilson, R.P. (1998) Effect of water temperature on growth and nutrient utilization of sunshine bass (*Morone chrysops* × *M. saxatilis*) fed diets containing different energy/protein ratios. *Aquaculture* 166, 151–162.

Kim, K.I., Kayes, T.B. and Amundson, C.H. (1992) Requirements for sulfuric amino acids and utilization of D-methionine by rainbow trout (*Oncorhynchus mykiss*). *Aquaculture* 101, 95–103.

Kissil, G.W. (1991) Gilthead seabream, *Sparus aurata*. In: Wilson, R.P. (ed.) *Handbook of Nutrient Requirements of Finfish*. CRC Press, Boca Raton, Florida, pp. 83–88.

Kocabas, A.M. and Gatlin, D.M., III (1999) Dietary vitamin E requirement of hybrid striped bass (*Morone chrysops* female × *M. saxatilis* male). *Aquaculture Nutrition* 5, 3–7.

Lall, S.P., Kaushik, S.J., LeBall, P.Y., Keith, R., Anderson, J.S. and Plisetskayu, E. (1994) Quantitative arginine requirement of Atlantic salmon (*Salmo salar*) reared in sea water. *Aquaculture* 124, 13–25.

Lopez, L.A. and Woods, L.C. (1999) Fatty acid profiles of eggs from domestic striped bass broodstock. In: *Aquaculture America '99 Book of Abstracts*, p. 102.

Millikin, M.R. (1983) Interactive effects of dietary protein and lipid on growth and protein utilization of age-0 striped bass. *Transactions of the American Fisheries Society* 112, 185–193.

Moon, H.Y. and Gatlin, D.M., III (1991) Total sulfur amino acid requirement of juvenile red drum, *Sciaenops ocellatus*. *Aquaculture* 95, 97–106.

Nematipour, G.R. and Gatlin, D.M., III (1993) Requirement of hybrid striped bass for dietary (n-3) highly unsaturated fatty acids. *Journal of Nutrition* 123, 744–753.

Nematipour, G.R., Brown, M.L. and Gatlin, D.M., III (1992a) Effects of dietary energy protein ratio on growth characters and body composition of hybrid striped bass, *Morone chrysops* × *M. saxatilis*. *Aquaculture* 107, 359–368.

Nematipour, G.R., Brown, M.L. and Gatlin, D.M., III (1992b) Effects of dietary carbohydrate lipid ratio on growth and body composition of hybrid striped bass. *Journal of the World Aquaculture Society* 23, 128–132.

Nose, T. (1979) Summary report on the requirement of essential amino acids for carp. In: Halver, J.E. and Tiews, K. (eds) *Finfish Nutrition and Fishfeed Technology*. Heenemann, Berlin, Germany, pp. 138–150.

NRC (National Research Council) (1993) *Nutrient Requirement of Fish.* National Academy Press, Washington, DC.

Ogino, C. (1980) Requirements of carp and rainbow trout for essential amino acids. *Bulletin of the Japanese Society for Scientific Fisheries* 46, 171–174.

Postel, R.T., La Douceur, M., Holbert, D. and Gallagher, M.L. (1996) Texture and flavor of hybrid striped bass fed soybean meal diets. *Journal of Aquatic Food Product Technology* 5(2), 83–91.

Rawles, S.D. and Gatlin, D.M., III (1998) Carbohydrate ultilization in striped bass (*Morone saxatilis*) and sunshine bass (*Morone chrysops* × *M. saxatilis*). *Aquaculture* 161, 201–212.

Robinette, H.R., Taylor, J.B., Gatlin, D.M., III and Craig, S. (1997) Effects of dietary catfish and menhaden oils on hybrid striped bass production. *Progressive Fish-Culturist* 59, 261–265.

Robinson, E.H., Wilson, R.P. and Poe, W.E. (1981) Arginine requirement and apparent absence of a lysine arginine antagonist in fingerling channel catfish. *Journal of Nutrition* 111, 46–52.

Rollin, X., Hidalgo, Y., Valdoz, M., Teller, E. and Vanbelle, M. (1994) Methionine requirement of Atlantic salmon (*Salmo salar*). *Aquaculture* 124, 61.

Rumsey, G.L., Page, J.W. and Scott, M.L. (1983) Methionine and cystine requirement of rainbow trout. *Progressive Fish-Culturist* 45, 139–143.

Santiago, C.B. and Lovell, R.T. (1988) Amino acid requirements for growth of Nile tilapia. *Journal of Nutrition* 118, 1540–1546.

Sealey, W.M. and Gatlin, D.M. III (1999) Dietary vitamin C requirement of hybrid striped bass *Morone chrysops* × *M. saxatilis*. *Journal of the World Aquaculture Society* 30, 297–301.

Small, B.C. and Soares, J.H. (1998) Estimating the quantitative essential amino acid requirements of striped bass *Morone saxatilis*, using fillet A/E ratios. *Aquaculture Nutrition* 4, 225–232.

Small, B.C., Woods, L.C., Brougher, D.S. and Soares, J.H. (1999) Establishing the optimal dietary amino acid profile for growth and reproduction of striped bass *Morone saxatilis*. In: *Aquaculture America '99 Book of Abstracts*, p. 176.

Smith, R.R., Peterson, M.S. and Allred, A.C. (1980) The effect of leaching on apparent digestion coefficients in determining digestibility and metabolizable energy of feedstuffs for salmonid. *Progressive Fish-Culturist* 42, 195–199.

Sullivan, J.A. and Reigh, R.C. (1995) Apparent digestibility of selected feedstuffs in diets for hybrid striped bass (*Morone saxatilis* × *M. chrysops*). *Aquaculture* 138, 313–322.

Thompson, K.R., Webster, C.D., Morgan, A.M. and Grisby, E.J. (2000) Effects of different feeding frequencies as growth, body composition, and fillet composition of juvenile sunshine bass, *Morone chrysops* × *M. saxatilis*, grown indoors. *Journal of Applied Aquaculture* 10(2), 55–65.

Tuncer, H. and Harrell, R.M. (1992) Essential fatty acid nutrition of larval striped bass (*Morone saxatilis*) and palmetto bass (*M. saxatilis* × *M. chrysops*). *Aquaculture* 101, 105–121.

Tuncer, H., Harrell, R.M. and Houde, E.D. (1990) Comparative energetics of striped bass (*Morone saxatilis*) and hybrid (*M. saxatilis* × *M. chrysops*) juveniles. *Aquaculture* 86, 387–400.

Webster, C.D. and Lovell, R.T. (1990) Response of striped bass larvae fed brine shrimp from different sources containing different fatty acid composition. *Aquaculture* 90, 49–61.

Webster, C.D., Tiu, L.G., Tidwell, J.H., Van Wyk, P. and Howerton, R.D. (1995) Effects of dietary protein and lipid levels on growth and body composition of sunshine bass (*Morone chrysops* × *M. saxatilis*) reared in cages. *Aquaculture* 131, 291–301.

Webster, C.D., Tiu, L.G. and Tidwell, J.H. (1997) Effects of replacing fish meal in diets on growth and body composition of palmetto bass (*Morone saxatilis* × *M. chrysops*) raised in cages. *Journal of Applied Aquaculture* 7(1), 53–67.

Webster, C.D., Tiu, L.G., Morgan, A.M. and Gannam, A. (1999) Effect of partial and total replacement of fish meal on growth and body composition of sunshine bass (*Morone chrysops* × *M. saxatilis*) fed practical diets. *Journal of the World Aquaculture Society* 30, 443–453.

Webster, C.D., Thompson, K.R., Morgan, A.M., Grisby, E.J. and Gannam, A.L. (2000) Use of hempseed meal, poultry by-product meal, and canola meal in practical diets without fish meal for sunshine bass (*Morone chrysops* × *M. saxatilis*). *Aquaculture* 188, 299–309.

Wilson, R.P. and Poe, W.E. (1985) Apparent digestible protein and energy coefficients of common feed ingredients for channel catfish. *Progressive Fish-Culturist* 47, 154–158.

Wilson, R.P., Harding, D.E. and Garling, D.L. Jr (1977) Effect of dietary pH on amino acid utilization and the lysine requirement of fingerling channel catfish. *Journal of Nutrition* 107, 166–170.

Wilson, R.P., Allen, O.W., Robinson, E. and Poe, W.E. (1978) Tryptophan and threonine requirements of fingerling channel catfish. *Journal of Nutrition* 108, 1595–1599.

Woiwode, J.G. and Adelman, I.R. (1991) Effects of temperatures, photoperiod, and ratio size on growth of hybrid striped bass × white bass. *Transactions of the American Fisheries Society* 120, 217–229.

Woods, L.C., Yust, D., McLeod, C. and Subramanyam, M. (1995) Effects of dietary protein : energy ratio on weight gain, body composition, serum glucose and triglyceride levels, and liver functions of striped bass. *Water Science and Technology* 31, 195–203.

Zhang, Q., Reigh, R.C. and Wolters, W.R. (1994) Growth and body composition of pond-raised hybrid striped bass, *Morone saxatilis* × *M. chrysops* and *M. saxatilis* × *M. mississippiensis*, fed low and moderate levels of dietary lipid. *Aquaculture* 125, 119–129.

Sturgeon, *Acipenser* spp. 24

Silas S.O. Hung[1] and Dong-Fang Deng[2]
[1]Department of Animal Science, University of California, One Shield Ave., Davis, CA 95616, USA; [2]Department of Biochemistry and Molecular Biology, Mississippi State University, PO Box 9650, Mississippi State, MS 39762, USA

Introduction

Sturgeon are primitive teleosts with a subcylindrical body, an extended hard snout and a ventral protrusible mouth (Moyle, 1976). They are excellent bottom-feeders because of their extremely sensitive barbels on the underside of the snout for detecting bottom animals and their extraordinarily long and protruding lip for sucking up the prey. These fish have a unique gastrointestinal tract because the pyloric stomach wall is hypertrophied to a gizzard-like organ (Buddington and Doroshov, 1986). Furthermore, the intestine of adult sturgeon has a functional ciliated epithelium and their hind-gut is modified into a spiral valve.

Following successful large-scale production of juvenile sturgeon in the early 1980s, sturgeon farms have been established in many countries. Major production in 1996 included 600 metric tons (t) of fish and less than 1 t of caviare from white sturgeon (*Acipenser transmontanus*) in Italy and the USA; 235 t of fish and a few tonnes of caviare from Siberian sturgeon (*Acipenser baeri*) in Belgium, France, Germany, Italy and Poland; 60 t from Adriatic sturgeon (*Acipenser naccarii*) in Spain; and 67 t of the hybrid bester (beluga × sterlet) sturgeon in Austria and Hungary (Bronzi *et al.*, 1999).

There are no specially designed systems and commercial feeds for sturgeon, and most production is either adapted or modified from existing systems. Sturgeon are mainly monocultured in circular or rectangular tanks and raceways, but cages and extensive grow-out ponds are sometimes used (Bronzi *et al.*, 1999). Most sturgeon farms use a flow-through water system with underground or surface water, but recirculation with tap water and warm water from power stations is used at a few facilities. Sturgeon juveniles are mainly grown in indoor hatcheries and market-size fish are grown in outdoor tanks, raceways, ponds or cages.

Information on nutrition and feeding is scarce for most sturgeon species, being limited and scattered for Adriatic (Randall *et al.*, 1992; Agradi *et al.*, 1993; McKenzie *et al.*, 1994, 1995, 1997, 1999), Atlantic (*Acipenser oxyrhynchus*) (Mohler *et al.*, 1996; Bardi *et al.*, 1998; Kelly and Arnold, 1999), Chinese (*Acipenser sinensis*) (Deng *et al.*, 1998a; Xiao *et al.*, 1999), lake (*Acipenser fulvescens*) (Moreau and Dabrowski, 1996; DiLauro *et al.*, 1998; Moreau *et al.*, 1999a,b) and hybrid (Gershanovich and Kiselev, 1993; Papp *et al.*, 1995, 1999) sturgeon, along with paddlefish (*Polyodon spathula*) (Kroll *et al.*, 1992, 1996). Most information is generated from nutrition studies conducted on white and Siberian sturgeon, but this information is still incomplete.

Nutrient Requirements

Information on nutrition and feeding of sturgeon is urgently needed because of increased interest from government hatcheries in producing juveniles for release into natural waters and from commercial farms in producing meat and caviare. Three reviews have been published on the nutrition of sturgeon (Hung, 1991a,b; Médale *et al.*, 1995), and the objective of this chapter is to update these reviews with emphasis on future research needs.

Protein and amino acids

Protein requirements for optimum and maximum growth of Siberian and white sturgeon were 40% and 50%, respectively (Hung, 1991a,b; Médale *et al.*, 1995). Médale *et al.* (1995) preferred to express protein requirements for growth per unit body weight and reported that Siberian sturgeon required about 300 g of dietary protein kg^{-1} body weight gain with a protein-to-energy ratio of 20–22 mg kJ^{-1}. The quality of dietary protein for white sturgeon was discussed by Hung (1991a,b). Kaushik *et al.* (1994) reported that the growth performance and apparent digestibility coefficient (ADC) of protein were higher with casein or casein and soybean diets than with a fish-meal diet in Siberian sturgeon. The availability of total essential amino acid (EAA) was lower in the fish-meal diets than in the casein diet.

Similarly to common carp (*Cyprinus carpio*) and channel catfish (*Ictalurus punctatus*), white sturgeon grew poorly when fed a diet with its intact protein substituted by crystalline amino acids (AA) (Ng *et al.*, 1996). This may have resulted from the poor diet intake and low protein synthesis in sturgeon fed the crystalline AA diet. The low protein synthesis may have resulted from the rapid absorption from the gastrointestinal tract and fast excretion through urine and gills of dietary amino acids. Ng *et al.* (1996) reported that plasma amino acids peaked 2–4 h after white sturgeon were fed the crystalline AA diet than in those fed the intact protein diet (6–8 h). Urinary amino acid excretion was also higher in sturgeon fed the crystalline AA diet than those fed the intact protein diet.

Neutralizing the acidity of the crystalline AA diet only slightly improved the growth and reduced the urinary amino acid excretion.

Due to the lack of a suitable crystalline AA diet to support adequate growth of sturgeon, EAA requirements of sturgeon have not been determined by the standard dose–response method. Kaushik *et al.* (1991) used the daily whole-body increment to estimate the EAA requirements of 22 g Siberian sturgeon, and they were (mg 100 g^{-1} body weight day^{-1}): arginine, 2.8; histidine, 1.1; isoleucine, 2.1; leucine, 3.2; lysine, 5.4; phenylalanine, 1.5; threonine, 2.2; and valine, 2.3. Ng and Hung (1995), using similar methods, found that the EAA requirements of 67 g white sturgeon (g 100 g^{-1} protein) were: arginine, 4.8; cystine, 0.2; histidine, 2.3; isoleucine, 3.0; leucine, 4.3; lysine, 5.4; methionine, 2.0; phenylalanine, 3.0; threonine, 3.3; tryptophan, 0.3; tyrosine, 2.3; and valine, 3.3 (based on a diet containing 40% protein). Some of these values are different from the whole-body and egg amino acid pattern of Siberian sturgeon (Kaushik *et al.*, 1991) and white sturgeon (Ng and Hung, 1994). The different requirement values between the two species are mainly attributable to different units of expression between the studies. Both requirement values, however, need to be validated by the standard dose–response method. The relationship between the requirement and tissue (whole-body, muscle and egg) amino acid compositions also needs to be established. Further research should be focused on developing a suitable crystalline AA diet for sturgeon.

Energy

There is little information on energy partition and utilization, energy value of feedstuffs and energy requirements of sturgeon under culture conditions. Médale and Kaushik (1991) conducted three experiments with a commercial trout diet (50% protein, 11% fat and 21.5 kJ g^{-1} dry diet) to study energy utilization of three age-groups of farmed Siberian sturgeon (3-, 10- and 24-month-old fish with 40, 230 and 1500 g body weight, respectively) at 18°C. Voluntary diet intake decreased with age from 1.9% in the 3-month-old fish to 0.5% body weight day^{-1} in the 24-month-old fish. Energy retention was higher in the 24-month-old (55% energy retained as lipid) than the 3- and 10-month-old fish, which retained energy primarily as protein. Endogenous nitrogen loss was 60 mg kg^{-1} body weight day^{-1}. Due to discrepancies between the two methods used to estimate the endogenous energy loss, energy requirements for maintenance were not established in Siberian sturgeon.

Cui *et al.* (1996) determined the effect of ration size (from starvation to maximum or satiation) of tubificid worms and body size (2.4, 11.1 and 22.5 g) on the energy budget of white sturgeon held at 18.5°C. Specific growth rate increased linearly with increasing ration size, and feed efficiency was highest at the maximum ration. Growth rate decreased with the increase in fish size at the maximum ration. Faecal production accounted for 3.2–5.2% of food energy and the proportion of food energy lost in nitrogenous excretion decreased with increased ration size. At the maximum ration, an average 64.9% of the metabolizable energy was spent on metabolism and 35.1% on growth.

The difficulty encountered in the Siberian sturgeon study was the discrepancy between the comparative slaughter method and indirect calorimetry. The discrepancy is most probably caused by the inaccurate measurement of oxygen consumption and ammonia excretion in the indirect calorimetry. The difficulty encountered in the white sturgeon study was the continuous feeding (nibbling) habits of the fish, making diet (energy) intake impossible to measure. Future studies should be focused on improving the indirect calorimetry method and developing a water-stable diet to allow the accurate determination of diet and energy intake.

Lipids and fatty acids

High-energy salmon diets containing 25.8, 30.4, 35.7 and 40.2% lipid and a protein-to-gross energy ratio (mg kJ^{-1}) of 22.7, 20.2, 17.8 and 14.4, respectively, were fed to 110 g white sturgeon in an 8-week growth trial (Hung *et al.*, 1997). Specific growth rate and feed efficiency did not differ for sturgeon fed diets with 25.8–35.7% lipid, but decreased when dietary lipid was increased to 40.2%. Sturgeon fed diets with 25.8–35.7% lipid showed a rapid growth and high feed efficiency, suggesting that these fish could efficiently utilize dietary lipid as high as 35.7%. Médale *et al.* (1991), however, reported that the ADC of lipid was lower in Siberian sturgeon fed a diet containing 22% lipid with raw starch than in the group fed a diet containing 12.5% lipid with gelatinized starch. The lower ADC may have resulted from the inclusion of a high level of raw starch. This remains to be clarified because a lipid ADC value of 95% was determined in white sturgeon fed diets with 27% of dextrin or raw maize starch (Herold *et al.*, 1995). The protein ADCs of sturgeon fed these diets were 92 and 95%, respectively. The optimal dietary lipid level has not been determined in any species of sturgeon.

The ability to utilize dietary lipids was studied in white and Adriatic sturgeon. White sturgeon utilized different lipid sources equally well when 15% lipid was added to the diets in an 8-week study (Hung, 1991b). Fatty acids of liver and muscle were more sensitive to dietary fatty acids than were those of the brain (Xu *et al.*, 1993). Xu *et al.* (1996) found that sturgeon could desaturate and elongate linoleic acid and linolenic acid, suggesting that Δ6-desaturase might not be a limiting step in the fatty acid metabolism of sturgeon. Similarly, McKenzie *et al.* (1994) observed no difference in growth between Adriatic sturgeon fed diets supplemented with fish-oil or coconut oil. Different sources of dietary lipids, however, resulted in significant differences in tissue fatty acid composition. Adriatic sturgeon fed the fish-oil diet had higher n-3 highly unsaturated fatty acids in the liver, muscle and heart than those fed the coconut-oil diet (Agnisola *et al.*, 1996; McKenzie *et al.*, 1997).

Deng (1996) and Deng *et al.* (1998b) suggested that white sturgeon may require both n-3 and n-6 fatty acids based on growth and the 20:3n-9/20:4n-6 and 20:3n-9/22:6n-3 ratio in liver phospholipids. Synthetic triacylglycerols were used in this study and the poor growth of sturgeon fed diets deficient in n-3 and/or n-6 fatty acid may have resulted from the low digestibility of the synthetic triacylglycerols when compared with those fed natural lipids that were not deficient in n-3 and n-6 fatty acids. There is no information on ADC and utilization of synthetic triacylglycerols by sturgeon. More studies are needed to develop a purified diet containing highly digestible synthetic triacylglycerol as a basal lipid for determining the essential fatty acid requirement of sturgeon.

Carbohydrates

The ability of white sturgeon to utilize carbohydrates differs from that of other species of fish (Hung *et al.*, 1989). White sturgeon utilized D-glucose and maltose more efficiently than dextrin, raw maize starch, fructose, sucrose and lactose. Furthermore, D-glucose and raw maize starch utilization was improved with continuous feeding as compared with meal feeding (Lin *et al.*, 1997).

The ADC of D-glucose, galactose, fructose, maltose, sucrose, lactose and raw maize starch in white sturgeon were 99.4, 99.8, 53.9, 99.9, 57.1, 35.8 and 31.8%, respectively (Herold *et al.*, 1995). The low ADC of sucrose and lactose was attributed to the low activities of the digestive enzymes sucrase and lactase in the brush-border membrane (Hung *et al.*, 1989). The low ADC of fructose may have resulted from the lack of an active fructose transport system in the brush-border membrane similar to those of D-glucose and galactose. Therefore fructose can only be absorbed through the brush-border membrane by passive diffusion, resulting in a lower ADC as compared with D-glucose and galactose.

Kaushik *et al.* (1989) concluded that Siberian sturgeon (90–150 g) did not utilize raw starch well. Replacing part of the dietary raw starch with gelatinized starch or extruded whole maize improved the growth of the sturgeon. These authors suspected that high levels of digestible carbohydrates might have adversely affect the liver size, morphology and function of Siberian sturgeon. Histological examination, however, revealed no such adverse effect because hepatocytes of white sturgeon fed a diet with 27% D-glucose, maltose, dextrin or raw maize starch appeared normal (Hung *et al.*, 1990). This was confirmed by a later study showing no adverse effect on growth and liver functions of white sturgeon fed a diet with 35% D-glucose (Fynn-Aikins *et al.*, 1993). The metabolic and histological impact of carbohydrate utilization in white sturgeon was reviewed by Hung and Fynn-Aikins (1993).

The optimum dietary D-glucose for white sturgeon was 21%, and lipogenesis and glycogenesis were higher in sturgeon fed a 21–35% than in those fed a 0–14% D-glucose diet (Fynn-Aikins *et al.*, 1992). A lower lipogenic effect, however, was observed in sturgeon fed diets with 0–35% hydrolysed potato starch (Deng, 1999).

Hung (1991c) used the traditional oral challenge test to study plasma glucose homoeostasis and found that maximum plasma glucose was higher and appeared earlier in sturgeon challenged with D-glucose than in those challenged with complex carbohydrates. Handling stress and animal variation may have confounded the traditional oral challenge test. A new technique combining oesophageal intubation, dorsal aorta cannulation and urinary catheterization was developed to overcome the above limitations (Deng *et al.*, 2000). The new technique also allows urine collection.

A recent study using this technique (Deng, 1999; Deng *et al.*, 2000) showed that plasma glucose peaked at 6 h post-intubation of simple and complex carbohydrates. The peak value was highest in the glucose groups and lowest in the raw starch groups. No difference in urinary glucose excretion was observed in sturgeon intubated with maltose, dextrins and starches, but those intubated with D-glucose showed a higher excretion, which accounted for less than 1% of the intubated dose. More studies using this technique are needed to determine the renal threshold of D-glucose in sturgeon and whether this fish behaves in the same manner as a diabetic mammal in response to dietary carbohydrates.

Vitamins and minerals

Currently the requirements of only three vitamins (choline, vitamin C and vitamin E) and one mineral (selenium) have been determined for sturgeon. It was recommended that 0.4–0.6% choline chloride be incorporated in white sturgeon diets (Hung, 1991b). Siberian (Moreau *et al.*, 1996) and lake (Moreau *et al.*, 1999a,b) sturgeon have been shown to be able to synthesize ascorbic acid and thus do not have a dietary requirement for this vitamin. Tissue storage of ascorbic acid increased with dietary vitamin C, which did not seem to inhibit renal L-gulono-1,4-lactone oxidase activity. Sturgeon hybrid did not require dietary vitamin C either, and their tissue level increased with dietary level (Papp *et al.*, 1995, 1999). Vitamin C and vitamin E interaction was studied in lake sturgeon (Moreau *et al.*, 1999a) and liver α-tocopherol level was markedly affected by dietary vitamin C and vitamin E, but no vitamin E requirement was determined. The effect of n-3 fatty acids and vitamin E supplements on growth was studied in Adriatic sturgeon, but the vitamin E requirement was not determined (Agradi *et al.*, 1993). The selenium requirement was determined in white sturgeon (S.S.O. Hung, unpublished). There was no difference in the growth of sturgeon fed purified diets supplemented with 0.05–1.2 mg selenium kg^{-1} diet in the form of sodium selenite in a 14-week growth trial, but those fed a diet without a selenium supplement grew less than those with 0.3, 1.0 and 1.2 mg of added selenium kg^{-1} diet. Vitamin and mineral premixes designed for salmonids are being used for Adriatic, Siberian and white sturgeon under research and production conditions. More studies are needed to determine vitamin and mineral requirements in sturgeon.

Practical Diets

Larval diet

Gawlicka (1995) and Herold (1996) followed the growth, survival and composition changes of white sturgeon larvae for the first 24 days after the larvae began to feed (12 days post-hatch). There was no difference in the growth, survival and body composition of larvae fed diets containing either 25%, 35% or 45% lipid. Histological and histochemical examination, however, revealed some adaptive changes in the spiral intestine epithelium and hepatocytes of larvae fed the 35% and 45% lipid diets. Feeding high-lipid diets appeared to interfere with lipid and glycogen storage and assimilation (S.S.O. Hung, unpublished). Furthermore, Herold (1996) and Hung *et al.* (1998b) suggested that the poor growth of larvae fed the high-lipid diet might have resulted from the low protein-to-gross energy ratio, a high percentage of linoleic acid and a low percentage of eicosapentaenoic acid. No differences in growth, survival and whole-body protein content were observed in larvae fed for 24 days at 25–30% body weight day^{-1} with diets containing 27% and 45% protein. There was, however, a decrease in liver glycogen reserves and an abnormal accumulation of lipid in larvae fed the low-protein diet (S.S.O. Hung, unpublished).

Another study was conducted to determine the interaction between low (LP = 25%) and high-protein (HP = 45%) diets and low (LFR), medium (MFR) and high feeding rates (HFR) on the survival, growth and body composition of sturgeon larvae during the first 24 days after initiation of feeding (S.S.O. Hung, unpublished). Each diet was fed at 10% (LFR), 20% (MFR) and 30% (HFR) body weight day^{-1} on days 1–7, and feeding rates were reduced by one-third and two-thirds on days 8–16 and days 17–24, respectively. Dietary protein levels and feeding rates had no effect on the survival and dry-matter content of the larvae. However, the weight gain of larvae fed the HP and LP diets at MFR and HFR was higher than that of those fed at LFR. An increase in feeding rate with the HP diet caused no changes in whole-body composition, whereas with the LP diet there was an increase of body protein content at day 16 and of lipid and glycogen content at day 16 and day 24. In general, feeding an LP diet for 24 days resulted in larvae with a higher body lipid and glycogen content. More studies are needed to determine whether the increased body lipid and glycogen are desirable for the larvae.

Five water-stable microdiets were evaluated for white-sturgeon larvae by Gawlicka *et al.* (1996, 1997). It was found that starch ticaloid microbound and starch ticaloid bound diets were most promising and worth further investigation. Free amino acids as diet stimulants were evaluated in Russian, Siberian, green (*Acipenser medirostris*), sevryuga (*Acipenser stellatus* Pallas) and beluga (*Huso huso*) sturgeon larvae (Kasumyan, 1994, 1999; Kasumyan and Taufik, 1994). Of the 20 amino acids tested, only glycine and L-alanine at the threshold of 1 μM were shown to induce food-searching behaviour. Moreau and Dabrowski (1996) tested 2% betaine, 2% betaine plus 3 or 6% krill and 2% betaine plus 3 or 6% fish-protein concentrate. The results showed that fish-protein concentrate improved the growth and survival of lake sturgeon larvae. Kuzmin *et al.* (1999) reported that lysine, methionine and alanine are the most attractive amino acids, and extruded diet components stimulated more bites than granulated ones in hybrid sturgeon. Among the diet components only fish-protein concentrate, dry milk and krill meal stimulated feeding behaviour. Flavour enhancers, such as concentrated low-molecular-weight nitrogenous substances, sodium glutamate and Finnstim all stimulated feeding behaviour of fry. The effectiveness of these stimulants in practical larval diets needs to be determined.

There is no commercial sturgeon larval diet available, but high growth (specific growth rate over 10% day^{-1}) and survival (95%) have been achieved with semimoist salmonid diets. Live organisms or moist, semimoist and dry microparticulated and crumbled pellets manufactured for salmonid and marine larvae are commonly used at the initiation of feeding of several species of sturgeon (Kroll *et al.*, 1992; Mohler *et al.*, 1996; Bardi *et al.*, 1998).

Grow-out diet

There is no standard practical grow-out diet for sturgeon. A few diets have been developed by fish farmers and feed mills but these diets are still under research and development. Most sturgeon farmers use existing commercially available diets, particularly those of salmonid diets, with or without modification. Hung *et al.* (1998a) fed 72 g white sturgeon one of seven commercial salmonid diets or a purified diet for 8 weeks. These diets contained 5.9–8.3% moisture, 31.5–51.4% crude protein, 7.6–19.4% lipid and 3.8–11.7% ash. Sturgeon fed the diet with 4.5% moisture, 51.4% crude protein, 18.0% lipid and 11.7% ash had the best growth and feed efficiency. Addition of 0.03% phytase, 0.05% deodorase or 0.33% carnitine to the diets did not improve growth and feed efficiency.

Brood-stock diet

Due to their large size, long lifespan and late sexual maturity, studies of the nutrient requirements of sturgeon brood-stock have rarely been conducted. It is, however, a common practice to include two to three times the amount of vitamin premix commonly used in salmonid feed for sturgeon brood-stock. A future challenge in sturgeon brood-stock nutrition is to develop an alternative method to the traditional growth trial to predict reproductive outcome accurately.

Feeding Practices

Larval feeding

Several studies on the nutrition of white (Gawlicka *et al.*, 1995) and Siberian (Gisbert and Williot, 1997; Gisbert *et al.*, 1997, 1999; Gisbert, 1999) sturgeon larvae have dealt mainly with the first feeding and/or gastrointestinal-tract ontogenic development. Gisbert and Williot (1997) recommended that Siberian sturgeon larvae kept at 18°C should be fed at 9 days post-hatch. Earlier initiation of feeding before completion of yolk-sac reserves (5–7 days post-hatch) provided no advantage. However, growth and survival rate decreased when larvae were fed later than 9 days post-hatch. In our experience, at 18–19°C, 12 days post-hatch is the best time to initiate the first feeding in white-sturgeon larvae.

Optimum feeding rates of white-sturgeon larvae at 19°C with a commercial semimoist salmonid diet containing 15% moisture, 42.5% protein, 12.8% lipid and 7.7% ash were 30%, 20%, 10% and 7.5% body weight day^{-1} for the first, second, third and fourth week after initiation of feeding (12 days post-hatch), respectively (S.S.O. Hung, unpublished). There is no information on the optimum feeding rate of other species of sturgeon.

Juvenile feeding

A series of studies were conducted to determine the optimal feeding rate of juvenile white sturgeon (Hung 1991a, b; Hung *et al.*, 1993, 1995). Cui and Hung (1995) used data from these studies to develop a prototype feeding–growth table based on an empirical model to predict the growth rate (G) from optimum feeding rate ($FR_{optimum}$), body weight (W) and water temperature (T):

$$G = -4.43 + 0.556T - 0.0151T^2 + 0.00849\ln(F + 0.1)\,T^2 - 0.001222\ln(F + 0.1)\,T\ln W$$

and

$$\ln FR_{optimum} = -2.88 - 0.25\ln W + 0.4T - 0.0077T^2$$

where F is the feeding rate (% body weight day^{-1}). These models, however, did not make biological sense and more studies are needed to modify the models so that they can be explained biologically.

Cui *et al.* (1997) compared the growth performance of 8.6 g white sturgeon kept under natural photoperiod but under six different feeding regimens: continuously for 24 h, 12.8 h during the day (light) and 12.8 h during the night (darkness) and two, four and six meals in 24 h during both light and darkness. It was concluded that 24 h continuous feeding was optimum for juvenile white sturgeon. Jatteau (1997), based on daily patterns of ammonia nitrogen output of Siberian sturgeon, also concluded that continuous feeding is suitable in sturgeon farms in terms of ammonia loadings. There is no information on the optimum feeding rate of sturgeon brood-stock. Continuous feeding with automatic (belt or electrical) feeders is commonly used for larvae, and automatic feeders or demand feeders are used for grow-out and brood-stock.

References

Agnisola, C., Mckenzie, D.J., Taylor, E.W., Bolis, C.L. and Tota, B. (1996) Cardiac performance in relation to oxygen supply varies with dietary lipid composition in sturgeon. *American Journal of Physiology* 27, 417–425.

Agradi, E., Abrami, G., Serrini, G., McKenzie, D., Bolis, C. and Bronzi, P. (1993) The role of n-3 fatty acid and vitamin E supplements in growth of sturgeon (*Acipenser naccarii*). *Comparative Biochemistry and Physiology* 105A, 187–195.

Bardi, R.W. Jr, Chapman, F.A. and Barrows, F.T. (1998) Feeding trials with hatchery-produced Gulf of Mexico sturgeon larvae. *Progressive Fish-Culturist* 60, 25–31.

Bronzi, P., Rosenthal, H., Arlati, G. and Williot, P. (1999) A brief overview on the status and prospects of sturgeon farming in Western and Central Europe. *Journal of Applied Ichthyology* 15, 224–227.

Buddington, R.K. and Doroshov, S.I. (1986) Structural and functional relations of the white sturgeon alimentary canal (*Acipenser transmontanus*). *Journal of Morphology* 190, 201–213.

Cui, Y. and Hung, S.S.O. (1995) A prototype feeding-growth table for white sturgeon. *Journal of Applied Aquaculture* 5, 25–34.

Cui, Y., Hung, S.S.O. and Zhu, X. (1996) Effect of ration and body size on energy budget of juvenile white sturgeon. *Journal of Fish Biology* 49, 863–876.

Cui, Y., Hung, S.S.O., Deng, D.F. and Yang, Y. (1997) Growth of white sturgeon as affected by feeding regimen. *Progressive Fish-Culturist* 59, 31–35.

Deng, D.F. (1996) Qualitative requirement of essential fatty acids for white sturgeon (*Acipenser transmontanus*). MS dissertation, University of California, Davis, California.

Deng, D.F. (1999) Carbohydrate utilization by white sturgeon (*Acipenser transmontanus*). PhD dissertation, University of California, Davis, California.

Deng, X., Cui, Y. and Hung, S.S.O. (1998a) Initial trials with feeding of Chinese sturgeon (*Acipenser sinensis*) larvae on artificial diet. *Acta Hydrobiologica Sinica* 22, 189–191.

Deng, D.F., Hung, S.S.O. and Conklin, D.E. (1998b) White sturgeon (*Acipenser transmontanus*) require both n-3 and n-6 fatty acids. *Aquaculture* 161, 333.

Deng, D.F., Refstie, S., Hemre, G.I., Crocker, C.E., Chen, H.Y., Cech, J.J. Jr and Hung, S.S.O. (2000) A new technique of feeding, repeated sampling of blood and continuous collection of urine in white sturgeon. *Fish Physiology and Biochemistry* 22, 191–197.

DiLauro, M.N., Krise, W.F. and Fynn-Aikins, K. (1998) Growth and survival of lake sturgeon larvae fed formulated diets. *Progressive Fish-Culturist* 60, 293–296.

Fynn-Aikins, K., Hung, S.S.O., Liu, W. and Li, H. (1992) Growth, lipogenesis and liver composition of juvenile white sturgeon (*Acipenser transmontanus*) fed different levels of D-glucose. *Aquaculture* 105, 61–72.

Fynn-Aikins, K., Hung, S.S.O. and Hughes, S.G. (1993) Effects of feeding a high level of D-glucose on growth and liver function in juvenile white sturgeon. *Fish Physiology and Biochemistry* 12, 317–325.

Gawlicka, A. (1995) Physiologie digestive et alimentation des larves de l'esturgeon blanc (*Acipenser transmontanus*). PhD dissertation, Université Laval, Quebec, Canada.

Gawlicka, A., Teh, S.J., Hung, S.S.O., Hinton, D.E. and de la Noüe, J. (1995) Histological and histochemical changes in the digestive tract of white sturgeon larvae during ontogeny. *Fish Physiology and Biochemistry* 14, 357–371.

Gawlicka, A., McLaughlin, L., Hung, S.S.O. and de la Noüe, J. (1996) Limitations of carrageenan microbound diets for feeding white sturgeon larvae. *Aquaculture* 141, 245–265.

Gawlicka, A., Gagnon, D., Dubeau, K., de la Noüe, J., Hung, S.S.O. and Barrows, R. (1997) Development and evaluation of water-stable microdiets for white sturgeon larvae. *Bulletin of Aquaculture Association Canada* 4, 26–32.

Gershanovich, A.D. and Kiselev, G.A. (1993) Growth and haematological response of sturgeon hybrids Russian sturgeon (*Acipenser gueldenstaedti* brandt) × beluga (*Huso huso* L.) to protein and lipid contents in the diet. *Comparative Biochemistry and Physiology* 106A, 581–586.

Gisbert, E. (1999) Early development and allometric growth patterns in Siberian sturgeon and their ecological significance. *Journal of Fish Biology* 54, 852–862.

Gisbert, E. and Williot, P. (1997) Larval behaviour and effect of the timing of initial feeding on growth and survival of Siberian sturgeon (*Acipenser baeri*) larvae under small scale hatchery production. *Aquaculture* 156, 63–76.

Gisbert, E., Rodriguez, A., Castello-Orvay, F. and Williot, P. (1997) A histological study of the development of the digestive tract of Siberian sturgeon (*Acipenser baeri*) during early ontogeny. *Aquaculture* 167, 195–209.

Gisbert, E., Sarasquete, M.C., Williot, P. and Castello Orvay, F. (1999) Histochemistry of the development of the digestive system of Siberian sturgeon during early ontogeny. *Journal of Fish Biology* 55, 596–616.

Herold, M.A. (1996) Lipid nutrition in white sturgeon (*Acipenser transmontanus*). PhD dissertation, University of California, Davis, California.

Herold, M.A., Hung, S.S.O. and Fynn-Aikins, K. (1995) Apparent digestibility coefficient of carbohydrates for white sturgeon. *Progressive Fish-Culturist* 57, 137–140.

Hung, S.S.O. (1991a) Nutrition and feeding of hatchery-produced juvenile white sturgeon (*Acipenser transmontanus*): an overview. In: Williot, P. (ed.) *Proceedings of the First International Symposium on the Sturgeon*. CEMAGREF, France, pp. 65–77.

Hung, S.S.O. (1991b) Sturgeon, *Acipenser* spp. In: Wilson, R.P. (ed.) *Handbook of Nutrient Requirements of Finfish*. CRC Press, Boca Raton, Florida, pp. 153–160.

Hung, S.S.O. (1991c) Carbohydrate utilization by white sturgeon as assessed by oral administration tests. *Journal of Nutrition* 121, 1600–1605.

Hung, S.S.O. and Fynn-Aikins, K. (1993) Carbohydrate utilization and its impact on some metabolic and histological parameters in white sturgeon. In: Kaushik, S.J. and Luquet, P. (eds) *Fish Nutrition in Practice. IVth International Symposium on Fish Nutrition and Feeding*, Biarritz, France. INRA, Paris, pp. 127–136.

Hung, S.S.O., Fynn-Aikins, F.K., Lutes, P.B. and Xu, R. (1989) Ability of juvenile white sturgeon (*Acipenser transmontanus*) to utilize different carbohydrate sources. *Journal of Nutrition* 119, 727–733.

Hung, S.S.O., Groff, J.M., Lutes, P.B. and Fynn-Aikins, K. (1990) Hepatic and intestinal histopathology of juvenile white sturgeon fed different carbohydrates. *Aquaculture* 87, 349–360.

Hung, S.S.O., Lutes, P.B., Shqueir, A.A. and Conte, F.S. (1993) Effect of feeding rate and water temperature on growth of juvenile white sturgeon (*Acipenser transmontanus*). *Aquaculture* 115, 297–303.

Hung, S.S.O., Conte, F.S. and Lutes, P.B. (1995) Optimum feeding rate of white sturgeon (*Acipenser transmontanus*) yearlings under commercial production conditions. *Journal of Applied Aquaculture* 5, 45–51.

Hung, S.S.O., Storebakken, T., Cui, Y., Tian, L. and Einen, O. (1997) High-energy diets for white sturgeon, *Acipenser transmontanus*, Richardson. *Aquaculture Nutrition* 3, 281–286.

Hung, S.S.O., Conte, F.S. and Frost, F. (1998a) Growth of white sturgeon fed seven commercial feeds. *Aquaculture* 161, 89.

Hung, S.S.O., Herold, M.A., Gawlicka, A. and de la Noüe, J. (1998b) Effects of dietary lipid on growth and fatty acid composition of white sturgeon (*Acipenser transmontanus*) larvae. *Aquaculture* 161, 333.

Jatteau, P. (1997) Daily patterns of ammonia nitrogen output of Siberian sturgeon *Acipenser baeri* (Brandt) of different body weights. *Aquaculture Research* 28, 551–557.

Kasumyan, A.O. (1994) Olfactory sensitivity of the sturgeon to free amino acids. *Biophysics* 39, 519–522.

Kasumyan, A.O. (1999) Olfaction and taste senses in sturgeon behavior. *Journal of Applied Ichthyology* 15, 228–232.

Kasumyan, A.O. and Taufik, L.R. (1994) Behaviour reaction of juvenile sturgeons (Acipenseridae) to amino acids. *Journal of Applied Ichthyology* 34, 90–103.

Kaushik, S.J., Luquet, P., Blanc, D. and Paba, A. (1989) Studies on the nutrition of Siberian sturgeon, *Acipenser baeri*. I. Utilization of digestible carbohydrates by sturgeon. *Aquaculture* 76, 97–107.

Kaushik, S.J., Breque, J. and Blanc, D. (1991) Requirement for protein and essential amino acids and their utilization by Siberian sturgeon (*Acipenser baeri*). In: Williot, P. (ed.) *Proceedings of the First International Symposium on the Sturgeon.* CEMAGREF, France, pp. 25–39.

Kaushik, S.J., Breque, J. and Balanc, D. (1994) Apparent amino acid availability and plasma free amino acid levels in Siberian sturgeon (*Acipenser baeri*). *Comparative Biochemistry and Physiology* 107A, 433–438.

Kelly, J.L. and Arnold, D.E. (1999) Effects of ration and temperature on growth of age-0 Atlantic sturgeon. *North American Journal of Aquaculture* 61, 51–57.

Kroll, K.J., van Eenennaam, J.P., Doroshov, S.I., Hamilton, E.J. and Russell, T.R. (1992) Effect of water temperature and formulated diets on growth and survival of larval paddlefish. *Transactions of the American Fisheries Society* 121, 538–543.

Kroll, K.J., van Eenennaam, J.P., Doroshov, S.I., Linares, J., Hamilton, E.J. and Russell, T.R. (1996) Growth and survival of paddlefish fry raised in the laboratory on natural and artificial diets. *Progressive Fish-Culturist* 56, 169–174.

Kuzmin, S., Mironov, S., Vostroushkin, D. and Shutov, V. (1999) Behavioural responses to various chemical incentives in hybrid beluga × Russian sturgeon (*Huso huso × Acipenser gueldenstaedtii*) fry. *Journal of Applied Ichthyology* 15, 233–236.

Lin, J.H., Cui, Y., Hung, S.S.O. and Shiau, S.Y. (1997) Effect of feeding strategy and carbohydrate source on carbohydrate utilization by white sturgeon and hybrid tilapia. *Aquaculture* 148, 201–211.

McKenzie, D.J., Piraccini, G., Taylor, E.W., Steffensen, J.F., Bronzi, P. and Bolis, C.L. (1994) Effects of dietary lipids on responses to stress in fish. In: MacKinlay, D.D. (ed.) *Proceedings of International Symposium on High Performance Fish.* Fish Physiology Society/American Fisheries Society, Vancouver, Canada, pp. 40–46.

McKenzie, D.J., Piraccini, G., Steffensen, J.F., Bolis, C.L., Bronzi, P. and Taylor, E.W. (1995) Effects of diet on spontaneous locomotor activity and oxygen consumption in Adriatic sturgeon (*Acipenser naccarii*). *Fish Physiology and Biochemistry* 14, 342–355.

McKenzie, D.J., Piraccini, G., Papini, N., Galli, C., Bronzi, P., Bolis, C.G. and Taylor, E.W. (1997) Oxygen consumption and ventilatory reflex are influenced by dietary lipids in sturgeon. *Fish Physiology and Biochemistry* 16, 365–379.

McKenzie, D.J., Piraccini, G., Agnisola, J.F., Steffensen, J.F., Bronzi, P., Bolis, C.L., Tota, B. and Taylor, E.W. (1999) The influence of dietary fatty acid composition on the respiratory and cardiovascular physiology of Adriatic sturgeon (*Acipenser naccarii*): a review. *Journal of Applied Ichthyology* 15, 265–269.

Médale, F. and Kaushik, S.J. (1991) Energy utilization by farmed Siberian sturgeon (*Acipenser baeri*) from 3 age classes. In: Williot, P. (ed.) *Proceedings of the First International Symposium on the Sturgeon*. CEMAGREF, France, pp. 13–23.

Médale, F., Blanc, D. and Kaushik, S.J. (1991) Studies on the nutrition of Siberian sturgeon, *Acipenser baeri*. II. Utilization of dietary non-protein energy by sturgeon. *Aquaculture* 93, 143–154.

Médale, F., Corraze, G. and Kaushik, S.J. (1995) Nutrition of farmed Siberian sturgeon. In: Gershanovic, A.D. and Smith, T.I.J. (eds) *Proceedings of the Third International Symposium on Sturgeons*. VNIRO Publishing, Moscow, Russia, pp. 289–298.

Mohler, J.W., Fynn-Aikins, K. and Barrows, R. (1996) Feeding trials with juvenile Atlantic sturgeons propagated from wild broodstock. *Progressive Fish-Culturist* 58, 173–177.

Moreau, R. and Dabrowski, K. (1996) Feeding stimulants in semipurified diets for juvenile lake sturgeon, *Acipenser fulvescens* Rafinesque. *Aquaculture Research* 27, 953–957.

Moreau, R., Kaushik, S.J. and Dabrowski, K. (1996) Ascorbic acid status as affected by dietary treatment in the Siberian sturgeon (*Acipenser baeri* Brandt): tissue concentration, mobilization and L-gulonolactone oxidase activity. *Fish Physiology and Biochemistry* 15, 431–438.

Moreau, R., Dabrowski, K., Czesy, S. and Cihla, F. (1999a) Vitamin C–vitamin E interaction in juvenile lake sturgeon (*Acipenser fulvescens* R.), a fish able to synthesise ascorbic acid. *Journal of Applied Ichthyology* 15, 250–257.

Moreau, R., Dabrowski, K. and Sato, P.H. (1999b) Renal L-gulono-1,4-lactone oxidase activity as affected by dietary ascorbic acid in lake sturgeon (*Acipenser fulvescens*). *Aquaculture* 180, 359–372.

Moyle, P.B. (1976) Sturgeon family, Acipenseridae. In: *Inland Fishes of California*. University of California Press, Berkeley, California, pp. 95–99.

Ng, W.K. and Hung, S.S.O. (1994) Amino acid composition of whole body, eggs, and selected tissues of juvenile white sturgeon. *Aquaculture* 126, 329–339.

Ng, W.K. and Hung, S.S.O. (1995) Estimating the ideal dietary indispensable amino acid pattern for growth of white sturgeon *Acipenser transmontanus* (Richardson). *Aquaculture Nutrition* 1, 85–94.

Ng, W.K., Hung, S.S.O. and Herold, M.A. (1996) Poor utilization of dietary free amino acids by white sturgeon. *Fish Physiology and Biochemistry* 15, 131–142.

Papp, Z.G., Jeney, Z. and Jeney, G. (1995) Effect of different levels of dietary vitamin C on growth, ascorbate concentration in some tissue and physiological status of sturgeon hybrid (*Acipenser ruthenus* L. × *Acipenser baeri* Brandt). In: Gershanovich, A.D. and Smith, T.I.J. (eds) *Proceedings of the Third International Symposium on Sturgeons*. VNIRO Publishing, Moscow, Russia, pp. 309–313.

Papp, G.Z., Saroglia, M., Jeney, Z., Jeney, G. and Terova, G. (1999) Effects of dietary vitamin C on tissue ascorbate and collagen status in sturgeon hybrids (*Acipenser ruthenus* L. × *Acipenser baeri* Brandt). *Journal of Applied Ichthyology* 15, 258–260.

Randall, D.J., McKenzie, D.J., Abrami, G., Bondiolotti, G.P., Natiello, F., Bronzi, P., Bolis, L. and Agradi, E. (1992) Effects of diet on responses to hypoxia in sturgeon (*Acipenser naccarii*). *Journal of Experimental Biology* 170, 113–125.

Xiao, H., Cui, Y., Hung, S.S.O., Zhu, X., Zou, Z. and Xie, S. (1999) Growth of juvenile Chinese sturgeon *Acipenser sinensis* Grey fed live and formulated diets. *North American Journal of Aquaculture* 61, 184–188.

Xu, R., Hung, S.S.O. and German, B.J. (1993) White sturgeon tissue fatty acid compositions are affected by dietary lipids. *Journal of Nutrition* 123, 1685–1692.

Xu, R., Hung, S.S.O. and German, J.B. (1996) Effects of dietary lipids on the fatty acid composition of triglycerides and phospholipids in tissues of white sturgeon. *Aquaculture Nutrition* 2, 101–109.

Silver Perch, *Bidyanus bidyanus* 25

Geoff L. Allan[1] and Stuart J. Rowland[2]
[1]New South Wales (NSW) Fisheries, Port Stephens Fisheries Centre, Taylors Beach, NSW 2316, Australia; [2]NSW Fisheries, Grafton Aquaculture Centre (GAC), Grafton, NSW 2460, Australia

Introduction

Silver perch, *Bidyanus bidyanus* (Mitchell) (Teraponidae) is a freshwater, temperate fish, endemic to the Murray–Darling river system west of the Great Dividing Range in south-eastern Australia. It is a schooling, omnivorous fish that feeds naturally on zooplankton, insects, crustaceans, molluscs, small fish and some aquatic plants, in particular filamentous algae. Silver perch is an attractive, small-scaled, laterally compressed fish with premium-quality white flesh, a mild, delicate flavour, a fine texture, few bones and about 40% meat recovery. A recent study of over 200 Australian seafood products found that farmed silver perch had the third highest levels of fatty acids beneficial to human health (Nichols *et al.*, 1998).

Hatchery techniques, based on hormone-induced spawning and extensive pond rearing of larvae, were developed in the early 1980s (Rowland *et al.*, 1983; Rowland, 1984) and up to 5 million fingerlings are produced annually for aquaculture, stock enhancement and conservation purposes. However, it was not until the early 1990s that research into the grow-out of silver perch commenced at the New South Wales (NSW) Fisheries' Grafton Aquaculture Centre (GAC), Australia (29° 41′ S; 152° 56′ E). The research demonstrated that silver perch is an excellent species for semi-intensive or intensive culture in earthen ponds. High survival rates (> 90%) and fast growth rates from fingerlings (2–5 g $fish^{-1}$ day^{-1}) to market-size (~500 g), at high stocking densities (20,000 fish ha^{-1}), have routinely produced high production rates of around 10 t ha^{-1} $year^{-1}$ in static, aerated, 0.1 ha earthen ponds (Rowland, 1995a; Rowland *et al.*, 1995).

Most silver perch cultured commercially are grown in static, aerated, earthen ponds that range in surface area from 0.1 to 0.5 ha. Recent research suggests that cages may also be suitable for culture. A three-phase production strategy is recommended for this species: I, hatchery; II, fingerling; III, grow-out (Rowland, 1995b). This strategy is usually combined with a single-batch system, where each pond only has fish of the same age or batch, which are totally harvested before the next batch is stocked. Fry (30 mm, 0.5 g) become available from hatcheries in mid- to late summer (January to February), and are stocked at densities up to 150,000 fish ha^{-1} in fingerling ponds for 3 or 4 months. The fingerlings are then harvested, graded and stocked at a density of 10,000–30,000 ha^{-1} for grow-out. The grow-out phase takes 10–18 months, depending on the temperature regime, to reach market size (400–800 g). Fish and ponds are closely managed because of the intensity of production. Major water-quality variables (temperature, dissolved oxygen, pH and ammonia) are monitored every 2–3 days. Fish are sampled monthly to estimate the mean weight and biomass, and the daily ration is adjusted accordingly. Sampled fish are checked for disease; there are no major disease problems in the industry at present. Fish are harvested using seine nets on most farms, and are placed live in clean water for 7 days to purge off-flavours and to ensure a uniform, high-quality product. Silver perch are sold live, principally into Asian communities, or whole chilled with prices ranging from A$7 to A$15 kg^{-1}. The industry is expected to have a large processing component in the future.

Although the silver perch farming industry is currently in its infancy, it has great potential for growth. Australia has limited wild fisheries, most of which are fully or overexploited, and approximately 70% of the white-fleshed finfish consumed is imported. There are abundant sites with high-quality water available for aquaculture and freshwater effluent is easily managed, making possible environmentally sound practices with no release of effluent to natural waterways. Considerable opportunities exist for the integration of silver-perch culture with established agricultural industries, especially where irrigation is used.

Silver perch are farmed predominantly in the states of NSW and Queensland, with small quantities also produced in Victoria, South Australia and Western Australia. Only 2.6 t were farmed in NSW in 1992/93, but production is now increasing rapidly as more farms are constructed and farming practices improve. The established culture techniques, the availability of sites and the premium quality of the product provide a basis for a dramatic increase in production over the next 5–10 years, which is predicted to exceed 1000 t by 2005. There is also much interest in the culture of silver perch in other countries – especially in the People's Republic of China, Taiwan and Israel. Fingerlings and broodfish have been exported to these countries from some hatcheries in Australia.

Nutrient Requirements

Protein and energy

Early research indicated that protein requirements of silver perch were similar to those reported for other omnivores, such as channel catfish, and lower than those reported for carnivores (Allan and Rowland, 1991; Allan *et al.*, 1994). Protein efficiency ratios increased with increasing digestible energy (DE) (in the range

12–15 MJ kg^{-1}) and tended to decrease with increasing digestible protein (in the range 25–45%) (Allan *et al.*, 1994). Harpaz *et al.* (1999) reported increased growth of 3 g silver perch fingerlings when fed 41% protein diets compared with the growth of fish fed 23% protein diets. Although gross energy was similar for Harpaz *et al.*'s diets, the DE content of the 23% protein diet was much lower than that of the 45% protein diet. Unfortunately, such differences in the DE content of experimental diets have confounded interpretation of this and other experiments that have tried to estimate the protein requirements of silver perch and other fish species.

Protein and energy need to be kept in balance (NRC, 1993). Research with pigs and poultry has led to the development of the concept of two phases in protein deposition: a protein-dependent phase and an energy-dependent phase. At a constant energy content, protein deposition will increase with increasing protein intake until a plateau is reached, after which there will be no response to increasing protein intake. At a higher energy intake, protein deposition will increase with a higher protein intake (Bikker, 1994).

A study was conducted at our laboratory to estimate requirements of silver perch for digestible protein using isocaloric diets (14 MJ of DE kg^{-1}) made using practical ingredients (fish-meal, maize gluten, peanut meal, wheat, wheat starch and fish-oil) with varying protein contents (11–40% digestible protein). Fish were fed to satiation. The minimum dietary protein content before growth was retarded was 28% (Fig. 25.1; Allan *et al.*, 2001).

The optimum protein requirements for juvenile silver perch at three different DE contents were also evaluated. Using similar methods to those described above, but with three series of diets, each with a different DE content, and with fish fed at about 90% of satiation level, protein retention efficiency was plotted against digestible dietary protein for each series of diets. Results indicate that optimum

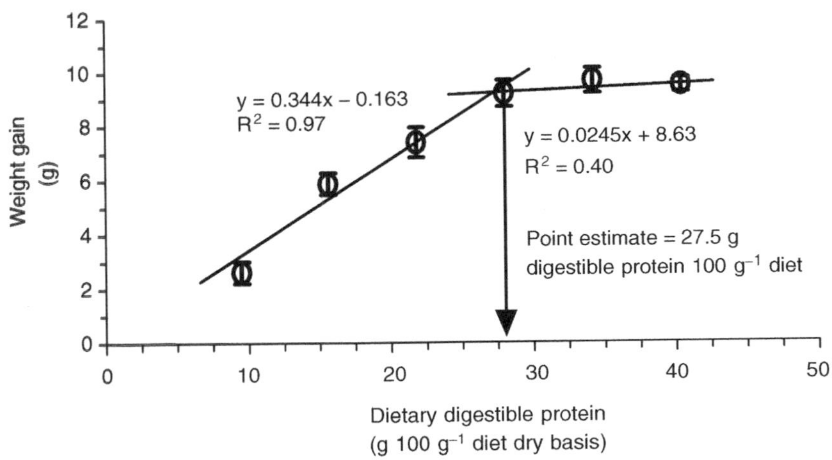

Fig. 25.1. Effect of dietary digestible protein content and a digestible energy content of 14 MJ kg^{-1} on individual weight gain of juvenile silver perch. Error bars indicate standard error of mean (SEM).

digestible protein requirements were 25% for diets with 12–14 MJ of DE kg^{-1}, 26% for diets with 14–16 MJ of DE kg^{-1} and 29% for diets with 16–17 MJ of DE kg^{-1} (G.L. Allan, unpublished data).

Requirements of protein and energy for maintenance were estimated by feeding juvenile silver perch restricted rations (approximately 90% of satiation) of a nutritionally adequate reference diet or diets with different substitution levels of an inert filler (diatomaceous earth). Using this approach, Booth *et al.* (2000a) estimated requirements of protein and energy for maintenance as approximately 2.1 g digestible protein kg^{-1} body weight day^{-1} and 116 kJ of DE kg^{-1} body weight day^{-1}. These amounts were slightly higher than those estimated for some other species (Lupatsch *et al.*, 1998), possibly because in the study by Booth *et al.* (2000a) fish were still expending energy for feeding activity, whereas in the other studies maintenance requirements were estimated by progressively reducing food intake, thereby also reducing feeding activity.

Protein is made up of 20 α-amino acids linked into chains by peptide bonds (NRC, 1993). The protein supplied to fish must contain sufficient quantities of the amino acids that fish cannot synthesize (there are ten of these essential amino acids) to provide for tissue protein synthesis.

The most common method for estimating amino acid requirements involves making a series of diets using an intact protein (e.g. casein plus gelatin) plus a mixture of crystalline amino acids. Graded amounts of the amino acid under investigation are supplied in the crystalline amino acid mix to give a series of diets. The total protein content of the crystalline amino acid mix is balanced in the series of diets using non-essential amino acids. Ngamsnae *et al.* (1999) used this method to estimate requirements of silver perch for arginine and phenylalanine and then estimated the requirements for each of the other indispensable amino acids by multiplying the requirement for arginine by the content of each amino acid (as a proportion of total amino acids) (see Table 25.1).

Growth rates of fish fed diets composed of purified sources of protein plus mixtures of crystalline amino acids are usually inferior to those of fish fed diets based on intact, practical protein sources, probably due to a relatively poor utilization of crystalline amino acids and purified protein sources (Nose and Murai, 1990; NRC, 1993; Cowey, 1994; Ngamsnae *et al.*, 1999). Allan *et al.* (2001) reported that, for diets with 14 MJ of DE kg^{-1}, requirements for essential amino acids could not be as high as those reported by Ngamsnae *et al.* (1999) because the growth of fish did not increase when additional amino acids were added past the contents in the 28% protein diet (see Fig. 25.1). They suggested that the amino acid content of the 28% protein diet could be considered as 'recommended values' for diets at 14 MJ of DE kg^{-1} (Table 25.1).

Lipids and fatty acids

Dietary lipids are an important source of energy and essential fatty acids and help with the absorption of fat-soluble vitamins (NRC, 1993). The DE content of

Table 25.1. Requirements and 'recommended' values of essential amino acids (% of diet) for juvenile silver perch.

Amino acid	Estimated requirements*	Recommended values†
Arginine	2.7	1.9
Histidine	1.0	0.7
Isoleucine	1.8	1.1
Leucine	3.6	2.0
Lysine	3.7	1.5
Methionine	1.5	0.6
Phenylalanine	2.2	1.2
Threonine	2.8	1.0
Valine	2.3	1.3
Tryptophan‡		

* Ngamsnae *et al.* (1999) estimated requirements of juvenile silver perch after fish were fed diets based on casein and gelatin plus a mixture of crystalline amino acids with graded contents of crystalline arginine or phenylalanine.
† Allan *et al.* (2001) fed juvenile silver perch diets with 14 MJ of DE kg^{-1} but graded digestible protein contents, all from intact protein sources rather than crystalline amino acids. The amino acid contents of the lowest-protein diet where growth or protein deposition was not retarded (28% protein diet – see Fig. 25.1) have been listed as the 'recommended' amino acid content for practical diets (digestible protein level 28% – see Fig. 25.1).
‡ Tryptophan was not measured in either study.

fish-oil and lard fed to silver perch was determined as described by Allan *et al.* (1999) (with revised calculations for ingredient digestibility recommended by Sugiura *et al.* (1998)). Digestibility coefficients and DE contents were 92% and 36 MJ of DE kg^{-1} for cod-liver oil and 52% and 21 MJ of DE kg^{-1} for lard. Practical diets for silver perch typically contain 6–10% crude lipid.

Anderson and Arthington (1989) measured the deposition of fatty acids in silver perch fed a soybean meal-based diet. They found that fatty acids accumulated in both the depot lipids and the phospholipids of the fish. When fish were transferred on to a fat-free diet, the preferential incorporation of both arachidonic (20:4n-6) and docosahexaenoic (22:6n-3) acids into the phospholipids during lipid turnover led the authors to conclude that silver perch require both n-3 and n-6 series fatty acids.

Silver perch were able to chain elongate and desaturate dietary linoleic acid (LA) (18:2n-6) and linolenic acid (LNA) (18:3n-3) to their longer-chain metabolites (Anderson and Arthington, 1992). This is in agreement with Hunter (2000), who found that silver perch require both LA and LNA and can chain-elongate and desaturate LA and LNA into their longer-chain polyunsaturated fatty acids. However, as growth was enhanced by the addition of longer-chain n-3 fatty acids to the diet, the conversion process alone is inadequate to optimize growth performance.

Dietary lipid has a major impact on fish lipid content (Buckley and Groves, 1979; Shearer, 1994). Measurement of lipid and fatty acid composition of whole silver perch and fillets confirms this and indicates that the fatty acid profile of farmed fish is strongly influenced by dietary lipid source (Hunter *et al.*, 1994, 2000). The whole-body lipid content of cultured silver perch is relatively high and increases with size. For silver perch grown in different experiments but all on the same diet (SP35–35% crude protein, 14–15 MJ of DE kg^{-1}), carcass contents were 24.6% lipid (dry basis) and 75.1% moisture for 2 g fish; 33.9% and 68.0% for 6 g fish; 41.9% and 59.7% for 72 g fish; and 50.9% and 58.8% for 395 g fish (Allan *et al.*, 2000b; Hunter, 2000; Stone *et al.*, 2000). Indicative fatty acid contents of farmed silver perch are listed in Table 25.2.

Hunter *et al.* (1994) reported that, for juvenile silver perch fed diets with protein contents ranging from 25 to 45% and 10.6 to 14.6 MJ of DE kg^{-1}, there was an inverse linear relationship between dietary protein : energy content and fish-fillet lipid content.

Carbohydrates

Although fish have no requirement for carbohydrate, provision of some carbohydrate is important in formulated diets to facilitate pellet binding, and to provide an alternative, lower-cost energy source to spare protein and/or lipid (Lovell, 1989; NRC, 1993). One of the most noticeable differences between diets for different warm-water fish species is that omnivores (e.g. channel catfish, common carp and tilapia) tend to be fed less protein in their diets than strictly carnivorous species (e.g. Asian sea bass = barramundi, gilt-head sea bream and red drum) (Wilson, 1991). However, as the overall protein composition of different fish species (especially on a lipid-free basis) does not vary greatly and net protein retention rates are similar (NRC, 1993), it is likely that this difference reflects the ability of different species to digest and utilize carbohydrate for energy.

One approach to investigating whether a species can tolerate different types or amounts of carbohydrate is to measure the uptake and clearance rates of carbohydrates (e.g. glucose) following intraperitoneal injection. Figure 25.2 shows the uptake and clearance rates of glucose for silver perch and barramundi, demonstrating that silver perch are more efficient at initial uptake and clearance of glucose than barramundi (A.J. Anderson, Z.S. Lipovsek and D.A.J. Stone, unpublished data; D.A.J. Stone, G.L. Allen and A.J. Anderson, unpublished data). Silver perch were also better able to utilize galactose than barramundi but both species were xylose-intolerant (A.J. Anderson, Z.S. Lipovsek and D.A.J. Stone, unpublished data; D.A.J. Stone, G.L. Allen and A.J. Anderson, unpublished data).

Unpublished data from our laboratory indicate that silver perch can efficiently digest diets containing 30% wheat starch, regardless of the degree of gelatinization, without adverse effects on liver enlargement, as indicated by the hepatosomatic index (HIS). However, when starch content was increased to 60%,

Table 25.2. Fatty acid composition of various sizes of silver perch (% of total fatty acids).

Fatty acid	Juvenile fish 2 g*	6 g*	72 g†	Adult fish (~ 500 g)‡
C12:0	0.1	0.1	Nm	Nm
C14:0	3.0	3.6	2.9	3.7
C15:0	0.2	0.3	0.2	Nm
C16:0	24.7	28.3	28.2	24.6
C17:0	0.3	0.2	Tr	Nm
C18:0	4.3	4.6	3.9	3.6
C20:0	0.2	0.1	Nd	Nm
C22:0	0.1	Tr	Nm	Nm
C14:1	0.1	0.1	Nm	Nm
C16:1	8.9	8.3	8.2	12.3
C17:1	0.2	Tr	Nd	Nm
C18:1n-9	27.5	32.0	34.6	31.5
C18:1n-7	2.2	2.2	1.9	2.5
C20:1n-11	0.1	0.6	0.2	Nm
C20:1n-9	0.8	2.7	1.4	1.1
C20:1n-12	Nm	Nm	Nm	0.7
C22:1n-11	0.2	1.6	1.0	Nm
C22:1n-9	0.1	0.3	Nm	Nm
C24:1n-9	0.4	0.4	Nm	Nm
C18:2n-6	9.3	7.3	8.2	7.7
C18:2n-9	Nm	Nm	Nm	0.8
C18:3n-6	0.9	0.4	0.5	Nm
C20:2n-6	0.3	0.2	Tr	Nm
C20:3n-6	0.7	0.3	0.2	Nm
C20:4n-6	0.6	0.2	Tr	Nm
C18:3n-3	1.2	0.7	0.9	1.0
C18:4n-3	0.6	0.5	0.5	Nm
C20:4n-3	0.4	0.3	Nm	Nm
C20:5n-3	2.1	0.8	1.5	1.8
C22:3n-3	0.2	Tr	Nm	Nm
C22:5n-3	1.8	0.6	1.2	1.8
C22:6n-3	5.8	2.3	4.6	6.3

* Ground whole fish (results are means for *n* = 4 fish) (Hunter, 2000).
† Ground, head-off, gutted fish (results are means for *n* = 4 fish) (Hunter *et al.*, 2000).
‡ Skinless fillets (results are means for *n* = 4 fillets) (B.J. Hunter, personal communication).
Tr, trace; Nd, not detected; Nm, not measured.

digestibility was positively correlated with the degree of gelatinization and HIS was elevated. Digestibility coefficients of various carbohydrate sources by silver perch and barramundi are presented in Table 25.3. The experimental diets were the same for both species but faeces of silver perch were collected by settlement

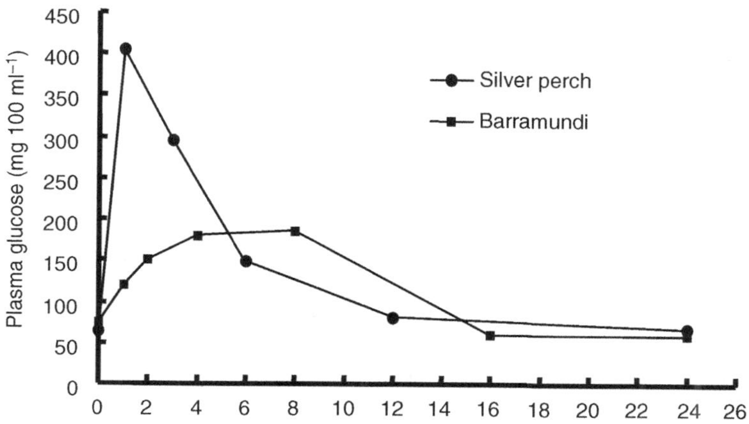

Fig. 25.2. Plasma glucose levels of silver perch and barramundi following an intraperitoneal injection of glucose (dose rate = 0.1% body weight).

Table 25.3. Apparent digestibility coefficients (%) for different carbohydrate sources for silver perch and barramundi (all ingredients were included at 30% with a basal diet of fish-meal 80.2%, wheat gluten 10%, fish oil 7%, vitamins and minerals 27%) (D.A.J. Stone, G.L. Allan and A.J. Anderson, unpublished data).

Starch/ingredient	Silver perch	Barramundi
Pea starch	75.1	39.6
Wheat starch	82.7	7.6
Pregelatinized wheat starch	95.2	–6.5
Dextrin (9%)	97.6	78.4
Dextrin (17%)	95.5	56.2
Dextrin (30%)	97.7	62.4
Maltose	98.4	59.0
Glucose	99.5	40.0

(Allan *et al.*, 1999), while those of barramundi were collected by stripping. Although this difference may have led to an overestimation of digestibility for silver perch and/or an underestimation for barramundi, previous results for protein-rich ingredients have been in much closer agreement (Williams *et al.*, 1998), indicating that the differences reflect differences between the species.

Vitamins and minerals

No research has been conducted on vitamin or mineral requirements for silver perch. The vitamin and mineral mixes used in experimental and commercial diets

Table 25.4. Recommended vitamin and mineral mixes in experimental and some commercial silver-perch diets.

Vitamin*	IU kg^{-1} diet*	mg kg^{-1} diet	Mineral*	mg kg^{-1} diet
Retinol (A)	8000		Calcium iodate (62% I)	0.47
Cholecalciferol (D_3)	1000		Manganese sulphate monohydrate (31% Mn)	10.0
D,L-α-Tocopherol acetate (E)		125.0	Zinc sulphate monohydrate	100.0
Menadione sodium bisulphite (K_3)		16.5	Copper sulphate pentahydrate	3.0
Thiamine hydrochloride (B_1)		10.0	Ferrous sulphate heptahydrate	29.9
Pyridoxine hydrochloride (B_6)		15.0	Sodium chloride	0.33
Folic acid		3.8	Magnesium sulphate (10% Mg)	500.0
Ascorbic acid (C)		1000.0	Lime	2.0
Calcium D-pantothenate		49.6		
Myoinositol		600.0		
Choline chloride		1500.0		
Niacin		200.3		
Cyanocobalamin (B_{12})		0.02		
Ethoxyquin (antioxidant)		150.0		
Calcium propionate (mould inhibitor)		250.0		

* Amount of active ingredient.

of silver perch are given in Table 25.4. Diets supplemented with these premixes have been used successfully for silver perch grown in static, earthen ponds. Premixes containing similar vitamin and mineral levels have been used in the commercial feeds for silver perch for around 5 years and deficiency symptoms have not been reported to date.

Practical Diets

The proximate composition and digestibility coefficients for dry matter, energy and protein of various ingredients for silver perch were determined by Allan *et al.* (2000a) and are presented in Table 25.5. Booth *et al.* (2000b) showed that, for silver perch > 17 g, grinding ingredients to ensure that 80% of particles were < 500 μm did not improve digestibility compared with when 80% of particles were between 710 and 1000 μm. They also demonstrated that steam conditioning or extrusion processing significantly improved digestibility, weight gain and feed conversion ratio (FCR) compared with when diets were pelleted without steam.

In Australia, several diets are commercially available for silver perch. Initially, formulations (e.g. SP35) were based on published requirements for other commonly cultured omnivorous species and on high-quality feed ingredients not readily available in Australia (Allan and Rowland, 1992). Following research on nutritional requirements and ingredient digestibility, diets with relatively low protein (35%–28% digestible protein) and energy (< approximately 15 MJ of DE kg^{-1}) were formulated (Allan and Rowland, 1999; Allan *et al.*, 2000b;

Table 25.5. Dry-basis proximate composition and apparent digestibility coefficients of ingredients for silver perch (Allan *et al.*, 2000a).

Test ingredient	Protein* (%)	Fat (%)	Ash (%)	GE (MJ kg^{-1})	Apparent digestibility coefficient (%)	
					Protein	Energy
Fish meals						
Australian fish-meal[†]	73.2	9.9	14.2	21.3	92.3	89.6
Danish fish-meal	72.9	11.4	13.0	21.5	94.2	98.0
Peruvian fish-meal	70.2	11.3	17.6	20.9	89.0	88.6
Animal meals						
Blood meal (spray dried)	94.9	–	3.1	23.9	90.2	99.9
Meat-and-bone meal (beef)	49.2	9.2	36.0	16.1	71.5	75.2
Meat-and-bone meal (lamb)	54.3	7.2	34.5	16.2	73.9	80.8
Poultry meal	60.3	18.2	15.0	22.7	85.4	93.7
Feather meal (hydrolysed, ring-dried)	84.3	11.2	3.0	24.9	92.8	101.0
Oilseeds						
Soybean meal (solvent)	47.8	3.7	8.0	17.0	94.8	78.0
Soybean meal (whole, expeller)	47.5	6.4	6.3	20.9	95.7	84.1
Soybean meal (dehulled, full-fat)	35.8	19.5	5.5	23.3	92.3	79.8
Canola meal (solvent)	36.6	2.6	7.4	19.9	83	58.1
Canola meal (whole, expeller)	31.8	12.5	6.6	21.8	79.6	58.6
Groundnut meal	41.2	1.3	5.2	19.7	98.2	7.0
Cottonseed meal (dehulled)	48.1	4.6	8.3	19.9	83.0	53.1
Linola (linseed meal)	29.8	11.3	6.1	21.2	77.7	44.3
Legumes (whole)						
Lupins, *Lupinus angustifolius* (cv. Gungurru)	34.1	5.7	2.8	17.9	97.1	51.2
Lupins, *Lupinus albus*	37.6	6.2	3.7	20.9	95.9	70.1
Field pea, *Pisum sativum* (cv. Dunn)	25.5	1.1	3.4	17.0	81.0	51.0
Faba bean, *Vicia faba* (cv. Fijord)	27.7	1.3	3.6	17.3	90.5	59.2
Chick-pea, *Cicer arietinum* (cv. Desi)	20.8	4.7	3.4	19.4	82.2	54.8
Vetch, *Vicia sativa* (cv. Blanch-fleur)	30.9	0.9	3.3	17.9	71.2	52.8
Cow-pea, *Vigna unguiculata*	25.2	2.3	3.7	18.8	96.5	44.7
Cereals						
Wheat gluten	76.9	–	–	23.1	99.8	94.0
Maize gluten meal	62.0	–	1.1	24.1	95.4	104.5
Wheat 1 (Aust. Std. Wheat)	12.2	–	–	18.3	99.5	53.2
Wheat 2 (high protein)	15.2	–	–	18.5	106.1	31.0
Mill-run	22.3	–	4.3	19.6	86.2	55.2
Sorghum	14.5	–	2.3	18.8	77.8	37.8

* N × 6.25.
[†] Sea Fish Pty Ltd, Triabunna, Tasmania, Australia.

Table 25.6). Lower-energy diets were used in an attempt to limit the excess carcass lipid deposition that some silver-perch farmers had reported to be a problem. Performances of silver perch on these diets were compared during a series of large-scale farming trials, with the fish being grown to over 400 g in replicate 0.1 ha earthen ponds at GAC, using very similar methods to those used by commercial farmers. A summary of the results of two experiments is given in Table 25.7. Sensory profiles of fish fed different diets were compared by accredited

Table 25.6. Composition of practical diets SP35,* 95LC1,* 95LC2,*† GRC2† and GRC3† used in two separate experiments, each using nine 0.1 ha static, aerated earthen ponds.

	Per cent in diet (dry matter basis)				
Ingredient*	SP35	95LC1	95LC2	GRC2	GRC3
Fish-meal (Danish)	27.0	10.0	5.0	0	0
Meat meal (lamb meal)	–	21.7	36.9	37.5	29.4
Blood meal (ring-dried)	2.0	2.1	–	1.7	0
Maize-gluten meal	4.0	3.8	5.2	0	0
Soybean meal (solvent-extracted)	20.0	–	–	5.0	0
Canola	–	–	5.0	0	0
Groundnut meal	–	–	5.0	5.0	0
Field peas (*Pisum sativum*)	–	14.9	10.4	0	0
Lupins (cv. Gungurru) dehulled	–	25.5	7.4	0	0
Wheat	26.9	–	–	10.6	25.9
Sorghum	11.0	4.7	–	0	0
Mill-run	2.0	10.0	17.7	20.0	20.0
Fish-oil (cod-liver oil)	1.0	2.9	3.2	3.2	7.6
D,L-Methionine	0.2	0.4	0.3	0.1	0
Vit./min. premix	4.0	4.0	4.0	1.5	1.5
Dicalcium phosphate	2.0	–	–	0	0

* Allan *et al.*, 2000b.
† 95LC2 was manufactured separately for both experiments. For the second experiment, minor changes to the vitamin and mineral premixes were made and an additional 2.5% mill-run was added (Allan and Rowland, 1999).

Table 25.7. Final weight, growth rate, food conversion ratio and production rate for silver perch fed different diets during two separate experiments in 0.1 ha earthen ponds.*

Experiment†	Diet	Final weight (g)	Growth rate (g fish^{-1} day^{-1})	FCR	Production (t ha^{-1})
1	SP35	395.4 ± 11.9^{a}	2.2 ± 0.07^{a}	2.2	5.8 ± 0.13^{a}
1	95LC1	433.8 ± 10.8^{b}	2.5 ± 0.09^{b}	2.0	6.3 ± 0.12^{b}
1	95LC2	441.7 ± 7.2^{b}	2.5 ± 0.03^{b}	1.9	6.5 ± 0.08^{b}
2	95LC2	461 ± 9.7^{a}	2.4 ± 0.05^{a}	1.6	6.5 ± 0.14^{a}
2	GRC2	453 ± 6.1^{a}	2.4 ± 0.03^{a}	1.7	6.5 ± 0.10^{a}
2	GRC3	433 ± 6.9^{a}	2.3 ± 0.05^{a}	1.7	6.2 ± 0.11^{a}

* Values are means ± SEM for three replicate ponds. For each experiment means in columns which share the same superscript were not significantly different ($P < 0.05$).
† Experiment 1 fish were stocked at 15,000 fish ha^{-1} at a mean weight of 81 g and cultured for 143 days (Allan *et al.*, 2000b). Experiment 2 fish with a mean weight of 58 g were stocked at 15,000 fish ha^{-1} and cultured for 187 days (Allan and Rowland, 1999).

taste panels. The most successful diets contained fish-oil but much less fish-meal (0 to 10%) and produced fish that tasted as good as or better than fish fed a fish-meal/soybean-meal-based diet (SP35). They were also much more cost-effective (Allan and Rowland, 1999; Allan *et al.*, 2000b) and several (95LC2, GRC3) have been adopted or are being considered for adoption by commercial feed-manufacturing companies. The replacement of expensive imported fish-meal and other protein sources with high-quality Australian agricultural protein sources has reduced ingredient costs for silver-perch diets by approximately 60% to around A$0.74 kg^{-1} without compromising performance. Silver-perch diets are currently the cheapest diets for any fish cultured in Australia.

Feeding Practices

Feed currently constitutes approximately 20% of production costs in silver-perch culture and efficient delivery of the diet is necessary for economic viability. Poor feeding practices increase the cost of production; overfeeding wastes feed and adversely affects water quality, while underfeeding results in reduced growth. Practices vary across the industry, from feeding to satiation to feeding a restricted amount based on a proportion of body weight. Satiation can be difficult to determine in the characteristically turbid silver-perch ponds where not all fish feed at the surface. Until recently, a feeding strategy specifically for silver perch had not been developed, and recommendations to industry were based on regimes used in other warm-water finfish industries, such as the channel catfish industry in the USA (Allan, 1995).

Russell *et al.* (1996) conducted a 4-week study in tanks and suggested that feeding between 5% and 10% body weight day^{-1} twice daily at water temperatures around 25°C would produce optimum growth and FCR (food conversion ratio) in small (1.3 g) fingerlings. A series of experiments was recently conducted at GAC to identify appropriate feeding rates and frequencies for fingerling and market-size silver perch at different ambient water temperatures. In each experiment, fish were stocked in floating cages in a 0.32 ha aerated pond and fed the commercial least-cost diet 95LC2 (35% crude protein, 5% fish-meal; extruded, slow-sinking) that had been specifically developed for silver perch (Allan *et al.*, 2000b). Silver perch performed well in these experiments, with very high survival (> 98%) and growth rates, similar to those reported in open pond culture (Rowland *et al.*, 1995). The new recommended feeding regimes for fingerling and market-size silver perch are given in Table 25.8. It is also recommended that the application of feed does not exceed 150 kg ha^{-1} day^{-1} in individual ponds.

Silver-perch broodfish are fed 2% body weight daily, commencing in early spring in the lead up to the breeding season when the water temperature reaches about 18°C; this rate is maintained during the summer. At lower water temperatures, rates of 0.5–1.0% are adequate for growth, health and gonadal development. First-feeding larvae (4.5 mm) are stocked in ponds, where they feed

Table 25.8. Suggested feeding rates and frequencies for fingerling and large silver perch at different water temperatures.

	Fingerlings (2–50 g)		Large fish (> 50 g)	
Water temperature(°C)	Feeding rate (% BW day^{-1})	Feeding frequency (no. of times day^{-1})	Feeding rate (% BW day^{-1})	Feeding frequency (no. of times day^{-1})
10–12	0.5	1	0.5	Alternate days
12–15	1	1	0.5	1
15–20	2.5	1	2	1
20–25	5	2	3	2 (50–500 g)
			1.5	2 (> 500 g)
25–30	7.5	2	2	2 (50–500 g)
			1	1 (> 500 g)

BW, body weight.

on zooplankton for several weeks before being weaned on to a commercially available diet containing 50% protein. Post-larvae and fry (up to 2 g) are fed this diet to satiation, four times daily.

A range of feeding techniques are used in the silver-perch industry. At the GAC and smaller farms fish are fed by hand, while on larger farms (> 10 ha of ponds) vehicle-mounted blowers are used to deliver the feed. Several farmers have recently started using automatic feeders.

Initially cold-pressed pellets were used (Allan and Rowland, 1992), but since the formulation and evaluation of 95LC2 (Allan *et al.*, 2000b) a majority of silver-perch farmers use extruded, slow-sinking or floating pellets. Although some silver perch feed aggressively at and near the surface, particularly in the warmer months, many fish feed mid-water. Thus, slow-sinking pellets are recommended for this species to ensure that all fish receive their daily ration.

Future Nutrition Research Priorities

Relatively low-cost production of silver perch is possible because the species performs well on low-protein diets composed of agricultural ingredients. An examination of the relative cost–benefit of using different proportions of protein, lipid and carboyhydrates as energy sources could also lead to more cost-effective diets. Further cost savings may be possible if reduction in the content of expensive vitamins can be achieved without compromising performance or immune response. Research to clarify amino acid requirements and the ability of silver perch to utilize crystalline amino acids is warranted. Recent reports of apparently stress-related 'winter diseases' indicate that research on immunostimulants may be of value. Further research to determine requirements for long-chain n-3 fatty acids to optimize performance is also needed. In all nutrition research with silver perch, it is necessary to investigate the impacts of diet on carcass composition and sensory characteristics.

Conclusions

1. Silver perch is a new species for aquaculture, is relatively easy to culture and performs well on low-protein (e.g. 28% digestible protein) and low-energy (e.g. 15 MJ of DE kg^{-1}) diets because of the efficient digestion and utilization of dietary carbohydrates, especially starch.
2. Maintenance requirements for protein and energy were estimated as 2.1 g digestible protein kg^{-1} body weight day^{-1} and 116 kJ of DE kg^{-1} body weight day^{-1}.
3. Recommended amino acid contents were estimated as those contents sufficient to ensure that growth was not retarded for diets made from intact protein sources with 15 MJ of DE kg^{-1}.
4. Silver perch are 'fatty' and, although they are able to chain-elongate and desaturate dietary linoleic and linolenic acid, growth is enhanced by the addition of longer-chain n-3 fatty acids in the diet. There is an inverse linear relationship between dietary protein : energy content and fish lipid content.
5. Australian agricultural ingredients, such as meat and poultry meals, lupins, peas and wheat, can be used in low-cost rations to replace most or all fish-meal without compromising performance or taste.

Acknowledgements

We would like to thank NSW Fisheries scientists and technicians, in particular David Stone, Mark Booth, Charlie Mifsud, Mark Nixon, Peter Boyd and David Glendenning, for providing data or help with conducting experiments that are reported here. Belinda Hunter (Newcastle University), Graham Bell (University of NSW) and Anne Ford (Queensland Department of Primary Industries) provided data on lipid requirements, fatty acid profiles and sensory profiles for silver perch fed different diets. We are grateful to Helena Heasman for assistance in preparing the manuscript and to Mike Heasman, Wayne O'Connor, Steve Kennelly, Carl Webster and Chhorn Lim for reviewing earlier drafts.

References

Allan, G.L. (1995) Development of artificial diets for silver perch. In: Rowland, S.J. and Bryant, C. (eds) *Silver Perch Culture: Proceedings of Silver Perch Workshops, Grafton and Narrandera, April 1994*. Austasia Aquaculture for NSW Fisheries, Sydney, pp. 77–87.

Allan, G.L. and Rowland, S.J. (1991) Growth of silver perch *Bidyanus bidyanus* on diets with different levels and sources of protein. *Proceedings Nutrition Society of Australia* 16, 211 [abstract].

Allan, G. and Rowland, S. (1992) Development of an experimental diet for silver perch (*Bidyanus bidyanus*). *Austasia Aquaculture* 6(3), 39–40.

Allan, G.L. and Rowland, S. (1999) *Consumer Sensory Evaluation of Silver Perch Cultured in Ponds on Meat Meal Based Diets*. Report to Meat and Livestock Australia PRCOP.009, NSW Fisheries Final Report Series No. 20, NSW Fisheries, Sydney, 21 pp.

Allan, G.L., Warner-Smith, R.J. and Frances, J. (1994) The effects of varying protein and energy concentrations on growth, protein efficiency ratio and food conversion ratio for silver perch *Bidyanus bidyanus* (Mitchell). *Proceedings Nutrition Society of Australia* 18, 53 [abstract].

Allan, G.L., Rowland, S.J., Parkinson, S., Stone, D.A.J. and Jantrarotai, W. (1999) Nutrient digestibility for juvenile silver perch *Bidyanus bidyanus*: development of methods. *Aquaculture* 170, 131–145.

Allan, G.L., Parkinson, S., Booth, M.A., Stone, D.A.J., Rowland, S.J., Frances, J. and Warner-Smith, R. (2000a) Replacement of fish meal in diets for Australian silver perch, *Bidyanus bidyanus*: I. Digestibility of alternative ingredients. *Aquaculture* 186, 293–310.

Allan, G.L., Rowland, S.J., Mifsud, C., Glendenning, D., Stone, D.A.J. and Ford, A. (2000b) Replacement of fishmeal in diets of silver perch *Bidyanus bidyanus*: V. Least-cost formulation of practical diets. *Aquaculture* 186, 327–340.

Allan, G.L., Johnson, R.J., Booth, M.A. and Stone, D.A.J. (2001) Estimating digestible protein requirements of silver perch, *Bidyanus bidyanus* Mitchell. *Aquaculture Research* 32, 1–11.

Anderson, A.J. and Arthington, A.H. (1989) Effect of dietary lipid on the fatty acid composition of silver perch (*Leiopotherapon bidyanus*) lipids. *Comparative Biochemistry and Physiology* 93B, 715–720.

Anderson, A.J. and Arthington, A.H. (1992) Influence of diet on the biosynthesis of lipid classes and fatty acids in silver perch, *Bidyanus bidyanus*. In: Allan, G.L. and Dall, W. (eds) *Proceedings of Aquaculture Nutrition Workshop, Salamander Bay, 15–17 April 1991*. NSW Fisheries, Brackish Water Fish Culture Research Station, Salamander Bay, Australia, pp. 41–45.

Bikker, P. (1994) Protein and lipid accretion in body components of growing pigs: effects of body weight and nutrient intake. PhD thesis, Agricultural University, Wageningen, the Netherlands, 203 pp.

Booth, M.A., Allan, G.L. and Stone, D.A.J. (2000a) Maintenance requirements of juvenile silver perch. In: *Book of Abstracts*. World Aquaculture Society, Baton Rouge, Louisiana, p. 83.

Booth, M.A., Allan, G.L. and Warner-Smith, R. (2000b) Effects of grinding, steam conditioning and extrusion of a practical diet on digestibility and weight gain of silver perch, *Bidyanus bidyanus*. *Aquaculture* 182, 287–299.

Buckley, J.T. and Groves, T.D.D. (1979) Influence of feed on the body composition of finfish. In: *Proceedings of World Symposium on Finfish Nutrition and Fishfeed Technology*. Heenemann, Berlin, pp. 335–343.

Cowey, C.B. (1994) Amino acid requirements of fish: a critical appraisal of present values. *Aquaculture* 124, 1–11.

Harpaz, S., Sklan, D., Karplus, I., Barki, A. and Noy, Y. (1999) Evaluation of juvenile silver perch *Bidyanus bidyanus* (Mitchell) nutritional needs using high and low-protein diets at two feeding levels. *Aquaculture Research* 30, 603–610.

Hunter, B.J. (2000) Dietary influences on the lipid composition of silver perch (*Bidyanus bidyanus*). A thesis submitted for PhD, University of Newcastle, Australia, 214 pp.

Hunter, B.J., Roberts, D.C.K. and Allan, G.L. (1994) The effects of varying dietary protein/energy ratios in the fat content and fatty acid composition of silver perch (*Bidyanus bidyanus*). *Proceedings of the Nutrition Society of Australia* 18, 54 [abstract].

Hunter, B., Roberts, D. and Allan, G. (2000) Meat meal replacement in diets for silver perch: effect on growth, protein, and lipid composition. *Journal of Applied Aquaculture* 10(3), 51–67.

Lovell, T. (1989) *Nutrition and Feeding of Fish.* Van Nostrand Reinhold, New York, 260 pp.

Lupatsch, I., Kissil, G.W., Sklan, D. and Pfeffer, E. (1998) Energy and protein requirements for maintenance and growth in gilthead seabream (*Sparus aurata* L.). *Aquaculture Nutrition* 4, 165–173.

Ngamsnae, P., De Silva, S.S. and Gunasekera, R.M. (1999) Arginine and phenylalanine requirement of juvenile silver perch *Bidyanus bidyanus* and validation of the use of body amino acid composition for estimating individual amino acid requirements. *Aquaculture Nutrition* 5, 173–180.

Nichols, P.D., Virtue, P., Monney, B.O., Elliot, N.G. and Yearsby, G.K. (1998) *Seafood the Good Food: the Oil Content and Composition of Australian Commercial Fishes, Shellfishes and Crustaceans.* CSIRO Marine Research, Hobart, Australia.

Nose, T. and Murai, T. (1990) Amino acid requirements in fishes. In: Yoshida, A., Naito, H., Niiyama, Y. and Suzuki, T. (eds) *Nutrition, Proteins and Amino Acids.* Japan Science Society Press, Tokyo, pp. 85–95.

NRC (National Research Council) (1993) *Nutrient Requirements of Fish.* National Academy Press, Washington, DC, 114 pp.

Rowland, S.J. (1984) The hormone-induced spawning of silver perch, *Bidyanus bidyanus* (Mitchell) (Teraponidae). *Aquaculture* 42, 83–86.

Rowland, S.J. (1995a) High density pond culture of silver perch, *Bidyanus bidyanus. Asian Fisheries Science* 8, 73–79.

Rowland, S.J. (1995b) Production of fingerling and market-size silver perch in earthen ponds. In: Rowland, S.J. and Bryant, C. (eds) *Silver Perch Culture: Proceedings of Silver Perch Workshops, Grafton and Narrandera, April, 1994.* Austasia Aquaculture for NSW Fisheries, Sydney, pp. 41–49.

Rowland, S., Dirou, J. and Selosse, P. (1983) Production and stocking of golden and silver perch in NSW. *Australian Fisheries* 42(9), 24–28.

Rowland, S.J., Allan, G.L., Hollis, M. and Pontifex, T. (1995) Production of the Australian freshwater silver perch, *Bidyanus bidyanus* (Mitchell), at two densities in earthen ponds. *Aquaculture* 130, 317–328.

Russell, A.J., Rowland, S.J. and McMeniman, N.P. (1996) Effects of feeding level and frequency on growth and food conversion ratio of juvenile silver perch (*Bidyanus bidyanus*). *Proceedings of the Australian Society of Animal Production* 21, 370.

Shearer, K.D. (1994) Factors affecting the proximate composition of cultured fishes with emphasis on salmonids. *Aquaculture* 119, 63–88.

Stone, D.A.J., Allan, G.L., Parkinson, S. and Rowland, S.J. (2000) Replacement of fish meal in diets for Australian silver perch, *Bidyanus bidyanus* III. Digestibility and growth using meat meal products. *Aquaculture* 186, 311–326.

Sugiura, S.H., Dong, F.M., Rathbone, C.K. and Hardy, R.W. (1998) Apparent protein digestibility and mineral availabilities in various feed ingredients for salmonid feeds. *Aquaculture* 159, 177–202.

Williams, K.C., Allan, G.L., Smith, D.M. and Barlow, C.G. (1998) Fishmeal replacement in aquaculture diets using rendered protein meals. In: *Proceedings Fourth International Symposium on Animal Nutrition, Protein, Fats and the Environment, Melbourne, 24–26 September, 1997.* Australian Renderers' Association, Sydney, Australia, pp. 13–26.

Wilson, R.P. (1991) *Handbook of Nutrient Requirements of Finfish.* CRC Press, Boca Raton, Florida, 196 pp.

Centrarchids: Largemouth Bass, *Micropterus salmoides*

26

James H. Tidwell, Shawn D. Coyle and Carl D. Webster
Aquaculture Research Center, Kentucky State University, Frankfort, KY 40601, USA

Introduction

The largemouth bass (*Micropterus salmoides*) is the largest member of the North American family of Centrarchidae, collectively known as the sunfish family (Fig. 26.1). Because of their popularity as a sport-fish, they have been transplanted across North, Central and South America, and Asia.

Research on this species has primarily focused on fisheries management or hatchery production to sizes of ≤ 5 cm (Simco *et al.*, 1986). In recent years interest has increased in the development of methods for growing largemouth bass to larger sizes used in correctional stocking of unbalanced sport-fish ponds, for fee fishing and increasingly as a foodfish. Although poorly documented, it is estimated that as much as 500,000 kg of 400–700 g (market-size) largemouth bass are being produced annually in the USA for sale as a foodfish. Primary outlets appear to be as a live product in urban, Asian markets, where they are highly regarded.

In 1984, the US Department of Agriculture Joint Subcommittee on Aquaculture (JSA, 1984) identified a lack of knowledge on nutritional requirements as a primary constraint on the profitable commercial production of the species. Even now, the number of studies conducted and our state of knowledge on largemouth-bass nutrition remain limited.

Nutrient Requirements

Protein and amino acids

The largemouth bass is a strict predator in nature. Carnivorous fish normally require high levels of protein. Using semipurified diets, Anderson *et al.*

Fig. 26.1. Photograph of largemouth bass.

(1981) determined minimum protein requirements of largemouth bass age 0 and age 1 to be 39.9% and 40.8%, respectively. Tidwell *et al.* (1996) found that largemouth bass reared under pond-production conditions and fed a practical diet containing 47% protein (Table 26.1) had higher harvest weight and survival than those fed 42% or 44% protein (Fig. 26.2). To date, protein : energy relationships for largemouth bass have not yet been adequately evaluated.

Requirements of largemouth-bass for individual amino acids have not yet been determined. Tidwell *et al.* (1996) reported that, based upon the amino acid profile of largemouth-bass muscle, their dietary requirements for lysine and methionine may be relatively high compared with other aquaculture species. To evaluate this, Coyle *et al.* (2000) fed juvenile largemouth bass experimental diets supplemented with L-lysine or L-methionine at levels sufficient to match largemouth muscle compositions. After 12 weeks, growth was not significantly affected by supplementation with lysine. The authors reported that the control diet, which contained a lysine concentration of 2.8% of diet (6% of protein), met dietary requirements for lysine. Supplementation with crystalline L-methionine did not significantly improve growth, but significantly improved feed conversion ratios, from 2.5 to 1.7. This may indicate that the methionine requirement for largemouth bass is less than 2.2%, but greater than the 0.6% and 1.2% of the diet required by channel catfish (*Ictalurus punctatus*) and common carp (*Cyprinus carpio*), respectively (Wilson, 1989).

Table 26.1. Ingredients and chemical composition of diets, containing three levels of protein, fed to largemouth bass.

Ingredient	Dietary protein (%)		
	42	44	47
Menhaden fish-meal (67%)	30.0	35.0	40.0
Soybean meal (44%)	25.5	33.0	41.0
Maize meal	35.9	26.9	17.4
Menhaden oil	7.0	3.5	0.0
Dicalcium phosphate	0.4	0.4	0.4
Vitamin mix[a]	0.5	0.5	0.5
Mineral mix[b]	0.6	0.6	0.6
Chemical analysis			
Protein[c]	41.7 ± 0.3	44.1 ± 0.4	46.9 ± 0.1
Lipid[c]	8.7 ± 0.2	5.9 ± 0.1	3.7 ± 0.0
Moisture	6.4 ± 0.0	9.7 ± 0.1	11.6 ± 0.1
Ash	11.5 ± 0.0	11.1 ± 0.1	11.5 ± 0.1
Lysine[d]	6.11 ± 0.00	6.28 ± 0.01	6.33 ± 0.07
Methionine[d]	1.98 ± 0.05	2.02 ± 0.05	2.07 ± 0.05
Arginine[d]	6.61 ± 0.06	6.76 ± 0.02	6.56 ± 0.02
Energy (kcal g^{-1})[e]	4.56 ± 0.01	4.32 ± 0.01	4.13 ± 0.01
Protein/energy (mg $kcal^{-1}$)	91.5 ± 0.4	102.1 ± 0.6	113.6 ± 0.4

[a] Vitamin mix supplied the following vitamins (IU or mg kg^{-1} of diet): vitamin A (as retinol palmitate), 6000 IU; vitamin C (as ascorbic acid), 780 mg; vitamin D (as cholecalciferol), 2200 IU; vitamin E (as α-tocopherol), 150 IU: vitamin K (as menadione), 10 mg; niacin, 200 mg; pantothenic acid, 60 mg; thiamine, 30 mg; riboflavin, 20 mg; pyridoxine, 20 mg; folic acid, 5 mg; B_{12}, 0.01 mg; biotin, 2 mg; choline, 2500 mg. [b] Mineral mix supplied the following (mg kg^{-1} of diet): manganese, 180 mg; copper, 8 mg; cobalt, 1.5 mg; iron, 66 mg; zinc, 150 mg; iodine, 6 mg; selenium, 0.3 mg. [c] Moisture-free basis. [d] Per cent of total amino acids. [e] Gross energy values (NRC, 1993).

Energy

As stated previously, energy requirements and optimum protein-to-energy ratios for largemouth bass have not been determined. Diets utilized in Coyle *et al.* (2000) contained approximately 4.2 kcal gross energy (GE) g^{-1} of diet and protein-to-energy ratios of approximately 118 mg protein $kcal^{-1}$. One problem experienced by largemouth-bass producers has been the occurrence of pale livers with some associated mortality, especially during the winter. It is possible that these problems are related to improper protein-to-energy ratios in the commercial salmonid diets typically fed to largemouth bass or an inability to digest or fully metabolize dietary carbohydrate (Goodwin *et al.*, 2000).

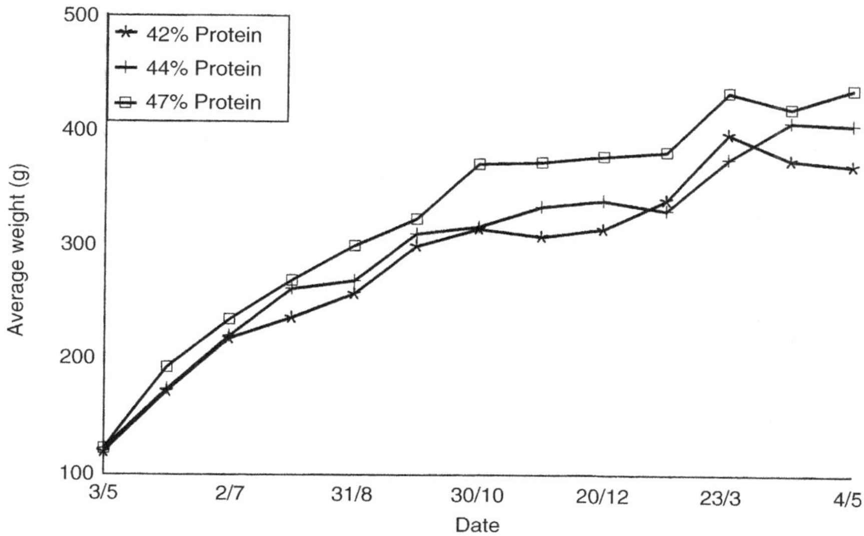

Fig. 26.2. Mean weights of largemouth bass, at different periods, fed three dietary protein levels (42, 44, 47%).

Lipids and fatty acids

Optimal lipid levels for largemouth-bass diets have not been determined. Tidwell *et al.* (1996) reported that, for practical diets containing 44–47% protein, lipid levels in excess of 8% should be evaluated for their potential protein-sparing effect. Coyle *et al.* (2000) included 10% lipid in diets containing 47% protein and reported good growth.

Although requirements for individual fatty acids in largemouth bass have not been evaluated, as a freshwater fish, largemouth bass are not likely to have requirements for long-chain polyunsaturated fatty acids. Most freshwater fish can elongate and desaturate shorter-chain precursors. However, Tidwell *et al.* (1996) reported that largemouth bass had very high levels of docosahexaenoic acid (DHA) (22:6n-3) in eggs and muscle tissues – almost 300% higher than levels found in the diet. The authors hypothesized that these levels represent extremely high metabolic demands and that providing higher levels of preformed DHA in the diet might represent significant metabolic savings and potentially improve growth. Based on this, Coyle *et al.* (2000) evaluated elevating dietary polyunsaturated fatty acid concentrations for largemouth bass, using squid oil. After 12 weeks, growth was not improved. However, increased levels of polyunsaturated fatty acids produced a significant shift in body composition toward a more lean body mass (i.e. decreased lipid and increased protein concentrations), from 20.2% lipid in fish fed a control diet to 17.3% lipid in fish fed diet with 4.0% squid oil. The mechanism controlling this shift in body composition is not currently understood.

Practical Diets

There are not enough data available to formulate diets specific to any life stage of largemouth bass. Unlike some aquaculture species, newly hatched largemouth bass fry (swim-up stage) cannot be trained to accept prepared diets (Williamson *et al.*, 1993). At this stage, they must be fed with live foods in fertilized ponds or in tanks until they reach 2.5–5.0 cm. At this size, the fingerlings are crowded into tanks at high densities and presented highly palatable feed items, such as fish eggs or freeze-dried krill at frequent intervals (Kubitza and Lovshin, 1997). Although carp eggs serve well as a training diet, their use as the sole source of nutrients has shown them to be nutritionally deficient (Willis and Flickinger, 1981). Gradually, a salmonid starter diet which has been moistened and softened with ground fish flesh or eggs is mixed in with the natural training diet. Over the period of about 10–14 days, the fish are switched to 100% dry diet. Addition of commercial attractants has produced some positive effects on the acceptance of dry diet but not soft, moist diets (Lovshin and Rushing, 1989). Juveniles are then stocked into ponds.

Currently salmonid diets are utilized during the first year of growth, due primarily to availability rather than suitability. During the second-year grow-out to harvest sizes (> 400 g), fish are also usually fed salmonid diets. Catfish diets have been utilized, but have been found to be unsuitable. Commercial diets produced for hybrid striped bass may be suitable if they contain more than 40% protein and less than 15% carbohydrates.

Feeding Practices

After feed-trained fingerlings are transferred to fingerling ponds. Williamson *et al.* (1993) recommended the following feeding rates for the first summer. Initially, pellets of 2–3 mm in diameter are fed four times daily to achieve a total daily allowance of 15% of estimated biomass. This feeding rate is reduced over the next week to 10% of estimated biomass daily, where it is maintained for 30 days. Thereafter, fish are fed at a rate of 5% of estimated biomass. All of these rates should be adjusted as dictated by fish consumption and water quality.

Brandt and Flickinger (1987) recommended that feeding of largemouth bass should continue as temperatures decrease, but at reduced rates. They recommended 2.5% of biomass daily at 20°C, gradually decreasing to 1% every 5–7 days at 5°C and below. In the spring, feeding resumes as temperatures increase above 8°C. Largemouth bass can be active feeders, especially in early summer.

For larger fish, Tidwell *et al.* (1996) recommended the use of floating extruded diets. Fish should be fed to satiation at least once daily. Field trials indicate that feeding twice per day may be advantageous at temperatures between 24 and 28°C. Figure 26.3 presents monthly diet consumption patterns over a year period for largemouth bass under production conditions in ponds.

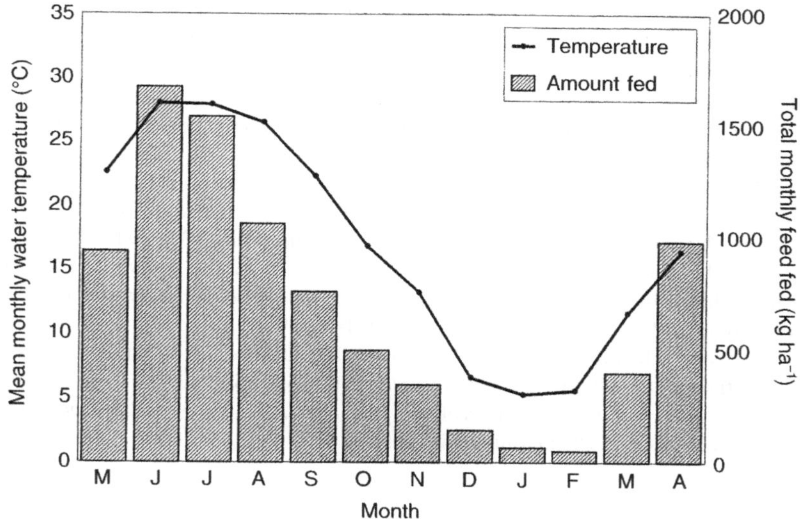

Fig. 26.3. Total monthly feed fed to largemouth bass in ponds over a 1-year culture period.

Interest in largemouth bass as an aquaculture species is increasing. However, as this chapter illustrates, our lack of knowledge of nutritional requirements and optimal feeding practices represents a current and future impediment to development.

References

Anderson, R.J., Kienholz, E.W. and Flickinger, S.A. (1981) Protein requirements of smallmouth bass and largemouth bass. *Journal of Nutrition* 111, 1085–1097.

Brandt, T.M. and Flickinger, S.A. (1987) Feeding largemouth bass during cool and cold weather. *Progressive Fish Culturist* 49, 286.

Coyle, S.D., Tidwell, J.H. and Webster, C.D. (2000) Response of largemouth bass *Micropterus salmoides* to dietary supplementation of lysine, methionine, and highly unsaturated fatty acids. *Journal of the World Aquaculture Society* 31, 89–95.

Goodwin, A.E., Tieman, D.M., Lochmann, R.T. and Mitchell, A.J. (2000) Massive hepatic necrosis and nodular regeneration in largemouth bass fed feeds high in available carbohydrate. In: *Proceedings of the 25th Annual Eastern Fish Health Workshop.* US Geologic Survey Biological Resources Division, National Fish Health Research Laboratory, Kearneysville, West Virginia, p. 16.

JSA (Joint Subcommittee on Aquaculture) (1984) Largemouth bass species plan. In: *National Aquaculture Development Plan*, Vol. II. US. Department of the Interior, Washington, DC, pp. 68–75.

Kubitza, F. and Lovshin, L.L. (1997) The use of freeze-dried krill to feed train largemouth bass (*Micropterus salmoides*): feeds and training strategies. *Aquaculture* 148, 299–312.

Lovshin, L.L. and Rushing, J.H. (1989) Acceptance by largemouth bass fingerlings of pelleted feeds with a gustatory additive. *Progressive Fish Culturist* 51, 73–78.

Simco, B.A., Williamson, J.H., Carmichael, G.J. and Tomasso, J.R. (1986) Centrarchids. In: Stickney, R.R. (ed.) *Culture of Nonsalmonid Freshwater Fishes*. CRC Press, Boca Raton, Florida, pp. 73–89.

Tidwell, J.H., Webster, C.D. and Coyle, S.D. (1996) Effects of dietary protein level on second year growth and water quality for largemouth bass (*Micropterus salmoides*) raised in ponds. *Aquaculture* 145, 213–223.

Webster, C.D., Tiu, L.G., Tidwell, J.H., Van Wyk, P. and Howerton, R.D. (1995) Effects of dietary protein and lipid levels on growth and body composition of sunshine bass (*Morone chrysops* × *M. saxatilis*) reared in cages. *Aquaculture* 131, 291–301.

Williamson, J.H., Carmichael, G.J., Graves, K.G., Simco, B.A. and Tomasso, J.R. (1993) Centrarchids. In: Stickney, R.R. (ed.) *Culture of Nonsalmonid Freshwater Fishes*, 2nd edn. CRC Press, Boca Raton, Florida, pp. 146–197.

Willis, D.W. and Flickinger, S.A. (1981) Intensive culture of largemouth bass fry. *Transactions of the American Fisheries Society* 110, 650–655.

Wilson, R.P. (1989) Amino acids and proteins. In: Halver, J.E. (ed.) *Fish Nutrition*, 2nd edn. Academic Press, San Diego, California, pp. 111–151.

Centrarchids: Hybrid Bluegill (*Lepomis cyanellus* × *Lepomis macrochirus*)

27

Carl D. Webster and James H. Tidwell
Aquaculture Research Center, Kentucky State University, Frankfort, KY 40601, USA

Introduction

The pay-lake industry (fee-fishing ponds) is an important sector of the aquaculture industry in many states in the USA. Pay lakes provide a source of income for the pond owner, a source of food and recreation for the public and a market for producers of live fish. The hybrid bluegill (female green sunfish, *Lepomis cyanellus* × male bluegill, *Lepomis macrochirus*) is a desirable fish for the pay-lake industry. Growth of hybrid bluegill is higher than that of either parental stock. Hybrid bluegill reach an acceptable catch size (100 g) quickly, and the fish can be fed a prepared diet due to its aggressive feeding response (Lewis and Heidinger, 1978). This response also increases the vulnerability to hook-and-line capture by anglers. From an angler's viewpoint, an aggressive fish translates into less effort expended to catch the fish. Thus, hybrid bluegill are popular with anglers. However, there have been few reports on the nutrient requirements, practical diet formulations and feeding practices of hybrid bluegill. It is hoped that this brief overview will assist persons interested in culturing this fish.

Nutrient Requirements

Protein and amino acids

Because protein is the most expensive component in a diet, knowledge about the protein requirements of the fish is essential for formulation of nutritious, economical diets. Information on protein, especially the amino acid, requirements of hybrid bluegill is limited. Tidwell *et al.* (1992) formulated three experimental practical diets to contain either 26, 31 or 37% protein, each diet being isocaloric at 4.0 kcal gross energy g^{-1} of diet. Juvenile hybrid bluegill (average weight of 5 g) were stocked into aquaria and fed one of the practical diets twice daily to apparent satiation for 10 weeks. At the conclusion of the study, fish fed a diet containing 37% protein had a significantly ($P < 0.05$) higher percentage weight gain (300%) than fish fed a diet containing 26% protein (226%), while the feed conversion ratio was significantly lower (1.9 compared with 2.6). The growth curves are shown in Fig. 27.1. Webster *et al.* (1997) fed juvenile hybrid bluegill (20 g), which had been stocked in cages, diets containing either 35%, 40%, 44% or 48% protein and reported no differences ($P > 0.05$) in final individual weight, specific growth rate (SGR), feed conversion ratio (FCR) or condition factor among treatments. Whole-body amino acid composition of hybrid bluegill indicated no differences ($P > 0.05$) among treatments for any amino acid analysed. When hybrid bluegill (14 g) were fed diets containing various percentages of protein (28, 32, 36 or 38%) in aquaria, final individual weight and percentage weight gain of hybrid bluegill fed a diet containing 38% protein were significantly ($P < 0.05$) higher (53 g and 265%, respectively) than fish fed diets containing 28% and 32% protein (45 g and 211%, respectively), but were not different from fish fed a diet containing 36% protein (51 g and 251%, respectively) (Webster *et al.*, 1997). Thus, it appears that feeding hybrid bluegill a diet containing between 35 and 36% protein is sufficient for good growth, FCR and survival percentages.

It may be that protein requirements are lower if fish are grown in ponds, where natural food organisms may play a role in supplying supplemental nutrients, or if grown to a larger size, when protein requirements may possibly be reduced compared with those of smaller fish. When hybrid bluegill were grown in

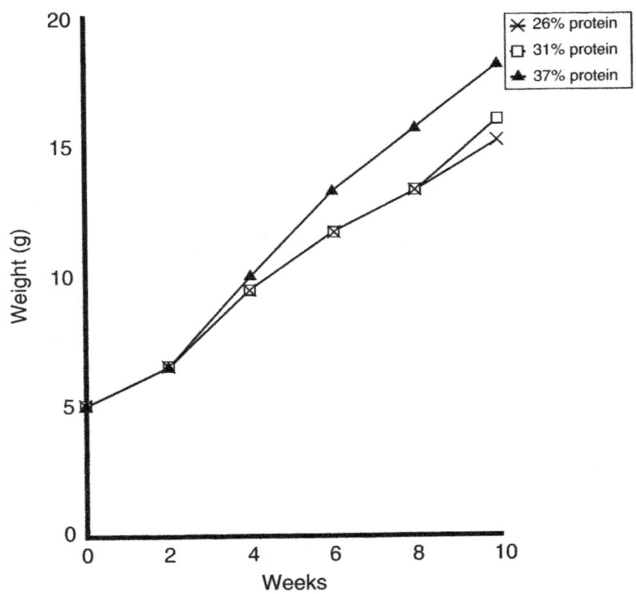

Fig. 27.1. Growth curves of hybrid bluegill grown in aquariums and fed diets containing various percentages of protein (26%, 31% or 37%) for 10 weeks.

ponds and fed diets containing either 32% or 38% protein, no difference in final individual weight, percentage weight gain, FCR or percentage survival was reported (Webster *et al.*, 1992). This is in agreement with Tidwell and Webster (1993), who reported that, when hybrid bluegill were fed during the winter in ponds, there were no differences in percentage weight gain, SGR and FCR.

Quantitative amino acid requirements have not been established for hybrid bluegill; however, the essential amino acid requirements of fish have been shown to correlate with the essential amino acid pattern of whole-body or muscle tissue of the fish. Whole-body amino acid compositions of hybrid bluegill from two feeding trials are presented in Table 27.1. If minimum requirements for lysine and the sulphur amino acids (methionine and cystine) are met, the requirements for the other essential amino acids should also be met. Practical diets containing lysine at 1.72% of the diet (4.70% of the protein) and methionine at 0.59% of the diet (1.68% of the protein) appear sufficient to meet requirements for these two amino acids for hybrid bluegill (Webster *et al.*, 1997).

Fish-meal is an important ingredient in fish diets because of its high protein quality and palatability; however, of all diet ingredients, fish-meal is one of the most expensive. There is a need to reduce the amount of fish-meal in diets to lower the cost of the diet and to decrease reliance. The long-term availability of fish-meal is unclear and it is important for the aquaculture industry to reduce the

Table 27.1. Whole-body amino acid composition (g 100 g^{-1} amino acids) of hybrid bluegill fed a diet with 37% protein (adapted from Tidwell *et al.*, 1992 – Study 1) and from hybrid bluegill fed a diet containing 48% protein (adapted from Webster *et al.*, 1997 – Study 2).

Amino acid	Study 1	Study 2
Alanine	6.7	7.3
Arginine	5.9	6.5
Aspartic acid	11.6	10.3
Cystine	0.9	0.9
Glutamic acid	14.9	14.1
Glycine	8.0	8.3
Histidine	2.1	3.1
Isoleucine	4.3	3.9
Leucine	7.3	7.4
Lysine	8.3	8.3
Methionine	3.1	3.0
Phenylalanine	4.1	4.2
Proline	5.1	5.5
Serine	4.1	4.8
Threonine	4.6	4.9
Tryptophan	1.4	n/a
Tyrosine	2.9	3.1
Valine	4.8	4.5

n/a, not available.

amount of fish-meal added to fish diets without adversely affecting the growth and health of the fish.

Vitamins and minerals

There are no published reports on vitamin or mineral requirements for hybrid bluegill, but there has been success in growing these fish with prepared diets. Levels reported in Webster *et al.* (1997) can be used until definitive requirements are elucidated (mg or IU kg^{-1} of diet): vitamin A, 6000 IU; vitamin D, 2200 IU; vitamin E, 150 IU; vitamin K, 10 mg; niacin, 200 mg; pantothenic acid, 60 mg; thiamine, 30 mg; riboflavin, 20 mg; pyridoxine, 20 mg; folic acid, 5 mg; B_{12}, 0.01 mg; biotin, 2 mg; manganese, 180 mg; copper, 8 mg; cobalt, 1.5 mg; iron, 66 mg; zinc, 150 mg; iodine, 6 mg; selenium, 0.3 mg.

Practical Diets

No data have focused on reducing fish-meal levels in a practical diet for hybrid bluegill, mainly due to the desire to formulate a practical diet that would be palatable to the fish and to ensure that diets would have high digestible-protein and lipid sources. Webster *et al.* (1992) fed commercial catfish diets with between 4 and 8% fish-meal to hybrid bluegill in ponds. Tidwell *et al.* (1992) fed practical diets with different percentages of protein and 10% fish-meal to hybrid bluegill in aquariums, while Tidwell and Webster (1993) fed diets containing between 6.5% and 10% fish-meal to hybrid bluegill grown in ponds during the winter. Webster *et al.* (1997) fed practical diets containing 15–21% fish-meal (29% of the protein) to hybrid bluegill grown in cages and diets with between 18% and 35% fish-meal (40–50% of the protein) to fish in aquariums (Table 27.2).

Practical diets fed to hybrid bluegill in most of the feeding trials have been shown to have some effect on body composition. Webster *et al.* (1992) reported that hybrid bluegill fed a diet containing 38% protein and 3.4% lipid had a significantly ($P < 0.05$) lower percentage of whole-body lipid (12.6% lipid on a dry-matter basis) compared with fish fed a diet containing 32% protein and 4.4% lipid (18.7% lipid on a dry-matter basis), but no differences ($P > 0.05$) in whole-body moisture or protein levels. Webster *et al.* (1997) stated that hybrid bluegill fed diets containing 40%, 44% and 48% protein had higher ($P < 0.05$) percentages of whole-body protein and lower percentages of whole-body lipid compared with fish fed a diet with 35% protein.

Feeding Practices

Results from feeding trials where hybrid bluegill are grown in aquariums or cages appear to indicate that, when feeding small (4–10 g) fish, a diet with 35–36% protein and where fish-meal comprises between 15% and 30% of the protein is

Table 27.2. Formulation and proximate composition of practical diets fed to hybrid bluegill.

	Diet no.		
	1	2	3
Ingredient			
Menhaden fish-meal	15.00	17.00	28.00
Soybean meal	41.00	49.00	14.85
Maize meal	36.95	27.45	20.00
Wheat	0.00	0.00	29.00
Hydrolysed feather meal	0.00	0.00	3.50
Monocalcium phosphate	1.00	1.00	0.75
Vitamin and mineral mix	2.00	2.00	0.85
Ascorbic acid	0.05	0.05	0.05
Menhaden (or cod-liver) oil	4.00	3.50	3.00
Proximate analysis (dry-matter basis)			
Protein (%)	35.2	40.2	36.6
Lipid (%)	7.8	7.2	11.7
Ash (%)	1.9	2.0	1.6
Protein-to-energy ratio	87	101	89

sufficient (Tidwell *et al.*, 1992; Webster *et al.*, 1997); however when hybrid bluegill (3.5 g) are stocked in ponds at low densities (12,350 fish ha^{-1}), it may be possible to feed a diet with a lower percentage of protein (32%) and lower levels of fish-meal (4–8% of diet) (Webster *et al.*, 1992). Hybrid bluegill do not appear to consume zooplankton, but can utilize benthic organisms from the taxa Oligochaeta, Chironomidae and Planorbiidae when grown in ponds (Brunson and Robinette, 1982, 1986).

Wang *et al.* (1998) reported that feeding small (4–7 g) hybrid bluegill to satiation three times per day increased growth and food consumption compared with fish fed once or twice daily, but was not different ($P > 0.05$) from fish fed four times per day. Feeding this often may be practical for fish grown indoors or in tanks, cages or outdoor raceways, but may not be practical or necessary if fish are grown in ponds. If fish are stocked in ponds at a low density (12,000 fish ha^{-1}), it may be possible to feed fish only once daily due to the fish consuming natural foods present in the pond. The FCR for fish stocked at 12,350 fish ha^{-1} was 1.3 when fish were fed a sinking diet during the winter in Kentucky, but fish stocked at 24,700 fish ha^{-1} had an FCR of 5.8 (Tidwell and Webster, 1993). A producer with a large number of ponds may be able to feed only once a day since it may take all day to feed the fish in all the ponds. Also, Webster *et al.* (1997) reported that, during the hot summer months, hybrid bluegill consumed less diet during the day so that feeding once a day was more desirable than feeding twice daily. Thus, it may be best to feed hybrid bluegill to apparent satiation and not to use a feeding chart when fish are fed only once during the day.

Feeding hybrid bluegill twice daily to satiation appears to be a viable method of feeding where practical. Feeding hybrid bluegill stocked in cages a floating diet twice daily (0800 and 1900 h) produced good growth rates, but high FCR values 3.5–4.4 (Webster *et al.*, 1997). This may have been due to some fish not being able to consume the diet because of the territorial nature of hybrid bluegill. Some fish (smaller, less aggressive fish) may have been prevented from getting to the surface to feed by larger, more aggressive fish. Likewise, Tidwell and Webster (1993) also reported a high (3.4–3.8) FCR when hybrid bluegill were fed in ponds during the winter when a sinking diet was fed according to a feeding chart used in feeding channel catfish, *Ictalurus punctatus*. Webster *et al.* (1992) reported FCR values of approximately 3.8 when hybrid bluegill grown in ponds were fed a floating diet.

For each size of fish desired for pay lakes (> 100 g) and for seafood markets (> 250 g), hybrid bluegill must be overwintered before a second-year grow-out in most regions of the USA. However, during the colder winter months, where water temperatures decline below those for optimal growth and fish feed aggressively, it does appear that feeding fish, at reduced levels, is advantageous. Brunson and Robinette (1982) reported that hybrid bluegill that were not fed a prepared diet during the winter had weight gains of 183% while fish fed a prepared diet had a significantly ($P < 0.05$) higher percentage weight gain (261%). In areas that have lower winter temperatures, hybrid bluegill fed a prepared diet during the winter had weight gains of 12% for fish stocked at 24,700 fish ha^{-1} and 26% of fish stocked at 12,350 fish ha^{-1} (Tidwell and Webster, 1993).

References

Brunson, M.W. and Robinette, H.R. (1982) Supplemental winter feeding of hybrid sunfish in Mississippi. *Proceedings of the Annual Conference of the Southeastern Association of Fish and Wildlife Agencies* 36, 157–161.

Brunson, M.W. and Robinette, H.R. (1986) Evaluation of male bluegill × female green sunfish hybrids for stocking Mississippi farm ponds. *North American Journal of Fisheries Management* 6, 156–167.

Lewis, W.M. and Heidinger, R.C. (1978) Use of hybrid sunfishes in the management of small impoundments. In: Novinger, J.D. and Dillard, J.C. (eds) *New Approaches to the Management of Small Impoundments.* Special Publication 5, North Central Division, American Fisheries Society, Bethesda, Maryland, pp. 104–108.

Tidwell, J.H. and Webster, C.D. (1993) Effects of stocking density and dietary protein on green sunfish (*Lepomis cyanellus*) × bluegill (*L. macrochirus*) hybrids overwintered in ponds. *Aquaculture* 113, 83–89.

Tidwell, J.H., Webster, C.D. and Clark, J.A. (1992) Growth, feed conversion, and protein utilization of female green sunfish × male bluegill hybrids fed isocaloric diets with different levels. *Progressive Fish-Culturist* 54, 234–239.

Wang, N., Hayward, R.J. and Noltie, D.B. (1998) Effect of feeding frequency on food consumption, growth, size variation, and feeding pattern of age-0 hybrid sunfish. *Aquaculture* 165, 261–267.

Webster, C.D., Tidwell, J.H., Goodgame, L.S., Clark, J.A. and Yancey, D.H. (1992) Effects of protein level or growth and body composition of hybrid sunfish (*Lepomis cyanellus* × *L.*

macrochirus) reared in ponds. *Transactions of the Kentucky Academy of Science* 53, 97–100.

Webster, C.D., Tiu, L.G. and Tidwell, J.H. (1997) Growth and body composition of juvenile hybrid bluegill *Lepomis cyanellus × L. macrochirus* fed product diets containing various percentages of protein. *Journal of the World Aquaculture Society* 28, 230–240.

Brazilian Species 28

Débora Machado Fracalossi
Departamento de Aquicultura, Centro de Ciências Agrárias, Universidade Federal de Santa Catarina, Caixa Postal 476, 88040-900 Florianópolis, Santa Catarina, Brazil

Introduction

Brazilian territory comprises several freshwater hydrographic basins which hold many native fish species with aquaculture potential. Four species have been selected for this review based on information available on nutritional requirements, feeding management practices, current production and market preferences. They are the omnivorous tambaqui (*Colossoma macropomum*) and pacu (*Piaractus mesopotamicus*, former *Colossoma mitrei*), both characins native to the Amazon (northern Brazil) and Prata (western-central Brazil) river basins, respectively, and the carnivores peacock bass (*Cichla* sp.) and pintado (*Pseudoplatystoma coruscans*). Peacock bass is a cichlid native to the Amazon basin and pintado is a silurid native to the Prata and São Francisco (north-eastern Brazil) river basins. Despite the increased interest in raising Brazilian freshwater native species in the past 20 years, the available information on their nutritional requirements and feeding practices is still scarce. A significant amount of information included in this chapter was retrieved from theses or dissertations, developed at Brazilian universities or freshwater fish research centres, thus having limited distribution. Most of the studies reviewed here were performed using practical-type diets, which can confound results due to uncontrollable variables and nutrient interactions. Additionally, the nutritional contribution of natural food in the pond studies was seldom accounted for. This could also lead to confounding results since some species, such as tambaqui, have numerous fine and long gill rakers, which enable them to efficiently consume food organisms besides the experimental diet.

Nutritional Requirements

Protein and energy

Pintado fingerlings, averaging 25.1 g, were fed practical-type diets containing 30, 36, 42 or 48% crude protein (CP) and two dietary digestible energy (DE) concentrations of 3500 and 4000 kcal kg^{-1} for 90 days. Highest weight gain and protein efficiency ratio (PER) were obtained for the fish fed the 30% CP with a DE of 4000 kcal kg^{-1} (Machado, 1999). However, at the lower dietary energy level (3500 kcal kg^{-1}), the optimum dietary CP concentration for best performance was 36%. Based on this information, good performance was obtained with diets containing a CP-to-DE ratio (CP/DE) of 75–102 mg $kcal^{-1}$. In a similar study, 10 g peacock-bass fingerlings were fed diets containing 30, 33, 37 or 41% CP and 3500 kcal DE kg^{-1} for 65 days. For this carnivorous species, highest weight gain and feed efficiency were observed when diets had a CP/DE between 105.7 and 117.1 mg $kcal^{-1}$ (Sampaio *et al.*, 2000).

Results of studies on the dietary protein and energy requirement for tambaqui are summarized in Table 28.1. The dietary CP-to-energy (DE or gross energy (GE)) ratios were not given, but they were calculated in an attempt to define a requirement range. The ratio that promoted the best performance for tambaqui was between 79 and 89 mg protein $kcal^{-1}$ of DE or GE. Optimum dietary protein and energy concentration varied from 24 to 50% and from 2700 to 4660 kcal kg^{-1}, respectively. However, Macedo (1979) reported a lower dietary protein requirement for tambaqui. She observed that 5 g fish had the highest weight gain when fed diets containing 22–26% CP, while there was no difference in weight gain among 18 g fish fed dietary protein levels of 18, 22 or 26%. However, Eckmann (1987) pointed out that the growth rate of fish in Macedo's study was far below the maximum growth rate of the species. Tambaqui showed that the highest growth rates when 100 or 75% of the dietary

Table 28.1. Dietary protein and energy requirements of different sizes of tambaqui.

Crude protein (%)	Energy* (kcal kg^{-1})	Fish initial weight (g)	Reference
18	3200 GE	18	Macedo (1979)
22	3200 GE	5	Macedo (1979)
24	3300 DE	30.2	Camargo *et al.* (1998)
25	3100 GE	37.5	Vidal *et al.* (1998)
30	2700 DE	30	Merola and Cantelmo (1987)
31	3800 DE	0.5	Hernández *et al.* (1995)
37	4660 GE	–	Eckmann (1987)
40	4613 GE	125	Van der Meer *et al.* (1995)
40	4493 GE	50	Van der Meer *et al.* (1995)
50	4589 GE	5	Van der Meer *et al.* (1995)

* DE = digestible energy (estimated or measured) or GE = gross energy.

protein originated from plant sources in a 30% CP diet (Werder and Saint-Paul, 1979).

Dietary CP and energy concentration for pacu ranged from 26 to 36% and 2600 to 4200 kcal kg^{-1}, respectively (Table 28.2). The optimum range of the dietary protein-to-energy (DE or GE) ratio, calculated by the author (D.M. Fracalossi, unpublished), which promoted best performance of pacu was 72–109 mg $kcal^{-1}$. Lower protein-to-energy ratios were found in two studies (Carneiro *et al.*, 1992a; Fernandes, 1998). The reasons for such a discrepancy are not clear but could be linked to an underestimation of protein intake in at least one of the studies (Carneiro *et al.*, 1992a), since the experiment was carried out in outdoor ponds for 1 year and natural pond food could have contributed to the protein requirement of the fish. Apparent net protein utilization (ANPU) by pacu fingerlings was high when the dietary DE concentration was 3000 kcal kg^{-1} in a 26% CP diet, although the highest weight gain was obtained with a dietary DE concentration of 2600 kcal kg^{-1} (Cantelmo, 1993). Pacu showed a 15% weight-gain increase when fed diets containing a mixture of meat meal, fish-meal and soybean meal, compared with diets containing only soybean meal (Borguetti *et al.*, 1991).

Lipids and carbohydrates

Hernández *et al.* (1995) found that carbohydrates seemed to be as effective as lipids as dietary energy sources for 0.5 g tambaqui fingerlings. At dietary protein concentrations of 11–48%, feed efficiency and protein efficiency ratio (PER) were similar for carbohydrate and lipid energy sources. Similarly, Gunther (1996) concluded that tambaqui can efficiently use dietary carbohydrates and that best growth and protein utilization were obtained with a diet containing 38% carbohydrate and 11% lipid. Van der Meer *et al.* (1997a), however, found that growth and ANPU increased with increasing dietary lipid from 5 to 20% in either 20 or 40% protein diets. Tambaqui fingerlings (14 g average weight) showed

Table 28.2. Dietary protein and energy requirements of different sizes of pacu.

Crude protein (%)	Energy* (kcal kg^{-1})	Fish initial weight (g)	Reference
22	4200 GE	112.1	Fernandes (1998)
22	4000 GE	240	Carneiro *et al.* (1992b)
23	3200 DE	200	Carneiro (1983)
26	4200 GE	7.9	Fernandes (1998)
26	3000 DE	39	Carneiro (1983)
26	2600 DE	27	Cantelmo (1993)
30	3600 DE	28	Carneiro (1990)
35	3225 DE	44.5	Borghetti *et al.* (1991)
36	3300 DE	22.2	Brener (1988)

* DE = digestible energy (estimated or measured) or GE = gross energy.

the highest weight gain when fed 6% lipid diets containing either maize oil or a mixture containing high concentration (3% or more) of palm (*Elaies guineensis*) oil and deodorization distillates of soybean oil (a by-product of soybean-oil extraction) as the dietary lipid sources (Viegas and Guzman, 1998). Despite the wide variation of the saturated-to-unsaturated fatty acid ratio in the diets (0.27 to 0.62), fish maintained a fairly steady saturated-to-unsaturated fatty acid ratio both in the whole body and in the fillet (0.61 to 0.69) except with the maize-oil diet. This indicates that tambaqui can use dietary saturated-fat sources, such as palm oil, without affecting its fatty acid composition.

Visceral lipid deposition of juvenile pacu increased with increasing dietary lipid concentrations from 8% to 24% of either lard or soybean oil (Pezzato, 1990). However, 8% of saturated fat (lard) in the diet promoted a higher visceral fat accumulation than the 24% soybean-oil diet. Additionally, it was demonstrated that young pacu can synthesize *de novo* medium-chain saturated and monoenoic fatty acids from their short-chain precursors. Alves (1999) reported that pacu fingerlings exhibited better weight gain, PER and ANPU when fed a 26% protein diet containing 7% lipid instead of 5 or 9% lipid. A dietary crude-fibre concentration of 16% promoted higher weight gain and feed conversion in 45 g pacu fingerlings than diets containing 4, 8 or 12% crude fibre among practical diets containing 30% CP and 3000 kcal DE (Zanoni, 1996). Feed consumption, PER and carcass composition, however, were not affected by crude-fibre concentrations.

Feeding Practices

Peacock bass fingerlings, conditioned to accept ground fish flesh (GF), were tentatively weaned from GF to dry pellets using gradual feed ingredient transition (GFIT) (Moura *et al.*, 2000). The fish, 0.5 g in weight, were fed a series of diets containing from 90% to 40% GF for 3 days. During that period, approximately 80% of the fish accepted the diet containing 40% GF. These fish were pooled and weaned to a 10% GF diet for the next 3 days with a series of diets containing from 30% to 10% GF combined with a fish-flavoured extruded commercial diet. In another study, 3.9 g pintado were initially trained to accept ground bovine heart, followed by a mixture of ground bovine heart and dry diet in different proportions until only dry diet was accepted (Machado *et al.*, 1998). After 84 days, fish showed a survival rate of 78.4%. Pintado larvae, 6.4 mm in length, were fed diets containing bovine heart, ground sardine and a mixture of bovine heart and ground sardine, four times a day (Oliveira *et al.*, 1998). All diets were supplemented with vitamin C, oxytetracycline and vitamin and mineral premixes; fish also received zooplankton twice a day. At the end of the 15-day weaning period, larval survival was 45%, 34% and 20% when fed the bovine heart, sardine and a mixture of the two, respectively. Hayashi *et al.* (1999) tested different fresh food items for weaning 0.18 g pintado fingerlings from fresh food to a 48% CP dry diet. Ground earthworm and ground bovine liver promoted higher weight gain than ground bovine heart and ground sardine, while survival rate was not affected by diet, during a 12-day weaning period. Fernandes *et al.* (1998) reported that small pintado fingerlings averaging 0.67 g had better growth (1.49 g against 0.89 g) and survival (82.5% against 65%) when fed moist rather than dry diets.

Van der Meer *et al.* (1997b) reported that feed consumption of tambaqui fingerlings, 0.87 g average weight, was lower at 0700 than at 1900 h. The authors suggested that growth and diet uptake could be improved by extending the feeding period after 1900 h. The feeding frequencies tested ranged from one to five meals per day and intervals between meals from 3 h to 24 h. Higher growth rate and diet intake and lower feed utilization efficiency were observed with higher feeding frequencies.

The optimum diet particle sizes for pacu and tambaqui fingerlings were determined by Cantelmo and Ribeiro (1994) and are summarized in Table 28.3. Pacu fingerlings, 1.5 g initial weight, were stocked in outdoor ponds at a density of 0.7 fish m^{-2} and subjected to three feeding regimes: complete diet (26% CP at 5% body weight), supplemental diet (30% CP at 3% body weight) plus pig manure or supplemental diet (30% CP at 2.5% body weight) plus agricultural by-products (Barros *et al.*, 1994). Weight gain was not affected by dietary treatment after 10 months of feeding, suggesting that either pacu can efficiently utilize agricultural by-products and manure or that pacu can benefit indirectly from the addition of these ingredients into the pond through an increase in the production of natural foods. However, Bernardino and Ferrari (1986) reported that monthly bovine-manure or inorganic pond fertilization improved weight gain of tambaqui fingerlings (average weight 8.2 g) by 13.2% and 13.8%, respectively, when compared with feeding only a 30% protein diet, after 1 year of feeding. Juvenile pacu, 102 g average weight, were stocked in outdoor ponds at a density of 1 fish m^{-2} and were fed the same diet in three different forms, namely meal, pelleted or extruded, for 1 year (Carneiro *et al.*, 1992b). Fish showed higher weight gain and PER and lower feed conversion when fed the extruded diet during the warmer

Table 28.3. Recomended feed particle sizes for various stages of pacu and tambaqui (adapted from Cantelmo and Ribeiro, 1994).

Species	Fish: Standard length (cm)	Fish: Mouth size (mm)	Particle size (mm): Min.	Particle size (mm): Max.
Pacu	1.6	1.92	0.35	0.42
	2.1	2.52	0.50	0.71
	3.1	3.72	0.71	1.00
	4.1	4.92	1.00	1.41
	4.5	5.40	1.00	1.41
Tambaqui	1.00	1.2	–	0.25
	1.34	1.6	–	0.25
	1.93	2.31	0.35	0.42
	2.85	3.42	1.00	1.41

season. During the cooler season, however, fish showed better performance when fed the pelleted (sinking) diet.

Borguetti and Canzi (1993) tested three feeding rates for juvenile pacu, average weight 95 g. Fish were stocked in cages at a density of 1.4 kg m^{-3} and fed the same diet (35% CP and 3225 kcal energy kg^{-1}) for approximately 8 months. Optimum feeding rate varied depending on water temperature: 1% body weight day^{-1} was sufficient for growth at a water temperature of 19–20°C, while 3% and 5% body weight day^{-1} were better at 24°C and 27–30°C, respectively.

References

Alves, J.M.C. (1999) Níveis de lipídios em dietas para o crescimento inicial do pacu, *Piaractus mesopotamicus* (Holmberg, 1887). MSc thesis, Universidade Estadual Paulista, Jaboticabal, Brazil.

Barros, M.M., Pezzato, L.E., del Carratore, C.R., Oliveira, M.C.B. and Salaro, A.L. (1994) Alevinos de pacu (*Piaractus mesopotamicus*) submetidos a diferentes manejos nutricionais. In: *Resumos do VIII Simpósio da Associação Brasileira de Aquicultura.* Associação Brasileira de Aquicultura, Piracicaba, p. 42.

Bernardino, G. and Ferrari, V.A. (1986) Efeitos da fertilização na produção de pacu *Colossoma mitrei* alimentado com ração. In: *Síntese dos Trabalhos Realizados com Espécies do Gênero* Colossoma. Centro de Pesquisa e Treinamento em Aquicultura, Pirassununga, p. 18.

Borghetti, J.R. and Canzi, C. (1993) The effect of water temperature and feeding rate on the growth rate of pacu (*Piaractus mesopotamicus*) raised in cages. *Aquaculture* 114, 93–101.

Borghetti, J.R., de Lepeleire, R.E.M. and Fernandez, D.R. (1991) Efeitos da origem da proteína no crescimento do pacu. *Revista Brasileira de Biologia* 51, 689–694.

Brener, M. (1988) Determinação da exigência de proteína do pacu (*Colossoma mitrei*, Berg, 1895). MSc thesis, Universidade Federal de Viçosa, Viçosa, Brazil.

Camargo, A.C.da S., Vidal, M.V. Jr, Donzele, J.L., de Andrade, D.R. and dos Santos, L.C. (1998) Níveis de energia metabolizável para tambaqui (*Colossoma macropomum*) dos 30 aos 180 gramas de peso vivo. 1. Composição das carcaças. *Revista da Sociedade Brasileira de Zootecnia* 27, 409–415.

Cantelmo, O.A. (1993) Níveis de proteína e energia em dietas para o crescimento do pacu *Piaractus mesopotamicus* (Holmberg, 1887). MSc thesis, Universidade Federal de Santa Catarina, Florianópolis, Brazil.

Cantelmo, O.A. and Ribeiro, M.A.R. (1994) Determinação do tamanho de partícula alimentar para o pacu *Piaractus mesopotamicus* Holmberg 1887 e tambaqui *Colossoma macropomum* Cuvier, 1818, no estágio de alevino. *Boletim Técnico do CEPTA* 7, 9–17.

Carneiro, D.J. (1983) Níveis de proteína e energia na alimentação do pacu, *Colossoma mitrei* (Berg, 1895). MSc thesis, Universidade Estadual Paulista, Jaboticabal, Brazil.

Carneiro, D.J. (1990) Efeito da temperatura na exigência de proteína e energia em dietas para alevinos de pacu, *Piaractus mesopotamicus* (Holmberg, 1887). PhD dissertation, Universidade Federal de São Carlos, São Carlos, Brazil.

Carneiro, D.J., Chaim, S.H.S. and Dias, T.C.R. (1992a) Efeito do processamento das dietas comerciais sobre o desenvolvimento produtivo de pacu, *Piaractus mesopotamicus*,

(Holmberg, 1887). In: *Anais do VII Simpósio da Associação Brasileira de Aquicultura.* Associação Brasileira de Aquicultura, Peruíbe, pp. 44–51.

Carneiro, D.J., Wagner, P.M. and Dias, T.C.R. (1992b) Efeito da densidade de estocagem e do nível de proteína bruta dietário no desempenho de produção do pacu (*Piaractus mesopotamicus*). In: *Anais do VII Simpósio da Associação Brasileira de Aquicultura.* Associação Brasileira de Aquicultura, Peruíbe, pp. 52–61.

Eckmann, R. (1987) Growth and body composition of juvenile *Colossoma macropomum* Cuvier 1818 (Characoidei) feeding on artificial diets. *Aquaculture* 64, 293–303.

Fernandes, E.B., Oliveira, C.Z.C., Oliveira, M.C.P., Carrazza, L.R., Senhorini, J.A. and Cecarelli, P.S. (1998) Desenvolvimento de alevinos de pintado, *Pseudoplatystoma coruscans*, (Agassiz, 1829), mediante alimentação artificial. In: *Book of Abstracts of the Aquicultura Brasil '98.* Associação Brasileira de Aquicultura/World Aquaculture Society, Recife, p. 59.

Fernandes, J.B.K. (1998) Fontes e níveis de proteína bruta em dietas para alevinos e juvenis de pacu, *Piaractus mesopotamicus* (Holmberg, 1887). PhD dissertation, Universidade Estadual Paulista, Jaboticabal, Brazil.

Gunther, J. (1996) Growth of tambaqui (*Colossoma macropomum*) juveniles at different carbohydrate–lipid ratios. *Journal of Aquaculture in the Tropics* 11, 105–112.

Hayashi, C., Gonçalves, G.S., Furuya, V.R.B., Nagae, M.Y. and Soares, C.M. (1999) Utilização de diferentes alimentos durante o treinamento alimentar de alevinos de pintado (*Pseudoplatystoma corruscans*) (Agassiz, 1829). In: *Proceedings of the Acuicultura '99.* Sociedade Venezolana de Acuicultura/World Aquaculture Society, Puerto La Cruz, pp. 258–267.

Hernández, M., Takeuchi, T. and Watanabe, T. (1995) Effect of dietary energy sources on the utilization of protein by *Colossoma macropomum* fingerlings. *Fisheries Science* 61, 507–511.

Macedo, E.M. (1979) Nível proteico ideal na nutrição do tambaqui, *Colossoma macropomum* Cuvier 1818 (Pisces, Characidae). MSc thesis, Universidade Estadual Paulista, Jaboticabal, Brazil.

Machado, J.H. (1999) Desempenho produtivo de juvenis de pintado (*Pseudoplatystoma coruscans*), arraçoados com diferentes níveis de proteína e energia. MSc thesis, Universidade Estadual Paulista, Ilha Solteira, Brazil.

Machado, J.H., del Carratore, C.R., Garossino, A.P.R., Mazeto, M.D. and Grechi, F.C.S. (1998) Treinamento alimentar para a aceitação de rações artificiais em alevinos de pintado (*Pseudoplatystoma coruscans*). In: *Proceedings of the Aquicultura Brasil '98.* Vol. 2. Associação Brasileira de Aquicultura/World Aquaculture Society, Recife, pp. 101–108.

Merola, N. and Cantelmo, O.A. (1987) Growth, feed conversion and mortality of cage-reared tambaqui, *Colossoma macropomum*, fed various dietary feeding regimes and protein levels. *Aquaculture* 66, 223–233.

Moura, M.A.M., Lovshin, L.L., Kubitza, F. and Cyrino, J.E.P. (2000) Feed training of peacock bass *Cichla* sp. *Revista Brasileira de Biologia* 60(4), 645–654.

Oliveira, M.C.P., Oliveira, C.Z.C., Fernandes, E.B., Carrazza, L.R., Senhorini, J.A. and Cecarelli, P.S. (1998) Efeito de diferentes fontes de alimento no desenvolvimento larval de pintado, *Pseudoplatystoma coruscans*, (Agassiz, 1829). In: *Book of Abstracts of the Aquicultura Brasil '98.* Vol. 2. Associação Brasileira de Aquicultura/World Aquaculture Society, Recife, p. 5.

Pezzato, L.E. (1990) Efeito de diferentes níveis de gordura de origem animal e vegetal sobre o desempenho e deposição de ácidos graxos em pacu (*Piaractus mesopotamicus*). PhD dissertation, Universidade Estadual Paulista, Jaboticabal, Brazil.

Sampaio, A.M.B. de M., Kubitza, F. and Cyrino, J.E.P. (2000) Relação proteína : energia na nutrição do tucunaré *Cichla* sp. *Scientia Agricola* 57(2), 213–219.

Van der Meer, M.B., Machiels, M.A.M. and Verdegem, M.C.J. (1995) The effect of dietary protein level on growth, protein utilization and body composition of *Colossoma macropomum* (Cuvier). *Aquaculture Research* 26, 901–909.

Van der Meer, M.B., Zamora, J.E. and Verdegem, M.C.J. (1997a) Effect of dietary lipid level on protein utilizaton and the size and proximate composition of body compartments of *Colossoma macropomum* (Cuvier). *Aquaculture Research* 28, 405–417.

Van der Meer, M.B., Van Herwaarden, H. and Verdegem, M.C.J. (1997b) Effect of number of meals and frequency of feeding on voluntary feed intake of *Colossoma macropomum* (Cuvier). *Aquaculture Research* 28, 419–432.

Vidal, M.V., Donzele, J.L., Camargo, A.C.da S., de Andrade, D.R. and dos Santos, L.C. (1998) Níveis de proteína bruta para tambaqui (*Colossoma macropomum*), na fase de 30 a 250 gramas. 1. Desempenho dos tambaquis. *Revista da Sociedade Brasileira de Zootecnia* 27, 421–426.

Viegas, E.M.M. and Guzman, E.C. (1998) Effect of sources and levels of dietary lipids on growth, body composition, and fatty acids of the tambaqui (*Colossoma macropomum*). *World Aquaculture* 29, 66–70.

Werder, U. and Saint-Paul, U. (1979) Experiência de alimentação com tambaqui (*Colossoma macropomum*), pacu (*Mylossoma* sp.), jaraqui (*Semaprochilodus theraponura*) e matrinchã (*Brycon melanopterus*). *Acta Amazonica* 9, 617–619.

Zanoni, M.A. (1996) Níveis de fibra bruta em dietas de crescimento do pacu (*Piaractus mesopotamicus*) Holmberg, 1887. MSc thesis, Universidade Federal de Santa Catarina, Florianópolis, Brazil.

Snakehead and *Pangasius* Catfish

29

Tippawan Paripatananont*
Institute of Agricultural Technology, Walailak University, Tasala, Nakhon Si Tamarat 80160, Thailand

Introduction

Snakehead and *Pangasius* catfish are found in South Africa, India, Burma, Indonesia, Malaysia, Cambodia and Thailand (Wee, 1982; Chuapohuk, 1994). The popularity of farming these fish is due to their fast growth, resistance to disease and handling and extreme tolerance to inferior water quality. In Thailand, they are two of 27 freshwater species that are cultivated with numerous types of systems, ranging from superintensive farming for commercial production to extensive farming, mainly for household consumption. Both snakehead and *Pangasius* catfish are farmed in cages and ponds. According to the official statistics, production of snakehead and *Pangasius* catfish together represents 15% of total freshwater aquaculture production in Thailand, with an annual value of US$18 million (Jantrarotai and Jantrarotai, 1994).

Snakehead

Snakehead belongs to the family Channidae. It is also known as murrel and serpent-headed fish. Snakehead is a very tasty and popular fish in most southern and south-eastern Asian countries (De Silva, 1989). It has long been regarded as a valuable foodfish and its flesh is claimed to be rejuvenating, particularly during recuperation from serious illness, and is therefore given to elderly people (Ling, 1977). Snakehead farming began around 1955 in Thailand and spread to other countries, including Hong Kong, India, Cambodia, Taiwan and Vietnam. The popular species for farming is *Channa striatus*. However, other species, such as *Channa micropeltes* and *Channa marulius*, are also farmed (De Silva, 1989).

*Deceased.

Pangasius *catfish*

There are two species of pangasids that are widely cultured in South-east Asian countries: *Pangasius sutchi* (striped catfish) and *Pangasius lamaudii* (black-ear catfish). They are found in rivers of India, Burma, Indonesia, Malaysia, Cambodia and Thailand (Ling, 1977). *Pangasius* catfish is a fast-growing fish and can tolerate poorer water quality than snakehead. It is also popular as a sport-fish. Due to its high dress-out weight percentage, which is comparable to that of channel catfish (*Ictalurus punctatus*), *Pangasius* catfish has great value in food processing.

Nutrient Requirements

Snakehead

Little is known about the nutrient requirement for snakehead. Snakehead is strictly carnivorous and is generally fed with diets of animal origin. The protein requirement is relatively high compared with other freshwater species. Boonyaratpalin (1980) reported that snakehead fry fed on 3.1 kcal digestible energy (DE) g^{-1} diet required 43% dietary protein, while 1-month-old fish consuming the same energy-containing diet required 36% dietary protein. The dietary lipid requirement of snakehead fingerlings for maximum growth and survival is 6% (Boonyaratpalin, 1981a). No information is available on the essential fatty acid requirements. Lipids rich in the n-6 or n-3 series of fatty acids are well utilized by snakehead. There are no studies on the vitamin requirement, except that of pantothenic acid. Boonyaratpalin (1981b) reported that dietary pantothenate is essential for normal growth and maximum survival of snakehead.

Pangasius *catfish*

Nutrient requirements for *Pangasius* catfish are even less known than those for snakehead. *Pangasius* catfish is an omnivore and requires lower dietary protein than snakehead. Jantrarotai *et al.* (1992) indicated that *P. sutchi* fingerlings require 27–29% of dietary protein for normal growth. Chuapoehak (1994) conducted a feeding study in *P. sutchi* and found that protein could be reduced to 18% without a reduction in growth performance and feed efficiency if high-quality protein sources were used in diets containing a digestible energy-to-protein ratio of 13 kcal g^{-1}. In a more recent study, Chutjareyaves *et al.* (1998) concluded that 35% of dietary protein is required by *P. lamaudii* fingerlings. The protein requirement is reduced to 20% for juvenile *P. lamaudii* (Chutjareyaves *et al.*, 1999). There are no studies on the requirements of other nutrients in *Pangasius* catfish.

Practical Diets

In general, farm-made diets and manufactured (commercial) diets are used in Thailand and other neighbouring countries. In the case of snakehead, farmers produce diets for their own use. Occasionally, snakehead farmers feed their fish with catfish diet manufactured by commercial feed-mills. Diet formulas used by snakehead farmers are shown in Table 29.1. Catfish diets produced in Thailand for fry, fingerlings, grow-out and finisher contain 40%, 35%, 30% and 25% crude protein, respectively. The diet ingredients used are similar to those of other aquadiets, including fish-meal, soybean meal, groundnut meal, rice bran, maize meal, broken rice, tapioca flour, vitamin and mineral premixes and binder.

The farm-made diet for *Pangasius* sp. is primarily composed of several kinds of locally available by-products. However, farmers occasionally used a formulated fish diet when local by-products are not available. Table 29.2 shows model diet formulae for *Pangasius* sp. Commercial diets for grow-out of *Pangasius* contain crude protein ranging from 15% to 26%.

Information on vitamin and mineral requirements of both snakehead and *Pangasius* catfish is practically unavailable. Therefore, the vitamin and mineral levels established for other tropical freshwater species (NRC, 1993) are recommended.

Feeding Practices

Snakehead

Ponds used for snakehead culture generally vary from 1600 to 3200 m^2 in area and 2 to 3 m in depth. A stocking density of 25 fingerlings (50 g each) m^{-2} is commonly used. Snakehead culture is totally dependent on animal sources of

Table 29.1. Traditional snakehead diet formulations used by farmers in Thailand.

Ingredient	(%)	References
Diet 1		Somseub (1994)
Trash fish	90	
Rice bran	10	
Diet 2		
Trash fish	80	
Rice bran	20	
Diet 3		S. Koonsomboon (personal communication)
Trash fish	50.0	
Fish-meal	17.5	
Soybean meal	7.5	
Broken rice	7.0	
Rice bran	17.0	
Vitamin and mineral premix	1.00	

protein, especially trash fish. In some areas where trash fish is easily obtained, minced trash fish is the sole diet for snakehead and is fed to satiation. The fish are fed three times a day from fry (15 g) to fingerling (50 g) stage. The inclusion of trash fish in a snakehead diet is reduced to 80% by the addition of 20% rice bran or cooked broken rice when the fish reach the fingerling size. The mixture of diet ingredients is minced through a meat mincer and placed on wooden platforms along the pond sides (S. Koonsomboon, personal communication). Fish are grown for 6–7 months to attain a weight of approximately 1 kg, which is the most common marketable size in Thailand.

Two species of snakehead, *C. micropeltes* and *C. striatus*, are customarily cultured in cages. Cages used for snakehead cultures are 15–200 m^3. Fish weighing approximately 50 g are stocked at a density of 400–500 fish m^{-3}. Trash fish are chopped before feeding. Fish grow to more than 1 kg in 8–12 months. Growth is dependent on the availability of diet and stocking density. The feed conversion ratio of pond and cage culture of snakehead is about 4 to 1 (Nuov and Nandeesha, 1994).

Pangasius *catfish*

Pangasius species are cultured in both ponds and cages. Pond culture is common with *P. sutchi*, whereas cage culture consists of *P. sutchi* and *P. lamaudii*. The ponds used in *Pangasius* culture generally vary from 600 to 2000 m^2 and are about 2–3 m in depth. Fingerlings, weighing approximately 50 g are stocked at a density of 3000 kg ha^{-1}. Most ponds have a stagnant water system and are either filled with rainwater or with water pumped from a nearby lake.

The feeding strategy adopted is largely based on the availability of feed-stuffs in the farming area rather than on any scientific consideration. In the first

Table 29.2. Model diet formulae for *Pangasius* catfish (P. Rimteerakul, personal communication).

Ingredient	%
Fingerling (30% protein)	
Fish-meal	30
Rice bran	45
Soybean meal	24
Vitamin and mineral premix	1
Grow-out (25% protein)	
Fish-meal	15
Soybean meal	15
Groundnut	24
Rice bran	30
Broken rice	15
Vitamin and mineral premix	1

2 months, fish are fed with rice bran. Some farmers use a mixture of rice bran, broken rice and a small amount of trash fish. From the third month onwards, a number of diet formulations are used in different localities, based on the ingredients that are locally available. For example, vegetable wastes, cooked broken rice, various by-products from fruit-processing factories (such as pineapple wastes, baby maize husk, etc.), trash fish, fish offal and the like are widely used in Thailand, Cambodia and Burma (New *et al.*, 1994). The daily feeding rate with a moist diet is about 10% of body weight from fry to fingerling stages, and this is decreased to 5% in juvenile and adult phases.

Fish are fed three times daily in the pond for 8–12 months and usually attain a weight of 1–1.5 kg. A net production of 25–37.5 tons ha^{-1} is very common in Thailand (P. Rimteerakul, personal communication). The feed conversion ratio varies from 4 to 6 : 1.

Pangasius sp. is also cultured in ponds for integrated farming (Somsueb, 1994). The stocking ratios of the animals suggested are 1250 chicken or ducks or 30–60 pigs together with 12,500 fish ha^{-1}.

Cage culture of *Pangasius* sp. is practised around the lake and river areas. Cage size ranges from 20 to 900 m^3. The common stocking rate is 5–10 kg m^{-3} with fingerlings weighing approximately 75–100 g. The fish are usually fed with trash fish and sometimes a pelleted diet. After 8–12 months, typically 35–65 kg m^{-3} is harvested.

Acknowledgements

The author would like to express her thanks to Chhorrn Lim, US Department of Agriculture, Aquatic Animal Health Research Laboratory, Auburn, Alabama, USA, and Mali Boonyaratpalin, Nutrition Specialist, Department of Fisheries, Thailand, for their valuable suggestions. Data from fish farmers would not have been possible without the help of Konggumnerd's family, whom the author wishes to thank deeply. Grateful appreciation is also extended to her parents and family for their continuous love and support.

References

Boonyaratpalin, M. (1980) Protein requirement of *Ophicephalus striatus*. In: *National Inland Fisheries Institute Annual Report*. Department of Fisheries, Bangkok, Thailand, pp. 37–38 (in Thai).

Boonyaratpalin, M. (1981a) *Lipid Requirements of Snakehead Fingerlings*. Progress Report of the Regional Project RAS/76/003, Network of Aquaculture Centers in Asia, Bangkok, Thailand, 30 pp.

Boonyaratpalin, M. (1981b) *Vitamin Requirements in Snakehead Diets*. Progress Report of the Regional Project RAS/76/003, Network of Aquaculture Centers in Asia, Bangkok, Thailand, 18 pp.

Chuapohuk, W. (1994) Development of a low protein high energy practical diet for Sutchi's catfish, *Pangasius sutchi* Fowler. *Thai Journal of Agricultural Science* 27, 19–25.

Chutjareyaves, S., Pongsirijun, S. and Janesirisak, S. (1998) *Protein Requirement of Juvenile Black Ear Catfish (*Pangasius lamaudii*)*. National Inland Fisheries Institute Annual Report, Bangkok, Thailand, 26 pp. (in Thai).

Chutjareyaves, S., Pongsirijun, S. and Pongjunyakul, T. (1999) *Protein Requirement of Black Ear Catfish (*Pangasius lamaudii *Bocourt) Fingerling*. National Inland Fisheries Institute Annual Report, Bangkok, Thailand, 21 pp. (in Thai).

De Silva, K.H.G.M. (1989) Cheap sources of alternate feed for the farming of snakehead fish (Teleostei: Channidae) in Sri Lanka. In: Huisman, E.A., Zonneveld, N. and Bouwmans, A.H.M. (eds) *Aquacultural Research in Asia: Management Techniques and Nutrition*. Proceedings of the Asian Seminar on Aquaculture, Malang, Indonesia, pp. 241–267.

Jantrarotai, W. and Jantrarotai, P. (1994) On-farm feed preparation and feeding strategies for catfish and snakehead. In: New, M.B., Tacon, A.G.J. and Csavas, I. (eds) *Farm-made Aquafeeds*. Proceedings of the FAO/AADCP Regional Expert Consultation on Farm-made Aquafeeds, FAO–RAPA/AADCP, Bangkok, Thailand, pp. 101–119.

Jantrarotai, W., Sitasit, P., Chumsumgnern, S. and Chinmoog, S. (1992) *Effect of Various Protein Levels in Isocaloric Diets on Growth and Fat Deposit of Striped Catfish (*Pangasius sutchi*)*. National Inland Fisheries Institute Annual Report, Bangkok, Thailand, 13 pp. (in Thai).

Ling, S.W. (1977) *Aquaculture in Southeast Asia: A Historical Overview*. University of Washington Press, Seattle, Washington, 108 pp.

New, M.B., Tacon, A.G.J. and Csavas, I. (eds) (1994) *Farm-made Aquafeeds. Proceedings of the FAO/AADCP Regional Expert Consultation on Farm-made Aquafeeds*. FAO–RAPA/AADCP, Bangkok, Thailand, 434 pp.

NRC (National Research Council) (1993) *Nutrient Requirements of fish*. National Academy Press, Washington, DC, USA.

Nuov, S. and Nandeesha, M.C. (1994) Aquafeeds and feeding strategies in Cambodia. In: New, M.B., Tacon, A.G.J. and Csavas, I. (eds) *Farm-made Aquafeeds. Proceedings of the FAO/AADCP Regional Expert Consultation on Farm-made Aquafeeds*. FAO–RAPA/AADCP, Bangkok, Thailand, pp. 181–200.

Somseub, P. (1994) Aquafeeds and feeding strategies in Thailand. In: New, M.B., Tacon, A.G.J. and Csavas, I. (eds) *Farm-made Aquafeeds. Proceedings of the FAO/AADCP Regional Expert Consultation on Farm-made Aquafeeds*. FAO–RAPA/AADCP, Bangkok, Thailand, pp. 365–385.

Wee, K.L. (1982) Snakeheads – their biology and culture. In: Muir, J.F. and Roberts, R.J. (eds) *Recent Advances in Aquaculture*. Croom Helm, London, pp. 179–213.

Baitfish 30

Rebecca Lochmann and Harold Phillips
Department of Aquaculture and Fisheries, University of Arkansas at Pine Bluff, PO Box 4912, Pine Bluff, AR 71611, USA

Introduction

Small fish produced specifically for anglers to attract desirable food- or game fish are referred to as 'baitfish'. Goldfish and the rosy red variety of fathead minnows may also be marketed as 'feeder' fish, which are meant for consumption by piscivorous pets. Baitfish and 'feeder' fish are raised under similar conditions, and nutrition and feeding practices for these species are considered together.

The total value of baitfish (including 'feeder' goldfish) sold in the USA in 1998 was $37.5 million (USDA, 2000). The golden shiner, *Notemigonus crysoleucas*, accounted for about half of this total. Goldfish, *Carassius auratus*, and the fathead minnow, *Pimephales promelas*, collectively comprised the remaining half. Baitfish value is determined primarily by size, and this feature dictates production and marketing strategies. Baitfish production is concentrated in the southern USA, especially in Arkansas, and the majority of fish are spawned and raised in ponds. Other fish species, such as white suckers, *Catostomus commersoni*, are sometimes preferred as bait outside the southern USA and in other countries. In some regions (such as the north-central USA), the distinction between wild-caught and cultured baitfish is not always clear and prepared diets may not be used (J.L. Gunderson, University of Minnesota, 2000, personal communication).

The known nutritional requirements of baitfish species are similar to those of other warm-water omnivores, such as channel catfish and common carp. However, quantitative requirements of many nutrients for baitfish species are still unknown. Feeding practices for baitfish also differ from those of foodfish in several important ways (Table 30.1).

Table 30.1. Major production factors that affect feeding and nutrition of baitfish and channel catfish in ponds.

Factor	Baitfish*	Channel catfish†
Feed cost	18% of total costs	≥ 50% of total costs
Fish growth	Rate is manipulated to achieve desired market sizes (variable)	Maximum rate is desirable throughout production cycle
Natural foods	Provide 40% or more of nutrition for non-larval stages in intensive culture in presence of prepared diets	Uncertain contribution; may provide some micronutrients for non-larval stages in intensive culture in presence of prepared diets
Body composition	Large amount of body fat does not reduce marketability/ may be advantageous for fish vigour	Large amount of body fat reduces dressing percentage and shelf-life

* Includes golden shiners, goldfish and fathead minnows (Stone *et al.*, 1997).
† From Robinson and Li (1996).

Nutrient Requirements

Protein and amino acids

The essential amino acid ratios of the whole body of golden shiners, goldfish and fathead minnows are similar to those of channel catfish, *Ictalurus punctatus*, and common carp, *Cyprinus carpio* (Gatlin, 1987), indicating possible similarities in essential amino acid requirements among these species. Lochmann and Phillips (1994a) determined that growth, survival and feed efficiency of golden shiners and goldfish fed semipurified diets with 29% protein in aquariums was similar to that of fish fed diets with higher protein levels when fed at 4–7% of body weight (BW).

A.L. Gannam and H. Phillips (unpublished) found that there were no differences in weight gain or yield of golden shiners in ponds fed practical diets with only vegetable proteins versus diets with 5, 10 or 20% fish-meal. The primary vegetable-protein source for all diets was soybean meal. Alternative protein feedstuffs, including fish silage and worm meal, have been tested for golden shiners in aquaria. Worm meal supported growth similar to that of fish fed diets containing an isonitrogenous amount of fish-meal, while diets containing fish silage reduced fish growth (R. Lochmann and H. Phillips, unpublished). However, golden shiners fed diets with only vegetable protein (soybean meal) performed as well as fish fed diets with fish- or worm meal. Therefore, there is no need for animal protein sources in practical diets for juvenile (0.5 g in weight) golden shiners in aquariums or ponds. In ponds, consumption of biota by baitfish probably further reduces the need for supplemental high-quality protein sources.

Golden shiners (1.3 g) in ponds were fed diets similar in energy : protein ratios but different in crude protein levels (22 or 28%) to satiation twice daily for 10 weeks. The weight gain of fish fed the diet with 28% protein was higher after 4 weeks, but by week 10 there were no significant differences in weight gain, feed conversion, net yield or yield of individual size classes of fish fed the two diets (R. Lochmann and H. Phillips, unpublished).

Energy

The optimal dietary protein : energy ratio for growth of golden shiners and goldfish is 103 mg protein kcal^{-1} (Lochmann and Phillips, 1994a). Energy requirements for maintenance of goldfish increase with temperature between 20 and 26°C (Pannevis, 1993). In general, both energy and protein requirements for maintenance of goldfish decrease with increasing fish size (Pannevis and Earle, 1994). Goldfish appear to select their diet based on energy density (Sanchez-Vazquez *et al.*, 1998). Available energy of feedstuffs for baitfish is estimated mostly from growth under optimal conditions (Stone *et al.*, 1997). De Silva *et al.* (1997) documented an alteration in apparent total dry matter and apparent protein digestibility in goldfish in the presence of a stressor (sublethal cadmium levels). Results varied depending on the digestibility marker used (chromic oxide or fibre). When chromic oxide (Cr_2O_3) was used as a marker, small quantities of chromium were absorbed by the goldfish and digestibility coefficients were underestimated (De Silva *et al.*, 1997). Stable carbon isotope data are another index of nutrient assimilation in baitfish. Whole cooked maize is not assimilated well by golden shiners, while maize-gluten meal, maize starch and lipids are readily assimilated (Lochmann and Phillips, 1996, 2000).

Lipids and fatty acids

The optimal dietary lipid level for juvenile golden shiners and goldfish was determined in feeding trials using graded levels of a mixture of 1 : 1 ratio of cod-liver and soybean oils. The lipid mixture contained 18-carbon, as well as 20- and 22-carbon fatty acids of the n-3 and n-6 families, which encompasses the essential fatty acid (EFA) requirements of most fish species (Watanabe, 1982). The weight gain of golden shiners fed diets containing 34% protein and 7–12% lipid was higher than that of fish fed diets with lower or higher lipid levels (SRAC, 1998). Feed efficiency and survival of golden shiners were not affected by dietary lipid levels ranging from 3 to 15%. Optimum weight gain and feed efficiency of goldfish were obtained with diets containing 3–6% lipid (SRAC, 1998). Survival increased with increasing dietary lipid level, but was 93% or higher in all treatments.

Golden shiners in aquariums fed isonitrogenous and isocaloric practical diets with either 4% or 13% poultry fat for 7.5 weeks had similar growth, but survival was higher in fish fed the diet with 13% lipid (Lochmann and Phillips, 2001).

Similar diets were also used in ponds (0.04 ha) and outdoor pool (4137 l) studies. Golden shiners (0.9 g) in ponds stocked at 796 kg ha^{-1} were fed practical diets containing 4 or 13% poultry fat or 13% menhaden fish-oil to satiation twice daily for 12 weeks (R. Lochmann and H. Phillips, unpublished). The weight gain of fish fed the diet with 4% poultry fat was higher than that of fish fed diets with 13% lipid. The net yield of fish fed the three diets was similar, implying a higher survival rate among fish fed the diets with 13% lipid. Whole-body lipid of the golden shiners fed the diet with 13% menhaden fish-oil was higher than that of fish fed the diets with 4 or 13% poultry fat ($P = 0.07$). This apparent difference in metabolism of poultry and fish-oils by golden shiners needs further study.

Goldfish (0.9 g) stocked at 600 fish per fertilized pool (3 m diameter × 1.5 m depth) were fed practical diets (3–6% BW) containing 4 or 13% lipid as poultry fat or menhaden fish-oil for 9 weeks. These supplemental diets contained 24% protein and no added vitamins or minerals. Average individual weight gain, feed efficiency, net yield and whole-body lipid were significantly higher ($P < 0.05$) in goldfish fed diets with 13% of either lipid source, compared with fish fed diets with 4% lipid (R. Lochmann and H. Phillips, unpublished). Lipid source did not affect goldfish performance. Presumably, the improved performance of goldfish fed the high-lipid diets was due to a protein-sparing effect of dietary lipid.

Studies have also been conducted in aquariums to determine the qualitative EFA requirements of golden shiners fed purified diets (Lochmann and Phillips, 2001). Results are summarized in Table 30.2. These trials have not consistently indicated a specific requirement for n-3 or n-6 fatty acids or for 18-carbon or longer-chain fatty acids of either family. The ratio of n-3 : n-6 fatty acids in the diets ranged from 0.06 (rice-bran oil) to 9.0 (cod-liver oil). Similar studies with common carp (Watanabe *et al.*, 1975a,b), channel catfish (Dupree, 1969; Stickney and Andrews, 1972; Satoh *et al.*, 1989a,b), and tilapia (Kanazawa *et al.*, 1980; Takeuchi *et al.*, 1983; Chou *et al.*, 2000) also yielded conflicting results. Pozernick and Wiegand (1997) found that growth and survival of larval goldfish fed diets with cod-liver or canola oils was equally good, indicating that a dietary source of n-3 highly unsaturated fatty acids (HUFAs) is not required. However, they did not consider n-6 fatty acids. Prostaglandins derived from arachidonic acid (20:4n-6) stimulate steroid production in goldfish (Wade *et al.*, 1994; Mercure and Van Der Kraak, 1996) and courtship behaviour in fathead minnows (Cole and Smith, 1987). Until the qualitative and quantitative EFA requirements of baitfish are established, dietary sources of both n-3 and n-6 fatty acids should be provided to support normal growth, health, appearance and reproduction in these species.

Phospholipid supplementation in diets for baitfish species may be beneficial. Practical diets supplemented with soybean lecithin enhanced growth but did not affect survival of juvenile goldfish (Lochmann and Brown, 1997) relative to diets containing lipid as triglyceride from either soybean or fish-oils. Phospholipid supplementation of semipurified diets improved both growth and survival of larval goldfish (Szlaminska *et al.*, 1993) and carp (Geurden *et al.*, 1995). Phospholipids may facilitate lipid digestion, absorption and transport in baitfish, as in other fish (Hertrampf, 1992).

Table 30.2. Summary of studies to determine qualitative essential fatty acid requirements of golden shiners (data from Lochmann and Phillips, 2001)*.

Study no./length	Initial fish weight (g)	Lipid sources†	Main results
I/9 weeks	0.21	Soybean, rice bran, canola, cod-liver, poultry	No differences in weight gain or survival between diets; whole-body lipid was higher in fish fed vegetable vs. animal lipids
II/11.5 weeks	0.19	Soybean, rice bran, canola, cod-liver, poultry, olive	No differences in weight gain or survival between diets; mortality of fish stressed with low dissolved oxygen was lowest in fish fed soybean oil
III/6 weeks‡§	0.35	Soybean, cod-liver, soy + cod-liver (50/50%), olive, canola	Weight gain of fish fed the soy + cod-liver (E), olive (U) and soybean (E) diets was highest; no differences in survival; whole-body fatty acid profiles reflected dietary patterns
IV/34 weeks	0.22	Same as study III	No differences in weight gain between diets; survival was higher in fish fed ethanol-extracted diets, regardless of lipid source. Fish fed canola had intact fins, opercula and integument; those fed olive oil had pronounced erosion of these structures; fish fed other lipids were intermediate in appearance

* Purified diets for all experiments contained 34% protein and 10 kcal energy g^{-1} of protein from casein and gelatin. Diets were supplemented with 10% lipid from various sources. Diets were fed to fish in triplicate tanks per treatment at 5–7% BW divided into two daily feedings.

† Lipids used were whole oils or fats extracted commercially from plants and animals (approximately 80% triglycerides).

‡ Experiment was terminated after 6 weeks due to massive mortality from an unidentified disease.

§ One series of five diets contained the lipid sources shown in table (U = unextracted diets); a second set of five diets contained the same lipids but the casein, gelatin, dextrin, cellulose and carboxymethylcellulose were extracted with boiling ethanol to minimize lipid content (E = extracted diets).

Carbohydrates

Weight gain and survival of golden shiners fed semipurified isocaloric and isonitrogenous diets with 15%, 30% or 45% starch were similar, indicating that they perform well over a wide range of dietary levels of carbohydrate : lipid ratios (Lochmann and Phillips, 2001). In another experiment, the weight gain of golden shiners fed diets with 15% carbohydrate from different sources improved with increasing complexity of the carbohydrate: starch > dextrin > sucrose = glucose (Lochmann and Phillips, 2001). Results are similar to those for other warm-water omnivores (NRC, 1993). Survival of golden shiners was lower in fish fed sucrose than in those fed dextrin or glucose, but the value was 93% or higher in all treatments.

Vitamins and minerals

Natural foods consumed by baitfish in ponds are rich sources of vitamins and minerals, but the amounts and types vary with the composition of the natural food supply. Therefore, commercial baitfish diets are supplemented with the same types and amounts of vitamins and minerals used in diets for channel catfish (Lovell, 1989). Few studies have addressed vitamin and mineral nutrition in baitfish. Weight gain and total net yield of golden shiners in ponds fed diets with or without a combination vitamin and mineral supplement for 8 weeks did not differ (Lochmann and Phillips, 1994b). Presumably, natural foods supplied sufficient vitamins and minerals to maintain overall fish production.

A preliminary aquarium study was performed to determine whether or not golden shiners (0.4 g) have a dietary requirement for ascorbic acid (AA). Fish-meal- or casein-based diets supplemented with either 0 or 250 p.p.m. AA (Stay-C, Roche Vitamins, Inc.) were fed to golden shiners for 12.5 weeks. Weight gain was higher but survival of golden shiners was not affected by AA supplementation of the fish-meal diets, while the reverse was true with the casein diets. No AA-deficiency signs (e.g. scoliosis, lordosis, severe fin erosion, etc.) were observed in golden shiners fed fish-meal diets with or without AA. There were significant differences in total AA, reduced AA, and % AA content of whole bodies of golden shiners fed AA-supplemented and AA-unsupplemented diets (Lehmann *et al.*, 2001a). Golden shiners appear to have a dietary requirement for AA, but the requirement is influenced by diet composition.

Goldfish fed purified diets without vitamin A displayed haemorrhaging, exophthalmia, scale loss and anorexia (Jones *et al.*, 1971). Nutritional myopathy attributed to dietary vitamin E deficiency was reported for goldfish fed practical diets with 26 mg kg^{-1} or less of vitamin E (Huerkamp *et al.*, 1988). Quantitative dietary requirements of these vitamins, however, have not been determined.

The metabolism of some minerals such as calcium, phosphorus, magnesium and selenium has been studied in baitfish species used in toxicological or biomedical studies (Ichii and Mugiya, 1983; Houston, 1985; Kleinow and Brooks, 1986). A preliminary study of mineral availability from fish-meal for goldfish was conducted by Sugiura *et al.* (1998), but the study was terminated early due to technical difficulties. These studies could serve as a basis for establishing dietary mineral requirements for baitfish, which are currently unknown.

Practical Diets

Some baitfish producers use only agricultural by-products (e.g. cottonseed meal, rice bran) as a diet, but semi-intensive and intensive producers use compounded diets to double or triple production. The composition of diets used for different stages of the baitfish production cycle are similar to those for channel catfish (Lovell, 1989; Stone *et al.*, 1997). However, pond fertilization to stimulate natural food production is considered critical for successful baitfish production, regardless of production intensity and prepared feeds used. Commercial minnow meal containing 48–50% protein is applied to ponds containing newly hatched fry. These small fish consume the meal, but newly hatched fry of some cyprinids do not utilize prepared feeds well (Dabrowski and Culver, 1991). Therefore, the meal may serve more as a fertilizer than as a feed. When the fish are approximately 2 cm in length, they are offered an extruded pellet (floating or 'slow-sink') or crumble containing 28–32% protein. This diet is used until the fish are harvested. A similar diet or one with a higher protein level (36%) is fed to baitfish brood-stock.

Soybean meal is the primary protein source in commercial baitfish diets and very little (2–5%) fish-meal is used in diets for juveniles and adults. Some producers believe that fish-meal supports superior baitfish performance in ponds, but there is no experimental evidence to support this. Due to the proximity of the baitfish and poultry industries in Arkansas, poultry fat is frequently added to baitfish diets. Poultry fat appears to be palatable and provides satisfactory performance.

Recently, high-fat (13%) diets with lower protein levels (24%) have generated some interest among baitfish producers. The cost of increased dietary lipid is offset by the reduction in cost due to the lower protein content, making the diet a viable alternative for most producers. For feeder fish, economical means of enhancing skin colour are needed. The synthetic carotenoids used to enhance colour in salmonids are too expensive for use in feeder-fish diets, and further research is needed to identify more economical alternatives.

Feeding Practices

Because of the large impact of natural foods on baitfish produced in ponds, the need for nutritionally complete feeds throughout the production cycle is debatable. When desirable plankton blooms can be maintained in production ponds, supplemental feeds may be sufficient to maintain production. However, in ponds where blooms are sparse or cannot be maintained, baitfish production may be improved by the use of nutritionally complete diets (Lochmann *et al.*, 2001b). The desired market size of the fish strongly influences the feeding frequency used, but the type of diet (supplemental or complete) could be altered to control size and also to maintain fish health.

Feeding practices by baitfish producers are highly variable. Some of the variability is related to species. Stocking, feeding and production rates are highest for goldfish, followed by golden shiners and fathead minnows. Most fathead-minnow producers use little or no prepared feed, while some goldfish producers feed up to 23 kg ha^{-1} day^{-1} (2% of BW). For golden shiners, 11 kg ha^{-1} day^{-1} is considered a high feeding rate. Most farms feed once daily, but fish are fed more often when rapid growth is desired. Diet is usually applied to ponds by blowers mounted on trucks.

Some producers in the USA are now hatching baitfish eggs indoors in tanks before releasing fry into ponds. The procedures used are similar to those described by Kestemont (1995) for goldfish larvae, except that live foods are not used. Feeding regimes for hatchery-produced baitfish fry are still experimental, but diets generally consist of a mixture of finely ground meals, microparticulate diets or other foods (yeast) with a small particle size (100–250 μm). After the fry are moved into ponds, they receive the same feeds as fish spawned in ponds.

Feeding strategies for baitfish change seasonally. Survival improves and the condition of golden shiners during winter is maintained by feeding at a rate of 1–2% BW at afternoon air temperature of 7°C or higher (McNulty *et al.*, 2000). Increased numbers of fathead minnows in good condition in ponds can be achieved with a feeding rate of 3% BW day^{-1} (using a 32% protein feed) from late summer to winter (Ludwig, 1996). In summer, many producers reduce feeding rates when water temperatures exceed 30°C.

Production of golden shiners in indoor recirculating systems has been attempted in some areas (north-central and north-eastern USA) with restricted outdoor growing seasons. Nutritionally complete feeds are essential for the successful production of fish in these systems. Commercial salmonid diets have been fed to golden shiners in these systems (G. Raisanen, Alexandria Technical College, Minnesota, 2000, personal communication). In some cases, growth and survival have been adequate, but fish developed pathological signs, such as 'bent backs' and other anomalies, which made them unmarketable. To date, culture of golden shiners in recirculating systems has not been successful and lack of an adequate diet may be a cause. Due to the lack of information on nutrient requirements of baitfish, diets developed for tilapia, carp or channel catfish may be used. However, the nutritional benefits derived from natural foods in ponds may be difficult to duplicate economically in indoor systems.

Acknowledgements

The authors thank Neil Anderson and Eric Park for providing fish and valuable information on raising baitfish species under commercial conditions. Nathan Stone and Hugh Thomforde reviewed and improved the manuscript. Numerous student workers have assisted in the successful completion of baitfish nutrition studies at the University of Arkansas at Pine Bluff. The research reviewed in this paper was supported in part by the Southern Regional Aquaculture Center (SRAC) and Cooperative State Research Education and Extension Service (CSREES) research programmes of the United States Department of Agriculture.

References

Chou, B.S., Hung, S.S.O. and Shiau, S.Y. (2000) Effect of dietary cod liver oil on growth and tissue fatty acids of juvenile hybrid tilapia. In: *Responsible Aquaculture in the New Millennium*. Special Publication No. 28, European Aquaculture Society, Ostend, Belgium, p. 131 [abstract].

Cole, K.S. and Smith, R.J. (1987) Release of chemicals by prostaglandin-treated female fathead minnow, *Pimephales promelas*, that stimulate male courtship. *Hormonal Behavior* 21, 440–456.

Dabrowski, K. and Culver, D. (1991) The physiology of larval fish. *Aquaculture Magazine* 17, 49–61.

De Silva, S.S., Deng, D.F. and Rajendram, V. (1997) Digestibility in goldfish fed diets with and without chromic oxide and exposed to sublethal concentrations of cadmium. *Aquaculture Nutrition* 3, 109–114.

Dupree, H.K. (1969) *Influence of Corn Oil and Beef Tallow in Growth of Channel Catfish*. Technical Paper No. 27, US Fish and Wildlife Service.

Gatlin, D.M. III (1987) Whole-body amino acid composition and comparative aspects of amino acid nutrition of the goldfish, golden shiner and fathead minnow. *Aquaculture* 60, 223–229.

Geurden, I., Radunz-Neto, J. and Bergot, P. (1995) Essentiality of dietary phospholipids for carp (*Cyprinus carpio* L.) larvae. *Aquaculture* 131, 303–314.

Hertrampf, J.W. (1992) *Feeding Aquatic Animals with Phospholipids II. Fishes*. Publication No. 11, Lucas Meyer, Hamburg, Germany.

Houston, A.H. (1985) Erythrocyte magnesium in freshwater fishes. *Magnesium* 4, 106–128.

Huerkamp, M.J., Ringler, D.H. and Chrisp, C.E. (1988) Vitamin E deficiency in goldfish fed a shellfish derived diet. *Laboratory Animal Science* 38, 178–182.

Ichii, T. and Mugiya, Y. (1983) Effects of a dietary deficiency in calcium on growth and calcium uptake from the aquatic environment in the goldfish, *Carassius auratus*. *Comparative Biochemistry and Physiology* 74A, 259–262.

Jones, J.H., Bullard, E.B. and Rodriguez, A. (1971) The essential nature of vitamin A in the diet of goldfish. *Transactions of the American Fisheries Society* 4, 676–681.

Kanazawa, A., Teshima, S.-T., Sakamoto, M. and Awal, M.A. (1980) Requirement of *Tilapia zillii* for essential fatty acids. *Bulletin of the Japanese Society of Scientific Fisheries* 46, 1353–1356.

Kestemont, P. (1995) Influence of feed supply, temperature and body size on the growth of goldfish *Carassius auratus* larvae. *Aquaculture* 136, 341–349.

Kleinow, K.M. and Brooks, A.S. (1986) Selenium compounds in the fathead minnow (*Pimephales promelas*) – II. Quantitative approach to gastrointestinal absorption, routes of elimination and influence of dietary pretreatment. *Comparative Biochemistry and Physiology* 83C, 71–76.

Lochmann, R.T. and Brown, R. (1997) Soybean-lecithin supplementation of practical diets for juvenile goldfish (*Carassius auratus*). *Journal of the American Oil Chemists' Society* 74, 149–152.

Lochmann, R.T. and Phillips, H. (1994a) Dietary protein requirement of juvenile golden shiners (*Notemigonus crysoleucas*) and goldfish (*Carassius auratus*) in aquaria. *Aquaculture* 128, 277–285.

Lochmann, R.T. and Phillips, H. (1994b) Vitamin and mineral additions to golden shiners diet tested. *Arkansas Farm Research* 43(3), 8–9.

Lochmann, R. and Phillips, H. (1996) Stable isotopic evaluation of the relative assimilation of natural and artificial foods by golden shiners *Notemigonus crysoleucas* in ponds. *Journal of the World Aquaculture Society* 27, 168–177.

Lochmann, R. and Phillips, H. (2000) A stable carbon isotope study of the assimilation of dietary lipids in golden shiners *Notemigonus crysoleucas*. In: *Responsible Aquaculture in the New Millennium*. Special Publication No. 28, European Aquaculture Society, Ostend, Belgium, p. 408 [abstract].

Lochmann, R. and Phillips, H. (2001) Nutritional aspects of health and related components of baitfish performance. In: Lim, C. and Webster, C. (eds) *Nutrition and Fish Health*. Food Products Press, Binghamton, New York, pp. 119–130.

Lochmann, R., Phillips, H., Dabrowski, K. and Moreau, R. (2001a) Responses of juvenile golden shiner *Notemigonus crysoleucas* fed semipurified and practical diets with and without supplemental ascorbic acid. *Journal of the World Aquaculture Society* 32, 202–209.

Lochmann, R., Phillips, H., Dasgupta, S., Gatlin, D. and Rawles, S. (2001b) Stable carbon isotope ratios and standard production data as indices of golden shiner, *Notemigonus crysoleucas*, performance in pond feeding trials. *Journal of Applied Aquaculture* 11, 21–34.

Lovell, R.T. (1989) *Nutrition and Feeding of Fish*. Van Nostrand Reinhold, New York.

Ludwig, G.M. (1996) Seasonal growth and survival of rosy red and normal-colored fathead minnows receiving different feed rations. *Progressive Fish-Culturist* 58, 160–166.

McNulty, E., Stone, N. and Lochmann, S. (2000) Effect of winter feeding rate, feed form and trigger temperature on yield and condition of golden shiners, *Notemigonus crysoleucas*. *Journal of Applied Aquaculture* 10, 69–82.

Mercure, F. and Van Der Kraak, G. (1996) Mechanisms of action of free arachidonic acid on ovarian production in the goldfish. *General and Comparative Endocrinology* 102, 130–140.

NRC (National Research Council) (1993) *Nutrient Requirements of Fish*. National Academy Press, Washington, DC.

Pannevis, M.C. (1993) Nutrition of ornamental fish. In: Burger, I. (ed.) *The Waltham Book of Companion Animal Nutrition*. Pergamon Press, Oxford, pp. 85–96.

Pannevis, M.C. and Earle, K.E. (1994) Maintenance energy requirement of five popular species of ornamental fish. *Journal of Nutrition* 124, 2616S-2618S.

Pozernick, M. and Wiegand, M.D. (1997) Use of canola oil in the feed of larval and juvenile goldfish, *Carassius auratus* (L.). *Aquaculture Research* 28, 75–83.

Robinson, E. and Li, M. (1996) *A Practical Guide to Nutrition, Feeds and Feeding of Catfish*. Bulletin 1041, Mississippi Agricultural and Forestry Experiment Station, Mississippi State, Mississippi, USA.

Sanchez-Vasquez, F.J., Yamamoto, T., Akiyama, T., Madrid, J.A. and Tabata, M. (1998) Selection of macronutrients by goldfish operating self-feeders. *Physiology and Behavior* 65, 211–218.

Satoh, S., Poe, E.E. and Wilson, R.P. (1989a) Studies on the essential fatty acid requirement of channel catfish, *Ictalurus punctatus*. *Aquaculture* 79, 121–128.

Satoh, S., Poe, E.E. and Wilson, R.P. (1989b) Effect of dietary n-3 fatty acids on weight gain and liver polar lipid fatty acid composition of fingerling channel catfish. *Journal of Nutrition* 119, 23–28.

SRAC (Southern Regional Aquaculture Center) (1998) *Dietary Protein and Lipid Requirements of Golden shiners and Goldfish*. SRAC Publication No. 124, Cooperative States Research, Education, and Extension Service, US Department of Agriculture, Stoneville, MS.

Stickney, R.R. and Andrews, J.W. (1972) Effects of dietary lipids on growth, feed conversion, lipid and fatty acid composition of channel catfish. *Journal of Nutrition* 102, 249–258.

Stone, N., Park, E., Dorman, L. and Thomforde, H. (1997) *Baitfish Culture in Arkansas: Golden Shiners, Goldfish, and Fathead Minnows.* Cooperative Extension Program, University of Arkansas at Pine Bluff, US Department of Agriculture, and County Governments Cooperating.

Sugiura, S.H., Dong, F.M. and Hardy, R.W. (1998) Effects of dietary supplements on the availability of minerals in fish meal: preliminary observations. *Aquaculture* 160, 283–303.

Szlaminska, M., Escafre, A.M. and Bergot, P. (1993) Preliminary data on semisynthetic diets for goldfish (*Carassius auratus* L.) larvae. In: Kaushik, S.J. and Luques, P. (eds) *Fish Nutrition in Practice.* INRA, Biarritz, France, pp. 607–612.

Takeuchi, T., Satoh, S. and Watanabe, T. (1983) Requirement of *Tilapia nilotica* for essential fatty acids. *Bulletin of the Japanese Society of Scientific Fisheries* 49, 1127–1134.

USDA (United States Department of Agriculture) (2000) *1997 Census of Agriculture: 1998*, Vol. 3, *Special Studies*, Part 3. National Agricultural Statistics Service, United States Department of Agriculture, Washington, DC, USA.

Wade, M.G., Van der Kraak, G., Gerrits, M.F. and Ballantyne, J.S. (1994) Release and steroidogenic actions of polyunsaturated fatty acids in the goldfish testis. *Biology and Reproduction* 51, 131–139.

Watanabe, T. (1982) Lipid nutrition in fish. *Comparative Biochemistry and Physiology* 73B, 3–15.

Watanabe, T., Utsue, O., Koybayashi, I. and Ogino, C. (1975a) Effect of dietary methyl linoleate and linolenate on growth of carp – I. *Bulletin of the Japanese Society of Scientific Fisheries* 41, 257–262.

Watanabe, T., Takeuchi, T. and Ogino, C. (1975b) Effect of dietary methyl linoleate and linolenate on growth of carp – II. *Bulletin of the Japanese Society of Scientific Fisheries* 41, 263–269.

Index